BIODIESEL – QUALITY, EMISSIONS AND BY-PRODUCTS

Edited by **Gisela Montero**
and **Margarita Stoytcheva**

INTECHOPEN.COM

Biodiesel – Quality, Emissions and By-Products
Edited by Gisela Montero and Margarita Stoytcheva

Published by InTech

Janeza Trdine 9, 51000 Rijeka, Croatia

Edition 2014

Publishing Process Manager Danijela Duric

Technical Editor Teodora Smiljanic

Cover Designer InTech Design Team

Image Copyright Jon Nightingale, 2011. Used under license from Shutterstock.com

Additional hard copies can be obtained from orders@intechopen.com

Biodiesel – Quality, Emissions and By-Products,
Edited by Gisela Montero and Margarita Stoytcheva
p. cm.
ISBN 978-953-307-784-0

INTECH
open science | open minds

Contents

Preface

The use of biodiesel worldwide is becoming increasingly important, because of the widespread shortage of oil resources, which has raised the prices of fossil fuels. This biofuel also offers an opportunity to meet the energy demands with less impact to the environment due its renewable characteristics.

This book entitled *"Biodiesel – Quality, Emissions and By-Products"* comprises 20 chapters and covers topics related to biodiesel quality, performance of combustion engines that use biodiesel, and the emissions they generate. It emphasizes the applications of glycerol, a byproduct of biodiesel production process. It is divided in three sections: i) Biodiesel Quality and Standards, ii) Biodiesel: Development, Performance and Combustion Emissions, and iii) Glycerol: Properties and Applications.

The first section, *Biodiesel: Quality and Standards* is integrated by seven chapters. Chapter 1 presents the main standards on commercial biodiesel quality adopted in different regions of the world and the importance and significance of the main properties of biodiesel. Chapter 2 discusses the determination of photochemical properties, thermal diffusion, thermal conductivity and thermal effusivity as a promising route to characterize biodiesel oils. Chapter 3 explains how to characterize and differentiate each type of biofuel respect to the other ones by pressure differential scanning calorimetry. Chapter 4 comments that raw material source, impurities and production practices, can affect the quality of the biodiesel, performance and commercial approval of the final product. Chapter 5 compares the performance of one commercial biodiesel and three laboratory-produced biodiesels to verify the effect of storage temperature, type of storage container, storage time, as well as the moisture content on the properties of the biodiesel. Chapter 6 describes a method for sample preparation and quantification of FAME in diesel, and Chapter 7 details a research about standardization of procedures used for metal trace in biodiesel.

The second section, *Biodiesel: Development, Performance and Combustion Emissions* include eight chapters. Chapter 8 is a study of biodiesel and food balance, and how they represent opportunities for agriculture and rural development. Chapter 9 describes several studies for producing biodiesel from raw materials native of Baja California, Mexico. Chapter 10 presents the development of multifunctional detergent-dispersant additives based on fatty acid methyl esters. Chapter 11 discusses the results of a research on hydrogenation of fatty acid methyl esters to fatty alcohol. Chapter 12 is a review of impacts of biodiesel use as a fuel for diesel engines. Chapter 13 focuses on the toxicology of the compounds produced by the combustion of biodiesel. Chapter 14 depicts an assessment of the main properties of binary mixtures and triple mixtures between biodiesel from rapeseed oil, commercial diesel fuel and bioethanol compared to diesel fuel when were used in combustion engines. Chapter 15 exhibits the results of a research activity aimed at studying the high-blending biodiesel use in the "torque controlled" automotive diesel engines. In particular, based on the employment of an innovative biodiesel blending detection methodology, the capability of closed loop combustion control to improve both pollutant emissions and full load engine performance was investigated.

The third section, *Glycerol: Properties and Applications,* focuses on this product obtained as a co-product of biodiesel. Chapter 16 is a review of reactions such as dehydration, hydrogenolysis, oxidation, etherification, and reforming including results obtained by the author. All these uses allow considering the glycerol as one key-compound in the environment of future biorefinery. Chapter 17 is a study about antioxidative and anticorrosive properties of glycerol. Chapter 18 summarizes the state-of-the-art glycerol-based biotechnological processes and discusses future developments. Chapter 19 presents the utilization of glycerol in biotechnological applications and Chapter 20 discusses some unusual applications of glycerol.

All the contributing authors are gratefully acknowledged for their time and efforts in preparing the different chapters, and for their interest in the present project.

Dr. Gisela Montero and Prof. Margarita Stoytcheva
Mexicali, Baja California
Mexico

Part 1

Biodiesel: Quality and Standards

1

Biodiesel Quality, Standards and Properties

István Barabás and Ioan-Adrian Todoruț
Technical University of Cluj-Napoca
Romania

1. Introduction

Quality is a prerequisite for the long-term success (successful use, without technical problems) of a biofuel. Biodiesel quality depends on several factors that reflect its chemical and physical characteristics. The quality of biodiesel can be influenced by a number of factors: the quality of the feedstock; the fatty acid composition of the parent vegetable oil or animal fat; the production process and the other materials used in this process; the post-production parameters; and the handling and storage. Given the fact that most current diesel engines are designed to be powered by diesel fuel, the physicochemical properties of biodiesel should be similar to those of diesel oil.
This chapter presents the main standards on commercial biodiesel quality adopted in different regions of the world and the importance and significance of the main properties that are regulated (cetane number, density, viscosity, low-temperature performances, flash point, water content, etc.) and unregulated (elemetal composition, fatty acid methyl and ethyl esters composition, heating value, lubricity, etc.). Properties of fatty acid methyl and ethyl esters obtained from different feedstocks[1] are presented based mainly on data published in the specialized literature, but also on personal research.

2. Biodiesel standardization world-wide

The main criterion of biodiesel quality is the inclusion of its physical and chemical properties into the requirements of the adequate standard. Quality standards for biodiesel are continuously updated, due to the evolution of compression ignition engines, ever-stricter emission standards, reevaluation of the eligibility of feedstocks used for the production of biodiesel, etc. The current standards for regulating the quality of biodiesel on the market are based on a variety of factors which vary from region to region, including

---

[1] ALME – algae methyl ester, CCEE – coconut oil ethyl ester; CCME – coconut oil methyl ester; CME – canola oil methyl ester; COME – corn oil methyl ester; CSOME – cottonseed oil methyl ester; FOEE – fish oil ethyl ester; FOME – fish oil methyl ester; JME – jatropha oil methyl ester; OEE – olive oil ethyl ester; OME – olive oil methyl ester; PEE – palm oil ethyl ester; PEEE – peanut oil ethyl ester; PEME – peanut oil methyl ester; PME – palm oil methyl ester; REE – rapeseed oil ethyl ester; RME – rapeseed oil methyl ester; SAFEE – safflower oil ethyl ester; SAFME – safflower oil methyl ester; SEE - soybean oil ethyl ester; SFEE – sunflower oil ethyl ester; SFME – sunflower oil methyl ester; SME – soybean oil methyl ester; TEE – tallow ethyl ester; TME – tallow methyl ester; WCOEE – waste cooking oil ethyl ester; WCOME – waste cooking oil methyl ester; YGME – yellow grease methyl ester; YMEE – yellow mustard oil ethyl ester; YMME – yellow mustard oil methyl ester.

characteristics of the existing diesel fuel standards, the predominance of the types of diesel engines most common in the region, the emissions regulations governing those engines, the development stage and the climatic properties of the region/country where it is produced and/or used, and not least, the purpose and motivation for the use of biodiesel (European Commission, 2007).

In Europe the fleet of cars equipped with diesel engines is considerable, while in the United States of America and Brazil diesel engines are specifically used in trucks. The most common feedstocks used are rapeseed and sunflower oil in Europe, soybean oil and waste vegetable oil in the USA and Canada, soybean oil in South America, palm, jatropha and coconut oil in Asia, palm oil and soybean oil in Australia and waste vegetable oil and animal fat in New Zealand. It is therefore not surprising that there are some significant differences among the regional standards, a universal quality specification of biodiesel is, and will be impossible. Table 1 presents a list of the most important biodiesel quality standards in the world, while in Tables 2-9 specifications of the imposed limits for the main properties of biodiesel and the required test methods are presented.

Country/Area	Specifications	Title
EU	EN 14213	Heating fuels - Fatty acid methyl esters (FAME) - Requirements and test methods
EU	EN 14214	EN 14214 Automotive fuels - Fatty acid methyl esters (FAME) for diesel engines - Requirements and test methods
U.S.	ASTM D 6751	ASTM D6751 - 11a Standard Specification for Biodiesel Fuel Blend Stock (B100) for Middle Distillate Fuels
Australia		Fuel Standard (Biodiesel) Determination 2003
Brazil	ANP 42	Brazilian Biodiesel Standard (Agência Nacional do Petróleo)
India	IS 15607	Bio diesel (B 100) blend stock for diesel fuel - Specification
Japan	JASO M360	Automotive fuel - Fatty acid methyl ester (FAME) as blend stock
South Africa	SANS 1935	Automotive biodiesel fuel

Table 1. Biodiesel standards

The biodiesel standards in Brazil and the U.S. are applicable for both fatty acid methyl esters (FAME) and fatty acid ethyl esters (FAEE), whereas the current European biodiesel standard is only applicable for fatty acid methyl esters (FAME). Also, the standards for biodiesel in Australia, Brazil, India, Japan, South Africa and the U.S. are used to describe a product that represents a blending component in conventional hydrocarbon based diesel fuel, while the European biodiesel standard describes a product that can be used either as a stand-alone fuel for diesel engines or as a blending component in conventional diesel fuel. Some specifications for biodiesel are feedstock neutral and some have been formulated around the locally available feedstock. The diversity in these technical specifications is primarily related to the origin of the feedstock and the characteristics of the local markets (European Commission, 2007; NREL, 2009; Prankl, et al., 2004).

The European standard EN 14214 is adopted by all 31 member states of the European Committee for Standardization (CEN): Austria, Belgium, Bulgaria, Croatia, Cyprus, the Czech Republic, Denmark, Estonia, Finland, France, Germany, Greece, Hungary, Iceland, Ireland, Italy, Latvia, Lithuania, Luxembourg, Malta, the Netherlands, Norway, Poland, Portugal, Romania, Slovakia, Slovenia, Spain, Sweden, Switzerland, and the United

Kingdom. Thus, there are no national regulations concerning biodiesel quality, but there is a separate section (not presented in the table), which provides cold flow property regulations. The national standards organizations provide the specific requirements for some regulations of CFPP (cold-filter plugging point, method EN 116), viscosity, density and distillation characteristics depending on the climate (6 stages for moderate climate and 5 for arctic climate). The regular diesel quality standard EN 590 specifies that commercial diesel fuel can contain 7% v/v biodiesel, compliant with the standard EN 14214. The standard ASTM D6751 describes the quality requirements and the methods of analysis used for biodiesel blended with diesel oil, applying to methyl esters as well as for ethyl esters. As the requirements for low-temperature properties can vary greatly, the standard foresees the indication of the cloud point. The standard ASTM D975 allows mixing commercial diesel oil with 5% biodiesel that meets the requirements of ASTM D6751, and ASTM D7467 specifies the quality requirements of mixtures with 5-20% of biodiesel.

Property	Test method	Limits		Units
		min	max	
Ester content	EN 14103	96.5	-	% (m/m)
Density at 15°C	EN ISO 3675, EN ISO 12185	860	900	kg/m^3
Viscosity at 40°C	EN ISO 3104, ISO 3105	3.5	5.0	mm^2/s
Flash point	EN ISO 3679	120	-	°C
Sulfur content	EN ISO 20846, EN ISO 20884	-	10.0	mg/kg
Carbon residue (in 10% dist. residue)	EN ISO 10370	-	0.30	% (m/m)
Sulfated ash content	ISO 3987	-	0.02	% (m/m)
Water content	EN ISO 12937	-	500	mg/kg
Total contamination	EN 12662	-	24	mg/kg
Oxidative stability, 110°C	EN 14112	4.0	-	hours
Acid value	EN 14104	-	0.50	mg KOH/g
Iodine value	EN 14111	-	130	g I/100 g
Polyunsaturated methyl esters (>= 4 double bonds)		-	1	% (m/m)
Monoglyceride content	EN 14105	-	0.80	% (m/m)
Diglyceride content	EN 14105	-	0.20	% (m/m)
Triglyceride content	EN 14105	-	0.20	% (m/m)
Free glycerine	EN 14105, EN 14106	-	0.02	% (m/m)
Cold-filter plugging point	EN 116	-	-	°C
Pour point	ISO 3016	-	0	°C
Net calorific value (calculated)	DIN 51900, -1, -2, -3	35	-	MJ/kg

Table 2. European standard EN 14213 for biodiesel as heating oil

Property	Test method	Limits min	Limits max	Unit
Ester content	EN 14103	96.5	-	% (m/m)
Density at 15°C	EN ISO 3675, EN ISO 12185	860	900	kg/m^3
Viscosity at 40°C	EN ISO 3104, ISO 3105	3.5	5.0	mm^2/s
Flash point	EN ISO 3679	120	-	°C
Sulfur content	EN ISO 20846, EN ISO 20884	-	10.0	mg/kg
Carbon residue (in 10% dist. residue)	EN ISO 10370	-	0.30	% (m/m)
Cetane number	EN ISO 5165	51	-	-
Sulfated ash	ISO 3987	-	0.02	% (m/m)
Water content	EN ISO 12937	-	500	mg/kg
Total contamination	EN 12662	-	24	mg/kg
Copper strip corrosion (3 hours, 50°C)	EN ISO 2160	-	1	class
Oxidative stability, 110°C	EN 14112	6.0	-	hours
Acid value	EN 14104	-	0.50	mg KOH/g
Iodine value	EN 14111	-	120	g I/100 g
Linolenic acid content	EN 14103	-	12	% (m/m)
Content of FAME with ≥4 double bonds		-	1	% (m/m)
Methanol content	EN 14110	-	0.20	% (m/m)
Monoglyceride content	EN 14105	-	0.80	% (m/m)
Diglyceride content	EN 14105	-	0.20	% (m/m)
Triglyceride content	EN 14105	-	0.20	% (m/m)
Free glycerine	EN 14105; EN 14106	-	0.02	% (m/m)
Total glycerine	EN 14105	-	0.25	% (m/m)
Alkali metals (Na + K)	EN 14108; EN 14109	-	5.0	mg/kg
Earth alkali metals (Ca + Mg)	EN 14538	-	5.0	mg/kg
Phosphorus content	EN 14107	-	10.0	mg/kg

Table 3. European biodiesel standard (EN 14214)

Property	Test Method	Limits min	Limits max	Units
Calcium & Magnesium, combined	EN 14538	-	5	ppm (µg/g)
Flash Point (closed cup)	D 93	93	-	°C
Alcohol Control (one to be met):				
1. Methanol Content	EN 14110	-	0.2	% (m/m)
2. Flash Point	D93	130	-	°C
Water & Sediment	D 2709	-	0.05	% (v/v)
Kinematic Viscosity, at 40 °C	D 445	1.9	6.0	mm^2/sec.
Sulfated Ash	D 874	-	0.02	% (m/m)
Sulfur: S 15 Grade S 500 Grade	D 5453 D 5453	- -	0.0015 0.05	% (m/m) % (m/m)
Copper Strip Corrosion	D 130	-	3	No.
Cetane	D 613	47	-	-
Cloud Point	D 2500	report		°C
Carbon Residue, 100% sample	D 4530	-	0.05	% (m/m)
Acid Number	D 664	-	0.05	mg KOH/g
Free Glycerin	D 6584	-	0.020	% (m/m)
Total Glycerin	D 6584	-	0.240	% (m/m)
Phosphorus Content	D 4951	-	0.001	% (m/m)
Distillation-Atmospheric equivalent temperature 90% recovery	D 1160	-	360	°C
Sodium/Potassium, combined	EN 14538	-	5	ppm (µg/g)
Oxidation Stability	EN 15751	-	3	hours
Cold Soak Filtration For use in temperatures below -12 °C	D7501 D7501	-	360 200	seconds seconds

Table 4. Biodiesel standard ASTM D6751 (United States)

Property	Test method	Limits min	Limits max	Unit
Sulfur	ASTM D5453	–	50 10	mg/kg
Density at 15 °C	ASTM D1298, EN ISO 3675	860	890	kg/m^3
Distillation T90	ASTM D1160	–	360	°C
Sulfated ash	ASTM D 874	–	0.20	% (m/m)
Viscosity at 40 °C	ASTM D445	3.5	5.0	mm^2/s
Flash point	ASTM D93	120	–	°C
Carbon residue - 10% dist. residue - 100% dist. sample	 EN ISO 10370 ASTM D4530	– – –	– 0.30 0.05	– % (m/m) % (m/m)
Water and sediment	ASTM D2709	–	0.050	% (v/v)
Copper strip corrosion (3 hours at 50°C) < 10 mg/kg of sulfur > 10 mg/kg of sulfur	 EN ISO 2160 ASTM D130	–	Class 1 No. 3	–
Ester content	EN 14103	96.5		% (m/m)
Phosphorus	ASTM D4951	–	10	mg/kg
Acid value	ASTM D664	–	0.80	mg KOH/g
Total contamination	EN 12662, ASTM D5452	–	24	mg/kg
Free glycerol	ASTM D6584	–	0.02	% (m/m)
Total glycerol	ASTM D6584	–	0.25	% (m/m)
Cetane number	EN ISO 5165, ASTM D613 ASTM D6890, IP 498/03	51	–	–
Cold-filter plugging point		report	–	°C
Oxidation stability 6 hours at 110°C	EN 14112, ASTM D2274 (as relevant for biodiesel)	–	–	hours
Metals: Group I (Na, K)	EN 14108, EN 14109 (Group I)	–	5	mg/kg
Metals: Group II (Ca, Mg)	EN 14538 (Group II)	–	5	mg/kg

Table 5. Australian biodiesel standard

Property	Test method	Limits min	Limits max	Units
Density at 15°C	ISO 3675 /P 32	860	900	kg/m^3
Kinematic viscosity at 40°C	ISO 3104 / P25	2.5	6.0	mm^2/s
Flash point (closed cup)	P21	120	-	°C
Sulphur	D5443/P83	-	50	mg/kg
Carbon resiue (Ramsbottom)	D4530	-	0.05	% (m/m)
Sulfated ash	ISO 6245/P4	-	0.02	% (m/m)
Water content	D2709 / P40	-	500	mg/kg
Total contamination	EN 12662	-	24	mg/kg
Copper corrosion 3 hr at 50°C	ISO 2160 / P15	-	1	-
Cetane number	ISO 5156/ P9	51	-	-
Acid value	P1	-	0.50	mg KOH/g
Methanol	EN 14110	-	0.20	% (m/m)
Ethanol		-	0.20	% (m/m)
Ester content	EN 14103	-	96.5	% (m/m)
Free glycerol, max	D6584	-	0.02	% (m/m)
Total glycerol, max	D6584	-	0.25	% (m/m)m
Phosphorous, max	D 4951	-	10.0	mg/kg
Sodium and potassium	EN 14108	To report		mg/kg
Calcium and magnesium	-	To report		mg/kg
Iodine value	EN 14104	To report		-
Oxidation stability at 110°C	EN 14112	6	-	hours

Table 6. Biodiesel standard in India

Property	Test method	Limits min	Limits max	Units
Ester content	EN 14103	96.5	-	% (m/m)
Density	JIS K 2249	0.86	0.90	g/ml
Kinematic Viscosity	JIS K 2283	3.5	5.0	mm^2/s
Flash Point	JIS K 2265	120	-	°C
Sulfur	JIS K 2541-1, -2, -6, -7	-	10	ppm
10% Carbon Residue	JIS K 2270	-	0.3	% (m/m)
Cetane number	JIS K 2280	51	-	-
Sulfated Ash	JIS K 2272	-	0.02	% (m/m)
Water	JIS K 2275	-	500	ppm
Total contamination	EN 12662	-	24	ppm
Copper strip corrosion (3 hours at 50 °C)	JIS K 2513	-	Class 1	rating
Total acid number	JIS K 2501, JIS K0070	-	0.5	mgKOH/g
Iodine Number	JIS K 0070	-	120	gI/100g
Methyl linolenate	EN 14103	-	12.0	% (m/m)
Methanol	JIS K 2536, EN 14110	-	0.20	% (m/m)
Monoglyceride	EN 14105	-	0.80	% (m/m)
Diglyceride	EN 14105	-	0.20	% (m/m)
Triglyceride	EN 14105	-	0.20	% (m/m)
Free glycerol	EN 14105, EN 14106	-	0.02	% (m/m)
Total glycerol	EN 14105	-	0.25	% (m/m)
Metals (Na + K)	EN 14108, EN 14109	-	5	ppm
Metals (Ca + Mg)	EN 14538	-	5	ppm
Phosphorous	EN 14107	-	10	ppm

Table 7. Japanese Biodiesel Specification

Property	Test method	Limits min	Limits max	Units
Ester content	EN 14103	96.5	-	% (m/m)
Density, at 15°C	ISO 3675, ISO 12185	860	900	kg/m^3
Kinematic viscosity at 40°C	ISO 3104	3.5	5.0	mm^2/s
Flash point	ISO 3679	120	-	°C
Sulfur content	ISO 20846, ISO 20884	-	10.0	mg/kg
Carbon residue (on 10% distillation residue)	ISO 10370	-	0.3	% (m/m)
Cetane number	ISO 5165	51.0	-	-
Sulfated ash content	ISO 3987	-	0.02	% (m/m)
Water content	ISO 12937	-	0.05	% (m/m)
Total contamination	EN 12662	-	24	mg/kg
Copper strip corrosion (3 hours at 50°C)	ISO 2160	-	No.1	rating
Oxidation stability, at 110°C	EN 14112	6	-	hours
Acid value	EN 14104	-	0.5	mg KOH
Iodine value	EN 14111	-	140	g I/100 g
Linolenic acid methyl ester	EN 14103	-	12	% (m/m)
Content of FAME with ≥4 double bonds		-	1	% (m/m)
Methanol content	EN 14110	-	0.2	% (m/m)
Monoglyceride content	EN 14105	-	0.8	% (m/m)
Diglyceride content	EN 14105	-	0.2	% (m/m)
Triglyceride content	EN 14105	-	0.2	% (m/m)
Free glycerol	EN 14105; EN 14106	-	0.02	% (m/m)
Total glycerol	EN 14105	-	0.25	% (m/m)
Group I metals (Na + K)	EN 14108; EN 14109	-	5.0	mg/kg
Group II metals (Ca + Mg)	EN 14538	-	5.0	mg/kg
Phosphorus content	EN 14107	-	10.0	mg/kg
Cold Filter Plugging Point (CFPP) Winter/Summer	EN 116	-	-4/+3	°C

Table 8. South African Biodiesel Standard

Property	Test method	Limits min	Limits max	Units
Flash point	ABNT NBR 14598, ASTM D93, EN ISO 3679	100	-	°C
Water and sediments	ASTM D2709	-	0.05	% (v/v)
Kinematic viscosity at 40 °C	ABNT NBR 10441, EN ISO 3104, ASTM D445	report		mm^2/s
Sulfated ash	ABNT NBR 9842, ASTM D874; ISO 3987	-	0.02	% (m/m)
Sulfur	ASTM D5453; EN/ISO 14596	-	0.001	% (m/m)
Copper corrosion 3 hours at 50 °C	ABNT NBR 14359, ASTM D130; EN/ISO 2160	-	No. 1	-
Ester content	EN 14103	report		% (m/m)
Distillation - atmospheric equivalent temperature 90% Recovery	D 1160	-	360	°C
Cetane number	ASTM D613; EN/ISO 5165	45	-	-
Cloud point	ASTM D6371	-	-	°C
Carbon Residue, 100% sample	ASTM D4530; EN/ISO 10370	-	0.05	% (m/m)
Acid number	ASTM D664; EN 14104	-	0.80	mg KOH/g
Total contamination	EN 12662	report		mg/kg
Free glycerin	ASTM D6854; EN 14105-6	-	0.02	% (m/m)
Total glycerin	ASTM D6854; EN 14105	-	0.38	% (m/m)
Distillation recovery 95%	ASTM D1160	-	360	°C
Phosphorus	ASTM D4951; EN 14107	-	10	mg/kg
Specific gravity	ABNT NBR 7148, 14065 ASTM D1298/4052	report		-
Alcohol	EN 14110	-	0.50	% (m/m)
Iodine number	EN 14111	report		gI/100g
Monoglycerides	ASTM D6584; EN 14105	-	1.00	% (m/m)
Diglycerides	ASTM D6584; EN 14105	-	0.25	% (m/m)
Triglycerides	ASTM D6584; EN 14105	-	0.25	% (m/m)
Metals: Group I (Na, K)	EN 14108, EN 14109	-	10	mg/kg
Metals: Group II (Ca, Mg)	EN 14538	report		mg/kg
Aspect	–	clear		-
Oxidation stability at 110°C	EN 14112	6	-	hours

Table 9. Brazilian biodiesel standard

3. Biodiesel fuel properties

The properties of biodiesel can be grouped by multiple criteria. The most important are those that influence the processes taking place in the engine (ignition qualities, ease of starting, formation and burning of the fuel-air mixture, exhaust gas formation and quality and the heating value, etc.), cold weather properties (cloud point, pour point and cold filter plugging point), transport and depositing (oxidative and hydrolytic stability, flash point, induction period, microbial contamination, filterability limit temperature, etc.), wear of engine parts (lubricity, cleaning effect, viscosity, compatibility with materials used to manufacture the fuel system, etc.).

3.1 Chemical composition of biodiesel

The elemental composition (carbon - C, hydrogen - H and oxygen - O), the C/H ratio and the chemical formula of diesel and biodiesel produced from different feedstocks is shown in Table 10 (Barabás & Todoruţ, 2010; Chuepeng &Komintarachat, 2010). The elemental composition of biodiesel varies slightly depending on the feedstock it is produced from. The most significant difference between biodiesel and diesel fuel composition is their oxygen content, which is between 10 and 13%. Biodiesel is in essence free of sulfur.

Fuel	C	H	O	C/H	Empirical formula
Diesel	86.5	13.5	0	6.24	$C_{15.05}H_{27.94}$
RME	77.2	12.0	10.8	6.45	$C_{19.03}H_{35.17}O_2$
SME	77.2	11.9	10.8	6.60	$C_{19.05}H_{34.98}O_2$
PME	76.35	11.26	12.39	6.16	$C_{18.07}H_{34.93}O_2$

Table 10. Elemental composition of diesel fuel and biodiesel, % (m/m)

Unlike fuels of petroleum origin, which are composed of hundreds of hydrocarbons (pure substances), biodiesel is composed solely of some fatty acid ethyl and methyl esters; their number depends on the feedstock used to manufacture biodiesel and is between 6 and 17 (Shannon & Wee, 2009). The fatty acid methyl and ethyl esters in the composition of biodiesel are made up of carbon, hydrogen and oxygen atoms that form linear chain molecules with single and double carbon-carbon bonds. The molecules with double bonds are unsaturated. Thus, fatty acid esters take the form C*nc*:*nd* (lipid numbers), where *nc* is the number of carbon atoms in the fatty acid and *nd* is the number of double bonds in the fatty acid (e.g., 18:1 indicates 18 carbon atoms and one double bond). The ester composition of biodiesel (methyl and ethyl esters) is shown in Table 11 (Bamgboye & Hansen, 2008; Barabás & Todoruţ, 2010; Chuepeng &Komintarachat, 2010). The highest concentrations are C18:1, C18:2, C18:3, followed by C18:0. A significant exception is biodiesel from coconut oil, in the case of which the highest concentration is C12:0, C14:0 and C16:0, hence this biodiesel is more volatile than the others. The physicochemical properties of biodiesel produced from a given feedstock are determined by the properties of the esters contained.

3.2 Cetane number

Cetane number (CN) is a dimensionless indicator that characterizes ignition quality of fuels for compression ignition engines (CIE). Since in the CIE burning of the fuel-air mixture is initiated by compression ignition of the fuel, the cetane number is a primary indicator of fuel quality as it describes the ease of its self-ignition.

Theoretically, the cetane number is defined in the range of 15-100; the limits are given by the two reference fuels used in the experimental determination of the cetane number:

Ester[2]	C8:0	C10:0	C12:0	C14:0	C16:0	C18:0	C18:1	C18:2	C18:3	C20:0	C20:1	Others	Obs.
ALME	–	–	–	0.6	6.9	3	75.2	12.4	1.2	0.4	–	0.3	–
RME	–	–	–	–	3.8	1.9	63.9	19	9.7	0.6	–	1.1	–
REE	–	–	–	–	4.9	1.6	33.0	20.4	7.9	–	–	22.2	22.2% C22:1
CME	–	–	–	–	4.2	2	57.4	21.3	11.2	1.2	2.1	0.60	–
SME	–	–	–	–	9.4	4.1	22	55.3	8.9	–	–	0.3	–
SEE	–	–	–	–	10.8	3	26.5	47.3	9	–	–	3.40	–
SFME	–	–	–	–	4.2	3.3	63.6	27.6	0.2	–	–	1.1	–
PME	–	–	0.2	0.5	43.4	4.6	41.9	8.6	0.3	0.3	–	0.2	–
COME	–	–	–	–	12.1	1.8	27.2	56.2	1.3	0.4	–	1	–
AME	–	–	–	–	11.6	4.4	49.6	33.7	0.7	–	–	–	–
OEE	–	–	–	–	11.6	3.1	74.9	7.8	0.6	–	–	2	–
TME	–	–	0.2	2.9	24.3	22.8	40.2	3.3	0.7	0.2	0.6	4.8	–
FOME	–	–	0.2	7.7	18.8	3.9	15	4.6	0.3	0.2	1.4	47.9	25.1% – C20:5
JME	–	–	–	–	12.7	5.5	39.1	41.6	0.2	0.2	–	0.7	–
JME	–	–	–	–	12.5	30.9	34.4	20.4	–	–	–	1.8	–
WCOME	–	–	0.1	0.1	11.8	4.4	25.3	49.5	7.1	0.3	–	1.4	–
WCOEE	–	–	2.0	–	15.7	3.1	29.6	41.5	1.0	–	–	7.1	–
SAFME	–	–	–	–	7.3	1.9	13.6	77.2	–	–	–	–	–
CCME	6.3	6	49.2	18.5	9.1	2.7	6.5	1.7	–	–	–	–	–
CCEE	7.5	6	53.3	17.1	7.3	1.9	5.5	1.4	–	–	–	–	–
YMME	–	–	–	–	2.6	1.2	20.6	20.6	13.3	0.9	10.7	30.1	25.6% – C20:1
YGME	–	–	0.1	0.5	14.3	8	35.6	35	4	0.3	–	2.2	-

Table 11. Fatty acid composition of different biodiesels (methyl and ethyl esters), % (m/m)

a linear-chain hydrocarbon, hexadecane ($C_{16}H_{34}$, also called n-cetane), very sensitive to ignition, having a cetane number of 100, and a strongly branched-chain hydrocarbon, 2,2,4,4,6,8,8-heptamethylnonane (HMN, also called isocetane), having the same chemical formula $C_{16}H_{34}$, with high resistance to ignition, having a cetane number of 15. The cetane number is the percentage by volume of normal cetane in a mixture of normal cetane and HMN, which has the same ignition characteristics as the test fuel. Thus the cetane number is given by the formula: CN = n-cetane [%, v/v] + 0.15*HMN [%, v/v]. Determination of the cetane number on the monocylinder engine specially designed for this purpose (EN ISO 5165, ASTM D613) is an expensive and lengthy operation. A cheaper and faster alternative is to determine the derived cetane number through ignition delay in a constant-volume combustion chamber (ignition quality tester - IQT), a widely accepted method described in ASTM D6890 and ASTM D7170, accepted by the biodiesel quality standard ASTM D6751. The cetane number indicates ignition delay, i.e. the time elapsed since the injection of fuel into the combustion chamber and self-ignition of the fuel-air mixture. Thus, ignition time lag

[2] C8:0 - caprylate, C10:0 - caprate, C12:0 - laurate, C14:0 - myristate, C16:0 - palmitate, C18:0 - stearate, C18:1 - oleate, C18:2 - linoleate, C18:3 - linolenate, C20:0 - arachidate, C22:1 - erucate.

means a low cetane number and vice versa. The upper and lower limits of the cetane number ensure the proper functioning of the engine. If the cetane number is too low, starting the engine will be difficult, especially at low temperatures and the engine will function unevenly and noisily, with cycles without combustion, it will warm more slowly, combustion will be incomplete and engine pollution will increase, especially hydrocarbon emissions. In case of a fuel with a very high cetane number, ignition will be carried out before a proper mix with air, resulting in incomplete combustion and the increase of the amount of exhaust smoke. Also, if the cetane number is too high the fuel will ignite close to the injector causing it to overheat, and unburned fuel particles can plug the injector nozzles. The optimal range of the CN (Fig. 1) is between 41 and 56, but must not be higher than 65 (Bățaga et al., 2003). The minimum cetane number of biodiesel is 51 in the European Union, 47 in the United States and 45 in Brazil. The minimum CN for diesel oil is 40 in the USA (ASTM D 975) and 51 in Europe (EN 590). The cetane numbers of the main pure methyl and ethyl esters are shown in Table 12 (Bamgboye & Hansen, 2008; Barabás & Todoruț, 2010).

Acid (*Cnc:nd*)	Cetane number		Heat of combustion, kJ/kg	
	Methyl ester	Ethyl ester	Methyl ester	Ethyl ester
Caprylate (C8:0)	n.d.	n.d.	34907	35582
Caprate (C10:0)	48.53	55.6	36674	37178
Laurate (C12:0)	61.99	73	37968	n.d.
Myristate (C14:0)	69.48	69.45	38431	n.d.
Palmitate (C16:0)	81.17	86.55	39449	n.d.
Palmitoleate (C16:1)	53.80	n.d.	39293	n.d.
Stearate (C18:0)	88.57	86.83	40099	n.d.
Oleate (C18:1)	62.39	64.57	40092	40336
Linoleate (C18:2)	42.10	40.37	39698	n.d.
Linolenate (C18:3)	32.20	26.7	39342	n.d.
Arachidate (C20:0)	100.00	n.d.	n.d.	n.d.
Erucate (C22:1)	76.00	n.d.	n.d.	n.d.

Table 12. Cetane number and heat of combustion for fatty acid esters

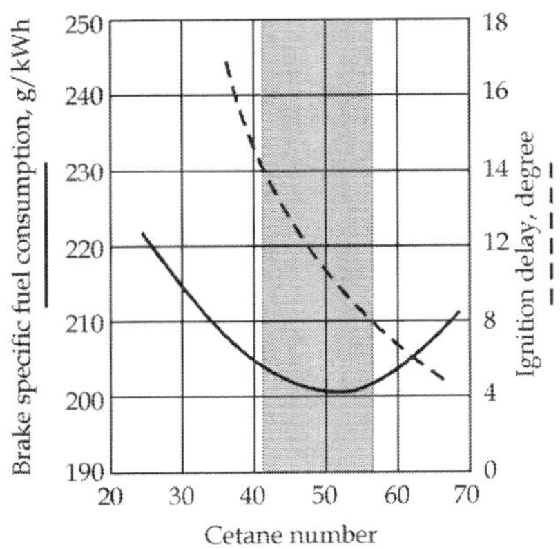

Fig. 1. Brake specific fuel consumption and ignition delay vs. fuel cetane number.

The cetane number of a substance depends on its molecular structure. The cetane number decreases with the number of double bonds, *nd*, in fatty acid ester molecules (degree of unsaturation, characterized by the iodine number) and increases with the number of carbon atoms, *nc*. Generally, the cetane number of ethyl esters is higher than that of methyl esters. Methyl- and ethyl palmitate as well as methyl- and ethyl stearate have a high cetane number, but methyl- and ethyl linoleate has a low cetane number. The cetane number of biodiesel depends on the cetane number and the concentration of the esters it is made up of. The cetane number of biodiesels is higher than that of the vegetable oils from which they are produced (34.6 < CN < 42), and is between 39 and 67. The cetane number values of biodiesel produced from various feedstocks are presented in Table 13 (Anastopoulos et al., 2009; Barabás & Todoruţ, 2010; Chuepeng &Komintarachat, 2010; Shannon et al., 2009; Fan et al., 2009).

Ester	CN	Q_g, kJ/kg	Q_n, kJ/kg	ν, mm²/s	ρ, kg/m³
AME	n.d.	n.d.	n.d.	4.52	879
CCEE	67.4	38158	n.d.	3.08	n.d.
CCME	57.4	n.d.	n.d.	n.d.	n.d.
CME	n.d.	n.d.	n.d.	4,34 - 4.84	883 - 888
COME	65	n.d.	38480	4.18 - 4.52	884
CSOME	45.5 - 51.2	40600	n.d.	4.06	874 - 884
FOEE	n.d.	n.d.	n.d.	3.98	887
FOME	51	n.d.	37800	4.96 - 5.76	850
JME	48	n.d.	38450	4.8 - 5.56	870 - 880.3
OEE	n.d.	38200	n.d.	4.0	881.5
OME	61	n.d.	37287	4.70	n.d.
PEE	56.2	39070	n.d.	n.d.	n.d.
PME	n.d.	39837	37500	3.70	864.4 - 870
REE	59.7 - 67.4	38300 - 40663	37820	4.84 - 6.17	876 - 881.2
RME	56 - 61.8	40540	37300 - 37780	4.83 - 5.65	880.2
SAFEE	62.2	39872	n.d.	4.31	n.d.
SAFME	49.8	40060	n.d.	4.03	880
SEE	48.2	40160	n.d.	4.40 - 5.03	833
SME	50.9	40400	37000	4.8	n.d.
SFEE	n.d.	38600	n.d.	4.43	882.7
SFME	45.5 - 58	n.d.	38472	4.03	878 - 884
SME	37 - 51.5	39871	37388	3.97 - 4.27	872 - 885
TEE	n.d.	n.d.	n.d.	5.20	n.d.
TME	58 - 61.8	39961-40200	37531	4.1 - 4.99	876 - 887
WCOEE	n.d.	37800 - 40500	37200	5.81	888.5
WCOME	n.d.	40110	n.d.	5.78 - 6.0	920
YMEE	54.9	40679	n.d.	5.66	n.d.

Table 13. Cetane number, gross and net heat of combustion, viscosity and density of biodiesels from different feedstoks

3.3 Heat of combustion

The heat of combustion (heating value) at constant volume of a fuel containing only the elements carbon, hydrogen, oxygen, nitrogen, and sulfur is the quantity of heat liberated when a unit quantity of the fuel is burned in oxygen in an enclosure of constant volume, the products of combustion being gaseous carbon dioxide, nitrogen, sulfur dioxide, and water, with the initial temperature of the fuel and the oxygen and the final temperature of the products at 25°C. The unit quantity can be mol, kilogram or normal square meter. Thus the units of measurement of the heating value are kJ/kmol, kJ/kg. The volumetric heat of combustion, i.e. the heat of combustion per unit volume of fuel, can be calculated by multiplying the mass heat of combustion by the density of the fuel (mass per unit volume). The volumetric heat of combustion, rather than the mass heat of combustion is important to volume-dosed fueling systems, such as diesel engines.
The gross (or high, upper) heating value (Q_g) is obtained when all products of the combustion are cooled down to the temperature before the combustion and the water vapor formed during combustion is condensed. The net or lower heating value (Q_n) is obtained by subtracting the latent heat of vaporization of the water vapor formed by the combustion from the gross or higher heating value. The net heat of combustion is related to the gross heat of combustion: $Q_n = Q_g - 0.2122 \cdot H$, where H is the mass percentage of hydrogen in the fuel. As in internal combustion engines the temperature of exhaust gases is higher than the boiling temperature of water (water vapor is discharged), for assessing the heating value of the fuel, the lower heating value of the biodiesel is more relevant. The heating value of fatty acid esters (Table 12) increases with molecular chain length (with the number of carbon atoms, *nc*) and decreases with their degree of unsaturation (the number of double bonds, *nd*). The mass heating value of unsaturated esters is lower than that of saturated esters, but due to their higher density, the volume heating value of unsaturated esters is higher than that of saturated esters. For example, methyl stearate (*nd*=0) has a mass heating value of 40099 kJ/kg, and methyl oleate (*nd*=1) has 40092 kJ/kg. Reported to the volume unit, the heating value of methyl stearate is 34070 kJ/L, while the volume heating value of methyl oleate is 34320 kJ/L. The presence of oxygen in the esters molecules (Table 1) decreases the heating value of biodiesel by 10...13% compared to the heating value of diesel fuel (see Table 13). Due to the fact that fuel dispensing in CIE is volumetric, the energy content of the injected dose will be more reduced in the case of biodiesel, therefore, the specific fuel consumption for biodiesel will be higher. This is partially compensated by the fact that the density of biodiesel is higher than that of diesel fuel.

3.4 Density of biodiesel

Fuel density (ρ) is the mass of unit volume, measured in a vacuum. Since density is strongly influenced by temperature, the quality standards state the determination of density at 15 °C. Fuel density directly affects fuel performance, as some of the engine properties, such as cetane number, heating value and viscosity are strongly connected to density. The density of the fuel also affects the quality of atomization and combustion. As diesel engine fuel systems (the pump and the injectors) meter the fuel by volume, modification of the density affects the fuel mass that reaches the combustion chamber, and thus the energy content of the fuel dose, altering the fuel/air ratio and the engine's power. Knowing the density is also necessary in the manufacturing, storage, transportation and distribution process of biodiesel as it is an important parameter to be taken into account in the design of these processes. The density of esters depends on the molar mass, the free fatty acid content, the water content and the temperature. Density values determined for pure esters are presented in Table 14 and for different biodiesel feedstock are listed in Table 13. The density of biodiesel is typically higher than that of diesel fuel and is dependent on fatty acid composition and purity. As biodiesel is made up of a small number of methyl or ethyl esters that have very similar densities, the density of biodiesel varies between tight limits. Contamination of the biodiesel significantly affects its density; therefore density can also be an indicator of contamination.

3.5 Viscosity of biodiesel

The viscosity of liquid fuels is their property to resist the relative movement tendency of their composing layers due to intermolecular attraction forces (viscosity is the reverse of fluidity). Viscosity is one of the most important properties of biodiesel. Viscosity influences the ease of starting the engine, the spray quality, the size of the particles (drops), the penetration of the injected jet and the quality of the fuel-air mixture combustion (Alptekin and Canakci 2009). Fuel viscosity has both an upper and a lower limit. The fuel with a too low viscosity provides a very fine spray, the drops having a very low mass and speed. This leads to insufficient penetration and the formation of *black smoke* specific to combustion in the absence of oxygen (near the injector) (Băţaga et al., 2003). A too viscous biodiesel leads to the formation of too big drops, which will penetrate to the wall opposite to the injector. The cylinder surface being cold, it will interrupt the combustion reaction and *blue smoke* will form (intermediate combustion product consisting of aldehydes and acids with pungent odor) (Băţaga et al., 2003). Incomplete combustion results in lower engine power. Too high viscosity leads to the increase of combustion chamber deposits and the increase of the needed fuel pumping energy, as well as the increased wear of the pump and the injector elements due to higher mechanical effort. Too high viscosity also causes operational problems at low temperatures because the viscosity increases with decreasing temperature (for temperatures at or below -20 °C viscosity should be at or below 48 mm^2/s). Viscosity also influences the lubricity of the fuel as some elements of the fuel system can only be lubricated by the fuel (pumps and injectors). Due to the presence of the electronegative oxygen, biodiesel is more polar than diesel fuel; as a result, the viscosity of biodiesel is higher than that of diesel fuel. The viscosity of pure ethyl esters are higher then viscosity of methyl esters (Table 14). The viscosities of biodiesels from different feedstoks are presented in Table 13.

3.6 Cold flow properties

Generally, all fuels for CIE may cause starting problems at low temperatures, due to worsening of the fuel's flow properties at those temperatures. The cause of these problems is the formation of small crystals suspended in the liquid phase, which can clog fuel filters partially or totally. Because of the sedimentation of these crystals on the inner walls of the fuel system's pipes, the flow section through the pipes is reduced, causing poor engine fueling. In extreme situations, when low temperatures persist longer (e.g. overnight), the fuel system can be completely blocked by the solidified fuel.

The cloud flow performances of the fuels can be characterized by the could point (CP), the pour point (PP), the cold filter plugging point (CFPP) and viscosity (ν). An alternative to CFPP is the low-temperature flow test (LTFT). Recently, the U.S. introduced a new method for assessing the cold flow properties of biodiesel, called cold soak filtration test (CSFT).

Acid (Cnc:nd)	Density, kg/m³				Dynamic and kinematic viscosity			
	Methyl ester		Ethyl ester		Methyl ester		Ethyl ester	
	15 °C	40 °C	15 °C	40 °C	mPa·s	mm²/s	mPa·s	mm²/s
Caprylate (C8:0)	881.5	859.6	871.6	850.0	1.0444	1.2150	n.d.	n.d.
Caprate (C10:0)	876.4	856.0	868.4	848.0	1.4773	1.7258	1.6000	1.8868
Laurate (C12:0)	873.7	853.9	866.4	846.8	2.0776	2.4331	2.2198	2.6214
Myristate (C14:0)	n.d.	852.2	864.8	845.8	2.8447	3.3381	2.9928	3.5384
Palmitate (C16:0)	n.d.	850.8	n.d.	n.d.	3.7551	4.4136	3.9558	n.d.
Palmitoleate (C16:1)	872.8	853.8	n.d.	n.d.	2.6162	3.0642	n.d.	n.d.
Stearate (C18:0)	n.d.	849.8	n.d.	844.8	4.9862	5.8675	5.0823	6.0160
Oleate (C18:1)	877.7	859.5	874.1	855.8	3.9303	4.5728	4.2137	4.9237
Linoleate (C18:2)	889.9	871.5	886.3	867.8	3.2270	3.7028	3.4060	3.9249
Linolenate (C18:3)	905.7	887.0	897.0	878.3	2.9253	3.2980	2.9750	3.3872
Erucate (C22:1)	874.3	856.5	n.d.	n.d.	5.9575	6.9556	n.d.	n.d.

Table 14. Density and viscosity of fatty acid esters

3.6.1 Cloud point (CP)

The cloud point (CP) is the temperature at which crystals first start to form in the fuel. The cloud point is reached when the temperature of the biodiesel is low enough to cause wax crystals to precipitate. Initially, cooling temperatures cause the formation of the solid wax crystal nuclei that are submicron in scale and invisible to the human eye. Further decrease of temperature causes these crystals to grow. The temperature at which crystals become visible (the crystal's diameter ≥ 0.5 μm) is defined as the cloud point because the crystals form a cloudy suspension. Below the CP these crystals might plug filters or drop to the bottom of a storage tank. The CP is the most commonly used measure of low-temperature operability of the fuel. The biodiesel cloud point is typically higher than the cloud point of conventional diesel. The cloud point of biodiesel depends on the nature of the feedstock it was obtained from (Table 15) (Barabás & Todoruţ, 2010; Fan et al., 2009), and is between -5 °C (ALME) and 17 °C (TME).

3.6.2 Pour point (PP)

The pour point is the temperature at which the fuel contains so many agglomerated crystals that it is essentially a gel and will no longer flow. This occurs if the temperature of the biodiesel drops below CP, when the microcrystals merge and form large clusters, which may disrupt the flow of the biodiesel through the pipes of the engine's fuel system. Similarly to the cloud point, the pour point values also depend on the feedstock the biodiesel was produced from (see Table 15). Pour point values are between -15 °C (REE and YMEE) and 16 °C (PME). Although CP and PP are relatively easily determined, they only provide indicative values for the minimum temperature at which the fuel can be used. While at cloud point the fuel can still be used in acceptable conditions, at pour point this is no longer possible. In other words, cloud point overestimates minimum operating temperature and pour point underestimates it.

3.6.3 Cold filter plugging point (CFPP)

The cold filter plugging point is the lowest temperature at which 20 mL of fuel passes through a filter within 60 s by applying a vacuum of 2 kPa. The CFPP test employs rapid cooling conditions. For this reason, CFPP does not reflect the actual limit of the fuel's operability temperature. The test does not take into account the fuel systems specially designed to operate at low temperatures (heavy-duty vehicles and some light-duty vehicles). Nevertheless, most standards require the determination of this parameter and its value is regulated depending on the climatic conditions of each region or country. The values of the CFPP of biodiesel produced from various feedstocks are listed in Table 15. CME has the lowest value, while TME has the highest. Biodiesel produced from the most common feedstocks has inferior cold flow properties compared to conventional diesel fuel (has a higher cloud point and pour point compared to petroleum diesel), which can lead to operational issues in cold climates, such as filter plugging due to wax buildup or reduced fuel flow. Conventional diesel blends with 10 % (v/v) biodiesels typically have significantly higher CP, PP and CFPP than petroleum diesel fuel.

Ester	CP, C	PP, °C	CFPP, °C	LTFT, °C	CSFT, s
ALME	-5	n.d.	-7	n.d.	85
CCEE	5	-3	n.d.	n.d.	n.d.
CCME	0	n.d.	-4	n.d.	49
CEE	-1	-6	n.d.	n.d.	n.d.
CME	-3...1	-9...-4	-13...-4	n.d.	113
COME	-3	-4	-3...-7	n.d.	131
CSOME	6	-4...0	3	n.d.	n.d.
FOME	4	n.d.	0	n.d.	68...81
JME	n.d.	-1	n.d.	n.d.	n.d.
OEE	7	-5	-3	n.d.	n.d.
OME	-2	-3	-6	n.d.	n.d.
PEE	8...16	6...12	n.d.	n.d.	n.d.
PME	13...16	14...16	10...14	n.d.	88
REE	-2...1	-15...-12	1	n.d.	n.d.
RME	-3...1	-9...-11	-10...-6	n.d.	233
SAFEE	-6	-6	n.d.	n.d.	n.d.
SAFME	n.d.	-6	n.d.	n.d.	n.d.
SEE	-2...15	-6...5	n.d.	n.d.	n.d.
SME	-2...3	-7...-1	-4...-2	-2...0	67
SFEE	-1...2	-6...-5	-3	n.d.	n.d.
SFME	0...4	-4...-3	-4...-2	n.d.	107
TEE	15	3...12	8	13	n.d.
TME	9...17	9...15	9...14	20	76
WCOEE	9	-1...8	3	n.d.	n.d.
WCOME	-2...3	-3...-6	-2...-9	n.d.	233
YGME	6	n.d.	2	n.d.	95
YMEE	1	-15	n.d.	n.d.	n.d.
YMME	4	n.d.	-5	n.d.	n.d.

Table 15. Cold flow properties of biodiesels from different feedstoks

3.6.4 Low-temperature flow test (LTFT)

Although CFPP is accepted almost worldwide as the minimum temperature at which fuel can be exploited, mainly because of the rapid cooling of the sample, the test does not entirely reflect real cooling conditions of the fuel. The Low-Temperature Flow Test (LTFT) is a similar attempt to the test determining the CFPP, the major difference being the cooling speed of the fuel sample, which in this case is 1 °C/h, reflecting more accurately the real conditions, when for example the fuel in the fuel system of a vehicle is cooled over a frosty night. In determining the low temperature flow temperature the sample volume is 180 mL, the filter is finer, and the vacuum filtration pressure is higher. Like CFPP, LTFT is defined as the lowest temperature at which 180 mL of fuel safely passes through the filter within 60 s. Since the LTFT is not included in biodiesel quality standards, currently there is very limited information about its values for biodiesel (see Table 15).

3.6.5 Cold soak filtration test (CSFT)

This test is the newest requirement under ASTM D6751, added in 2008 in response to data indicating that in blends with petroleum diesel of up to 20% some biodiesels could form precipitates above the cloud point. Some substances that are or seem to be soluble at ambient temperature come out of the solution if temperature decreases or biodiesel is stored at ambient temperature for a longer period. This phenomenon was observed both in the case of pure biodiesel and its blends with diesel fuel.

Solid or semi-liquid substances can, in turn, cause filter clogging. The CSFT allows highlighting this danger and improving biodiesel due to this phenomenon. Cold soak consists of chilling a 300 ml sample for 16 hours at 4 °C, then warming it up to ambient temperature (68-72 °F, 20-22 °C) and filtering with a 0.7 micron glass fiber filter with a stainless steel filter support. The result of this test is filtering time. There are two time limits for filtration: in the case of net biodiesel for use throughout the year, the filtration time is 360 seconds or less; if the seller claims the post-blended biodiesel is fit for use in temperatures below 10 °F (-12 °C) the filtration time is 200 seconds or less. The test result depends mainly on the type and quality of the used feedstock, the purity of biodiesel, the soap value, the total glycerin, etc. The higher the soap value, the higher the cold soak filtration results. In addition it was found that total glycerin can also negatively influence the cold soak filtration results. When the total glycerin is within the ASTM D 6751 standard's limits ($\leq 0.24\%$), it will show no negative effect on the cold soak filtration results (Fan et al., 2009). Because CSFT has only recently been included in biodiesel quality standards, at present there is very little reported data on this parameter (Table 15).

3.7 Biodiesel lubricity

Lubricity describes the ability of the fuel to reduce the friction between surfaces that are under load. This ability reduces the damage that can be caused by friction in fuel pumps and injectors (Schumacher, 2005). Lubricity is an important consideration when using low and ultra-low sulfur fuels (ULSD). The fuel lubricity can be measured with High Frequency Reciprocating Rig (HFRR) test methods as described at ISO 12156-1. The maximum corrected wear scar diameter (WS 1.4) for diesel fuels is 460 μm (EN 590). Reformulated diesel fuel has a lower lubricity and requires lubricity improving additives (which must be compatible with the fuel and with any additives already found in the fuel) to prevent excessive engine wear. The lubricity of biodiesel is excellent. Biodiesel may be used as a lubricity improver. The lubricity of some biodiesels and the influence of biodiesel concentration on this parameter in blends with diesel fuel are shown in Table 16 (Barabás & Todoruţ, 2010; Schumacher, 2005). The lubricity of biodiesel depends on the feedstock it is produced from. Biodiesel from jatropha oil has the highest and biodiesel sunflower oil has the lowest lubricity. Generally, it can be stated that 1% (v/v) biodiesel mixed with ultra-low sulfur diesel fuel (ULSD) already provides lubricity that meets the requirements of the commercial diesel fuel's lubricity quality standards.

Biodiesel	Biodiesel concentration, % (v/v)					
	0	0.25	0.5	1	2	100
CME	735	n.d.	n.d.	n.d.	351	n.d.
COME	735	n.d.	n.d.	n.d.	366	n.d.
JME	570	345	325	265	165	95
PME	570	490	380	265	185	135
RME	520	518	517	395	n.d.	n.d.
SFME	735	n.d.	n.d.	n.d.	429 (2 % w/w)	n.d.
SME	735	n.d.	n.d.	n.d.	375	n.d.
SME	n.d.	485	525	325	n.d.	n.d.

Table 16. The effect of the biodiesel additive in ULSD on WS1.4 (μm)

3.8 Flash point (FP)

The flash point is the minimum temperature calculated to a barometric pressure of 101.3 kPa at which the fuel will ignite (flash) on application of an ignition source under specified conditions. It is used to classify fuels for transport, storage and distribution according to hazard level. The flash point does not affect the combustion directly; higher values make fuels safer with regard to storage, fuel handling and transportation. FP varies inversely with the fuel's volatility. For biodiesel the minimum flash point is 93 °C in the United States, 100 °C in Brazil and 120 °C in Europe. Biodiesel's flash point decreases rapidly as the amount of residual (un-reacted) alcohol increases (methanol's flash point is 11–12 °C, and ethanol's is 13–14 °C). Thus, measuring the biodiesel flash point helps indicate the presence of methanol or ethanol. For example, the presence of 0.5% methanol in biodiesel reduces biodiesel flash point from 170 °C to 50 °C. If flash point is used to determine the methanol content, the ASTM standard imposes for it a minimum value of 130 °C. This limit may be considered too severe, because at the maximum permissible concentration of methanol of 0.2% w/w biodiesel flash point drops below 130 °C. The flash point of biodiesel produced from various feedstocks are presented in Table 17 (Anastopoulos et al., 2009; Barabás & Todoruţ, 2010; Barabás et al., 2010; Chuepeng &Komintarachat, 2010; Pinyaphong et al., 2011; Shannon et al., 2009; Fan et al., 2009).

3.9 Acid value

The acid value (AV), also called neutralization number or acid number is the mass of potassium hydroxide (KOH) in milligrams that is required to neutralize the acidic constituents in one gram of sample. The acid value determination is used to quantify the presence of acid moieties in a biodiesel sample. In a typical procedure, a known amount of sample dissolved in organic solvent is titrated with a solution of potassium hydroxide with known concentration and with phenolphthalein as a color indicator. The acidic compounds that could possibly be found in biodiesel are: 1) residual mineral acids from the production process, 2) residual free fatty acid from the hydrolysis process or the post- hydrolysis process of the esters and 3) oxidation byproducts in the form of other organic acids (Berthiaume & Tremblay, 2006). This parameter is a direct measure of the content of free fatty acids, thus the corrosiveness of the fuel, of filter clogging and the presence of water in the biodiesel. A too high amount of free glycerin can cause functioning problems at reduced temperatures and fuel filter clogging. This parameter can also be used to measure the freshness of the biodiesel. Fuel that has oxidized after long-term storage will probably have a higher acid value.

Ester	FP, °C	AV, mg KOH/g	IV, g Iodine/100 g FAME	Oxidation stability, hours
ALME	>160	n.d.	n.d.	8.5 - 11.0
CCEE	190	n.d.	n.d.	n.d.
CCME	115	n.d.	n.d.	35.5
COME	111 - 170	0.15	101.0 – 119.18	n.d.
CSOME	110	n.d.	n.d.	n.d.
FOME	>160	1.11	n.d.	0.2
JME	170	0.38	105.0	2.3
OEE	182	n.d.	n.d.	n.d.
OME	178	0.13	84.0	3.3
PME	176	0.12	57.0 – 59.0	n.d.
REE	170 – 181	n.d.	99.7	2.0. – 7.6
RME	166 - 179	0.14 – 0.16	97.4 – 109.0	n.d.
SAFEE	178	n.d.	n.d.	n.d.
SAFME	149..180	n.d.	139.83	n.d.
SEE	55	0.28	123.0	n.d.
SME	171	0.14	120.52 – 133.2	n.d.
SFEE	178	0.15	132.0 – 136.0	n.d.
SFME	85 – 177	0.15	n.d.	0.8...0.9
SME	120 – 190	0.30	n.d.	2.1
TME	96 – 188	n.d.	n.d.	1.6
WCOEE	124	n.d.	n.d.	0.33
WCOME	141	0.14 - 0.69	n.d.	>6
WCOME	110...160	0.14	n.d.	1.0
YGME	>160	n.d.	n.d.	5.2
YMEE	183	n.d.	n.d.	n.d.
YMME	n.d.	n.d.	n.d.	1.1

Table 17. Flash point, acid value, iodine value and oxidation stability of biodiesels from different feedstoks

3.10 Iodine value

The iodine value (IV) or iodine number was introduced in biodiesel quality standards for evaluating their stability to oxidation. The IV is a measurement of total unsaturation of fatty acids measured in g iodine/100 g of biodiesel sample, when formally adding iodine to the double bonds. Biodiesel with high IV is easily oxidized in contact with air. The iodine value highly depends on the nature and ester composition of the feedstocks used in biodiesel production. Therefore the IV is limited in various regions of the world depending on the specific conditions: 120 in Europe and Japan, 130 in Europe for biodiesel as heating oil, 140 in South Africa, in Brazil it is not limited and in the U.S., Australia and India it is not included in the quality standard (it would exclude feedstocks like sunflower and soybean oil). Biodiesel with high IV tends to polymerize and form deposits on injector nozzles, piston rings and piston ring grooves. The tendency of polymerization increases with the degree of unsaturation of the fatty acids.

3.11 Biodiesel stability

Biodiesel quality can be affected by oxidation during storage (in contact with air) and hydrolytic degradation (in contact with water). The two processes can be characterized by the oxidative stability and hydrolytic stability of the biodiesel. Biodiesel oxidation can occur during storage while awaiting distribution or within the vehicle fuel system itself.

The stability of biodiesel can refer to two issues: *long-term storage stability or aging* and *stability at elevated temperatures or pressures* as the fuel is recirculated through an engine's fuel system (NREL 2009).

For biodiesel, storage stability is highly important. Storage stability refers to the ability of the fuel to resist chemical changes during long term storage. These changes usually consist of oxidation due to contact with oxygen from the air (Gerpen, 2005).

Biodiesel composition greatly affects its stability in contact with air. Unsaturated fatty acids, especially the polyunsaturated ones (e.g. C18:2 and C18:3) have a high tendency to oxidation. After oxidation, hydroperoxides (one hydrogen atom and 2 oxygen atoms) are attached to the fatty acid chain. Oxidation reactions can be catalyzed by some of the materials present (the material the reservoir is produced from) and light. After the chemical oxidation reactions hydroperoxides are produced that can, in turn, produce short chain fatty acids, aldehydes, and ketones. Hydroperoxides can polymerize forming large molecules. Thus, oxidation increases the viscosity of biodiesel. In addition, oxidation increases acid value, the color changes from yellow to brown, solid deposits can form in the engine fuel system (pipes and filters), the lubricity and heating value of the biodiesel is reduced.

When water is present, the esters can hydrolyze to long chain free fatty acids, which also cause the acid value to increase (Gerpen, 2005). These acids can catalyze other degradation reactions such as reverse trans-esterification and oxidation. The water required for hydrolysis can be present as a contaminant (Engelen, 2009). For determining the oxidation stability of biodiesel two types of tests are currently used: the Rancimat test, contained in EN 14214 and the oxidative stability index (OSI) included in ASTM D6751.

The Rancimat test method (EN 14112, EN 15751) is an accelerated oxidation test in which the biodiesel to be tested is run at elevated temperatures (110 °C) whilst exposing the sample to a stream of purified air (10 L/hour) accelerating the oxidation process of the oil. After passing through the biodiesel, the air is fed into a collection flask containing distilled water and a probe to measure conductivity. As the biodiesel sample degrades, the volatile organic acids produced are carried to the collection flask, and the conductivity of the solution is recorded by the probe. Oxidation stability will be given by the induction period, defined as the time between the start of the test and the sudden conductivity increase of the solution in the collection flask. This results in auto-oxidation in a few hours, instead of months.

The oxidative stability index (OSI) is another measurement method of the conductivity increase caused by the formation of secondary products in the oxidation process. The OSI is defined as the time until the conductivity of a biodiesel sample rises most rapidly during an accelerated oxidation test. The oxidation of biodiesel is influenced by its composition (increases with the level of unsaturation of fatty acids in its composition), i.e. the feedstock used to manufacture the biodiesel. For example, the content of oleic acid methyl ester in the case of biodiesel produced from sunflower oil may vary between 48 and 74%. In addition, the induction period of biodiesel made from rapeseed oil is 12 times greater than those obtained from soybean oil and 25 times higher than those produced from linseed oil. The presence of metals (the tank walls and metals contained in the biodiesel) can accelerate the oxidation process, whereas sulfur is an antioxidant (Berthiaume & Tremblay, 2006). Oxidation stability can be improved by using the appropriate additives. Additives such as tert-butylhydroquinone (TBHQ), butylated hydroxyanisole (BHA), butylated hydroxytoluene (BHT), propyl gallate (PrG) and alpha-tocopherol (vitamin E) have been found to enhance the storage stability of biodiesel. Biodiesels produced from some feedstocks (e.g. soybean oil) naturally contain some antioxidants. Any fuel that will be stored for more than 6 months, whether it is diesel fuel or biodiesel, should be treated with an antioxidant additive (Gerpen, 2005).

3.12 Water and sediments

Water content is a purity indicator for the biodiesel. Biodiesel should be dried after water washing to get the water specification below 500 ppm (0.050 %). Even when biodiesel is dried properly by the producer, water can accumulate during storage and transportation.

The moisture accumulated in biodiesel leads to the increase of free fatty acid concentration, which can corrode metal parts of the engine's fuel system. Biodiesel is much more hygroscopic (it attracts water) than diesel oil. The biodiesel absorbs water during storage when the temperature is higher and the water absorbed is precipitated at lower temperatures. Following these repeated processes, the accumulated water is deposited on the bottom of the tank. Water in biodiesel facilitates microbial growth and the formation of sediments. To measure the water and sediment content, a 100 mL sample of undiluted fuel is centrifuged at a relative centrifugal force of 800 for 10 minutes at 21 to 32°C (70 to 90°F). After centrifugation, the volume of water and sediment which has settled into the tip of the centrifuge tube is read to the nearest 0.005 mL and reported as the volumetric percent of water and sediment.

3.13 Other properties

Sulfated ash is a measure of ash formed from inorganic metallic compounds. After the burning of biodiesel, in addition to CO_2 and H_2O a quantity of ash is formed consisting of unburned hydrocarbons and inorganic impurities (e.g. metal impurities). Metallic ash is very abrasive and may cause excessive wear of the cylinder walls and the piston ring.

Carbon residue indicates the presence of impurities and deposits in the engine combustion chamber, and is also an indicator of the quantity of glycerides, free fatty acids, soaps and transesterification reaction catalyst residues.

Copper-strip corrosion is an indicator of the corrosiveness of biodiesel, of the presence of fatty acids derived from materials which did not enter into reaction during the production process.
Content of metals (Ca, Na, Mg, K and P) can lead to combustion chamber deposits, filter- and fuel injection pump clogging, and can harm the catalyst.

4. Monitoring the quality of biodiesel

Biodiesel quality can be provided efficiently if its entire manufacturing process is monitored: from monitoring feedstock acidity, assuring complete separation of biodiesel from glycerin, to removing the excess of alcohol and contaminants before its marketing. Quality assurance and monitoring should include storage, testing, blending and distribution. Fuel quality monitoring is conducted by independent laboratories that can accredit manufacturers, distributors and quality analysis laboratories. One example is the BQ-9000® program in the United States of America, a program based on voluntary cooperation, which accredits manufacturers, marketers and biodiesel quality analysis laboratories. Monitoring the quality of biodiesel contributes to its promotion and public acceptance.

5. Conclusions

An adequate and constant quality of biodiesels can only be assured by respecting the biodiesel quality standards. To achieve this goal it is necessary to monitor the quality throughout the biodiesel manufacturing process, from the feedstock to the distribution stations. The physicochemical properties of biodiesels are strongly influenced by the nature and the composition of the feedstocks used in their production. Therefore, quality requirements for the marketing of biodiesel vary from region to region. The largest differences are found in cetane number, oxidation stability, iodine value, density and viscosity. Other reasons for these differences are the weather conditions, reflected in the regulations of properties describing performances of biodiesel at low temperatures. Due to these major differences, unifying the standards for biodiesel is not possible. This could be a serious impediment for both biodiesel imports and exports among different regions of the world, as well as automotive producers, who must adapt their engines to the quality of biodiesel in the region where the vehicles will be used.

6. References

Anastopoulos, G.; Zannikou, Y.; Stournas, S. & Kalligeros, S. (2009). Transesterification of Vegetable Oils with Ethanol and Characterization of the Key Fuel Properties of Ethyl Esters. *Energies*, Vol.2, No.2 (June 2009), pp. 362-376.

Bamgboye, A.I. & Hansen, A.C. (2008). Prediction of cetane number of biodiesel fuel from the fatty acid methyl ester (FAME) composition. *International Agrophysics*, (January, 2008), Vol.22, No.1, pp. 21-29. ISSN 0236-8722. 17.06.2011, Available from: http://www.international-agrophysics.org/artykuly/international_agrophysics/IntAgr_2008_22_1_21.pdf.

Barabás, I. & Todorut, A. (2010). *Combustibili pentru automobile: testare, utilizare, evaluare.* UT PRESS, 978-973-662-595-4, Cluj-Napoca, Romania.

Barabás, I., Todorut, A. & Baldean, D. (2010). Performance and emission characteristics of an CI engine fueled with diesel-biodiesel-bioethanol blends. *Fuel*, Vol.89, No.12, (December, 2010)pp. 3827-3832.

Băţaga, N., Burnete, N. & Barabás, I. (2003). Combustibili, lubrifianţi, materiale speciale pentru autovehicule. Economicitate şi poluare. Alma Mater, ISBN 973-9471-20-X, Cluj-Napoca, Romania.

Berthiaume, D. & Tremblay, A. (2006) Study of the Rancimat Test Method in Measuring the Oxidation Stability of Biodiesel Ester and Blends. NRCan project No. CO414 CETC-327, OLEOTEK Inc., Québec, Canada. 17.06.2011, Available from: http://www.technopolethetford.ca/Industrial-oleochemistry/info_observatoiredeloleochimie_etudes-et-recherches_187_ang.cfm.

Chuepeng, S. & Komintarachat, C. (2010). Thermodynamic Properties of Gas Generated by Rapeseed Methyl Ester-Air Combustion Under Fuel-Lean Conditions. *Kasetsart Journal: Natural Science*, Vol.044, No.2, (March 2010- April 2010), pp. 308-317, ISSN: 0075-5192.

Engelen, B., Guidelines for handling and blending FAME. (2009). Fuels Quality and Emissions Management Group. CONCAWE report no. 9/09. Prepared for the CONCAWE Fuels Quality and Emissions Management Group by its Special Task Force, FE/STF-24. 17.06.2011, Available from: www.concawe.org.

European Commission (2007). White paper on internationally compatible biofuel standards. 17.06.2011, Available from: http://ec.europa.eu/energy/renewables/biofuels/ standards_en.htm.

Fan, X., Burton, R. & Austic, G. (2009). Preparation and Characterization of Biodiesel Produced from Recycled Canola Oil. *The Open Fuels & Energy Science Journal*, Vol.2, pp. 113-118. ISSN: 1876-973X. 17.06.2010. Available from: www.benthamscience.com/open/ /toefj/articles/V002/113TOEFJ.pdf.

Gerpen, J.V. (January 2005). Biodiesel Production and Fuel Quality, 17.06.2011, Available from: http://www.uiweb.uidaho.edu/bioenergy/biodieselED/publication/.

NREL, (2009). Biodiesel Handling and Use Guide - Fourth Edition. National Renewable Energy Laboratory, NREL/TP-540-43672. Revised December 2009. 17.06.2011, Available from: http://www.osti.gov/bridge.

Pinyaphong, P., Sriburi, P. & Phutrakul, S. (2011). Biodiesel Fuel Production by Methanolysis of Fish Oil Derived from the Discarded Parts of Fish Catalyzed by Carica papaya Lipase. *World Academy of Science, Engineering and Technology*, Vol. 76. p.p. 466-472. 17.06.2011, Available from: http://www.waset.org/journals/ waset/v76/v76-91.pdf.

Prankl, H., Körbitz, W., Mittelbach, M. & Wörgetter, M. (2004). Review on biodiesel standardization world-wide. 2004, BLT Wieselburg, Austria. Prepared for IEA Bioenergy Task 39, Subtask "Biodiesel".

Schumacher, L. (January 2005). Biodiesel Lubricity. 17.06.2011, Available from: http://www.uiweb.uidaho.edu/bioenergy/biodieselED/publication/.

Shannon, D. S., White, J.M., Parag, S.S., Wee, C, Valverde, M.A. & Meier, G.R. (2009). Feedstock and Biodiesel Characteristics Report. 2009, Renewable Energy Group, Inc., Ames, Iowa, U.S.

2

Characterization of Biodiesel by Unconventional Methods: Photothermal Techniques

Maria Castro, Francisco Machado, Aline Rocha, Victor Perez, André Guimarães, Marcelo Sthel, Edson Corrêa and Helion Vargas
State University of the North Fluminense Darcy Ribeiro (UENF)
Brazil

1. Introduction

Atmospheric Pollution is one of the most concerning problems of modern society. The pollution of air can cause local and global impacts (Rockstrom et al., 2009, Steffen et al., 2003), modifying the climate (Hansen *et al.*, 2008, Rosenzweig *et al.*, 2008, Solomon et al., 2009), damaging the human healthy and the environment, have at the moment an international prominence. The Intergovernmental Panel on Climate Change (IPCC) report (IPCC, 2007a, 2007b) concluded that emissions of greenhouse effect gases (GEE) has increased due to anthropic actions resulting a rise of the Earth's average temperature. An alternative to minimize these effects is the use of renewable sources, as an attempt to reduce global warming (Meinshausen et al., 2009, Allen et al., 2009).
However, access to energy underpins our current way of life and promotes hopes among peoples around the world for improved lives. Mobility is a core component of these aspirations, as transport has become the main factor in increasing global primary oil demand, and is expected to grow by 1.3% per year up to 2030, reaching 116 million barrels a day (Meinshausen & Meinshausen, 2009) (from 84 million barrels per day in 2005). The transport sector in particular relies almost entirely on oil, which is to become increasingly scarce and costly in the next few decades, and supplies may be prone to interruption. Biofuels (Bauen et al., 2008, Tilman et al., 2006) - fuels derived from plant materials - have the potential to address these two issues. At first sight they appear to be carbon neutral, renewable (fresh supplies can be grown as needed) and can be cultivated in many different environments. In addition they are an integral part of the emerging 'bio-economy', where plant material is used to produce specific chemicals and bulk industrial chemicals. In the future biofuels may increasingly replace chemicals derived from fossil fuel. The full picture, however, is much more complex as different biofuels have widely differing environmental, social and economic impacts. The predicted shortage of fossil fuel has encouraged the search for substitutes for petroleum derivatives. This search resulted in an alternative fuel called "biodiesel".
Biodiesel obtained from different vegetable oils is considered to be as an attractive option. From environmental standpoint, biodiesel is carbon neutral since all carbon dioxide released during consumption is sequestered from the atmosphere for growth of vegetables oil crops. The other environmental advantages in using biodiesel are: it is easy to use, as well as being biodegradable, non-toxic, reduces emissions of particulate,

reduces emissions of carbon dioxide, its emissions causes 50% less ozone to form than conventional diesel fuel, and it is essentially free of sulfur and aromatics (Ragauskas et al., 2006). According to the Brazilian Biodiesel Program, biodiesel is defined as a biodegradable fuel derived from renewable sources such as vegetable oils and animal fats. Technically speaking, biodiesel is the alkyl ester of fatty acids, produced by chemical or enzymatic transesterification of vegetal oils or animal fats with short-chain alcohol such as methanol and ethanol (Barnwal & Sharma, 2005). Glycerin is consequently a by-product from biodiesel production (Pinto et al., 2005, Schuchardt et al., 1998). In addition, the flash point of biodiesel (around 150^{0}C), being significantly higher than that petroleum diesel (around 50^{0}C), is much safer for transportation than petroleum diesel. Vegetable biodiesel can be obtained from different sources, as soybean, sunflower, castor bean, cotton, African oil palm (*Elaeis guianensis*), babassu palm, animal fats or simply oil of domestic fry. In addition, biodiesel can be a total or partial substitute for petroleum diesel to diesel engines through preparation of blends diesel/ biodiesel with different proportions. Thus, petrodiesel blended with 2%, 5%, 10% and 20% of biodiesel are known as B2, B5, B10 and B20, respectively, up to pure biodiesel (B100).

As a consequence of the fast increasing use of biodiesel one of the major concerns is the lack of standardization and certification of the products purity. Any remaining reactants from the extraction procedure can induce changes in the physicochemical properties, which may result in erosion or failure of the fuel injection components of the diesel engine. A promising area of studies to characterize this biomaterial is the measurement of the thermal properties. Thermal properties reveal important information on the physicochemical processes in the material and may certainly be an additional route to characterize biodiesel oils (Demirbas, 2003). Therefore, the use of non-conventional and advantageous methods to measure the biodiesel thermal properties appears to be relevant. Photothermal methods include a wide range techniques and phenomena which are based on the conversion of absorbed optical energy into heat.

Since the middle of the 1970's, photothermal techniques have proved to be a powerful tool to investigate physical properties of materials, with multiple branches of application. In the beginning these techniques were devoted to spectroscopic studies, especially in non-conventional systems, such as highly light diffusive and opaque materials, biological samples, etc (Rosencwaig, 1973; Cesar et al., 1979). Further applications were directed to the thermal characterization of solid, gaseous, pasty and liquid materials. Studies of diverse systems are reported in literature, such as semiconductors, conductive polymers, optical fibers, ceramics, foodstuff, non-electrolytic and ionic liquids, as well as magnetic and organic materials (Guimarães et al., 2009; Vargas & Miranda, 1988; Almond & Patel, 1996). The photothermal techniques are based on the detection of very small temperature variations produced by the absorption of radiation. The periodic or transient heating generates thermal waves, and their detection can provide information about thermal and optical properties of materials. There are many techniques for detecting such thermal diffusion waves, and the choice for each one is determined by taking into account the specificity of the sample being investigated and the kind of study to be performed.

Among the photothermal techniques, two of them are very suitable for the characterization of liquid samples, which are The Thermal Lens (TL) and Photopyroelectric (PPE) techniques. In the thermal lens experiments, an excitation laser beam, with a Gaussian profile intensity, crosses the sample to be investigated. A portion of energy is absorbed and its conversion into heat generates a radial temperature profile into the sample, which consequently gives rise to a refraction index profile. The temporal evolution of such thermal lens is strictly related to optical and thermal properties of the sample and is detected by a probe beam, which impinges in a photodiode after passes through the sample (Shen et al., 1992). In the photopyroelectric technique, the thermal waves are produced by the absorption of radiation with modulated intensity and detected by a pyroelectric transducer, which is a polymeric or ceramic material, in contact with the sample. The temperature oscillations in the pyroelectric sensor induce changes in the dielectric polarization charge over its electrodes, which are detected as an ac electric voltage or current signal by a lock-in analyzer (Coufal, 1984).

Recently, both the Thermal Lens and the photoacoustic techniques have proved to be useful for the characterization of biodiesel. Using thermal lens, it was possible to measure the thermal diffusivity and the temperature coefficient of refraction index of biodiesel samples obtained from distinct precursor oils, using both ethanol and methanol for the transesterification processes (Castro et al., 2005). Furthermore, it was applied to soybean biodiesel samples in order to evaluate the influence of residues and antioxidants on its thermo-optical properties (Lima et al., 2009). The photopyroelectric was used to the complete thermal characterization of biodiesel, meaning the determination of the thermal diffusivity, effusivity, conductivity and the heat capacity per unit volume (Guimarães et al., 2009).

2. Photothermal methodology

The conduction of heat, for a stationary temperature regime, can be described by the Fourier's Law, which relates the temperature gradient and the heat flux density, with the thermal conductivity being the proportionality factor (Carslaw & Jaeger, 1959). Nevertheless, the photothermal approach is based on the detection of thermal waves, i.e., it involves time dependent distribution of temperature. In this case, the local energy conservation is necessary in order to write the heat diffusion equation, given by:

$$\nabla^2 T(\vec{r},t) - \frac{1}{\alpha}\frac{\partial T}{\partial t}(\vec{r},t) + \frac{s(\vec{r},t)}{k} = 0 \qquad (1)$$

The term $s(\vec{r},t)$ represents the heat sources distribution and the thermal diffusivity, α, is the relevant transport parameter. It is defined in terms of the thermal conductivity (k), density (ρ) and specific heat (c) as $\alpha=k/\rho c$ (Almond & Patel, 1996). The thermal diffusivity (α) reveals how fast the heat flows in a given material, since it simultaneously depends on the way how the heat is conducted (dependence with k) and absorbed (dependence with ρc).

The photothermal measurements usually deal with the propagation of thermal waves in layered systems. Thus, the mathematical approach consists in writing the heat diffusion equations (1) for all the media involved, and so coupling them by imposing boundary conditions, such as temperature and heat flux continuity at the interfaces. This procedure gives rise to another important parameter, the reflection coefficient of the thermal wave between two adjacent media, given by

$$R_{ij} = (b_{ij} - 1)/(b_{ij} + 1) \qquad (2)$$

where $b_{ij}=e_i/e_j$ is the ratio of the thermal effusivities of these media. The thermal effusivity and the thermal diffusivity are defined in terms of the thermal conductivity (k), density (ρ) and specific heat (c) as $e=(\rho ck)^{1/2}$ (Almond & Patel, 1996). This property is essential for

describing the behavior of heat flux, when going through different media, as it plays an important role in the detection of thermal wave.
In general, the photothermal techniques allow the measurement of thermal diffusivity and effusivity. The thermal conductivity and the heat capacity per unit volume are thus indirectly determined, by inverting the expressions written above. In some special cases, depending on the sample thickness and the photothermal configuration used, it is possible to obtain directly the thermal conductivity or the specific heat (Glorieux et al., 1995; Menon et al., 2009). The thermal properties can be measured by photothermal techniques from experiments presenting good reproducibility, with uncertainties being around 1-5 %. It is important to stress that such of liquid samples are not easily investigated with conventional methods, mainly because of convection currents caused by great stationary temperature gradients. In photothermal devices, the temperature oscillations generated by the laser beam absorption are quite small, as commented above, suppressing the convection effects and also avoiding changes in the sample's properties during the measurements.

2.1 Photopyroelectric technique

The photopyroelectric (PPE) detection was introduced in the eighties (1984), as a powerful tool for a measurement of thermal properties of materials (Coufal, 1984, Mandelis, 1984, Dadarlat et al., 1984). Many studies with edible oils, foodstuff, molecular associations in binary liquids, fatty acids and automotive fuels have been reported in the recent past (Coufal, 1984, Nockemann et.al, 2009, Bicanic et al., 1992, Longuemart et al., 2002, Dardalat & Neamtu, 2006, Dadarlat et al., 1995, Cardoso et al., 2001). Although this technique is most suitable for liquid and pasty systems, it can also be applied for solid samples, if a good care is taken concerning the sample-sensor coupling (Mandelis & Zver, 1985, Salazar, 2003).
This technique is based on the use of a pyroelectric transducer to convert thermal waves to an ac electric voltage or ac current signal via induced changes in the dielectric polarization charge over the transducer electrodes. The pyroelectric response $S(t)$ of the detector due to a periodic (frequency f) temperature variation is given by equation 3 (Chirtoc & Miháilescu, 1989):

$$S(t) = \frac{i2\pi f \tau_p p L_p}{\mathrm{K}_p(1 + i2\pi f \tau_p)} \theta_p(f) e^{i2\pi f t} \tag{3}$$

where $\tau_p=RC$ is the electrical time constant, considered the equivalent resistance-capacitance circuit, and p, L_p and K_p are the pyroelectric coefficient, the thickness and the dielectric constant of the pyroelectric sensor, respectively. The term $\theta_p(f)$ is the spatially averaged temperature field over the sensor thickness, which is obtained by solving the coupled one-dimensional heat diffusion equations. For simplicity, the pyroelectric response can be written as:

$$S(t) = A(f)\Gamma(f)e^{i2\pi f t} \tag{4}$$

$A(f)$ is considered a transfer function, it represents the global frequency response of the used electric circuitry that converts the oscillating temperature into the ac voltage or current signal that is synchronously detected by the lock-in analyzer. In general, the detection of the pyroelectric signal is performed using current mode, which reduces the resistive and capacitive influences of sensor and cables (Chirtoc et al., 2003). $\Gamma(f)$ is a dimensionless response factor containing relevant information about the thermal properties and thickness of the different layers, among which is the sample layer of interest.

The two main PPE configurations for thermal characterization of liquid samples are called standard (SPPE) and inverse (IPPE) configurations. In the so called standard photopyroelectric (SPPE) configuration, the incident light is absorbed at the sample's surface and the thermal wave reaches the sensor in contact with the other sample's surface, allowing the determination of the thermal diffusivity (α).

For the SPPE configuration, the one-dimensional heat diffusion is considered in a four-layer system, constituted by air (g), sample (s), pyroelectric sensor (p) and backing (b). Since the interest here is getting information about thermal properties, a very thin cooper foil is used in the gas-sample interface, in order to guarantee the superficial absorption of the incident radiation. Besides the opaque sample assumption, it is considered that the thermal waves are completely attenuated in the gas and backing layers. Thus, the general form of $\Gamma(f)$ (Chirtoc & Miháilescu, 1989) is reduced to the following expression:

$$\Gamma(f) = \frac{2(b_{gp}+1)}{(b_{gs}+1)(b_{sp}+1)} e^{-\sigma_s L_s} P(f) \tag{5}$$

where $P(f)$ can be considered as a perturbation factor that becomes effective only at low frequency. If the frequency is high enough to consider sample and sensor as thermally thick ($\mu_s < L_s$, $\mu_s > L_s$), the perturbation factor is $P(f)=1$ (Delencos et al., 2002), where:

$$\mu_s = \sqrt{\alpha_s / \pi f} \tag{6}$$

is the thermal diffusion length. Then, amplitude and phase of $\Gamma(f)$ can be written as:

$$|\Gamma(f)| = \frac{2(b_{gp}+1)}{(b_{gs}+1)(b_{sp}+1)} e^{\frac{L_s}{\mu_s}} \tag{7a}$$

$$\varphi(f) = -\frac{L_s}{\mu_s} \tag{7b}$$

One can notice that both amplitude and phase explicitly depend on the sample thickness and implicitly depend on the modulation frequency (f) besides the sample's thermal diffusivity, via $\mu_s=(\alpha/\pi f)^{1/2}$. Moreover, $ln\,|\,\Gamma(f)\,|$ as well as $\varphi\,(f)$ present a linear dependence with $(f)^{1/2}$, and both curves have the same slope S_{freq}. This behavior allows the determination of the sample's thermal diffusivity, from amplitude and phase, by performing frequency scan measurements, knowing the sample's thickness L_s:

$$\alpha_s = \frac{\pi L_s^2}{S_{freq}^2} \tag{8}$$

Since the term $A(f)$ in equation (4) also plays a role in a frequency scan, a calibration measurement is necessary in order to guarantee that the fitting data contains only information about the sample. In a different approach, the sample's thermal diffusivity is determined by performing thickness scan, with the modulation frequency kept constant.

The dependence of both $ln\,|\,\Gamma(f)\,|$ and $\varphi\,(f)$ with the sample's thickness is linear, with the curves having the same slope S_{thick}. In this case the thermal diffusivity is determined by

$$\alpha_s = \frac{\pi f}{S_{thick}^{\;2}} \tag{9}$$

As presented above, the standard SPPE configuration allows one to determine the thermal diffusivity of materials from two different ways. In recent studies the thickness scan approach has been adopted, in which thickness specific values do not play a hole, being important the values variation. Moreover, in such approach no calibration is necessary in order to cancel the transfer function $A(f)$. Typical relative errors using thickness scan are around 0.8-3 % (Dadarlat & Neamtu, 2009a, Balderaz-López & Mandelis, 2002, Dadarlat & Neamtu, 2009b)

The inverse IPPE configuration is similar to the standard one, but the modulated light beam reaches first the sensor, and the thermal wave is completely attenuated in the sample. Thus, we can consider almost the same schematic layer-system of SPPE, just replacing "s" by "p" and neglecting the "b" layer. With these assumptions, the general form of $\Gamma(f)$ is reduced to (Dadarlat & Neamtu, 2009a):

$$\Gamma(f) = \frac{1 - e^{-\sigma_p L_p} + R_{sp}(e^{-2\sigma_p L_p} - e^{-\sigma_p L_p})}{(b_{gp}+1)(1 - R_{gp} R_{sp} e^{-2\sigma_p L_p})} \tag{10}$$

Considering the extreme and opposite thermal effusivities of air and sensor, one can find $b_{sp}=0$ and $R_{sp}=-1$, and the equation (10) is simplified. From this point, we can choose for two different approaches, depending on the relation between the sensor thermal diffusion lenght and thickness.

For the first approach, the frequency is low enough to assume that the sensor is very thermally thin and the sample very thermally thick, then the amplitude of $\Gamma(f)$ becomes (Pittois et al., 2001):

$$|\Gamma(f)| = \frac{\sqrt{\pi f} L_p c_p}{e_s} \tag{11}$$

while the phase assumes the constant value of $-\pi/4$. Thus it has reached a special case that allows the determination of the thermal effusivity of the sample. Besides the dependence with sensor properties, the IPPE signal amplitude also depends on the transfer function $A(f)$. In order to eliminate these contributions and get the sample's effusivity e_S, the experiments consists in normalizing the signal amplitude of the sample to some reference material, with known effusivity e_{ref}. This normalized signal S_n is given by:

$$S_n = \frac{e_{ref}}{e_s} \tag{12}$$

Such method was widely used in the nineties; the experimental errors are around 1-10 % (Dadarlat et al., 1990, 1996 and 1997) mainly related to instabilities on the laser intensity.

For the second approach, the frequency is high enough to assume that the sensor is thermally thick and the sample very thermally thick. For this configuration, thicker sensors are better to manipulate and the information is obtained from the phase data, avoiding those laser instabilities (Longuemart et al., 2002).

Moreover, the pyroelectric signal for the sample must be normalized by the signal for the empty cell, i. e., the sample being air. With these assumptions, the normalized $\Gamma_n(f)$ is given by (Sahraoui & Longuemart, 2002):

$$\Gamma_n(f) = 1 - (1 + R_{sp})e^{-\sigma_p L_p} \tag{13}$$

and its phase can be written as:

$$\tan\varphi = \frac{(1 + R_{sp})\sin(L_p / \mu_p)e^{-L_p/\mu_p}}{1 - (1 + R_{sp})\cos(L_p / \mu_p)e^{-L_p/\mu_p}} \tag{14}$$

It depends on the modulation frequency, properties of the sensor and the reflection coefficient R_{sp}, which carries information about the sample's effusivity. The inversion of equation (14) leads to:

$$R_{sp} = \frac{\tan\varphi e^{L_p/\mu_p}}{\sin(L_p / \mu_p) + \cos(L_p / \mu_p)\tan\varphi} - 1 \tag{15}$$

The normalized phase is an oscillating function, which passes through 0 (zero) when $L_p/\mu_p = n\pi$, with $n=1,2,3\ldots$ Thus, performing a frequency scan, this parameter can be obtained from the signal phase and used in equation (15) in order to get R_{sp}. This equation is not mathematically defined for $\varphi=0$, but one can choose a frequency range where R_{sp} is constant and get and average value for that. Knowing the sensor effusivity e_p, one can obtain the sample's effusivity e_s from:

$$e_s = \frac{(1 + R_{sp})}{(1 - R_{sp})} e_p \tag{16}$$

The reported errors for this approach are around 1-3 % (Longuemart et al., 2002, Dadarlat & Neamtu, 2006, Dadarlat et al., 2009).

2.2 Thermal lens technique

The Thermal Lens effect was first observed by Gordon et al., when transient power and beam divergence alters the output of Helium-Neon laser, after placing transparent samples in a laser cavity (Gordon et al., 1964, 1965). Thermal Lens Spectrometry (TLS) is one of a family of photothermal techniques that can be used to measure spectroscopy and thermo-optical properties of transparent materials (Cruz et al., 2011, Franko 2002, Sampaio et al., 2002, Bialkowski, 1996, Shen et al., 1992, Baesso et al., 1994). This technique has proved to be high sensitive and accurate, it is approximately three orders more sensitive than conventional transmission methods. The Thermal Lens effect was created when the sample is exposed to an excitation laser beam with a Gaussian intensity profile. A fraction of absorbed energy is converted into heat, generating a radial temperature profile T(r,t). As a result of this local temperature increase a lens-like optical element in the heated region is created. There are many experimental configurations of Thermal Lens (Gordon et al., 1965,

Hu et al., 1973). Nowadays the dual beam mode-mismatched configuration is the most widely used because of its high sensitivity (Baesso et al., 1994). In the figure 1 we present an schematic figure for a typical beam configuration of dual beam mode-mismatched configuration. This arrangement consists of two lasers beams with a different spot sizes in the sample position.

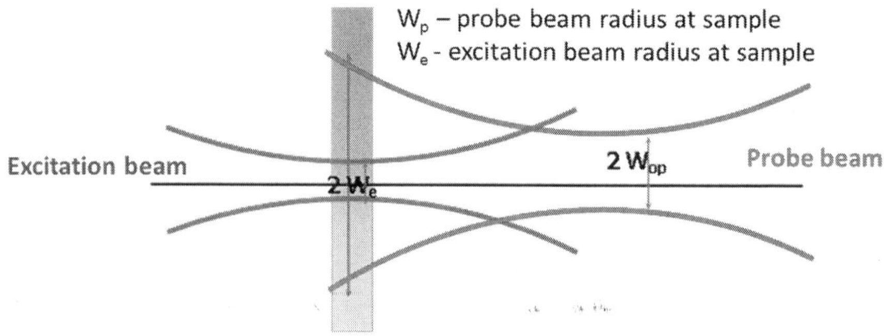

Fig. 1. Schematic diagram of dual beam mode-mismatched configuration, where w_p and w_e are, respectively, the probe and excitation beam radius at the sample

The presence of such thermal lens is detected by its effect on the propagation of a probing beam passing through the sample. The temporal evolution of the on-axis probe beam intensity, I(t), is measured in the far field using a circular aperture in front of a photodiode. Shen et al. using the Fresnel diffraction theory, determined an analytical expression for the probe beam intensity at the detector (Shen et al., 1992):

$$I(t) = I(0)\left[1 - \frac{\theta}{2}\tan^{-1}\left(\frac{2mV}{\left[(1+2m)^2 + V^2\right]\left({}^{t_c}\!/_{2t}\right) + 1 + 2m + V^2}\right)\right]^2 \tag{17}$$

where

$$m = \left(\frac{\omega_p}{\omega_e}\right)^2 \tag{18}$$

w_p and w_e are, respectively, the probe and excitation beam radius at the sample, $V = z_1/z_c$, z_1 is the distance between the sample and probe beam waist, and

$$z_c = \pi\omega_0^2 / \lambda_p \tag{19}$$

is the probe beam Rayleigh range, ω_0 is the minimum probe beam radius, λ_p is the probe beam wavelength, I(0) is the value of I(t) when the transient time t or the phase shift, h, is zero. The TL transient signal amplitude is proportional to its phase shift, h, given by :

$$\theta = -\frac{P_e A L}{k\lambda_p}\left(\frac{dn}{dT}\right)_p \tag{20}$$

in which P_e is the excitation laser power, k is the thermal conductivity, $(dn/dT)_p$ is the temperature coefficient of refractive index at the probe beam wavelength, A_e is the optical absorption coefficient at the excitation beam wavelength, $L_{eff} = (1-\exp(-A_eL))/A_e$ is the effective length and L is the sample thickness. The TL temporal signal evolution depends on the characteristic thermal time constant, t_c, which is given by (Baesso, 1994)

$$t_c = w_e^2 / 4\alpha \tag{21}$$

3. Potentialities of photothermal methods on biodiesel characterization

In order to explain the potentialities of Photothermal Methods in the Biodiesel characterization, a set of thermal properties studies using Thermal Lens and Photopyroelectric techniques have been done. Thus, the section 3.1 presents a complete characterization of thermal properties of Biodiesel using only PPE technique. In the section 3.2 we present a complete characterization of thermal properties of Biodiesel using both techniques.

3.1 Thermal properties of biodiesel using photopyroelectric technique

We first show the results for the SPPE configuration, related to thickness scan measurements in order to obtain thermal diffusivity. Water and ethylene glycol were chosen as reference samples to calibrate the cell (Guimarães et al., 2009).

For the photopyroelectric technique. the experimental arrangement is schematically shown in Fig. 2. For the standard (SPPE) configuration (2a), the pyroelectric sensor used was a PVDF (polyvinylidenedifluoride) foil, with 110 µm thickness and around 1 cm² area. The sample is hold by a cooper cylinder, glued to the sensor with silicone. An aluminum mask provides both the superficial absorption (80 µm Cu foil on the bottom) and the sample's thickness control, by means of an attached micrometer. The radiation source was an 80 mW argon laser (514 nm), chopped by an acoustic-optical modulator. Measurements were performed at a fixed frequency (3 Hz), scanning the sample's thickness from 600 to 200 µm, with a 20 µm step. The data acquisition was done as function of time, with the values of amplitude and phase of the signal being averaged for each thickness. For the inverse (IPPE) configuration (2b), two different sensors were used, a PVDF with 9 µm thickness, for the thermally thin sensor approach, and PZT (lead-titanium-zirconate) ceramic, with 210 µm,

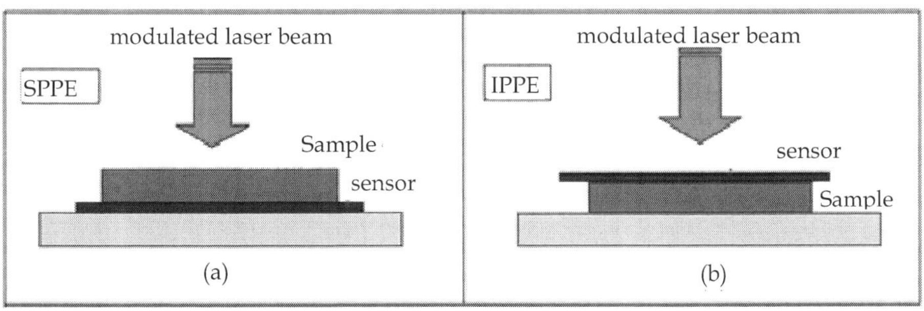

Fig. 2. Schematic figure for t he SPPE (a) and IPPE (b) setups.

for the thermally thick sensor approach. For both sensors, the surface which absorbs the laser beam was black inked. The sample is hold as in SPPE, but in this case having around 5 mm thickness, fulfilling the sample's thermally thick condition. The radiation source was a 15 mW diodo laser, electronically self modulated. In all the setups, the pyroelectric signal was measured by a SR830 lock-in analyzer, using the current mode detection.

The complete data analysis is presented for a Canola *(Brassica napus)* biodiesel, as an example, and the diffusivity values for the whole set of samples are in Table 3. As expected, the amplitude seams to decrease exponentially with the sample's thickness, while the phase presents a linear behavior (Eq. 7). For each thickness this signal is averaged and the results for the natural logarithm of amplitude and phase versus sample's thickness are shown in Fig. 3. The signal has a good stability, with a signal-to-noise ratio of 500, at the minimum. It is clear the accordance of both amplitude and phase with the predicted behavior in (Eq. 7).

With these data, according to Eq. 9, we calculate the thermal diffusivity from both amplitude (α_{ampl}) and phase (α_{phase}) of the signal. The data fitting present a good regression coefficient, which is confirmed by the small uncertainties in the diffusivity values (Fig. 3), around 0.5 %. Nevertheless, many repetitions were performed, which allowed us to check the very good reproducibility of the experiments, and determine a medium value for the thermal diffusivity of the samples Canola *(Brassica napus)* biodiesel (Table 1). The statistical error, considering the many repetitions, is more realistic, around 1 %, and it was also observed for all the samples, as it can be seen in (Table 1). The obtained thermal diffusivity values for water and ethylene glycol is in very good agreement with data reported in literature (Dadarlat & Neamtu, 2006, Delencos et al., 2002, Balderáz-Lopes et al., 2000, Bindhu et al., 1998). It is not easy to compare the results for the biodiesel samples, since their thermal properties are strictly related to the manufacturing and storage processes. However, we can notice that the thermal diffusivity present good precisions and assume values similar to results obtained for other biodiesel samples (Castro et al., 2005).

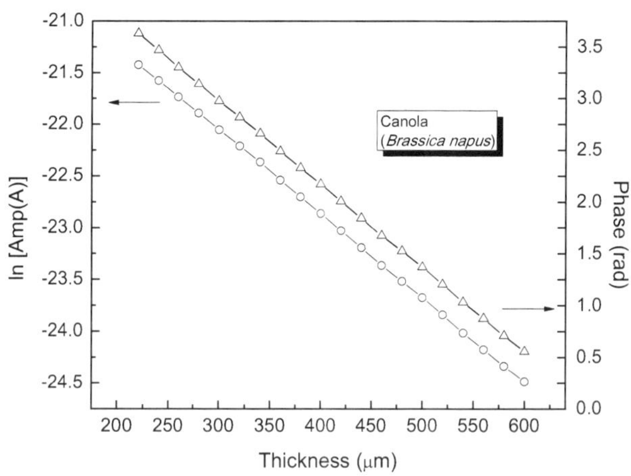

Fig. 3. Natural logarithm of amplitude and phase for SPPE signal as function of the sample's thickness.

The thermal effusivity determination, in any way, depends on a reference material, with known thermal effusivity. In this work we chose Ethylene Glycol as the standard, assuming an averaged value (Dadarlat & Neamtu, 2006, Delencos et al., 2002, Sahraoui et al., 2002, Menon et al., 2009), 810 $(Ws^{1/2}\ m^{-2}K^{-1})$. We first present the results for the thermally thin sensor approach, in which the thermal effusivity is obtained from the signal amplitude. The measurements were performed using frequencies from 0.01 to 2 Hz. For this frequency range, the normalized signal was averaged and calculations based on Eq. 12 lead to the thermal effusivity determination for water, macaw, canola, pequi *(Caryocar brasiliense) and* babassu (Table 1). In addition, the Figure 4 shows the normalized amplitude as a function of modulation frequency for the canola *(Brassica napus)* biodiesel. The results for water and the biodiesel samples are in good agreement with other reported data (Dadarlat & Neamtu, 2006, Delencos et al., 2002, Castro et al., 2005, Balderaz-López & Mandelis, 2003) and present reasonable experimental errors, between 2-5 %, considering the sensibility of the signal amplitude with fluctuations in the sensor response.

Concerning the thermally thick approach, the modulation frequency was scanned in the 1-70 Hz range. Measurements were performed for water, ethylene glycol and tree biodiesel samples. The results for normalized phase versus frequency are shown in Fig. 5.

As expected, all the curves cross the abscissa zero point at the same frequency, since it depends only on properties of the sensor. Thus, using the thickness of the sensor, we obtain get its thermal diffusivity, α_p=(4.82±0.08)x10^{-7} m^2/s. This value is used as an initial point for the data fitting which leads to the R_{sp} evaluation (Eq. 15). Knowing the average value of R_{sp} we can determine the thermal effusivity (Eq. 16). From the curve for Ethylene Glycol we get the thermal effusivity of the sensor, e_p=(1940±20)$Ws^{1/2}m^{-2}K^{-1}$, and this value is used to determine the thermal effusivity of the other samples. The measurements were repeated for all samples, and the results, shown in Table 1, present better precisions (0.6-2 %) compared to the values obtained from the signal amplitude.

Considering the values of thermal diffusivity, obtained from SPPE configuration, and effusivity, obtained from IPPE, we could determine the thermal conductivity of all the investigated materials, using the relation $k=e(\alpha)^{1/2}$, and these results are also presented in

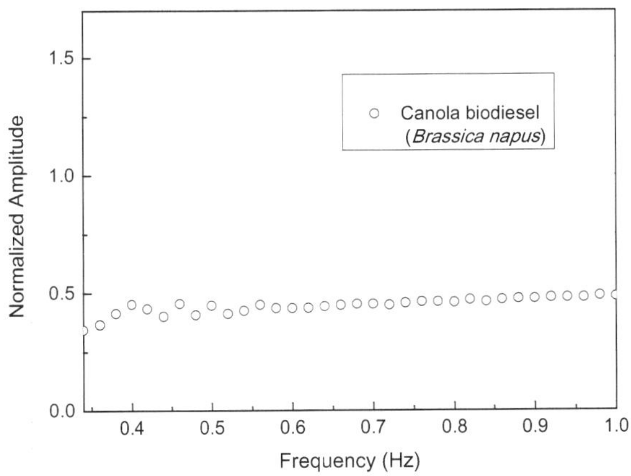

Fig. 4. Normalized amplitude as function of modulation frequency

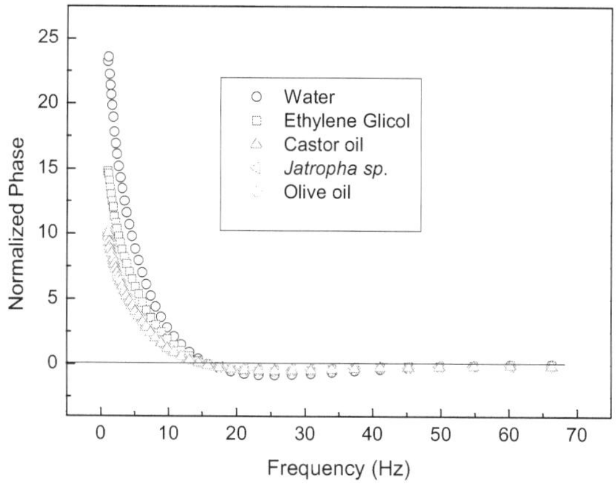

Fig. 5. Frequency scans of the normalized phase for various liquid samples, using the IPPE setup with the thermally thick sensor approach.

Biodiesel	***Thermal Diffusivity ×10⁷ (m²/s)**	***Thermal Effusivity ($W\,s^{1/2}\,m^{-2}\,K^{-1}$)**		***Thermal Conductivity (W/mK)**
		IPPE	SPPE	
Canola (*Brassica napus*)	0.860±0.010	516±11	—	0.153±0.008
Macaw (*Acrocomia aculeate)*	0.820±0.010	522±27	—	0.148±0.003
Olive oil *(Olea europaea L.)*	0.840±0.010	—	505±3	0.146±0.002
Castor oil *(Ricinus communis L.)*	0.790±0.007	—	532±5	0.150±0.002
Pinhão Manso (*Jatropha curcas*)	0.882±0.007	—	530±6	0.157±0.002
Babaçu (Orbignya phalerata)	—	556±16	—	—
Pequi (*Caryocar brasiliense*)	—	562 ± 15	—	—

*Thermal diffusivity, effusivity, conductivity were estimated by Eqs. (10, 17, 21) and $k = e\sqrt{\alpha}$, respectively.

Table 1. Thermal Properties of all the Investigated Samples

Table 1 The uncertainties are quite small, around 1-2 %, except for Canola, which is close to 5 %, certainly due to the error in the effusivity measurement.

According to attained results may be considered that photopyroelectric methodology allows the investigation of thermal properties of liquid and pasty materials. In this work we have shown specific applications to the characterization of biodiesel. With the technique, thermal diffusivity and effusivity can be obtained from different configurations (SPPE and IPPE) and the thermal conductivity can thus be calculated. Experiments present a good reproducibility, with uncertainties less than 5% and the setups are relatively simple.

3.2 Thermal properties of biodiesel using thermal lens and photopyroeletric techniques

The samples analyzed in this study were obtained by transesterification of macaw palm, canola oil, pequi and babassu oils using ethanol. The first technique used to determine the thermal diffusivity of biodiesel was the Thermal Lens. The Thermal Lens measurements were performed in the dual beam mode-mismatched configuration, the Fig. 6 shows the schematic diagram of the experimental set-up for thermal diffusivity measurements. An Ar^+ ion laser (Coehrent innova 300C) was used as excitation beam at 488nm and either an He-Ne laser was used as probe beam at 632,8nm. The samples were placed in a quartz cuvette (L=2mm). The samples were positioned at the waist of excitation beam; in Table 2 we present the experimental parameters used in the experimental set-up.

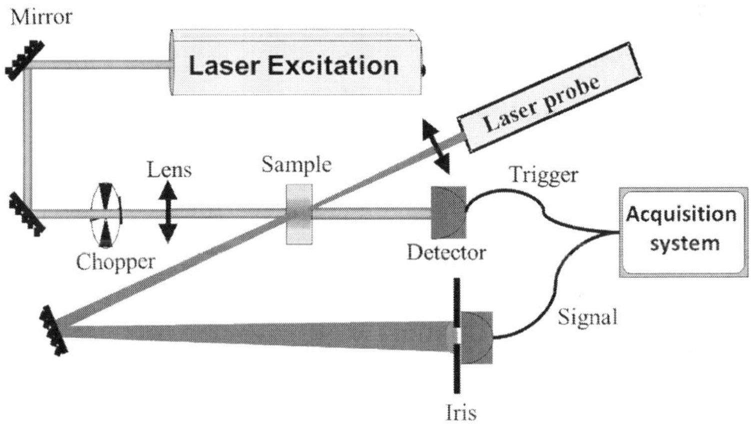

Fig. 6. Experimental set-up *TL* technique

Laser	Beam Wavelength ($\times10^{-9}$ m)	Probe Beam Rayleigh Range (cm)	Minimum Probe Radius ($\times10^{-3}$ cm)	Ratio of beam waists in the sample	Ratio distances	Excitation laser spot size at the sample ($\times10^{-3}$ cm)
	-					-
He-Ne	632.8	5.23	10.27	-	-	-
Ar	488.0	0.63	3.14	43.68	1.76	3.12

Table 2. Experimental parameters

In the transient Thermal Lens measurements, the parameters θ and t_c are determined directly by fitting the experimentally observed time profile of developing Thermal Lens

effect to Figure 7 shows a normalized time resolved Thermal Lens signal obtained for canola sample, where excitation beam power was 30mW. From the curve fitting using the equation (17), we obtained $\theta = 0.10005 \pm 0.00007$ and $\alpha = 1.090 \pm 0.005 \times 10^{-3}$ cm^{-2}/s.
We used the same procedure in order to determine the thermal diffusivity for other samples and the results are shown in the Table 2. The effusivity measurements were obtained using the Photopyrolectric technique with the IPPE configuration, in the section 3.1 we were present the details. The results of thermal conductivity can be determined using the relation $\alpha = k/\rho c$. The values obtained are in agreement with the results from literature (Castro et al. 2005, Lima et al. 2009).

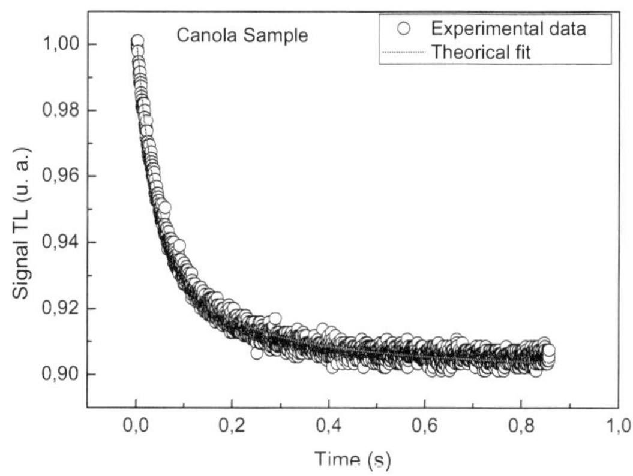

Fig. 7. TL transient signal for canola oil sample (circle) and the excitation beam power was 30mW.

Biodiesel	Thermal Diffusivity *(10^7 m^2/s)*	Thermal Effusivity *($W s^{1/2} m^{-2} K^{-1}$)*	Thermal Conductivity *(W/mK)*
Sunflower *(Helianthus annus)*	1.090±0.009	534±9	0.176±0.004
Macaw *(Acrocomia aculeate)*	1.160±0.009	522±10	0.177±0.001
Babassu *(Orbignya phalerata)*	1.120±0.007	556±16	0.186±0.001
Canola *(Brassica Napus)*	1.100±0.005	516±16	0.171±0.004
Pequi *(Caryocar brasiliense)*	1.200±0.009	562±15	0.195±0.002

Table 3. Thermal Properties of all the investigated Samples

Several factors, such as presence air, light, heat (high temperature), light traces of metal, antioxidants, peroxides as well as the nature of the storage containers can trigger oxidative processes, thus will be altered the quality of biodiesel (Knothe et al., 2005). The physical properties of biodiesel are strictly related to kind of fatty acid present in their composition and consequently the autoxidation is due to the presence of double bonds in the chains in the unsaturated fatty acids. Therefore, the properties such as viscosivity, density and thermal diffusivity are a function of the temperature and fatty acids composition, among other parameters. Thus, thought an analysis of these properties it is possible to verify of the biodiesel quality. In the sections 3.1 and 3.2 we presented the thermal characterization of biodiesel from several sources using the Thermal Lens and Photopyrolectric techniques, respectively, and the attained results showed differences between calculated thermal diffusivity values, e.g., for the macaw oil was obtained thermal diffusivity values of 1.16×10^{-7} m^2/s by TL method, in despite of 0.86×10^{-7} m^2/s by PPE technique. This variation may be related to oxidative processes because the thermal diffusivity measurements between the TL and PPE were carried out to set at the intervals of 15 months, approximately. Similar results were observed by Lima et al. 2009 and Castro et al., (2005) which observed a reduction in the thermal diffusivities in oxidated samples.

On the other hand, a inversely exponential correlation between the thermal diffusivity and iodine value (IV) for biodiesel produced from several oils source was observed (Fig. 8) reveling that thermal diffusivity for each biodiesel is a function from fatty acid composition, essentially for unsaturated compound. Further studies should be carried out to verify this behavior for other oils whose composition is preferentially of saturated fatty acids.

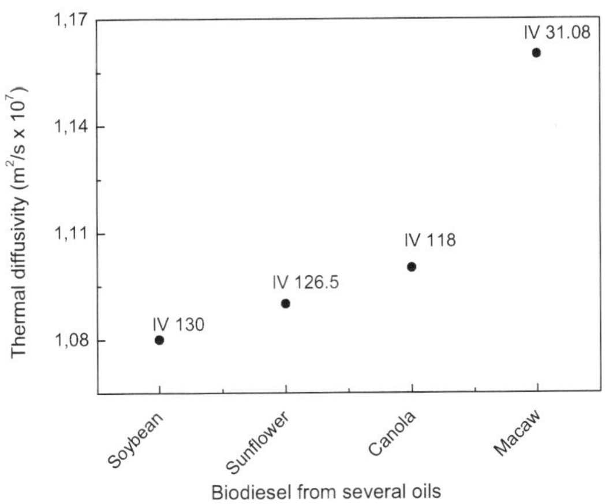

Fig. 8. Correlation between the thermal diffusivity and iodine value (IV) for biodiesel produced from several oils source.

4. Acknowledgements

The authors want to thank to The State of Rio de Janeiro Research Foundation (FAPERJ) and the National Council for Scientific and Technological Development (CNPq) for financial support. Also, thank to Prof. Dr. Donato Aranda goes providing kindly the biodiesel samples used in this work.

5. References

Allen, M. R.; Frame, D. J.; Huntingford, C.; Jones, C. D.; Lowe, J. A.; Meinshausen, M. and Meinshausen, N. (2009) Warning Caused by Cumulative Carbon Emissions Towards the Trillionth tonne, *Nature*, Vol. 458, 30 April 2009. ISSN 00280836

Almond, D. P. & Patel, P. M. (1996). Photothermal Science and Techniques, Chapman & Hall. ISBN: 0412153300.

Baesso, M. L.;Shen, J.;Snook, R. D.; (1994), Mode-mismatched thermal lens determination of temperature coefficient of optical path length in soda lime glass at different wavelengths, *Journal of Applied Physics* , 75, 8, 3732 - 3737, ISSN: 0021-8979

Barnwal, B. K. and Sharma, M. P. (2005) Prospects of biodiesel production from vegetable oils in India. *Renewable and Sustainable Energy Reviews*, 9, pp. 363–378. ISSN 13640321

Bauen, A.; Beal, J.; Parker, R.; Black, M.; Bright, I.; Chalmers, J.; Davidson, P.; Gaines, P.; Gameson, T.; Sustainable Biofuels Prospects and Challenges, Report, The Royal Society, UK, January 2008, ISBN 978 0 85403 662 2

Bialkowski, S. E. (1996). Photothermal Spectroscopy Methods for Chemical Analysis. John Wiley & Sons INC., ISBN 0-471-57467-8, Canada

Carslaw, H. S. & Jaeger, L. C. (1959). Conduction of Heat in Solids, Oxford Clarendon. . ISBN: 0198164599

Castro, M. P. P., Andrade, A. A., Franco, R. A. W., Miranda, P. C. M. L., Sthel, M. S., Vargas, H., Constantino, R., Baesso, M. L. (2005). Thermal properties measurements in biodiesel oils using photothermal techniques, *Chem. Phys. Lett.* 411: 18-22. ISSN 00092614

Cesar, C. L., Vargas, H., Meyer, J. A. & Miranda, L. C. M. (1979). Photoacoustic effect in solids, *Phys. Rev. Lett.* 42 (23): 1570-1573 ISSN 0031-9007 (print) 1079-7114 (online)

Coufal, H. (1984). Photothermal spectroscopy using a pyroelectric thin-film detector, *Appl. Phys. Lett.* 44 (1): 59.ISSN 00036951

Demirbas, A.; (2003) Biodiesel fuels from vegetable oils via catalytic and non-catalytic supercritical alcohol transesterifications and other methods: a survey. Energy Conversion & Management. 44 (2003) pp. 2093–2109. ISSN 01968904

Franko, M. (2001) Recent applications of thermal lens spectrometry in food analysis and environmental research *Talanta*, 54, 1, 30 March 2001, pp.1-13, ISSN 00399140

Glorieux, C., Thoen, J., Bednarz, G., White M. A. & Geldart, D. J. W. (1995). Photoacoustic investigation of the temperature and magnetic-field dependence of the specific-heat capacity and thermal conductivity near the Curie point of gadolinium, *Phys. Rev. B* 52 (17): 12770. ISSN 1050-2947 Print; ISSN 1094-1622 Online

Guimarães, A. O., Machado, F. A. L., Zanelato, E. B., Sthel, M. S., da Silva, E. C. & Aranda, D. A. G.. (2009). Photopyroelectric methodology applied to thermal characterization of biodiesel, *International Review of Chemical Engineering* 1 (6): 623-631 ISSN 20351763

Guimarães, A. O., Soffner, M. E., Mansanares, A. M., Coelho, A. A., Carvalho, A. M. G., Pires, M. J. M., Gama, S. & da Silva, E. C. (2009). Acoustic detection of the magnetocaloric effect: Application to Gd and Gd5.09Ge2.03Si1.88, *Phys. Rev. B* 80: 134406. ISSN 1050-2947 Print; ISSN 1094-1622 Online

Hansen, J.; Sato, M.; Kharecha, P.; Beerling, D.; Berner, R.; Masson-Delmotte, V. Pagani, M.; Raymo, M.; Royer, D. L. and Zachos, J. C. (2008) Target atmosphere CO2: Where Should Humanity Aim, *The Open Atmospherric Science Journal*, 2, 217-231, ISSN: 1874-2823

HU, C. e WHINNERY, J. R.; *Appl. Opt.* 1973, 12, 72. ISSN 0003-6935

IPCC 2007a: Climate Change 2007: The Physical Science Basis. Contribution of Working Group I to the Fourth Assessment Report of the Intergovernmental Panel on Climate Change. Solomon, S., Qin, D., Manning, M., Chen, Z., Marquis, M., Averyt, K. B., Tignor, M. and Miller H. L. (eds.). Cambridge University Press, Cambridge, United Kingdom and New York, NY, USA.

IPCC 2007b: Contribution of Working Group III to the Fourth Assessment Report of the Intergovernmental Panel on Climate Change. Metz, B., Davidson, O. R., Bosch, P. R., Dave, R. and Meyer L. A. (eds) Cambridge University Press, Cambridge, United Kingdom and New York, NY, USA.

Sampaio, J. A.; Lima, S. M.; Catunda, T. ; Medina, A. N.; Bento, A. C. and. Baesso, M. L. (2002) Thermal lens versus DTA measurements for glass transition analysis of fluoride glasses,*Journal of Non-Crystalline Solids*, Volume 304, Issues 1-3, June 2002, Pages 315-321 ISSN 00223093

Gordon, J. P.; Leite, R. C. C.; Moore, R. S.; Porto, S. P. S. and Whinnery, J. R. (1964) "Long transient effects in lasers with inserted liquids samples," Bull. *Am. Phys. Soc.* 9, 501. ISSN 00218979.

Gordon, J. P.; Leite, R. C. C.; Moore, R. S.; Porto, S. P. S. and Whinnery, J. R. (1965) "Long transient effects in lasers with lasers with inserted liquid samples," *J. Appl. Phys.* 36, 3-8.ISSN 00218979.

James Hansen, et al., Target atmosphere CO2: Where Should Humanity Aim, *The Open Atmospherric Science Journal*, 2, 217-231, (2008) ISSN 18742823

Knothe, G., Gerpen, J.V., Krahl, J. The biodiesel handbook. *AOCS PRESS*, Urbana, Illinois, 2005. ISSN: 0003-021X

Lima, S. M., Figueiredo, M. S., Andrade, L. H. C., Caíres, A. R. L., Oliveira S. L. & Aristone, F. (2009). Effects of residue and antioxidant on thermo-optical properties of biodiesel, *Applied Optics* 48: 5728 - 5732. ISSN 01436228

Meinshausen, M. and Meinshausen, N. (2009) Warning Caused by Cumulative Carbon Emissions Towards the Trillionth tonne, *Nature*, Vol. 458, 30 April 2009 ISSN 00280836

Meinshausen, M.; Meinshausen, N.; Hare, W.; Raper, S. C. B.; Frieler, K.; Knutti, R.; Frame, D. J. and Allen, M. R. (2009) Greehouse - Gas Emission Targets for Limiting Global Warning to 2 o C. *Nature*, Vol. 458, 30 April pp.1158-1163. ISSN 00280836

Menon, P.M., Rajesh, R. N. & Glorieux, C. (2009). High accuracy, self-calibrating photopyroelectric device for the absolute determination of thermal conductivity and thermal effusivity of liquids, *Rev. Sci. Instrum.* 74 (1): 054904-1. ISSN 0034-6748

Pinto, A. C.; Guarieiro, L. L. N.; Rezende, M. J. C.; Ribeiro, N. M.; Torres, E. A.; Lopes, W. A.; Pereira, P. A. P. and Andrade, J. B. (2005) Biodiesel: An Overview. *J. Braz. Chem. Soc.*, Vol. 16, No. 6B, pp. 1313-1330. ISSN 16784790

Ragauskas, A. J.; Williams, C. K.; Davison, B. H.; Britovsek, G.; Cairney, J.; Eckert, C. A.; Frederick Jr., W. J.; Hallett, J. P.; Leak, D. J.; Liotta, C. L.; Mielenz, J. R. ; Murphy, R.; Templer, R. and Tschaplinski, T. (2006) The Path Forward for Biofuels and Biomaterials. *Science* , Vol. 311, 27 January (2006) pp. 484-489. print ISSN 0036-8075; online ISSN 1095-9203.

Rocha, A. M.; Silva, W. C.; Sangiorgio, L. F. M.; Sthel, M. S.; Andrade, A. A. C.; Aranda, D. A. G. and Castro, D. A. G. (2009) Thermal diffusivity measurements of biodiesel fuel using Thermal Lens Technique. *International Review of Chemical Engineering.* November 2009,Vol. 1 No 6 pp. 336-639 ISSN 20351763

Rockström, J.; Steffen, W.;Noone, K.; Persson, A.; Chapin , F. S. and Lambin, E. F.; Lenton, T.M.; Scheffer, M. ; Folke, C.; Schellnhuber, H.J.; Nykvist, B.; de Wit, C A. ; Hughes, T.; Leeuw,S. V.; Rodhe, H.; Sörlin, S. ; Snyder, P.K.; Costanza, R.; Svedin, U.; Falkenmark, M. ; Karlberg, L.; Corell, R.W.; Fabry, V.J.; Hansen, J.; Walker, B. ; Liverman, D. ; Richardson, K.; Crutzen, P. and Foley, J.(2009) A safe operating space for humanity, *Nature*, Vol. 461, 24 September 2009, pp. 472 - 475. ISSN 00280836

Rosencwaig, A. (1973). Theory of the photoacoustic effect with solids, *J. Opt. Commun.* 7 (305). ISSN 01734911.

Rosenzweig, C.; Karoly, D.; Vicarelli, M., Neofotis, P., Wu, Q.; Casassa, G.; Menzel, A. ; Root, T. L.; Estrella, N; Seguin, B.; Tryjanowski, P.; Liu, C.; Rawlins, S. and Imeson, A. Z. (2008) Attributing physical and biological impacts to antropogenic climate change, *Nature*, Vol 453, 15 May 2008, pp.353-358. ISSN 00280836

Schuchardt, U.; Sercheli, R. and Vargas, R.M.; (1998) Transesterification of Vegetable Oils: a Review. *J. Brazil. Chem. Soc.* Vol. 9, N0 1, pp. 199-210 ISSN 0103-5053 printed version. ISSN 1678-4790 online version.

Shen, J., Lowe, R. D., Snook, R. D. (1992). A model for cw laser induced mode-mismatched dual-beam thermal lens spectrometry, *Chemical Physics*, 165 (1-2): 385-396. ISSN 03010104

Solomon, S.; Plattner, G.; Knutti, R. and Friedlingstein, P. (2009) Irreversible climate change due to carbon dioxide emissions, *PNAS*, Vol. 106, no. 6, 10 February 2009,pp. 1704-1709. ISSN 1091-6490

Steffen, W.; Sanderson, A.; Tyson, P. D.; Jäger, J.; Matson, P. A.; Moore, B.; Oldfield, F.; Richardson, K.; Schellnhuber H. J.; Turner, B. L. and Wasson R. J. (2003) Global Change and the Earth System: A Planet under Pressure, Springer - Verlag, Editors: Steffen, W.; Eliott, S. and Bellamy, J., pp.1-41. published by Springer-Verlag Berlin Heidelberg New York.. ISBN 91-631-5380-7, New York.

Tilman, D.; Hill, J. and Lehman, C. (2006) Carbon-Negative Biofuels from Low-Input High-Diversity Grassland Biomass. *Science*, Vol. 314, 8 December 2006 ISSN 09320776

Vargas, H. & Miranda, L. C. (1988). Photoacoustic and related photothermal techniques, *Phys. Rep.* 161 (2): 43-101. ISSN 03701573

3

Thermooxidative Properties of Biodiesels and Other Biological Fuels

Javier Tarrío-Saavedra[1], Salvador Naya[1], Jorge López-Beceiro[1],
Carlos Gracia-Fernández[2] and Ramón Artiaga[1]
[1]*Escola Politécnica Superior, University of A Coruña, Ferrol*
[2]*TA Instruments, Madrid*
Spain

1. Introduction

The aim of this chapter is to show how thermooxidative properties of biological fuels can be evaluated by pressure differential scanning calorimetry (PDSC) and used to correctly classify the fuels studied. The onset oxidation temperature (OOT) is an important parameter for estimating the oxidation stability that can be evaluated by the ASTM method. Nevertheless, in addition to the OOT, other meaningful information can be extracted from the PDSC tests. That additional information provides a better understanding of the thermooxidative process, allowing for identifying subtle differences between similar fuels. In fact, the following lines show that the features extracted from heat flow curves obtained by PDSC allow to characterize and to differentiate each type of fuel respect to the other ones if the adequate statistical tools are applied. Thus, the proposed statistical analysis of the PDSC curves allows to classify the different fuels types chosen for this study: two types of biodiesel, seven different classes of edible oils and two wood species. The statistical study consisted of the application of Analysis of Variance (ANOVA) procedures and the implementation of a simulation study, using parametric bootstrap and methods of multivariate supervised classification as Linear Discriminant Analysis (LDA), Logistic Regression and Naïve Bayes classifier.

Studying the thermooxidative properties of a fuel is important attending to various reasons. For example, vegetable oils are protected against oxidation thanks to antioxidants that precisely removed during the production process of biodiesel. For this reason, biodiesel is not stable, being susceptible to oxidation to a greater or lesser extent due to several factors including the presence of air, temperature, light, presence of hydroperoxides and antioxidants (Dunn, 2005; Knothe & Dunn, 2003; Knothe, 2007). The products resulting from the oxidation of biodiesel can damage internal combustion engines, it is therefore essential to study the oxidation stability of biodiesels. In the case of vegetable oils, they can produce significant changes in the salubriousness of food when the same oil is used repeatedly to fry due to the possible oxidation processes produced at the relatively high temperatures (Vorria et al., 2004). For this reason, the thermal stability to oxidation is an important parameter for oils. The study of thermooxidative characteristics of the species of wood is not as common as in oils or biodiesel. However, this is justified as it would allow to estimate the resistance to combustion in an oxidizing atmosphere, under similar conditions to wildfire.

The thermal analysis techniques used to measure thermooxidative stability are: thermogravimetry (TG), differential scanning calorimetry (DSC) and PDSC. The oxidation stability can be obtained using the TG technique (a) measuring the increase in sample weight due to absorption of oxygen, (b) measuring the temperature corresponding to maximum weight and (c) the temperature at the beginning of oxidation (Van Aardt et al., 2004). DSC and PDSC techniques can be applied to study the exothermic oxidation process. The PDSC provides results in a shorter time than DSC, further reducing evaporation in the sample. It is also important to note that using PDSC we can estimate the oxidation stability under pressures similar to those operating in a diesel engine. The values that are determined to study the oxidation stability by DSC and PDSC are the oxidation induction time (OIT) and the the onset oxidation temperature (OOT). High values of both parameters are related to a high oxidative stability. The two methods have been used by several authors to study the oxidation stability of biodiesel (Knothe, 2007; Moser et al., 2007; Dunn, 2006; Xu et al., 2007, Polavka et al., 2005), findding correlations with other procedures (Dunn, 2005; Tan, 2002). The OOT parameter measures the degree of oxidative stability of a substance at a constant heating rate, both at high pressure and high temperature. It is a non isothermal dynamic method. The procedure for calculating the OOT is explained in ASTM E2009 (2008). Recent results concerning the characterization of thermooxidative fuels such as biodiesel or edible oils can be found in (Tarrío-Saavedra et al., 2010; Artiaga et al., 2010; López-Beceiro et al., 2011).

2. Materials

In the present chapter, three different types of fuels are tested:

1. Two types of biodiesel: obtained from the soybeam and from the palm.
2. Four classes of vegetable oils: soy, sunflower, corn and two olive oil spanish varieties named hojiblanca and picual.
3. Two species of comercial wood: *Pinus sylvestris* (Scots pine) and *Eucalyptus globulus*.

2.1 Biodiesel

Biodiesel is a liquid biofuel made from natural fats such as vegetable oils or animal fats through a process of esterification and transesterification. The resulting substance of these transformations can be applied as a partial replacement of petroleum products. The reaction of the base oils with a low molecular weight alcohol and a catalyst (usually sodium hydroxide), resulting in fatty acids formed by long chains of mono-alkyl esters which are very similar to "diesel " derived from petroleum. The commercial biodiesel used today are mixed with other fuels. In this paper we have studied two types of pure biodiesel, obtained from the soybeans and, on the other hand, from palm oil. They have been supplied by Entaban Biofuels Galicia, SA (Ferrol, Spain). See Table 1.

2.2 Vegetable oils

Table 2 shows the chemical composition retrieved from the USDA National Nutrient Database for Standard Reference-22 (USDA, 2009). This table is an indication to compare

Test	Soybean	Palm	Standard
Humidity	193 µg/g	216 µg/g	EN ISO 12937
Neutralization number	0.37 mg KOH/g	0.31 mg KOH/g	EN 14104
Ester content	99.3 %	97.3%	EN 14103
Methyl esters of linolenic acid	6.45 %	1.0%	
Iodine index	124.5 g I2/100g	66.1 g I2/100g	EN 14111
Free Glycerol	0.002 %	0.003 %	EN 14105 EN 14106
Total Glycerol	0.14 %	0.19 %	EN 14105
Monoglycerides	0.47 %	0.62 %	EN 14105
Diglycerides	0.10 %	0.15 %	EN 14105
Triglycerides	0.003 %	0.05 %	EN 14105
Stability to oxydation	6.76 h	12.5 h	EN 14112
Cold filter plugging point	-3 °C	9 °C	EN 116
Methanol	0.04 %	0.1 %	EN 14110
Sulphated ashes	0.003 %	0.002 %	ISO 3987
Density	877.7 kg/m^3 at 23 °C	872 kg/m^3 at 21 °C	------------
Density at 15°C	883.5 kg/m^3	876.3 kg/m^3	ISO 3575 EN ISO 12185
Kinematic viscosity at 40°C	4.1 mm^2/s	4.4 mm^2/s	EN ISO 3104
Na+K	< 1 µg/g	< 1 µg/g	EN 14538
Ca+Mg	< 1 µg/g	< 1 µg/g	EN 14538
Phosphorous	< 1 µg/g	< 1 µg/g	EN 14107

Table 1. Characteristics of soybean and palm based biodiesel studied.

different types of oils. While poliinsaturated acids like linoleic acid are abundant in corn and sunflower oils, the monoinsaturated oleic acid is the predominant fatty acid in olive oil. The levels of palmitic and estearic acids, which are saturated fatty acids in sunflower oil are lower than the others. Other components of vegetable oils are the acilglycerides, phospholipids, and non-glycerides compounds as vitamin E, vitamins D and A, sterols, carotenoids, methyl sterols and squalene.

2.3 Wood species

The wood is mainly composed by three components that conditon the degradation of wood in an inert atmosphere (Alén et al., 1996). These are hemicellulose, cellulose and lignin (Yang et al., 1999; Alén et al., 1996; Grønli et al., 2010). Cellulose represents about 40 and 60% in the overall weight of dry wood, while lignin represents the 23–33% in softwoods and the 16–25% in hardwoods; finally, the hemicellulose represents the 25–35% (Miller et al., 1999; Grønli et al., 2010). The hemicellulose decomposes at 200-260 °C, the cellulose at 240-350 °C and the ligning in a temperature range between 280-500 °C (Alén, 1996; Wang, 2009; Mohan, 2006). Therefore, changes in the thermooxidative stabiliy are expected due to percentage differences in these 3 componens. Percentages that characterize the different wood species.

	Common name	Sunflower mid-oleic	Soybean	Corn
Total fat/g		100	100	100
Saturated fat/g		9	15.6	12.9
14:00	Myristic acid	57	0	24
16:00	Palmitic acid	4219	10455	10580
17:00	Margaric acid	37	34	67
18:00	Stearic acid	3564	4436	1848
20:00	Arachidic acid	297	361	431
22:00	Behenic acid	836	366	0
Monounsaturated fat/g		57.3	22.8	27.6
16:1 undifferentiated	Palmitoleic acid	95	0	114
16:1 c		95	N.A.	114
17:01		N.A.	0	N.A.
18 undifferentiated	Oleic acid	57024	22550	27335
18:1 c		57024	22550	27335
20:01	Gadoleic acid	211	233	129
Polyunsaturated fat/g		29	57.7	54.7
18:2 undifferentiated	Linoleic acid	28925	50960	53510
18:2 n-6 cc		28703	50422	53510
18:2 tt		N.A.	533	N.A.
18:2 i		219	N.A.	286
18:03	Linolenic acid	37	6789	1161
18:3 n-3 ccc		37	6789	1161
Total trans fatty acids/g		0.2	0.5	0.3
Total trans-polyenoic fatty acids/g		0.2	0.5	0.3
Total omega-3 fatty acids		37	6789	1161
Total omega-6 fatty acids		28925	50422	53510
Tocopherols		41.08	94.64	14.3

Table 2. Chemical composition of sunflower, soybean, corn and olive oil, retrieved from the USDA National Nutrient Database for Standard Reference-22 (USDA, 2009).

3. Data collecting

A design of experiments consisting of 1 factor (type of fuel) at 9 different levels (soy biodiesel, palm biodiesel, sunflower oil, soy oil, corn oil, hojiblanca olive oil, picual olive oil, eucalyptus wood and Scots pine) was done. Three samples per each fuel type were considered, capturing the existing variability. In fact, this sampling process seeks to obtain a compromise between capturing the existing variability and the minimization of the time of the experimental test. The tests are carried out by PDSC to study the oxidation stability of the fuels and to compare these materials according to this concept. The PDSC tests were performed in a TA Instruments pressure cell mounted on a Q2000 modulated DSC. The experimental conditions were the following: T-zero open aluminum pan, a heating rate of 10 °C min^{-1} from room temperature to 300 °C -taking into account the recommendations to obtain a better oxidation peak (Riesen & Schawe, 2006)-, sample mass in the 3-3.30 mg range, and an oxygen pressure of 3.5 MPa, applying a flow rate of 50 mL min^{-1} according to the ASTME2009 method. The experiments were manually stopped once the end of the exotherm was reached.

The Universal Analysis software supplied by the company TA was utilized to calculate the OOT using the standard E2009. The standard determines that the OOT corresponds to the temperature assigned to the crossover point between the tangent to the curve of heat flow at the point of maximum slope and tangent to the curve just before the occurrence of the peak corresponding to oxidation (which coincides with the baseline).

4. Analysis of variance (ANOVA)

The analysis of variance is a statistical tool performed to study the dependence of a quantitative variable with respect to one or more qualitative variables. In this chapter, an experimental design consisting of an nine-level factor was performed. The quantitative variable is the parameter OOT, an indicator of oxidation stability, while the factor is the type of fuel. The F test allows testing whether the mean OOT values for each fuel type are statistically equal or, conversely, there are at least one mean different (Maxwell, 2004).

H_0: $m_1 = m_2 = ... = m_9 = \mu$

H_1: *at least one* $m_i \neq \mu$, *where* μ *is the global mean*

If the before mentioned test is significant, Tukey's test can answer the question of which means are really different. Tukey's test applied to this case, provides information about what levels or types of fuel present OOT values statistically different (Maxwell, 2004). The significance level used in this work is 0.05.

5. Classification methods

The process of assigning a p-dimensional observation to one of several groups predetermined is called supervised classification. The principal aim is to obtain a discriminant function that summarizes the information corresponding to the different p variables that define a sample according to an indicator, with which each observation can be correctly classified as belonging to a group. In the statistical literature can be found several methods developed to address the classification problem.

5.1 Linear discriminant analysis

One of the most popular techniques in classification was proposed by Fisher (Fisher, 1936), this approach is called linear discriminant analysis (LDA) and basically divides the sample space into subspaces through the use of hyperplanes that allow to better separating the groups studied. The assumptions for the use of LDA are: multivariate normality and equal covariance matrices between groups. Under these assumptions, the LDA is based on finding a linear combination of features that describe or separate two or more classes of objects or events. The resulting combination can be used as a linear classifier, or more commonly, to reduce the dimension of the problem before a subsequent classification.

LDA is closely related to other statistical techniques such as analysis of variance (ANOVA) and regression analysis, however, in these two techniques, the dependent variable is a number, while in LDA is a categorical variable (class labels). Other statistical procedures related to LDA are the Principal Component Analysis (PCA) and Factor Analysis (FA), used when you look for linear combinations of variables that better explain the data.

Although the terms LDA and Fisher linear discriminant analysis are commonly used to indicate the same procedure of supervised classification, in fact, the early work of Fisher (Fisher, 1936) does not imply the assumptions of normality and equal covariances, undertaken by LDA.
The LDA has been successfully applied in fields as diverse as engineering, economics, computing science, biology, etc. Recently, the LDA has been applied in some works related to the classification of weeds (Lopez-Granados et al, 2008) and the classification of different species of wood through the use of features extracted by image processing (Mallik et al., 2011).

5.2 Logistic regression

The Logistic model is currently applied to these cases where the explicative variables (or set of different features) do not have a multivariate normal distribution (McLachlan, 2004).
Considering only two classes (*C1* and *C2*), the Logistic regression equation used to solve this classification problem is the following[1]:

$$p = \frac{e^{\alpha+\beta' x}}{1+e^{\alpha+\beta' x}} \tag{1}$$

where $p = P(Y = C1 \mid x)$ is the posterior probability of Y equal to *C1* class, $log\,(p/(1\text{-}p)) = a+\beta' x$ is the logit transformation, x is the p-dimensional vector of features or explanatory variables, β is a vector of p parameters and $p/(1-p)$ the odds ratio. Nevertheless, in the present study is necessary to use a classification model that could be applied in the case of the existence of multiples classes. This can be solved using the logit model generalized to more than two populations, i.e. for qualitative response with more than two possible classes. If G population are supposed, then, defining p as the probability that the observation i belongs to the class g, it is possible to write[2]:

$$p_{ig} = \frac{e^{\beta_{0g}+\beta'_{1g}x_i}}{1+\sum_{j=1}^{G-1} e^{-\beta_{0j}-\beta'_{1j}x_i}} \tag{2}$$

This equation[2] is an estimator of the posterior probabilities, i.e. the probabilities of belonging to a specific class, given the values of a vector of features (values of x). The $p_{ig,}$ or posterior probabilities, satisfy a multivariate logistic distribution. The following expression is used to do the different possible comparisons[3]:

$$\frac{p_{ig}}{p_{ij}} = \frac{e^{\beta_{0g}+\beta'_{1g}x_i}}{e^{-\beta_{0j}-\beta'_{1j}x_i}} \tag{3}$$

The logistic regression has been applied for classifying species of wood through the use of features extracted by image segmaentation (Mallik et al., 2011).

5.3 Bayes Naïve classifier

Naïve Bayes classifier is a supervised multivariate classification technique based on Bayes theorem, particularly suitable when the dimension of the vectors of features or inputs is considerably high. Calculating the posterior probability for an event among a group of possible outputs, $X = \{x_1, x_2, ..., x_d\}$, is intended. That is, using Bayes rule we intend to calculate the probability that a sample belongs to a particular class, C_j, from a group of possible classes $C = \{c_1, c_2, ..., c_k\}$, given some particular values corresponding to the characteristics that define the sample. Using Bayes rule, the probability that X belongs to C_j or posterior probability is[4]:

$$p(C_j \mid x_1, x_2, ..., x_d) \propto p(x_1, x_2, ..., x_d \mid C_j) p(C_j) \tag{4}$$

Using Bayes' rule, we estimate the class of the event or sample using the class corresponding to the largest posterior probability obtained. Since Naïve Bayes assumes that the conditional probabilities of the independent variables are statistically independent, the posterior probabilities can be rewrite as[5]:

$$p(C_j \mid X) \propto p(C_j) \prod_{k=1}^{u} p(x_k \mid C_j) \tag{5}$$

In addition, due to the assumption that the predictor variables are statistically independent, we can reduce the size of the estimated density function using a kernel estimation consisting of one dimension.

The Naïve Bayes classifier can be modeled with normal, log-normal, Gamma and Poisson density functions.

Naïve Bayes method appears in the 80's and is the supervised classification method most popular based on the Bayes rule. Several variants and extensions of the Naïve Bayesian classifier have been developed, for example, Cestnik (Cestnik, 1990) developed the m-estimations of the posterior probabilities and Kononenko (Kononenko, 1991) designed a semi-naïve Bayesian classifier that goes beyond the "naive" and detects dependencies between attributes. The advantage of fuzzy discretization of continuous attributes in the Naïve Bayesian classifier is described in the work of Kononenko (Kononenko, 1992). Langley (Langley, 1993) studied a system that uses the Naïve Bayesian classifier at the nodes of decision trees. Other recent works are those for Webb et al. (Webb et al, 2005) and Mozina et al. (Mozina et al., 2004). This technique has been used successfully in classification problems of spam and in areas such as medicine (to resolve, among other tasks, medical diagnosis), acoustic (automatic classification of sound and voice), image classification (Kononenko, 2001; Tóth et al., 2005; Mallik et al., 2011).

5.4 *K* nearest neighbors (KNN)

K Nearest Neighbors (KNN) is a non-parametric supervised classification method, which has been used successfully in populations where the assumption of normality is not verified. This assumption is required by traditional techniques such as linear discriminant analysis. We can summarize the KNN operation in the following three points:

1. A distance is defined between samples (represented by feature vectors), usually the Euclidean or Mahalanobis distances.
2. The distances between the test sample, x_0, and the other samples are calculated.
3. The k nearest samples to those that we want to classify are selected. Then, the proportion of these k samples belonging to each of the studied populations is calculated. Finally, the sample x_0 is classified within the population corresponding to

the highest existing frequency. Among the different methods available for choosing the value of k, the minimization of error of cross validation is one of the most used.

KNN method was introduced by Fix and Hodges (Fix & Hodges, 1951). Later, he shown some of the formal properties of this procedure, for example, that the classification error rate is bounded by twice the Bayes error value when you have an infinite number of samples for classifying and *k* is equal to 1 (Cover & Hart, 1967). Once developed the formal properties of this classifier, he established a line of research that goes up today, highlighting the work of Hellman (Hellman, 1970), which show a new approach to rejection, Fukunaga and Hostetler (Fukunaga & Hostetler, 1975), which sets out refinements with respect to the Bayes error rate, or those developed by (Dudani, 1976) and Bailey and Jain (Bailey & Jain, 1978), in which new approaches were established to the use of weighted distances. Other interesting work on the subject is related to soft computing (Bermejo & Cabestany, 2000) and fuzzy methods (Jozwik, 1983, Keller et al., 1985). Recent interesting papers are those of Bremner et al. (Bremner et al., 2005), Nigsch et al. (Nigsch et al., 2006), Hall et al. (Hall et al., 2008) and Toussaint (Toussaint, 2005). They are also very interesting applications of this algorithm to the analysis of functional data (Ferraty & Vieu, 2006).

The development of computer tools in recent years and the creation of the information society have led to that the technique KNN be used successfully in such diverse fields as chemistry, biology, medicine, computer science, genetics and materials science (Tarrío-Saavedra et al., 2011; Mallik et al, 2011).

5.5 Validation procedure: Leave-one-out cross validation

When we want to classify samples using supervised classification methods, working with training and testing data, extracted from the observed instances, is necessary. Each instance in the training set consists of the corresponding class label and a vector of several sample features. The aim of the classification methods applied is to produce a model, using the training sample, to estimate the class labels corresponding to each data instances corresponding to the testing set for which we only know the features. Leave-one-out cross-validation is the procedure used to obtain the probabilities of correct classification for each test sample and, therefore for comparing the different classification methods proposed. This is a technique widely used for the validation of an empirical model, especially suitable for working with small samples sizes. This procedure consists on the following steps:

1. One instance is leaving out: the testing sample.
2. Then, a model is obtained using the remaining samples (the training sample).
3. Finally, the developed model is used for classifying the left out instance.

This sequence is repeated until all the instances are left out once. The percentages (measured as per one) of correct classification are obtained using this procedure.

All the classification methods have been implemented using R statistical package

6. Parametric bootstrap resampling

In the case of the parametric bootstrap, the model from which data was generated of the original sample is known or assumed, ie the type of distribution is known. Therefore, successive resamplings are obtained by substituting the parameters of the distribution of probability corresponding to the studied variables by the maximum likelihood estimators, calculated from the original sample. In the present chapter the normal distribution of the features is assumed. In addition, we suppose a model where the observations (the chosen features) are independent, i.e. a diagonal covariance matrix is assumed. Taking into account these assumptions, knowing the mechanism that generates the data, generating new data from the parameters of the original sample is possible using the sample means and variances. This allows to do a simulation study to evaluate the discrimination power of the heat flow PDSC curves and their extracted features.

7. Results

The PDSC curves obtained using ASTME2009 are shown in Fig. 1. They represent the heat flow vs. temperature signals corresponding to the 9 different fuels, obtained using a heating ramp.

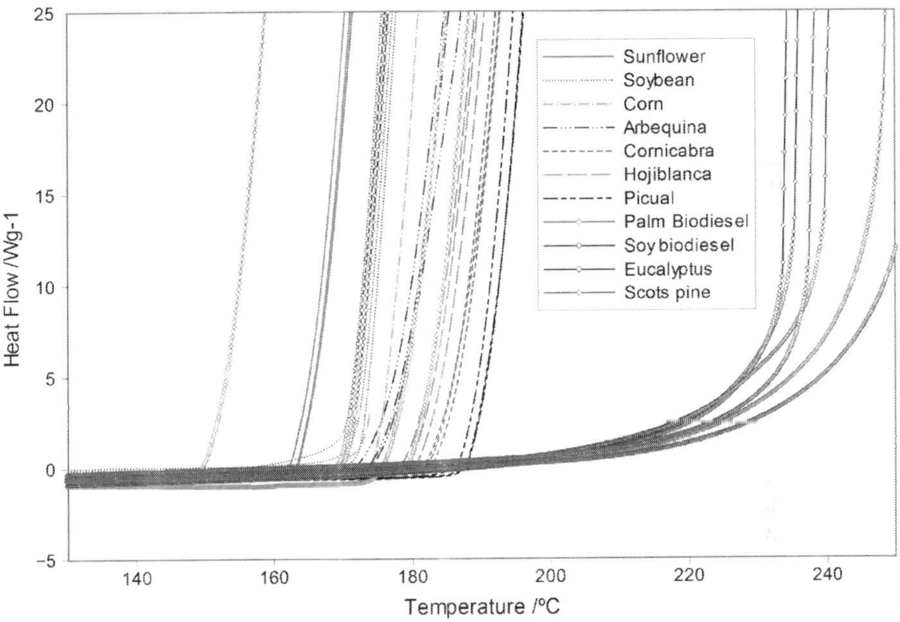

Fig. 1. PDSC curves corresponding to the studied fuel samples.

The Fig. 1 shows that the curves are different at lesser or major extend depending on the class of fuel tested. At a first glance, it seems there are two main groups. The first group is corresponding to the different studied oils and biodiesel types, and the second one consists of the wood species. It is clear that the OOT values corresponding to the first group are significantly lower than the OOT ones of the second group. But there are differences in oxidation stability (measured by OOT parameter, according to the Fig. 2) within these main groups? For answering this question we have used well known statistical tools as the F test and Tukey test. Using F test we can confirm that at least one class of fuel presents different OOT mean value than the others with statistical significance (p-value $\approx 0 < 0.05$). By means the Tukey test we can know which fuels are statistically different, observing the OOT variable. Table 3 shows the result of Tukey test. Each column represents a group of fuels different from the others, on the basis of the OOT value. For example, in the group number 1 there are three species that present no different OOT values (p-value = 0.154 < 0.05). We can observe that there are not differences between soy and soy biodiesel OOT. However, they are different to all the olive oil varieties tested and to the wood fuels studied. In fact, olive varieties form an independent group. On the other hand, palm biodiesel OOT is statistically different from the sunflower, picual and wood species OOT. It is important to note that the high OOT values obtained for wood species may condition the results for the remaining fuel classes. Attending to the means, the following fuels are sorted from largest to smallest OOT: Scots Pine > Eucalyptus > Picual > Hojiblanca > Palm biodiesel > Corn > Soy > Soy biodiesel > Sunflower

Fuel class	N	Different Groups					
		1	2	3	4	5	6
Sunflower	3	165.8100					
Soy biodiesel	3	172.9450	172.9450				
Soy	3	173.0967	173.0967				
Corn	3		175.7933				
Palm biodiesel	3		181.6833	181.6833			
Hojiblanca	3			184.9933	184.9933		
Picual	3				192.2767		
Eucalyptus	3					235.6567	
Scots pine	3						245.1867
p-value		0.154	0.054	0.911	0.154	1.000	1.000

Table 3. Mean values of OOT in the homogeneous subsets for the fuel class factor. Different groups obtained by applying the Tukey test (with significance lever $\alpha = 0.05$).

Foccusing to the soy and soy biodiesel OOT values, the OOT mean corresponding to soy oil is higher than soy biodiesel OOT, according to the theory. But when we want to compare an important quantity of fuels that presents a wide range of OOT values, the variance of the OOT measurements can prevent to distinguish the different fuels. The OOT is an important parameter that contains much information about the oxidation stability of a fuel. But, as we have observed, the OOT by itself is not enough to distinguish between all studied fuels. Obtaining more information about the PDSC curves is necessary to classify correctly among the different fuels. Therefore, additional features are chosen: the maximum slope of heat flow versus time (slope max, V) obtained in each case, the temperature at that point of maximum slope (T at max slope, H) and the slopes of the heat flow curves vs. temperature in the range from 5 to 10 Wg^{-1} (slope between 5 and 10, m). The Fig. 2 and 3 show the additional features extracted from the PDSC signals.

Moreover, having a large number of samples in a supervised classification problem is recommended. There are three samples of each fuel but, as shown in Fig. 1, the PDSC

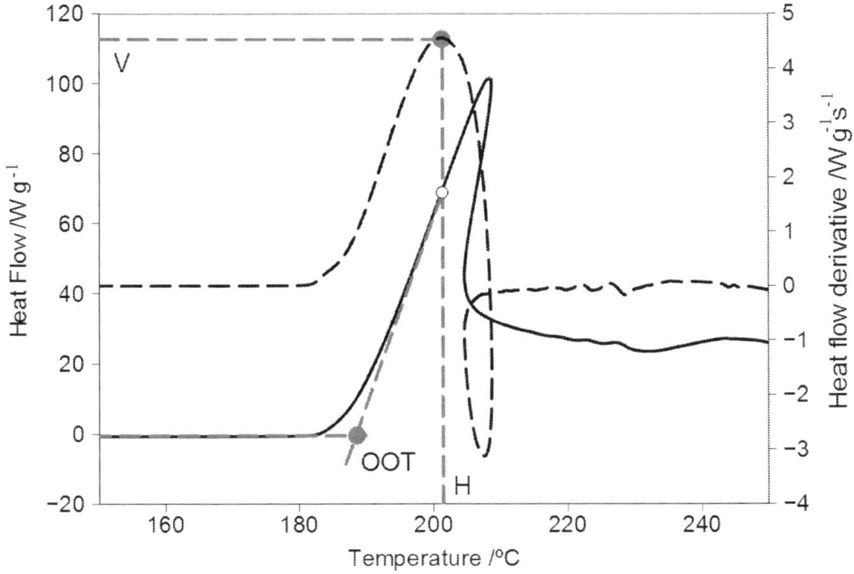

Fig. 2. OOT, H and V features extracted from the PDSC heat flow signal and its derivative.

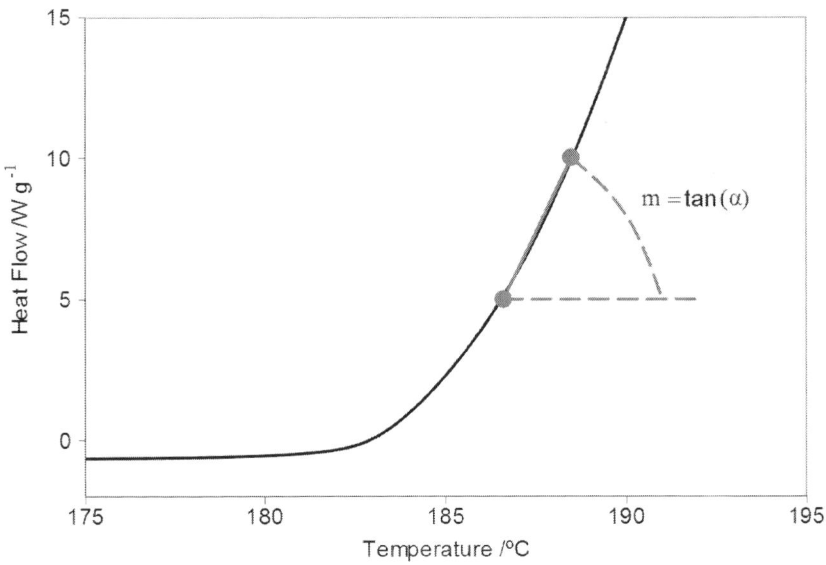

Fig. 3. The slopes corresponding to the heat flow curves vs. temperature in the range from 5 to 10 Wg^{-1} (slope between 5 and 10, m).

curves variability has been properly represented. So we can do a simulation study taking into account the sample parameters (mean and variance of the features measured in each fuel class). Then, a simulation study is presented to evaluate the power of classification of the chosen features. A parametric bootstrap resampling is chosen to increase the sample size until 100 items per fuel class. The parametric bootstrap is implemented for generating new values from OOT and the other chosen features, assuming that are independently distributed according a Gaussian distribution where the mean is the sample mean and the variance is the sample variance. The leave-one-out cross-validation method technique is used for the validation of the empirical model. It allow to estimate the probabilities of correct classification corresponding to the different classification methods. It works by leaving out one sample (represented by the features above mentioned); then a model is trained with the remaining parameter samples and, finally, the developed model is used for classifying the sample left out. This is repeated 900 times, until all the vectors have been left out once. Table 4 shows the probability of correct classification obtained by the above mentioned classification methods. These probabilities are very high, regardless of the method used. The best result corresponds to the use of logistic regression (99.7%) through almost all the samples are correctly classified. Table 5 shows the confusion matrices corresponding to the application of logistic regression, LDA, Bayes Naïve and KNN classification methods. The percentage of simulated samples correctly classified is shown in the diagonal of the matrices. The percentages of confusion obtained between the fuel types, two by two, are presented outside the diagonal. The little confussions existing between the two types of wood and between palm biodiesel and hojiblanca olive oil are solved using the logistic regresion method. According to these results, the OOT and the other characteristics are very useful parameters for classification purposes.

Classification method	Percentage of correct classification/ %
LDA	94.2
Logistic regression	99.7
Bayes Naïve Classifier	98.0
KNN	98.1

Table 4. Percentages of correct classification obtained by the three proposed methods. The best results are obtained by Logistic regression.

		Actual								
Method	Estimated	Corn	Eucal.	Hojib.	Palm Biod.	Picual	Scots p.	Soy	Soy Biod.	Sunfl.
Logistic Regress.	Corn	100	0	0	0	0	0	0	0	0
	Eucal.	0	100	0	0	0	0	0	0	0
	Hojib.	0	0	100	0	1	0	0	0	0
	PalmBiod.	0	0	0	99	0	0	0	0	0
	Picual	0	0	0	1	98	0	0	0	0
	Scots p.	0	0	0	0	0	100	0	0	0
	Soy	0	0	0	0	1	0	100	0	0
	Soy Biod.	0	0	0	0	0	0	0	100	0
	Sunfl.	0	0	0	0	0	0	0	0	100

LDA	Corn	75	0	0	0	0	0	0	0	0
	Eucal.	0	98	0	0	0	19	0	0	0
	Hojib.	0	0	100	4	2	0	0	0	0
	PalmBiod.	1	0	0	96	0	0	0	0	0
	Picual	0	0	0	0	98	0	0	0	0
	Scots p.	0	2	0	0	0	81	0	0	0
	Soy	7	0	0	0	0	0	100	0	0
	Soy Biod.	17	0	0	0	0	0	0	100	0
	Sunfl.	0	0	0	0	0	0	0	0	100
Bayes Naïve	Corn	99	1	0	0	0	0	0	0	0
	Eucal.	0	94	0	0	0	8	0	0	0
	Hojib.	0	0	100	3	0	0	0	0	0
	PalmBiod.	1	0	0	97	0	0	0	0	0
	Picual	0	0	0	0	100	0	0	0	0
	Scots p.	0	5	0	0	0	92	0	0	0
	Soy	0	0	0	0	0	0	100	0	0
	Soy Biod.	0	0	0	0	0	0	0	100	0
	Sunfl.	0	0	0	0	0	0	0	0	100
KNN	Corn	100	0	0	0	0	0	0	0	0
	Eucal.	0	96	0	0	0	10	0	0	0
	Hojib.	0	0	100	3	0	0	0	0	0
	PalmBiod.	0	0	0	97	0	0	0	0	0
	Picual	0	0	0	0	100	0	0	0	0
	Scots p.	0	4	0	0	0	90	0	0	0
	Soy	0	0	0	0	0	0	100	0	0
	Soy Biod.	0	0	0	0	0	0	0	100	0
	Sunfl.	0	0	0	0	0	0	0	0	100

Table 5. Confusion matrix or prediction percentages obtained by each classification method and leave-one-out cross-validation, using the features extracted from PDSC signals. The feature data set was tested with 9 classes or types of fuels. The results are shown as percentages.

8. Conclusion

The thermooxidative stability of 9 different types of fuels (including two types of biodiesel, soy and palm oil) has been measured using the OOT parameter. The use of the OOT parameter and ANOVA techniques allows to differentiate various groups of fuels: the varieties of olive oil, the two types of wood and finally the remaining fuels (although the sunflower oil is slightly different). But the OOT by itself is not enough to distinguish between all studied fuels with statistical significance.

The classification of the 9 fuels according to the thermooxidative properties has been possible using multivariate supervised classification method and additional features extracted from the PDSC curves as dataset: the maximum slope of heat flow versus time (slope max) obtained in each case, the temperature at that point of maximum slope (T at max slope) and the slopes of the heat flow curves vs. temperature in the range from 5 to 10 Wg^{-1} (slope between 5 and 10). That additional information provides a better understanding of the thermooxidative process, allowing for identifying subtle differences between similar fuels.

The evaluation of the discriminant power of the extracted thermooxidative features has been possible using parametric bootstrap resampling.

The overall percentages of correct classification are very high, in particular when Logistic regression classifier is used (99.7%); it seems to work better than LDA, Bayes Naïve and KNN.

9. Acknowledgment

This research has been partially supported by the Spanish Ministry of Science and Innovation, Grant MTM2008-00166 (ERDF included) and Grant MTM2011-22392.

10. References

Alén, R.; Kuoppala, E. & Pia, O. (1996). Formation of the main degradation compound groups from wood and its components during pyrolysis. *Journal of Analytical and Applied Pyrolysis*. Vol. 36, pp. 137–148.

Artiaga, R.; López-Beceiro, J.; Tarrío-Saavedra, J.; Mier, J. L.; Naya, S. & Gracia, C. (2010). Oxidation stability of soy and palm based biodiesels evaluated by Pressure Differential Scanning Calorimetry. *Journal of ASTM International*. Vol. 7, No. 4.

ASTM. Standard Test Method for Oxidation Onset Temperature of Hydrocarbons by Differential Scanning Calorimetry. E2009-08. *Annual Book of ASTM Standards*, ASTM Internacional, West Conshohocken, PA, 2008.

Bailey, T. & Jain, A. (1978). A note on distance-weighted k-nearest neighbor rules. *IEEE Transactions on Systems, Man, and Cybernetics*. Vol. 8, pp. 311-313.

Bermejo, S. & Cabestany, J. (2000). Adaptive soft k-nearest-neighbour classifiers. *Pattern Recognition*. Vol. 33, pp. 1999-2005.

Bremner, D.; Demaine, E.; Erickson, J.; Iacono, J.; Langerman, S.; Morin, P. & Toussaint, G. (2005). Output-sensitive algorithms for computing nearest-neighbor decision boundaries. *Discrete and Computational Geometry*. Vol. 33, pp. 593-604.

Cover, T.M. & Hart P.E. (1967). Nearest neighbor pattern classification. *IEEE Transactions on Information Theory*. Vol. 13, pp. 21-27.

Dudani, S.A. (1976) The distance-weighted k-nearest-neighbor rule. *IEEE Transactions on Systems, Man, and Cybernetics*. Vol. 6, pp. 325–327.

Dunn, R. O. (2005). Effect of antioxidants on the oxidative stability of methyl soyate (biodiesel). *Fuel Processing Technology*. Vol. 86, pp. 1071-1085.

Dunn, R. O. (2006). Oxidative stability of biodiesel by dynamic mode pressurized-differential scanning calorimetry (P-DSC). *Transactions of the ASABE*. Vol. 49, pp. 1633-1641.

Ferraty, F. & Vieu, P. (2006). Nonparametric Functional Data Analysis: Theory and Practice. Springer-Verlag, New York.

Fisher, R. A. (1936). The use of multiple measurements in taxonomic problems. *Annual Eugenics*. Vol. 7, pp. 179-188.

Fix, E. & Hodges, J.L. (1951). Discriminatory analysis, nonparametric discrimination: Consistency properties. Technical Report 4, USAF School of Aviation Medicine, Randolph Field, Texas.

Fukunaga, K. & Hostetler, L. (1975). k-nearest-neighbor bayes-risk estimation. *IEEE Transactions on Information Theory*. Vol. 21, pp. 285-293.

Grønli, M. G.; Várhegyi, G. & Blasi, C. (2002). Thermogravimetric analysis and devolatilization kinetics of wood. *Industral & Engineering Chemistry Research*. Vol. 41, pp. 4201–08.

Hall, P.; Park, B.U. & Samworth, R.J. (2008). Choice of neighbor order in nearest-neighbor classification. *Annals of Statistics*. Vol. 36, pp. 2135–2152.

Hellman, M.E. (1970). The nearest neighbor classification rule with a reject option. *IEEE Transactions on Systems, Man, and Cybernetics*. Vol. 3, pp. 179-185.

Jozwik, A. (1983). A learning scheme for a fuzzy k-nn rule. *Pattern Recognition Letters. Vol.* 1, pp. 287-289.

Keller, J.M.; Gray, M.R. & Givens, J.A. (1985). A fuzzy k-nn neighbor algorithm. *IEEE Transactions on Systems, Man, and Cybernetics*. Vol. 15, pp. 580-585.

Knothe, G. & Dunn, R. (2003). Dependence of oil stability index of fatty compounds on their structure and concentration and presence of metals. *J American Oil Chemists'Society*. Vol. 80, pp. 1021-1026.

Knothe, G. (2007). Some aspects of biodiesel oxidative stability. *Fuel Processing Technology*. Vol. 88, No.7, pp. 669-677.

Kononenko, I. (1991). Semi-naive Bayesian classifier, *Proceedings of the European Working Session on Learning-91*, pp. 206-219, Springer, Berlin.

Kononenko, I. (1992). Naive Bayesian classifier and continuous attributes. *Informatica*. Vol. 16, pp. 1–8.

Kononenko, I. (2001). Machine learning for medical diagnosis: history, state of the art and perspective. *Artificial Intelligence in Medicine*. Vol. 23, pp. 89-109.

Langley, P. (1993). Induction of recursive Bayesian classifiers. *Proceedings of the European Conference on Machine Learning*, Vienna.

López-Granados, F.; Peña-Barragán, J. M.; Jurado-Expósito, M.; Francisco-Fernández, M.; Cao, R.; Alonso-Betanzos, A. & Fontenla-Romero, O. (2008). Multispectral discrimination of grass weeds and wheat (*Triticum durum*) crop using linear and nonparametric functional discriminant analysis and neural networks. *Weed Research*. Vol. 48, pp. 28-37.

López-Beceiro, J.; Pascual-Cosp, J.; Artiaga, R.; Tarrío-Saavedra, J. & Naya, S. (2011). Comparison of olive, corn, soybean and sunflower oils by PDSC. *Journal of Thermal Analysis and Calorimetry*. Vol. 104, pp. 169-175.

Mallik, A.; Tarrío-Saavedra, J.; Francisco-Fernández, M & Naya, S. (2011). Classification of Wood Micrographs by Image Segmentation. *Chemometrics and intelligent laboratory systems*. Vol. 107, pp. 351-362.

Maxwell, S. E. & Delaney, H. D. (2004). Designing experiments and analyzing data. A model comparison perspective, Lawrence Erlbaum Associates, New Jersey.

McLachlan, G.L. (2004) Discriminant Analysis and Statistical Pattern Recognition, Wiley-Interscience, Hoboken, New Jersey.

Miller, R. B. (1999). Structure of wood. In: Wood handbook: *Wood as an engineering material*, Woodhead Publishing Limited, Department of Agriculture, Forest Service, Forest Products Laboratory, Madison, WI.

Mohan, D.; Pittman, J. C. U. & Steele, P. H. (2006). Pyrolysis of wood/biomassfor bio-oil: a critical review. *Energy Fuel*.Vol. 20, pp. 848–89.

Moser, B. R.; Haas, M. J.; Winkler, J. K.; Jacksona, M. A.; Erhana, S. Z.; List, G. R.. (2007). Evaluation of partially hydrogenated methyl esters of soybean oil as biodiesel. *European Journal of Lipid Science and Technology*. Vol. 109, pp. 17-24.

Mozina, M.; Demsar, J.; Kattan, M. & Zupan, B. (2004). Nomograms for Visualization of Naive Bayesian Classifier, *Proccedings of PKDD-2004*, pp. 337-348.

Nigsch, F.; Bender, A.; Van Buuren, B.; Tissen, J.; Nigsch, E. & Mitchell, J.B.O. (2006). Melting Point Prediction Employing k-nearest Neighbor Algorithms and Genetic Parameter Optimization. *Journal of Chemical Information and Modeling*. Vol. 46, pp. 2412-2422.

Polavka, J.; Paligová, J.; Cvengros J, et al. (2005). Oxidation stability of methyl esters studied by differential thermal analysis and rancimat. *American Oil Chemists' Society*. Vol. 82, pp. 519-524.

Riesen, R. & Schawe, E. K. (2006). Description of the O_2 pressure influence on the oxidation time of motor oils. *J Thermal Analysis and Calorimetry*. Vol. 83, pp. 157-161.

Tan, C. P.; Che Man, Y. B.; Selamat, J. & Yusoff, M.S.A. (2002). Comparative studies of oxidative stability of edible oils by differential scanning calorimetry and oxidative stability index methods. *Food Chemistry*. 2002, Vol. 76, p. 385- 389.

Tarrío-Saavedra, J. ; López-Beceiro, J.; Naya, S. ; Gracia, C. ; Mier, J. L. & Artiaga, R. (2010). Factores influyentes en la estabilidad a la oxidación del biodiesel. Estudio estadístico. *Dyna*. Vol. 85, No. 4, pp. 341-350.

Tóth, L.; Kocsor; A. & Csirik, J. (2005). On naive bayes in speech recognition. *International Journal of Applied Mathematics and Computer Science*. Vol. 2, pp. 287-294.

Toussaint, G.T. (2005). Geometric proximity graphs for improving nearest neighbor methods in instance-based learning and data mining. *International Journal of Computational Geometry and Applications*. Vol. 15, pp. 101–150.

U.S. Department of Agriculture, Agricultural Research Service (2009). USDA National Nutrient Database for Standard Reference, Release 22. In: *Nutrient Data Laboratory*, 11.09.2010, Avaliable from http://www.nal.usda.gov/fnic/foodcomp/Data/SR15/dnload/pk260w32.exe.

Van Aardt, M.; Duncan, S. E.; Long, T. E.; O'Keefe, S. F.; Marcy, J. E. & Sims, S.R. (2004). Effect of antioxidants on oxidative stability of edible fats and oils: thermogravimetric analysis. *Journal of Agricultural and Food Chemistry*. Vol. 52, pp. 587-591. 13.

Vorria, E.; Giannou, V. & Tzia, C. (2004). Hazard analysis and critical control point of frying-safety assurance of fried foods. *European Journal of Lipid Science and Technology*. Vol.106, pp. 759–65, ISSN 1438-9312

Wang, S.; Wang, K.; Liu, Q.; Gu, Y; Luo, Z; Cen, K & Fransson, T. (2009). Comparison of the pyrolysis behavior of lignins from different tree species. *Biotechnology Advances*. Vol. 27, pp. 562–7.

Webb, G.I.; Boughton, J. & Wang, Z. (2005). Not So Naive Bayes: Aggregating One-Dependence Estimators. *Machine Learning*. Vol. 58, pp. 5-24.

Xu, Y. X.; Hanna, M. A. & Josiah S J. (2007). Hybrid hazelnut oil characteristics and its potential oleochemical application. *Industrial Crops and Products*. Vol. 26, pp. 69–76.

Yang, H.; Lewis, I. R. & Griffiths, P. R. (1999). Raman spectrometry and neural networks for the classification of wood types. 2. Kohonen self-organizing maps. *Spectrochimica Acta Part A: Molecular and Biomolecular Spectroscopy*. Vol. 55, pp. 2783–2791.

4

Effects of Raw Materials and Production Practices on Biodiesel Quality and Performance

Jose M. Rodriguez
Mississippi State University
USA

1. Introduction

The demand for transportation fuels is increasing around the world, especially the demand for petroleum-based fuels. To cope with rising demand and dwindling petroleum reserves, alternative motor fuels such as biodiesel are at the forefront of commercialization. Biodiesel is an environmental renewable clean burning fuel. Biodiesel is a replacement for diesel in compression-ignition engines. Biodiesel is composed of mono-alkyl esters of long chain fatty acids. These esters are produced when virgin vegetable oils, i.e., soy, canola, palm and rapeseed oil, animal fats from tallow, poultry offal and fish oils or used cooking oils and trap grease from restaurants are reacted with an alcohol. The major chemical components of vegetable oils, fats and greases are triacylglycerols. The chemical reaction of converting triacylglycerols into methyl esters is termed transesterification. A stochiometric excess of alcohol and a catalyst is required for the effective transesterification of triacylglycerols into alkyl esters. The transesterification reaction is depicted in Figure 1. The alcohol used for producing biodesel is usually methanol. Methanol is the least expensive alcohol and therefore the alcohol of choice. The catalyst can be an acid or a base depending on the amount of free fatty acids present. The catalyst bases most commonly used are NaOH or KOH. The acid catalyst is usually H_2SO_4. In order to be commercially available in the United States and Canada, biodiesel must meet the specifications in ASTM D6751, Standard Specification for Biodiesel Fuel (B100) Blend Stock for Distillate Fuels. In Europe they follow the requirements and test methods for fatty acid methyl esters (FAME). The requirements are specified in EN 14214. The requirements for these two standards are given in Table 1. These specifications are designed to meet the requirements necessary for the proper performance of compression-ignited engines. Feedstock, feedstock quality and production practices can influence the quality of the biodiesel and therefore, the performance and commercial approval of the final product.

Feedstock

As previously stated, the feedstock sources can be virgin vegetable oils, animal fats and greases. The virgin vegetable oils that are commonly used are soybean, canola, rapeseed, sunflower and palm. Soybean vegetable oil, fats and yellow grease are mainly used in the United States [1]. Canola is used in Canada. Rapeseed and sunflower oil are the primary feedstock in Europe [2]. Palm oil, which is mainly produced in the tropics, is the main feedstock used there [3, 4]. The feed stock source can influence the cetane number, oxidation stability, cold soak filterability (deposition), and cold flow properties.

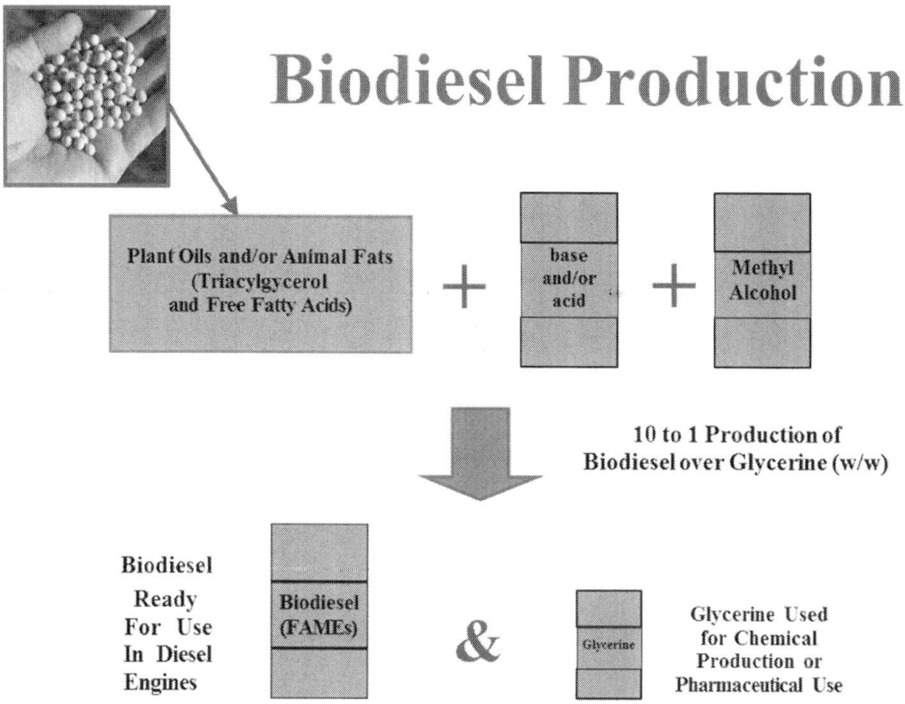

Fig. 1. The transesterification reaction for the production of Biodiesel from triacylglycerol.

Cetane Number

The performance of diesel engines depends on the compression ratio, injection timing, fuel/air mixture and ignition delay. The cetane number is a measurement based on the ignition delay of compression-ignition engines (the lower the ignition delay, the higher cetane number). ASTM D613 and EN ISO 5165 are the standard procedures for determining cetane number. The lower the ignition delay, the better the compression-ignition engines performs. The low ignition delay increases power, engine efficiency and the engine's ability to start at lower temperatures. The composition of the biodiesel influences the cetane number. The minimum acceptable cetane number necessary for acceptable performance in modern compression-ignition engines is 40 [5].

	ASTM D6751-10			EN 14214:2008	
Properties	Test Method	Grade S15	Grade S500	Test Method	
FAME				EN 14103	96.5 % (m/m)min.
Ca & Mg (total)	EN 14538	5 ppm (μg/g) max.	5 ppm (μg/g) max.	EN 14538	5 ppm (μg/g) max.
Density, 15°C				EN ISO 3675/EN ISO 12185	860-900 kg/m^3
Flash Point (close cup)	ASTM D93	93°C min.	93°C min.	EN ISO 2719/EN ISO 3679	>101°C
Alcohol Control					
One of the following must be met: 1. Methanol Content 2. Flash Point	EN 14110 ASTM D93	0.2 % mass max. 130°C min.	0.2 % mass max. 130°C min.		
Methanol				EN 141101	0.2% mass max.
Water and Sediment	ASTM D2709	0.050 % volume max.	0.050 % volume max.	EN ISO 12937	500 mg/kg max.water
Total contamination				EN 12662	24 mg/kg max.
Kinematic viscosity, 40°C	ASTM D445	1.9-6.0 mm^2/s	1.9-6.0 mm^2/s	EN ISO 3104	3.5-5.0
Sulfated Ash	ASTM D874	0.020 % mass max.	0.020 % mass max.	ISO 3987	0.02 %(m/m) max.
Sulfur	ASTM D5453	0.0015 % mass max.	0.05 % mass max.	EN ISO 20846/EN ISO 20884	10 mg/kg max.
Copper strip corrosion	ASTM D130	No. 3 max.	No. 3 max.	EN ISO 2160	Class 1 rating
Cetane Number	ASTM D613	47 min.	47 min.	EN ISO 5165	51.0 min.
Cloud point	ASTM D2500	Report	Report		
Carbon Residue	ASTM D4530	0.050 % mass max.	0.050 % mass max.	EN ISO 10370	0.3 % (m/m) max. at 10% dist. remnant
Acid Number	ASTM D664	0.50 mg KOH/g Max.	0.50 mg KOH/g max.	EN 14104	0.50 mg KOH/g max.
Cold soak filterability	ASTM D 6751 Annex A1	360 seconds max.	360 seconds max.		
Free glycerin	ASTM D6584	0.020 % mass max.	0.020 % mass max.	EN 14105/EN 14106	0.02 %(m/m) max.
Total glycerin	ASTM D6584	0.240 % mass max.	0.240 % mass max.	EN 14105	0.25 % (m/m) max.
Phosphorous Content	ASTM D4951	0.001 % mass max.	0.001 % mass max.	EN 14107	4 mg/kg max.
Distillation temperature A.E. temp., 90% recovery	ASTM D1160	360°C max.	360°C max.		
Na & K (total)	EN 14538	5 ppm (μg/g) max.	5 ppm (μg/g) max.	EN 14108/EN 14109/EN 14538	5 mg/kg max
Oxidation stability	EN 15751	3 hrs. min.	3hrs. min.	prEN 15751/EN 14112	6hrs. min.
Iodine Value				EN 14111	120 max.
Linolenic Acid Methylester				EN 14103	12 % max
Polyunsatureted Methyl Esters >=4double bonds				EN 14103	1 % (m/m) max.
Mono, Di and Tri glyceride content				N14105E	0.8/0.2/0.2 %(m/m) max.

Table 1. Biodiesel Standard Specifications for North America (ASTM D6751) and Europe (EN14214).

The chemical composition of the triacylglycerols from different feedstocks varies in chemical composition. Therefore, the methyl esters produced from different feedstocks varies according to the source. The cetane numbers of the methyl esters from different feedstocks are given in Table 2.

Feedstock	Cetane Number (Average of Lit. Values)
Soybean	48.8
Rapeseed	52.2
Sunflower	53.4
Beef Tallow	56.2
Palm	62.3
Yellow Grease	62.6

Table 2. Comparison of average cetane numbers from published data [6].

Oxidation Stability

All fuels, including biodiesel, have stability problems. Biodiesel is susceptible to oxidative degradation of the fuel quality. The oxidation degradation of the fuel is determined by the amount and position of the olefinic unsaturation in the fatty acid methyl ester molecular chains. All of the biodiesel feedstocks have polyunsaturated chains that are methylene-interrupted in their triacylglycerols molecules. The oxidation proceeds at different rates depending on the number and position of the olefinic unsaturation [7]. The fatty acids chemical composition of triacylglycerols used as feed stocks is given in Table 2. EN 15751 specifies a procedure to measure the propensity of biodiesel to oxidation.

Oxidation stars by attacking the methylene carbons between the olefinic carbons. Hydrogen is removed and a hydroperoxide and conjugated dienes are formed. The hydroperoxide decomposes and interacts to form aldehydes, alcohols, carboxylic acids and high molecular weight polymers [9]. Aldehydes detected in the oxidation process include hexenals [10], heptenals, propanal [11,12] and 2,4-heptadienal [12]. Short chain aliphatic acids and alcohols have also been detected [13, 14]. Increase acidity due to formation of organic acids increases corrosion. Polymerization products from oxidation will increase viscosity of the fuel and therefore it will influence the performance.

Cold Soak Filterability

In cold weather, the most common problem associated with biodiesel or biodiesel blends is the plugging of the fuel filter. In 2008, a cold soak filtration test was added to the ASTM specifications, to address this problem. Cold soak filterability is a measurement of how well biodiesel flows when chilled and poured through a filter. Previous studies showed that the formation of precipitates during cold weather conditions depends on the feedstock, blend concentration and storage time [15, 16]. Most of the precipitate formed at lower temperatures will be re-dissolved when they are warmed to room temperature [17]; however, minor precipitate components remain as precipitates after warming to room temperature.

Insoluble precipitates from soybean biodiesel can be attributed to sterols present in the soybean oil feedstock. Soybean oil contains approximately 0.36% sterols. Sterols are composed of a group of steroid alcohols present in plants. The culprit sterol was found to be sterol glucoside(SG) [15]. Soybean oil may contain up to 0.23 % SG [16].

The insoluble precipitates from palm biodiesel are due to both sterol glucoside and monoacylglycerols; while, the precipitates from poultry fat biodiesel are due only to monoacylglycerols [15].

Feedstock	Unsat./Sat. ratio	Saturated					Mono unsaturated	Poly unsaturated	
		C10:0	C12:0	C14:0	C16:0	C18:0	C18:1	C18:2	C18:3
Beef Tallow	0.9			3	24	19	43	3	1
Canola Oil	15.7				4	2	62	22	10
Lard	1.2			2	26	14	44	10	
Palm	1.0			1	45	4	40	10	
Soybean	5.7				11	4	24	54	7
Sunflower	7.3				7	5	19	68	1

Table 3. Composition of tracylglycerols used as feedstock in biodiesel production. Percent by weight of total fatty acids [8].

Cold flow properties

All diesel fuels, as well as biodiesel are subject to performance problems when they are subjected to cold temperatures. As a fuel is cooled, high molecular weight components present in the fuel begin to precipitate and this causes the fuel to start to solidify or gel.

The cold flow properties of the biodiesel are dependent on the fatty acids composition of the triacylglycerol feedstock. The transesterification does not change the chemical compositions of the fatty acids; it just makes methyl esters of these acids. Therefore, biodiesel made from triacylglycerol feedstock composed of high concentration of high molecular weight fatty acids will have poor cold flow properties. Tallow and palm biodiesel are the worst offenders. They start to have cold flow problems between 18 to 10°C. Canola, rapeseed, sunflower and soybean biodiesels start having problems around 0°C [18].

Feedstock Quality

Pure triacylglycerols feedstock is easy to convert to biodiesel. However, impurities that may be present in the feedstock can impact quality and cost of the final product. Common impurities present with the triacylglycerol feedstock are water, solids, free fatty acids and sulfur [19].

Water

In the production of biodiesel, it is important to keep water below 1%. The presence of water in the feedstock will produce soaps during the transesterication process and affect the completeness of the reaction. The soap and water can form a water in oil emulsion which will affect the final biodiesel fuel quality; since, it will create deposits, viscosity and engine performance problems.

These water emulsions can be broken by heating. Therefore, the oil can be heated and the water allowed settling to the bottom of the container. Water removal is performed by pumping the water out from the bottom of the container from under the oil.

Solids

Insoluble particles can be present with the feedstock. This is a particular problem with yellow and trap grease. These particles can create fuel filter plugging and engine deposits. Therefore, it is recommended to filter the feedstock before transesterification.

Free fatty acids

Base catalyzed transesterefication of high free acid feedstock will react with the catalyst and produce soaps. Feedstock with more than 2% free fatty acid needs to be caustic striped before being used in base catalyzed transesterification. Feedstocks with characteristic high amounts of free fatty acids are tallow and yellow grease. These feedstocks usually contain over 15% free fatty acids.

On the other hand, acid catalyzed transesterification produces water as a byproduct of the reaction. Water needs to be removed in order to drive the reaction to completion. This reaction also requires higher temperatures and a higher ratio of alcohol to free fatty acids, usually around 20:1 to 40:1.

A combination of acid catalyzed esterification followed by a base catalyzed reaction offers a good alternative for biodiesel production from high free fatty acid feedstocks. In this case, the acid catalyst of choice is phosphoric acid, H_3PO_4. After esterication, the H_3PO_4 is reacted with excess KOH. Finally, at the end of the process, the remaining KOH is reacted H_3PO_4. The K_3PO_4 is dried and sold as fertilizer.

Sulfur

The EPA regulates the amount of sulfur in fuels. For on road fuels, the EPA mandates 15 ppm sulfur maximum. In Europe, the sulfur level in biodiesel has to be lower than 10 ppm. Biodiesel made from pure feedstocks has virtually no sulfur. However, sulfur levels in waste grease can reach to 200 - 400 ppm. During production, the final sulfur concentration can be reduced by approximately 40 to 50%. Vacuum distillation can also reduce sulfur by 50%. Treatment with activated carbon can reduce sulfur in biodiesel to acceptable low levels.

Production Practices

Quality of the final product is also dependent on production practices. Good practices will insure completeness of the reaction, good separation of the glycerol from the reaction product, stripping of the alcohol, splitting of soaps and water and catalyst removal.

Reaction completeness

The trasesterication of triacylglycerols into biodiesel occurs by first producing a diacylglycerol, which in turn is converted to a monoacylglycerol and finally a glycerol molecule. Each of the reaction steps produces a molecule of fatty acid methyl ester. If left with the final product, they can produce cold flow problems and engine deposits and the biodiesel may not pass ASTM or EN specifications. However, there are absorbents in the marketplace that through filtration can selectively remove acylglycerols and glycerol.

Glycerol

Glycerol is an undesirable product in biodiesel production. It is insoluble in biodiesel and could be easily removed by settling to the bottom of the tank or by centrifugation. Excess methanol and high concentration of soaps will inhibit the separation. Glycerol in the biodiesel will create viscosity, engine combustion and filter plugging problems. Water washing or absorbents can reduce the concentration of glycerol in biodiesel to acceptable levels.

Alcohol

Biodiesel may contain up to 4% after glycerol separation. Excess methanol in the fuel will provide a dangerous explosive mixture in compression-ignited engines. The methanol present in the fuel influences the flash point. The change in flash point of fatty acid methyl ester biodiesel versus methanol and ethanol concentrations is given in Figure 2. Water washing or vacuum stripping will reduce alcohol to acceptable levels and meet ASTM and EN specifications.

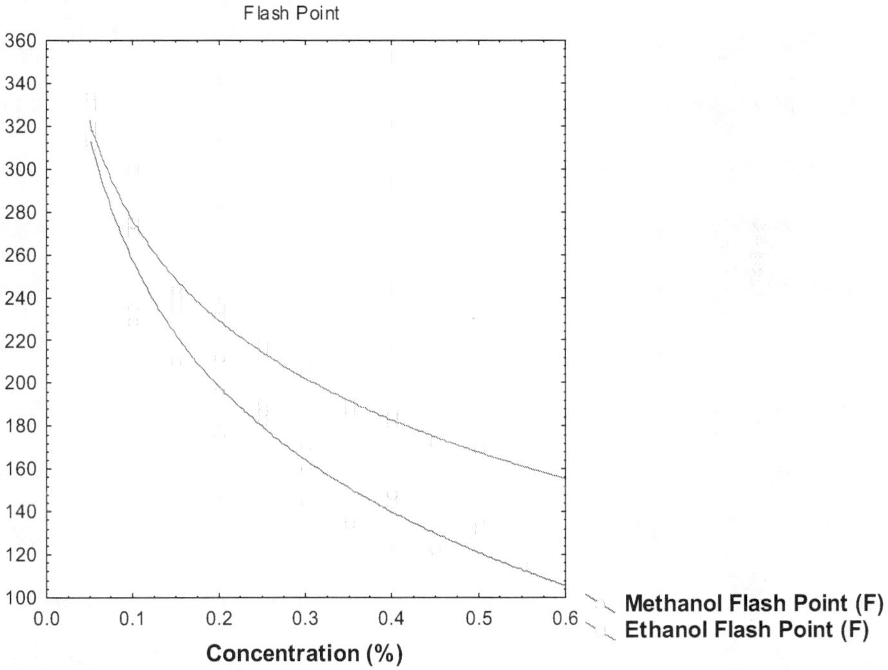

Fig. 2. Flash point of methanol and ethanol versus concentration in biodiesel.

Soaps

Soaps have been previously discussed. They can form microemulsions and influence the performance of the fuel. Soaps can be removed by water washing of the final product.

Water and catalyst removal

Water can be present as microemulsion or dissolved in the fuel. Biodiesel can contain up to 0.15% dissolved water. Water can contribute to corrosion, microbiological grows, sedimentation, etc.

Water can be removed by allowing it to settle to the bottom of the tank, boiling it off or by using solid absorbers.

Residual catalysts can form engine deposits and abrasion and wear of the fuel engine parts. Catalyst is usually removed with the glycerol and with the final water wash of the fuel.

BQ-9000 (Quality Assurance Program)

Finally, we could not leave this subject without mentioning BQ-9000.
The National Biodiesel Accreditation Program is a cooperative and voluntary program for the accreditation of producers and marketers of biodiesel fuel called BQ-9000. The program is a unique combination of the ASTM standard for biodiesel, ASTM D 6751, and a quality systems program that includes storage, sampling, blending, shipping, distribution and fuel management practices. BQ-9000 is open to any biodiesel manufacturer, marketer or distributor of biodiesel blends in the U.S. and Canada.

2. Conclusion

Biodiesel is a renewable fuel manufactured from feedstocks such as virgin and used vegetable oils, animal fats and recycled restaurant greases. It serves as a substitute for conventional diesel.
Feedstocks, feedstock quality and production practices can influence the quality of the final product. However, by taking appropriate steps in the production of biodiesel, a high quality fuel can be produced.

3. References

[1] Jewett, B., Inform 14: 528-530 (2003).
[2] Harold, S., Lipid Technol. 10: 67-70 (1997).
[3] Masjuki, H.H., Sapuan, S. M., J. Am. Oil Chem. Soc. 72: 609-612 (1995).
[4] Sii, H. S., Masjuki, H., Zaki, A. M., J. Am. Oil Chem. Soc. 72: 905-909 (1995).
[5] Clerc, J. C., "Cetane Number Requirements of Light Duty Diesel Engines at Low Temperatures," Report No. 861525, Society of Automotive Engineers, Warrendale, PA 1986.
[6] Gopinath, A., Puhan, S., Nagarajan, G., Proc. IMechE Vol. 223 Part D: J. Automobile Engineering, 211(4), 565-583 (2009).
[7] Frankel, E. N. , Lipid Oxidation, The Oily Press, Dundee, Scotland, 1998.
[8] http://www.scientificpsychic.com/fitness/fattyacids1.html
[9] Waynick, J. A., SwRI Project No. 08-10721. Task 1Results. August 2005.
[10] Andersson, K., Lingnert, H., J. Am. Oil Chem. Soc. 75(8), 1041-1046 (1998).
[11] Neff, W.E, Mounts, T. L., Rinsch, W. M., Konishi, H., J. Am. Oil Chem. Soc. 70(2), 163-168 (1993).
[12] Neff, W.E, El-Agaimy, M. A., Mounts, T. L., J. Am. Oil Chem. Soc. 71(10), 1111-1116 (1994).
[13] Loury, M., Lipids, 7, 671-675 (1972).
[14] DeMan, J. M, Tie, F., deMan, L., J. Am. Oil Chem. Soc. 64(7), 993-996 (1987).
[15] Tang, H. Y., Salley, S. O., Ng, K. Y. S., Fuel 87: 3006-3017 (2008).
[16] Tang, H. Y., De Guzman R. C., Salley, S. O., Ng, K. Y. S., J Am Oil Chemo c 85: 1173-1182 (2008).
[17] http://biodieselmagazine.com/article.jsp?article_id=196
[18] Dunn, R. O. in The Biodiesel Handbook, edited by G. Knothe, J.Van Gerpen and J. Krahl, AOCS Press, Champaign, Illinois 1962.
[19] Van Gerpen, J., Pruszko, R., Clements, D., Shanks, B., Knothe, G., Building a Successful Biodiesel Business, January 2005.

5

The Effect of Storage Condition on Biodiesel

Yo-Ping Wu, Ya-Fen Lin and Jhen-Yu Ye
Department of Chemical and Materials Engineering, National Ilan University
Taiwan, R.O.C.

1. Introduction

Biodiesel is an alternative diesel fuel derived from the varied processes of vegetable oils, animal fats, or waste frying oils to give the corresponding fatty acid methyl esters (Chang et al., 1996; Schmidt & Van Gerpen, 1996). In the transport sector it can be used blended with fossil diesel fuel and in pure form. The major chemically bound oxygen component in the biodiesel fuel has the effect of reducing the pollutant concentration in exhaust gases due to better burning of the fuel in the engine (Kahn et al., 2002). It is also described as an alternative fuel which improves environmental conditions and contributes to gaining energy sustainability (Edlund et al., 2002).

As biodiesel fuels are becoming commercialized and with its biodegradability, it is important to examine their properties as respect to transport, storage, or processing. Demirbas (Demirbas, 2007) has summarized the biodegradability data of petroleum and biofuels available in the literature and showed heavy fuel oil has low biodegradation of 11%, in 28 day laboratory studies while biodiesels have 77%-89% biodegraded, and diesel fuel was only 18% biodegraded.

Some studies have been conducted focusing on how biodiesel stimulated the degradation of petrol-diesel in varied environments. However, there are very few studies concentrated on biodiesel degradation under different storage temperatures and storage environments such as in a sealed or ambient environment, and in an environment with or without the presence of water moisture. Mittelbach and Gangl (Mittelbach & Gangl, 2001) studied the degree of physical and chemical deterioration of biodiesel produced from rapeseed and used frying oil under different storage conditions. They found there has severe effects when the fuel was exposed to daylight and air. But they found there were no significant differences between undistilled biodiesel made from fresh rapeseed oil and used frying oil. In their study, the viscosity and neutralization numbers rose during storage and did not reach the specified limits for over 150 days.

Zullaikah et al. (Zullaikah et al., 2005) examine the effect of temperature, moisture and storage time on the accumulation of free fatty acid when they used a two-step acid-catalyzed process to produce the biodiesel from rice bran oil. Their results showed rice bran stored at room temperature showed that most triacylglyceride was hydrolyzed and free fatty acid (FFA) content was raised up to 76% in six months. Leung et al. (Leung et al., 2006) divided twelve biodiesel samples into 3 groups and stored at different temperatures and environments to monitor the regular interval over a period of 52 weeks. Their results showed that the biodiesel under test degraded less than 10% within 52 weeks for those samples stored at 4 and 20 ^{o}C while nearly 40% degradation was found for those samples stored at 40 ^{o}C.

Bouaid et al. (Bouaid et al., 2007) used four different vegetable oils: high oleic sunflower oil (HOSO), high and low erucic Brassica carinata oil (HEBO and LEBO) respectively and used frying oil (UFO) to produce biodiesel through the process of transesterification. These biodiesels were then used to determine the effects of long storage under different conditions on oxidation stability. Their samples were stored in white (exposed) and amber (not exposed) glass containers at room temperature for a 30-months study period. Their results showed that acid value, peroxide value, viscosity and insoluble impurities increased, while iodine value decreased with increasing storage time. They also found there has slight differences between biodiesel samples exposed and not exposed to daylight before a storage time of 12 months and after this period the differences were significant.

Karavalakis et al. (Karavalakis et al., 2011) investigated the impact of various synthetic phenolic antioxidants on the oxidation stability of biodiesel blends with the employment of the modified Rancimat method. Their experimental results revealed Butylated hydroxytoluene (BHT) and butylated hydroxyanisol (BHA) showed the lowest effectiveness in neat biodiesel, whereas their use in biodiesel blends showed a greater stabilizing potential. Propyl gallate (PG) and pyrogallol (PA) additives showed the strongest effectiveness in both the neat biodiesel and the biodiesel blends. They conducted an ageing process-- a naturally ageing process of the biodiesel blends for a period of 10 weeks; samples were taken every 2 weeks to simulate the automotive biodiesel stored in the fuel tank of a vehicle. Their results showed a sharp decrease in fuel stability, significantly increased in acid value but limited effects in viscosity over time. The addition of antioxidants resulted in some increases in viscosity and acid value of the biodiesel blends.

In this study, one commercial biodiesel and three laboratory-produce biodiesels were used to verify the effect of storage temperature, type of storage container, storage time as well as the moisture content on the properties of the biodiesel. The major properties analyzed in this study include acid value, iodine value, viscosity, flash point, and heating value. The variation of the chemical species in the tested biodiesel were also analyzed and compared.

2. Materials and methods

The laboratory-produce biodiesel fuel used in this study were produced from the transesterification of vegetable oil with methanol (CH_3OH, Malliuckrodt Baker Inc., USA) catalyzed by sodium hydroxide (NaOH, Shimakyu, Osaka, Japan). The reaction scheme of the methanolysis of triacyloglycerols can be found elsewhere (Komers et al., 1998; Wu et al., 2007). Three types of vegetable oil, soybean oil, peanut oil, and sunflower seed oil, were converted into biodiesels- soybean oil methyl ester (SBM), sunflower seed oil methyl ester (SFM), and peanut oil methyl ester (PNM). A titration was performed to determine the amount of NaOH needed to neutralize the free fatty acids in each vegetable oil. The amount of NaOH needed as catalyst for every liter of soybean oil, sunflower seed oil, and peanut oil were determined as 4.4g, 4.3g, and 4.1g, respectively. For transesterification, 200mL CH_3OH plus the required amount of NaOH were added for every liter of cooking oil, and the reactions were carried out at 65°C. A total of 50L of each vegetable oil was used to produce biodiesel. The water wash process was performed by using a sprinkler which slowly sprinkled water into the biodiesel container until there was an equal amount of water and biodiesel in the container. The water/biodiesel mixture was then agitated gently for 10 min., allowing the water to settle out of the biodiesel. After the mixture had settled, the water was drained out.

A series of tests were performed to characterize the properties of the produced biodiesel. These properties include density (ASTM D 1298), kinematic viscosity (ASTM D445), acid value (ASTM D664), iodine value (CNS 15060), flash point (ASTM D 93), water and sediment (ASTM D 2709), and heating value (ASTM D 240). The heating values of biodiesels were measure by bomb calorimeter (PAAR 6200, USA). The chemical components in the biodiesels were also analyzed by a gas chromatograph/mass spectrometry (ThermoQuest Trace MS) with a 1.0□□m, 0.25mm × 30m DB-1 column (J & W Scientific).
There was another 50L of commercial biodiesel (NJC) obtained from Taiwan NJC Corp. used for this study. The experiments use NJC, SFM, SBM, and PNM as the biodiesels to examine the effects of storage condition. The biodiesels were stored in the polypropylene bottle (PP) and stainless steel container (ST), respectively. Each container contains 500mL of biodiesels, there were three groups of water contents contained in two different containers, which include the pure biodiesel storage in the PP bottle (ppB100) and ST cup (stB100), 98% biodiesel + 2% distilled water storage in PP bottle (ppB98) and ST cup (stB98), and 95% biodiesel + 5% distilled water storage in PP bottle (ppB95) and ST cup (stB95), and stored at 0°C, 25°C, and 40°C, respectively. The major properties of every sample were measured at the time interval of 0, 1st, 2nd, 4th, 8th, 16th, and 32nd week.

3. Results and discussions

As similar in our previous study (Wu et al., 2007), after the transesterification process, there was a nearly 90% volume ratio of methyl ester phase to a 10% volume ratio of glycerol phase during the separation process. In this study, the volume ratios of methyl ester phase were 91.98%, 89.63%, and 91.33% for SBM, SFM, and PNM, respectively, which yielded nearly 45L of each biodiesel for the use in this study. A total of 50 L of each vegetable oil was used and converted into biodiesel. After the transesterification process, there was a nearly 90% volume ratio of methyl ester phase to a 10% volume ratio of glycerol phase during the separation process, which gave nearly 45L of sunflower seed oil biodiesel (SFM) for the use in this study. Table 1 gives some of the major properties of the NJC, SFM, SBM and PNM biodiesels. As observed form these data, the NJC has higher acid value and gross heating value but lower iodine value, kinematic viscosity, and density than the laboratory-produce biodiesels.

Property	NJC	SFM	SBM	PNM	Method
Acid value (mg KOH/g)	0.685	0.145	0.267	0.154	ASTM D664
Iodine value (g I_2/100g)	103.37	137.81	129.1	130.52	CNS 15060
Kinematic Viscosity (mm^2/s) at 40°C	4.103	5.251	5.171	5.390	ASTM D445
Density (g/cm^3) at 15°C	0.884	0.897	0.896	0.896	ASTM D1298
Flash point (°C)	172	172	184	176	ASTM D93
Heating value (cal/g)	9529	9464	9460	9420	ASTM D240
Water content (ppm)	—	—	1699	1768	CNS 4446

Table 1. Original major properties of the NJC, SFM, SBM, and PNM used in this study.

Table 2 compares the initial major chemical composition of the commercialized biodiesel (NJC) and the laboratory-produce biodiesels. NJC contained 19.69% saturated fatty acid methyl esters (FAMEs) and 79.92% unsaturated FAMEs, SFM contained 12.38% saturated

Name	Formula	NJC	SFM	SBM	PNM
2-Methyl pentane	C_6H_{14}		0.373%		
3-Methyl pentane	C_6H_{14}		0.254%		
Hexane	C_6H_{14}		0.585%		
Methylcyclopentane	C_6H_{12}		0.172%		
Cyclohexane	C_6H_{12}		0.033%		
Methyl enanthate	$C_8H_{16}O_2$	0.001%			
Methyl caprylate	$C_9H_{18}O_2$	0.033%	0.022%		
2,4-Decadienal	$C_{10}H_{16}O$	0.011%	0.023%		
Methyl caprate	$C_{11}H_{22}O_2$	0.012%			
9-Oxo-nonanoic acid methyl ester	$C_{10}H_{18}O_3$	0.004%			
Butylated Hydroxytoluene (BHT)	$C_{15}H_{24}O$	0.168%			
Methyl laurate	$C_{13}H_{26}O_2$	0.114%			
Methyl 8-(2-furyl)octanoate	$C_{13}H_{20}O_3$	0.002%			
Methyl Z-11-tetradecenoate	$C_{15}H_{28}O_2$	0.008%			
Methyl myristate	$C_{15}H_{30}O_2$	0.626%	0.064%	0.005%	0.010%
Methyl cis-4-octenoate	$C_9H_{16}O_2$		0.002%		
Methyl (9E)-9-dodecenoate	$C_{13}H_{24}O_2$		0.005%		
Methyl pentadecanoate	$C_{16}H_{32}O_2$	0.030%	0.011%		
3,7,11,15-Tetramethyl-2-hexadecen-1-ol	$C_{20}H_{40}O$	0.034%			
Palmitoleic acid methyl ester	$C_{17}H_{32}O_2$	0.437%	0.094%	0.004%	0.011%
Methyl hexadecanoate	$C_{17}H_{34}O_2$	14.568%	6.911%	9.640%	10.579%
Ethyl palmitate	$C_{18}H_{36}O_2$	0.020%			
Methyl 8-(2-hexylcyclopropyl)octanoate	$C_{18}H_{34}O_2$	0.046%	0.025%		
Methyl margarate	$C_{18}H_{36}O_2$	0.071%	0.036%		0.022%
Methyl linileate	$C_{19}H_{34}O_2$	36.450%	57.762%	64.183%	61.332%
Methyl oleate	$C_{19}H_{36}O_2$	42.631%	26.186%	22.268%	22.386%
Methyl stearate	$C_{19}H_{38}O_2$	3.777%	4.129%	2.835%	3.986%
Methyl cis-9,cis-15-linoleate	$C_{19}H_{34}O_2$		0.014%		
Methyl linolelaidate	$C_{19}H_{34}O_2$		0.099%		
Dimethyl 9-oxoheptadecanedioate	$C_{19}H_{34}O_5$		0.027%		
Methyl linolenate	$C_{19}H_{32}O_2$	0.031%	0.034%	0.016%	
Methyl (6E,9E,12E)-6,9,12-octadecatrienoate	$C_{19}H_{32}O_2$		0.046%		
Methyl ricinoleate	$C_{19}H_{36}O_3$	0.022%			
Methyl (11E,14E)-11,14-icosadienoate	$C_{21}H_{38}O_2$	0.031%			
Methyl (11E)-11-icosenoate	$C_{21}H_{40}O_2$	0.274%	0.152%	0.106%	0.122%
Methyl 9-hydroxystearate	$C_{19}H_{38}O_3$	0.016%			
Methyl arachisate	$C_{21}H_{42}O_2$	0.256%	0.235%	0.224%	0.281%
Methyl heneicosanoate	$C_{22}H_{44}O_2$			0.014%	
Methyl 9,10-dihydroxystearate	$C_{19}H_{38}O_4$		0.008%		
2-Monopalmitin	$C_{19}H_{38}O_4$		0.024%		
Methyl behenate	$C_{23}H_{46}O_2$	0.136%	0.693%	0.255%	0.299%
Methyl tricosanoate	$C_{24}H_{48}O_2$	0.011%	0.024%	0.015%	
Glycerol 1-monolinolate	$C_{21}H_{38}O_4$	0.030%	0.896%	0.193%	0.442%

Name	Formula	NJC	SFM	SBM	PNM
Monoolein	$C_{21}H_{40}O_4$	0.057%	0.537%	0.127%	0.311%
Monostearin	$C_{21}H_{42}O_4$		0.035%		
Methyl (15E)-15-tetracosenoate	$C_{25}H_{48}O_2$	0.009%			
Methyl lignocerate	$C_{25}H_{50}O_2$	0.047%	0.253%	0.072%	0.105%
Squalene	$C_{30}H_{50}$	0.011%	0.018%		
Methyl pentacosanoate	$C_{26}H_{52}O_2$	0.004%			
Methyl hexacosanoate	$C_{27}H_{54}O_2$	0.004%			
Vitamin E	$C_{29}H_{50}O_2$		0.039%		
Stigmasterol	$C_{29}H_{48}O$		0.041%	0.013%	
β-Sitosterol	$C_{29}H_{50}O$	0.016%	0.138%	0.028%	0.117%

Table 2. Initial chemical compositions of NJC, SFM, SBM, and PNM.

FAMEs and 84.42% unsaturated FAMEs, SBM contained 13.06% saturated FAMEs and 86.58% unsaturated FAMEs, and PNM contained 15.28% saturated FAMEs and 83.85% unsaturated FAMEs. The major species in NJC's saturated FAMEs were Methyl hexadecanoate (Hexadecanoic acid methyl ester, $C_{17}H_{34}O_2$, 14.57%) and Methyl stearate (Octadecanoic acid methyl ester, $C_{19}H_{38}O_2$, 3.78%), while Methyl linoleate (9,12-Octadecadienoic acid methyl ester, $C_{19}H_{34}O_2$, 36.45%) and Methyl oleate (9-Octadecenoic acid, methyl ester$C_{19}H_{36}O_2$, 42.63%) were the major species in NJC's unsaturated FAMEs. These four species were also the major species in SFM, SBM, and PNM. SFM contained 6.91% $C_{17}H_{34}O_2$, 4.13% $C_{19}H_{38}O_2$, 57.76% $C_{19}H_{34}O_2$, and 26.19% $C_{19}H_{36}O_2$. SBM contained 9.64% $C_{17}H_{34}O_2$, 2.84% $C_{19}H_{38}O_2$, 64.18% $C_{19}H_{34}O_2$, and 22.27% $C_{19}H_{36}O_2$. PNM contained 10.58% $C_{17}H_{34}O_2$, 3.99% $C_{19}H_{38}O_2$, 61.33% $C_{19}H_{34}O_2$, and 22.39% $C_{19}H_{36}O_2$. Also, there was one species, Butylated hydroxytoluene ($C_{15}H_{24}O$, BHT), which can be uses as antioxidant, detected in NJC.

3.1 Acid value

The freshness of the oil is related to acid value of the oil, while the oil may generate the free fatty acids during longer storage due to the hydrolysis reactions. Hence the acid value becomes one of the important quality targets to determine the purity of oil. Acid value, which defined as the amount of the free fatty acids contained in each gram of the oil determined by the neutral reagent, KOH, denoted as mgKOH/g.

The acid values determined from the four tested biodiesels stored in varied types of container under different temperatures as well as the various moisture contents as a function of storage time are shown in Figure 1. As shown in Figures 1(A), 1(B), 1(C), and 1(D), the acid values of NJC, SFM, SBM, and PNFM remain at a stable range at the lowest storage temperature (0°C). As the storage temperature rose, as shown in Figure 1(E) to Figure 1(L), there shows an acid value increases for these four tested biodiesels with the storage time increased, which suggests the storage temperature did affect but will increase the acid value of the biodiesel if storage temperature increased. Also, as the storage time increased, there shows obvious acid value differences for laboratory-produce biodiesels which stored in ST cups, while the acid value of NJC in ST cups only shows a slightly increase for storage temperatures at 25°C and 40°C. The NJC biodiesel is a commercial product with antioxidant BHT added while SFM, SBM, and PNM are laboratory products with no additive. However, there was a small amount of Vitamin E (0.039%), which can be considered as a natural antioxidant, detected in SFM. This tiny amount of Vitamin E may have the retardation effect on oxidation of SFM at the beginning of study. Another difference is the storage container, the PP bottle has a screw-on lid which has a better sealing than the ST cups did. The effect of antioxidant can be observed from the results shown in Figures 1(A), 1(E), and 1(I). These three figures showed NJC kept in varied oxygen exposure environments (PP and ST) but show similar trend on the change of the acid value, which suggest the antioxidant can slow the degradation of biodiesel. These results show the oxygen in the air performed the oxidation and do affect the acidification of the biodiesel with the temperature acceleration. Among the three laboratory-produce biodiesels, there has a higher non-FAMEs observed for SFM (as shown in Table 2) which might caused the higher acid value change of SFM under oxygen contact environments. Also, the results observed, as shown in Figures 1, show the moisture content has no obvious effect on the acid value of tested biodiesels during the tested period in this study.

3.2 Iodine value

The iodine value is an index used to express the number of unsaturated bonds of the oil, the oil with higher iodine value implies it has higher number of unsaturated bonds. The measured iodine value from the NJC, SFM, SBM, and PNM which stored in PP and ST containers under varied temperatures and moisture contains as function of the storage time are shown in Figure 2. As observed from Figures 2(A), 2(E), and 2(I), the iodine value were stayed in a stable range regardless with the temperature changes, storage time or storage bottles. For iodine value observed from the sample of SFMs, as shown in Figures 2(B), 2(F), and 2(J), the iodine value stably remains for SFM stored at $0^{o}C$, and also stably remains for those kept in PP bottles as storage temperature increased, but decreased with increasing storage temperature and time for those stored in ST cups. The decreased iodine values of SFM stored in ST with increased storage time imply the effect from oxidation. As previous discussed, the NJC biodiesel contains the antioxidants which can prevent the oxidation of the fuel and kept the iodine value in a stable range. The laboratory-produce biodiesels with no oxidation inhibitor in the ST cups contact more oxygen at higher temperature and hence shows the effect on the difference of iodine value.

3.3 Viscosity

Viscosity is the important property that affects engine performance. Higher viscosity interferes with injector operation, resulting in poorer atomization of the fuel spray, and has been associated with increased engine deposits.

The kinematic viscosity of the NJC, SFM, SBM, and PNM stored in PP and ST bottles with varied moisture contents and stored at $0^{o}C$, $25^{o}C$, and $40^{o}C$ with respect to storage time were shown in Figure 3. Similar with the results of acid value and iodine value sections, the viscosity of NJC biodiesel, as shown in Figures 3(A), 3(E), and 3(I), remain constant with respect to storage time. The viscosities of laboratory-produce biodiesels shows varied results. For SFM, SBM, and PNM stored in PP bottles, the viscosity remains constant as storage time changed. However, the viscosity of SFM, SBM, and PNM kept in ST bottles increased as storage temperature increased and also increased as storage time increased. Allena et al. (Allena et al., 1999) has point out the viscosity of biodiesel fuels reduce considerably with increase in unsaturation. Prankl and Worgetter (Prankl & Worgetter,

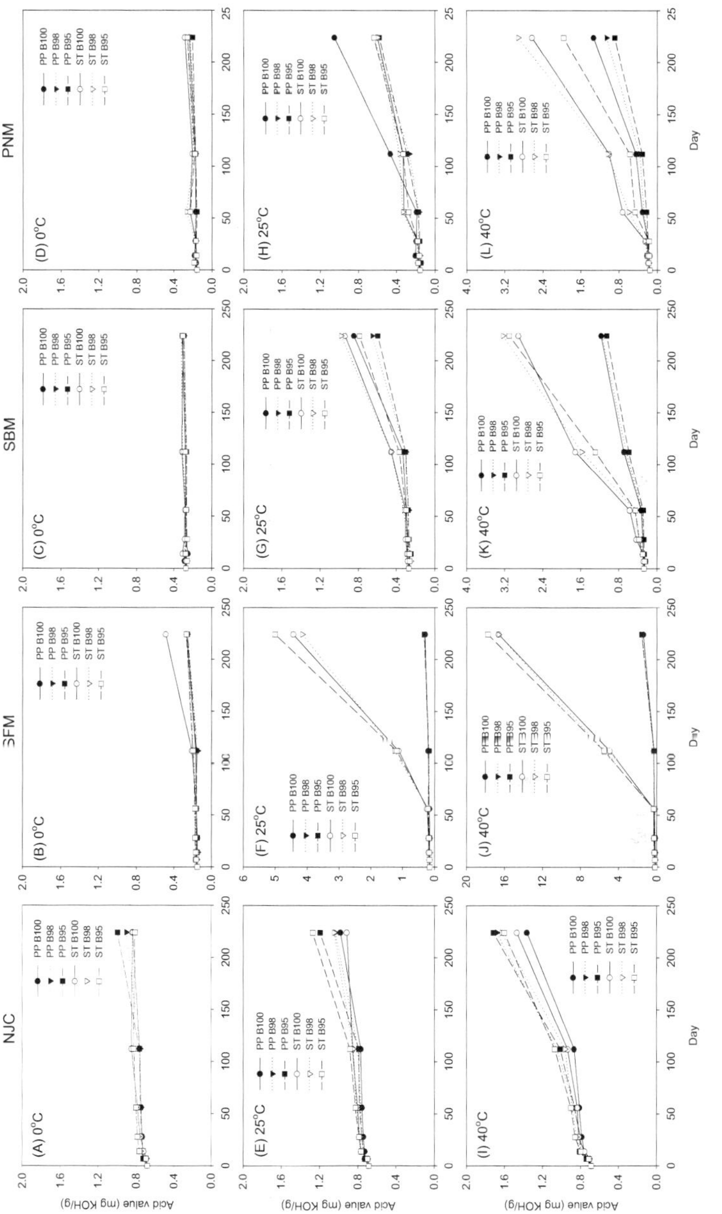

Fig. 1. The acid values determined from the NJC, SFM, SBM, and PNM stored in varied types of container under different temperatures and the various moisture contents as a function of storage time.

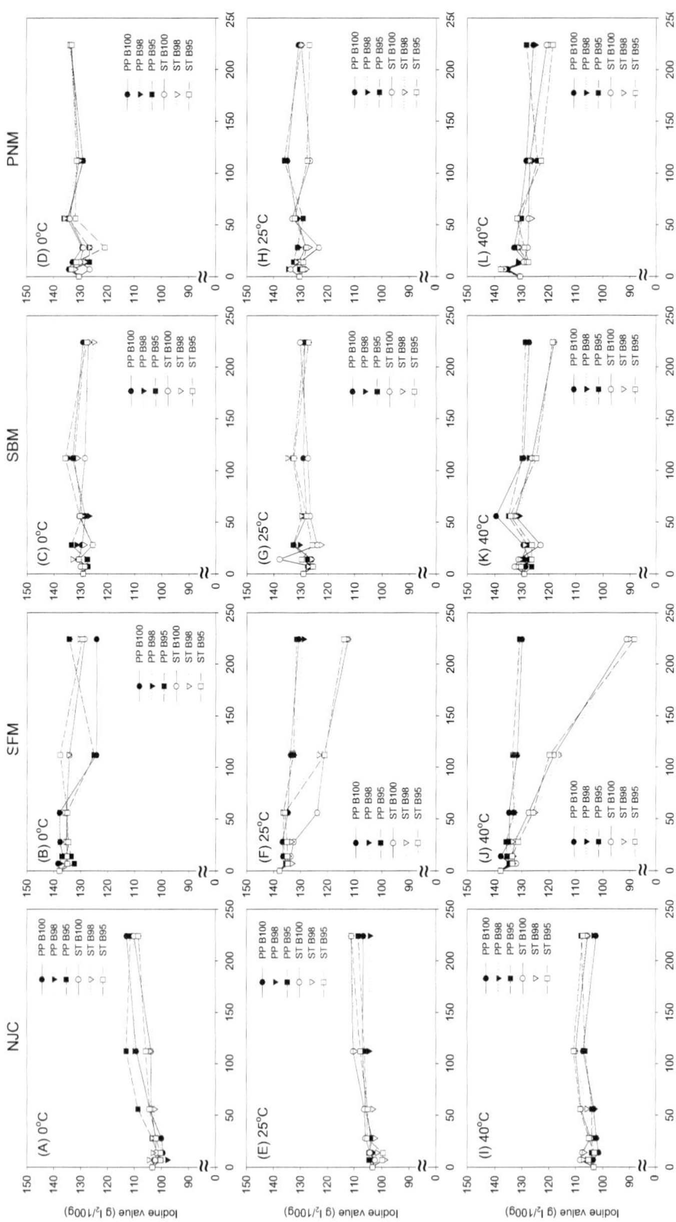

Fig. 2. The iodine values determined from the NJC, SFM, SBM, and PNM stored in varied types of container under different temperatures and the various moisture contents as a function of storage time.

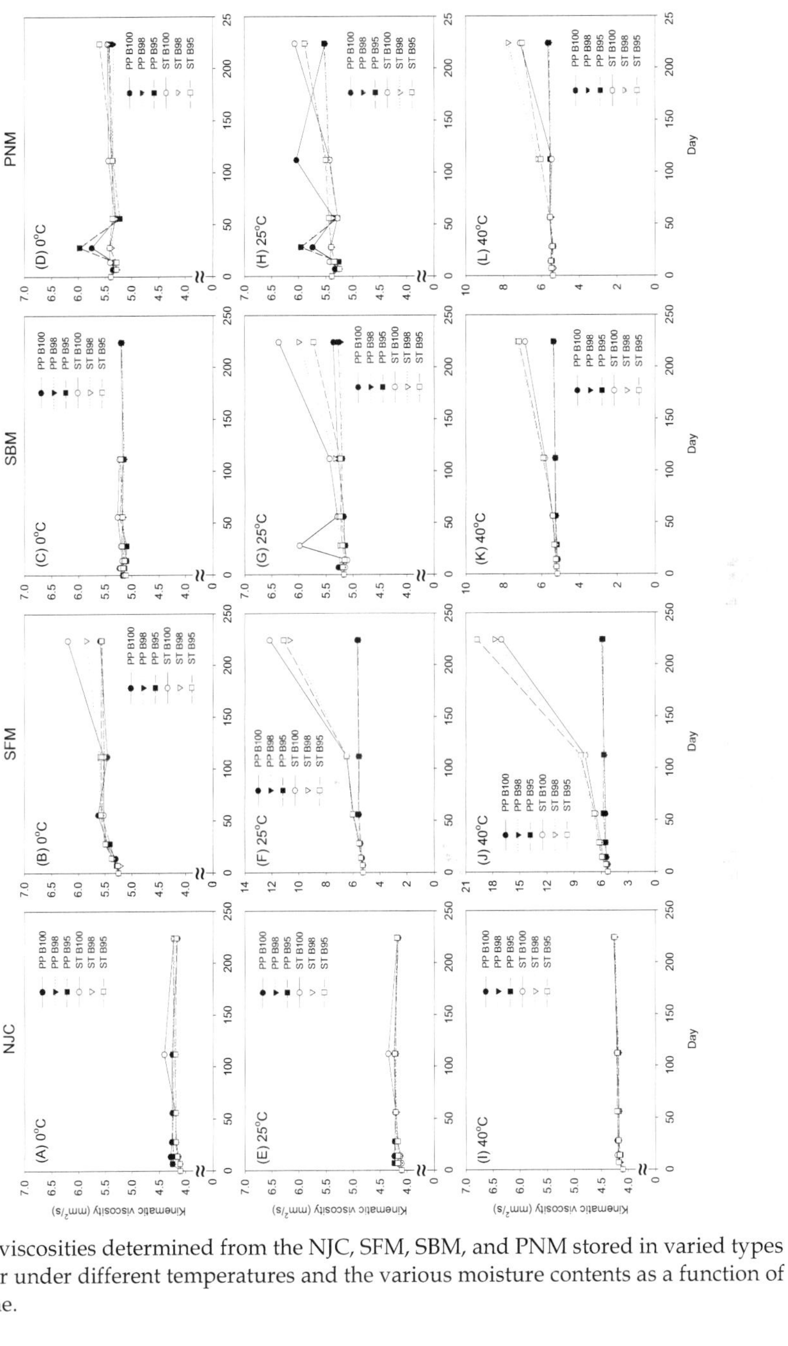

Fig. 3. The viscosities determined from the NJC, SFM, SBM, and PNM stored in varied types of container under different temperatures and the various moisture contents as a function of storage time.

1996) showed that the lower iodine value of biodiesel fuel gives higher viscosity. Compare the results show in Figures 2(F) and 2(J), and Figures 3(F) and 3(J), the viscosity of SFM in ST cups increased match up with the decreased iodine value with increasing storage time. Similarly, there are same trend can be observed from Figures 2(G) and 2(K), and Figures 3(G) and 3(K) for SBM, and from Figures 2(H) and 2(L), and Figures 3(H) and 3(L) for PNM. Here, again, the results show that storage temperature accelerates the degradation and adding antioxidants can prevent the biodiesel degradation.

Viscosity increases with chain length and with increasing degree of saturation (Knothe, 2007). Free fatty acids have higher viscosity than the corresponding methyl or ethyl esters. Since oxidation processes lead to the formation of free fatty acids, double bond isomerization, saturation and products of higher molecular weight, viscosity increases with increasing oxidation.

3.4 Density

Density has importance in diesel-engine performance and is required for the estimation of the Cetane index since fuel injection operates on a volume metering system. (Demirbas, 2008a, as cited in Song, 2000; Srivastava & Prasad, 2000). This study also measured the density of NJC, SFM, SBM, and PNM stored in PP and ST bottles with varied moisture contents and stored at 0°C, 25°C, and 40°C with respect to storage time and shown in Figure 4. Similar with the results of the properties discussed earlier, the density of tested biodiesels remain stable at 0°C. The density of NJC remains constant with regardless with the difference of storage temperature, time, or containers. The density of the SFM, SBM, and PNM kept in ST cups increased with increased storage temperature and storage time.

3.5 Flash point

Flash point is an important property for determining the flammability of a fuel and can be used as a safety indicator for the storage and transportation of a fuel. Biodiesel with a higher flash point indicates the methyl esters transesterification have been properly treated, eliminating any remaining alcohols.

The flash point of the NJC, SFM, SBM, and PNM stored in PP and ST bottles with varied moisture contents and stored at 0°C, 25°C, and 40°C with respect to storage time were shown in Figure. 5. Among these figures, the flash point of the tested biodiesels tends to decrease with increasing storage time at higher storage temperature. The flash point of SFM stored in ST cups show a larger difference with respect to storage time (Figures 5(F) and 5(J)). The FAMEs in biodiesels will get hydrolyzed to alcohols and acids in contact with air. Sharma et al. (Sharma et al., 2008) has point out the oxidation of biodiesel is dependent on the total number of bis-allylic sites. The conversion of FAMEs into alcohols will lead to the reduction of flash point which can easily observed from the results showed in Figures 5(F), 5(G), 5(J), 5(K), and 5(L).

3.6 FAMEs

The major FAME species of the NJC, SFM, SBM, and PNM with 5% water (ST95) in ST cups stored at 40°C was measured by GC/MS during the storage study. Figure 6(A) presents the major species detected in NJC at the beginning (0 week) and the end of study (48th week). This figure shows the shorter chain saturated FAME, Methyl hexadecanoate ($C_{17}H_{34}O_2$), slightly increased after the 48 weeks of durations, but the longer chain saturated FAME,

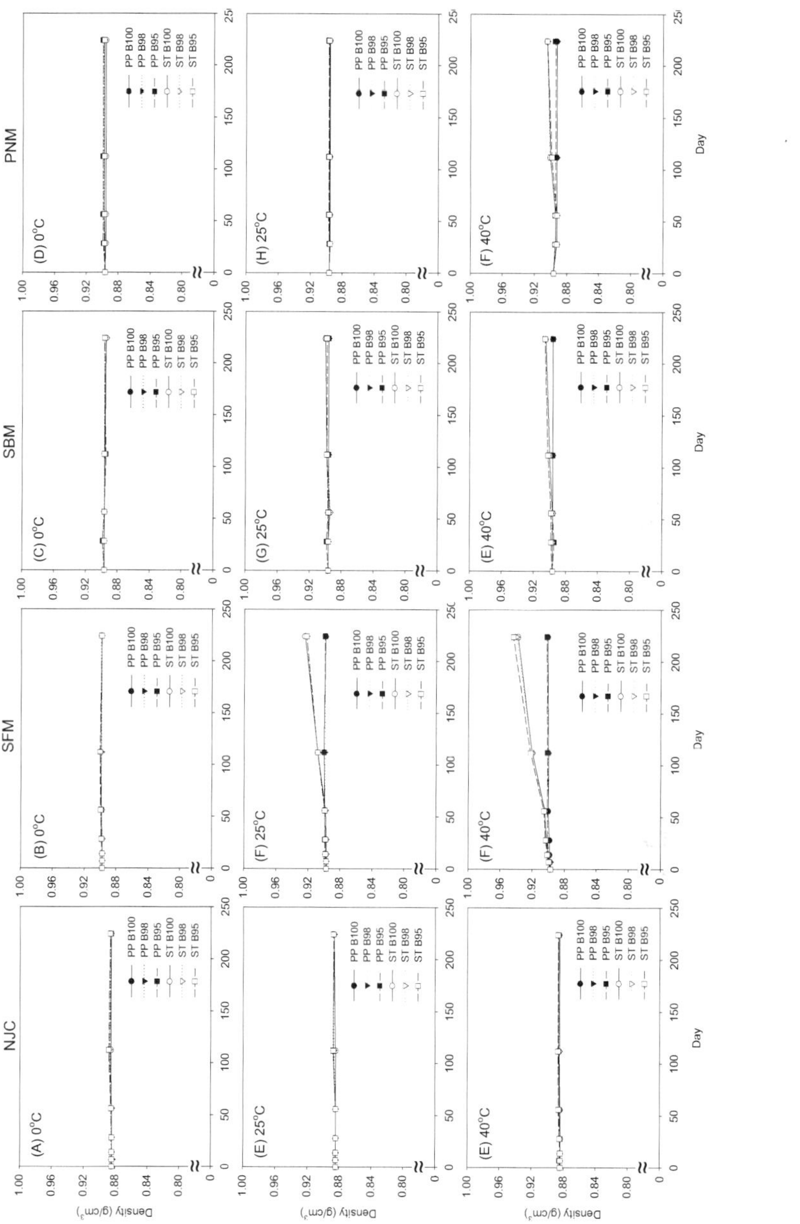

Fig. 4. The densities (15°C) determined from the NJC, SFM, SBM, and PNM stored in varied types of container under different temperatures and the various moisture contents as a function of storage time.

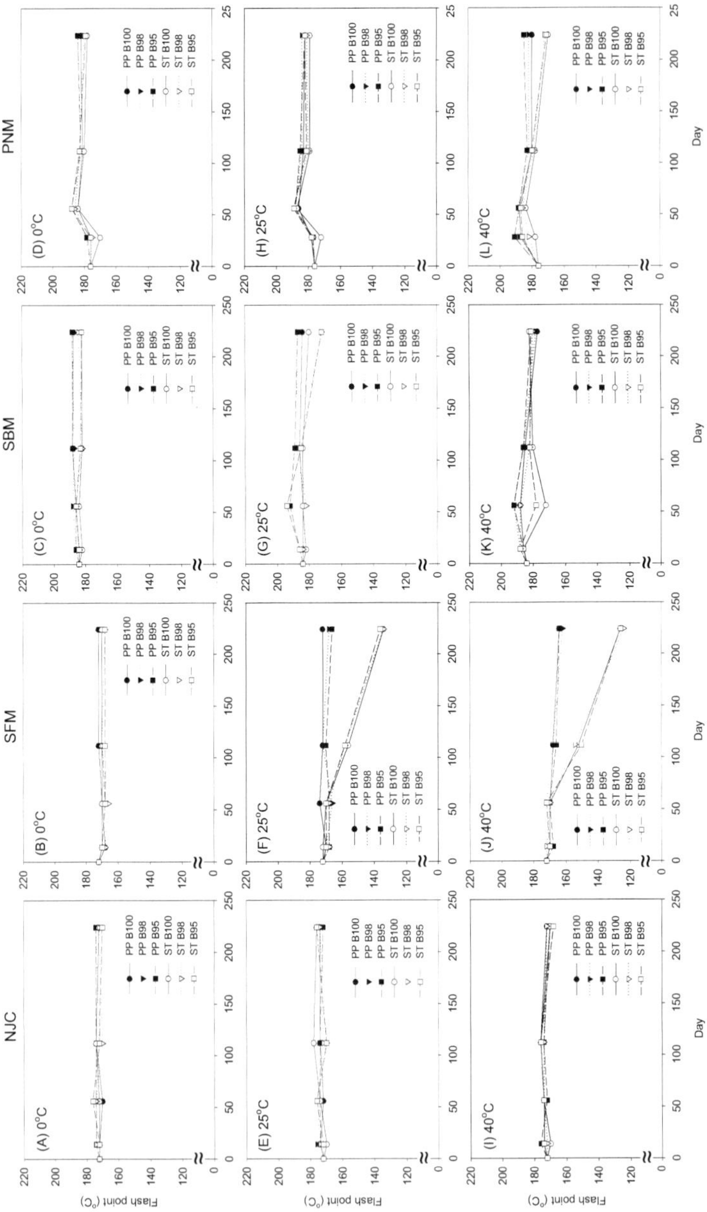

Fig. 5. The flash points determined from the NJC, SFM, SBM, and PNM stored in varied types of container under different temperatures and the various moisture contents as a function of storage time.

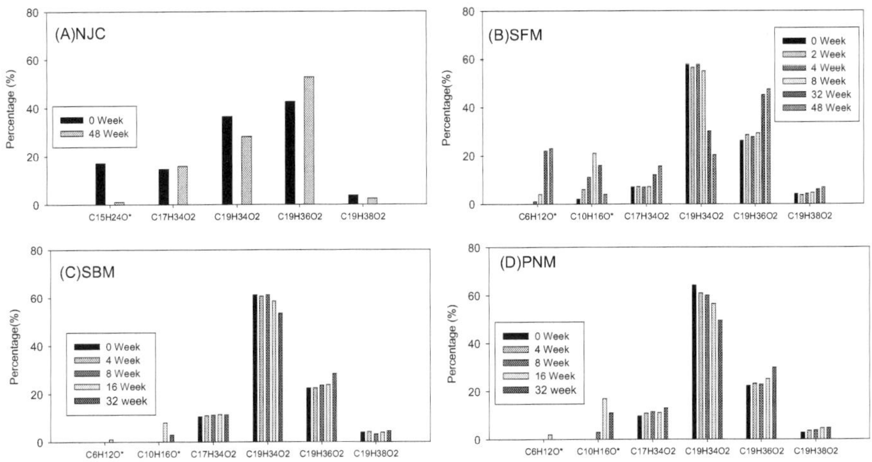

Fig. 6. The major FAME species of the NJC, SFM, SBM, and PNM with 5% water (ST95) in ST cups stored at 40°C.

Methyl stearate ($C_{19}H_{38}O_2$) decreased. For unsaturated FAMEs, the di-double bonds species, Methyl linoleate ($C_{19}H_{34}O_2$), decreased at the end of study and the mono-double bond species, Methyl oleate ($C_{19}H_{36}O_2$) increased. Figure 6(A) also presents the concentration difference of BTH between the beginning and the end of study. It shows the BTH concentration also decreased with increased storage time. Figure 6(B) shows the difference of the major FAME species during the study. As shown in the figure, the trends of the difference for the major FAMEs are similar with those in NJC. There were also have a small amount of alcohols, hexanal ($C_6H_{12}O$) and 2, 4-Decadienal ($C_{10}H_{16}O$), detected as storage time increased. It supports the hydrolysis reaction did occurs within the storage. Figures 6(C) and 6(D) present the major FAMEs analysis results from SBM and PNM but with a shorter period. The difference trends of the major FAMEs were similar with those of NJC and SFM.

3.7 Higher heating value (HHV)

The higher heating value (HHV) is an important property defining the energy content of fuels. The HHVs of the NJC, SFM, SBM, and PNM stored in PP and ST bottles with varied moisture contents and stored at 0°C, 25°C, and 40°C with respect to storage time were shown in Figure 7. The HHV remains at a stable level for NJC and for SFM, SBM, and PNM which stored at lower temperatures. The HHV of SFM stored in ST cups at higher temperatures, Figures 7(F) and 7(J), shows a larger decrease with respect to storage time. The HHV is the same as the thermodynamics heat of combustion with enthalpy change for the reaction of the compounds before and after combustion. Here, as refer to the species of the FAMEs from Figure 6, the formation of enthalpy ($H_{f,298}$) at 25°C were estimated by using the "THERM" computer code which is based on modified Group Additivity (Benson, 1976; Ritter & Bozzelli, 1991). The gross $H_{f,298}$ of each biodiesel at varied period were then calculated by assuming the $H_{f,298}$ is proportional to the fraction (x_i) of the species, that is

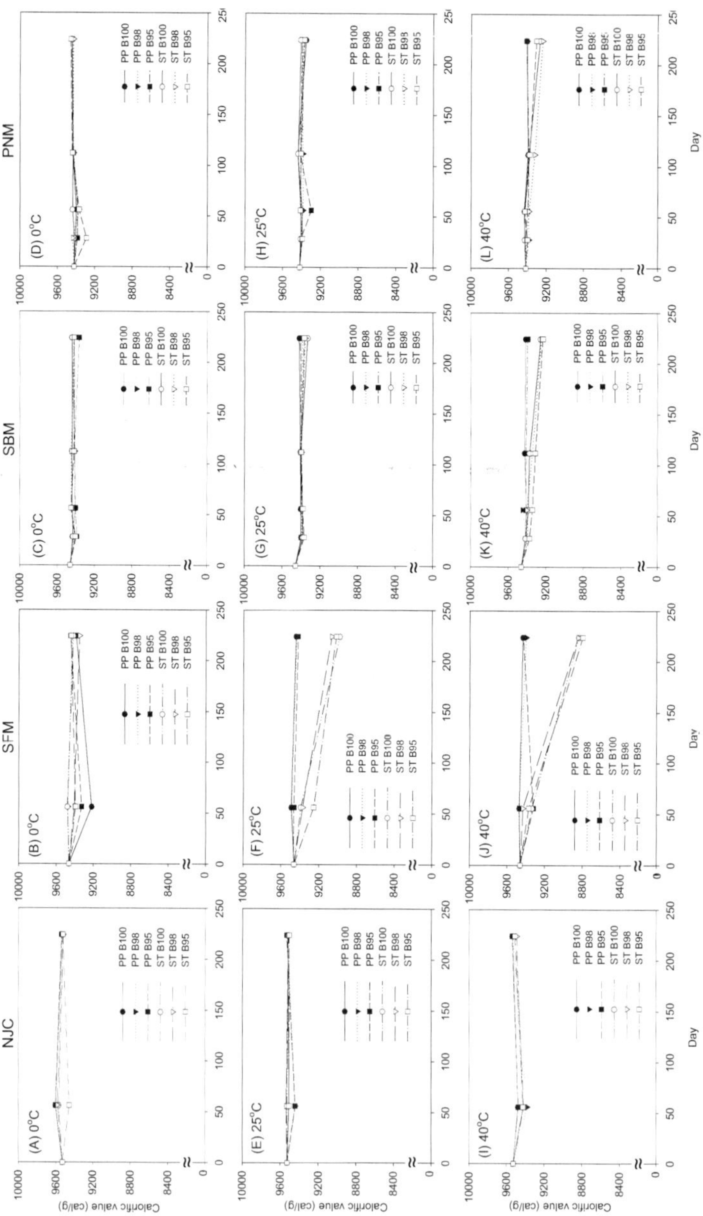

Fig. 7. The high heating values determined from the NJC, SFM, SBM, and PNM stored in varied types of container under different temperatures and the various moisture contents as a function of storage time.

$$H_{f,298} = \sum_i x_i H_{f,298'i} \cdot \quad (1)$$

The HHVs were then calculated by assuming the species is combusted completely to CO_2 and H_2O. Fig. 8 presents the calculated HHV of NJC, SFM, SBM, and PNM at varied storage time. Demirbas (Demirbas, 2008) presents several estimate methods for HHV in his study. One is using the correlation between flash point and HHV for biodiesels. The equation between flash point (FP) and HHV is

$$\text{HHV (MJ/Kg)} = 0.021\ \text{FP (K)} + 32.12 \quad (2)$$

Another equation from regression is between viscosity (VS) and HHV for biodiesels. The equation between VS and HHV is

$$\text{HHV (MJ/Kg)} = 0.4625\ \text{VS (cSt)} + 39.450 \quad (3)$$

There is a modified Dulong's formula may used to calculate the HHV for biomass fuels such as coal (Demirbas, 2008b, as cited in Perry & Chilton, 1973; Demirbas, 2008b, as cited in Demirbas *et al.*, 1997) as a function of the carbon (C%), hydrogen (H%), oxygen (O%), and nitrogen (N%) contents (wt.%).

$$\text{HHV (MJ/Kg)} = 0.335(\text{C\%}) + 1.423(\text{H\%}) - 0.154(\text{O\%}) - 0.145(\text{N\%}) \quad (4)$$

Figure 8 compares the HHV of NJC, SFM, SBM, and PNM from the methods described above with the experimental values. Among this figure, all the estimating HHVs were higher than experimental results. As shown in Figure 8(A), the estimated HHVs of NJC stay in a small variance range. For the other biodiesels, as shown in Figures 8(B), 8(C), and 8(D), the estimated methods giving an order of VS > Flash pt > Dulong's > THERM on giving the value of HHV. There has an exception on these estimations, as shown in Figure 8(B), there has large HHV difference for SFM at the end of storage study. The method using VS equation gives an extreme high HHV due to the high viscosity detected in the sample.

4. Conclusions

The acid values of tested biodiesels remain at a stable range at lowest temperature (0°C). As the storage temperature raised the acid value increases with the storage time increased, which suggests the storage temperature did affect but will increase the acid value of the biodiesel. Also, the results show the oxygen performed the oxidation and do affect the acidification of the biodiesel with the temperature acceleration. The laboratory-produce biodiesels with no oxidation inhibitor in the ST cups contact more oxygen at higher temperature and hence shows the effect on the difference of iodine value. The viscosity of SFM, SBM, and PNM kept in ST cups increased as storage temperature increased and also increased as storage time increased. These results suggesting the storage temperature and the degree of oxygen contact will affect the degradation of biodiesel. The water content shows no significant effect on the biodiesel storage in this study. There still have some research topics, for example, the ageing effect of natural or artificial antioxidants on the chemical composition of biodiesels, the proper amount of antioxidants if needed, and the quick analysis method for determining the degree of degradation, can be considered for helping us has better understanding on biodiesel storage.

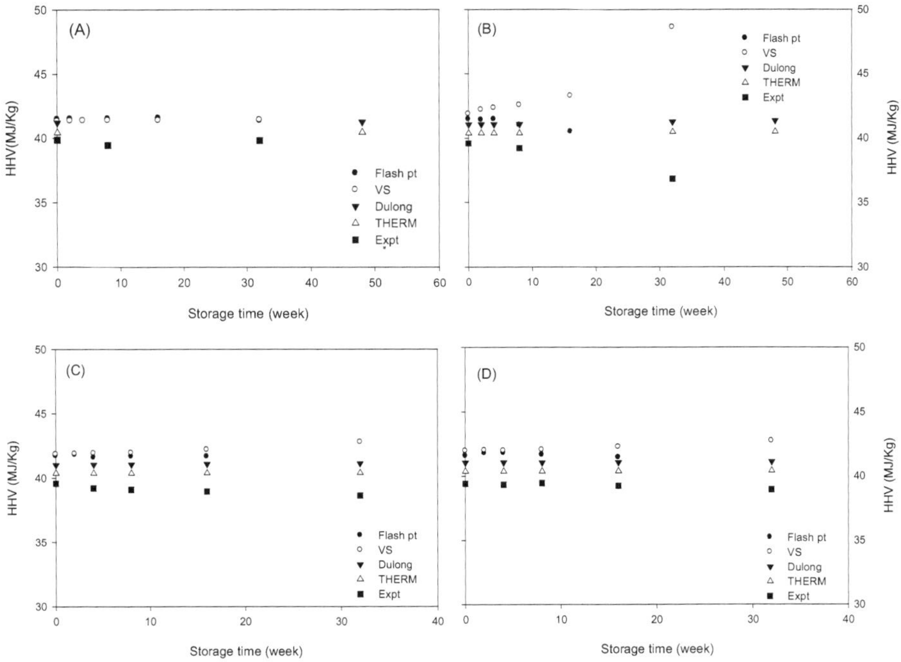

Fig. 8. Comparison of estimated higher heating values of NJC, SFM, SBM, and PNM with experimental results.

5. Acknowledgement

The authors wish to express their appreciation for the funding (NSC96-2221-E-197-010) from the National Science Council of the R.O.C. We also want the thank Taiwan NJC Corp. for the commercial biodiesel supplement.

6. References

Allena, C.A.W.; Wattsa, K. C., Ackmanb, R. G. & Peggc, M. J. (1999) Predicting the viscosity of biodiesel fuels from their fatty acid ester composition. Fuel, Vol. 78, pp. 1319–1326, ISSN 0016-2361

Benson, S. W. (1976) Thermochemical Kinetics; Methods for the Estimation of Thermochemical Data and Rate Parameters. 2nd ed. Wiley, ISBN 0471067814, New York.

Bouaid, A.; Martinez, M. & Aracil, J. (2007) Long storage stability of biodiesel from vegetable and used frying oils. Fuel, Vol. 86, pp. 2596–2602, ISSN 0016-2361

Chang, D. Y.; Van Gerpen, J. H., Lee, I., Johnson, L. A., Hammond, E. & Marley, S. J. (1996) Fuel properties and emissions of soybean oil esters as diesel fuel. J. Am. Oil Chem. Soc., Vol. 73, pp. 1549-1555, ISSN 0003-021X.

Demirbas, A. (2007) Importance of biodiesel as transportation fuel, Energy Policy, Vol. 35, pp. 4661–4670, ISSN 0301-4215

Demirbas, A. (2008a) Biodiesel: a realistic fuel alternative for diesel engines. Springer-Verlag London Limited, ISBN 978-1-84628-994-1.

Demirbas, A. (2008b) Relationships derived from physical properties of vegetable oil and biodiesel fuels. Fuel, Vol. 87, pp. 1743–1748, ISSN 0016-2361

Edlund, M.; Visser, H. & Heitland, P. (2002) Analysis of biodiesel by argon-oxygen mixed-gas inductively coupled plasma optical emission spectrometry. J. Anal. At. Spectrom., Vol.17, pp. 232-235, ISSN 0267-9477

Kahn, J.; Rang, H. & Kriis, J. (2002) Advance in biodiesel fuel research. Proc. Estonian Acad. Sci. Chem., Vol. 51, pp. 75-117, ISSN 1406-0124

Karavalakis, G.; Hilari, D., Givalou, L., Karonis, D. & Stournas, S. (2011) Storage stability and ageing effect of biodiesel blends treated with different antioxidants. Energy, Vol. 36, pp. 369-374, ISSN 0360-5442

Knothe, G. (2007) Some aspects of biodiesel oxidative stability. Fuel Processing Technology, Vol. 88, pp. 669–677, ISSN 0378-3820

Komers, K.; Stloukal, R., Machek, J., Skopal, F. & Komersova, A. (1998) Biodiesel fuel from rapeseed oil, methanol, and KOH. Analytical methods in research and production. Fett/Lipid, Vol. 100, no. 11, pp. 507-512, ISSN 1521-4133

Leung, D.Y.C.; Koo, B.C.P. & Guo, Y. (2006) Degradation of biodiesel under different storage conditions. Bioresource Technology, Vol. 97, pp. 250–256, ISSN 0960 8524

Mittelbach, M. & S. Gangl S. (2001) Long Storage Stability of Biodiesel Made from Rapeseed and Used Frying Oil. JAOCS, Vol. 78, no. 6, pp. 573-577, ISSN 0003-021X

Prankl, H. & Worgetter, M. (1996) Influence of the iodine number of biodiesel, to the engine performance. Third Liquid Fuel Conference, "Liquid Fuels and Industrial Products from Renewable Resources", Nashville, Tennessee, USA, September 15-1 7, 1996.

Ritter, E. R. & Bozzelli, J. W. (1991) THERM: Thermodynamic property estimation for gas-phase radicals and molecules. *Int. J. Chem. Kinetics,* Vol. 23, pp.767-778, ISSN 1097-4601

Schmidt, K & Van Gerpen, J. (1996) The effect of biodiesel fuel composition on diesel combustion and emissions. International Fuels & Lubricants Meeting & Exposition, Dearborn, MI, USA, May 1996.

Sharma, Y.C.; Singh, B. & Upadhyay, S.N. (2008) Advancements in development and characterization of biodiesel: A review. Fuel, Vol. 87, pp. 2355–2373, ISSN 0016-2361.

Srivastava, A. & Prasad R. (2000) Triglycerides-based diesel fuels. Renew Sust Energy Rev., Vol. 4, pp. 111–33, ISSN 1364-0321

Wu Y. P.; Lin, Y. F. & Chang, C. T. (2007) Combustion Characteristics of Fatty Acid Methyl Esters Derived from Recycled Cooking Oil. Fuel, Vol. 86, pp. 2810–2816, ISSN 0016-2361

Zullaikah, S.; Lai, C. –C., Vali R. S., & Ju, Y. –H. (2005) A two-step acid-catalyzed process for the production of biodiesel from rice bran oil. Bioresource Technology, Vol. 96, pp. 1889–1896, ISSN 0960-8524

6

Analysis of FAME in Diesel and Heating Oil

Vladimir Purghart
Intertek (Switzerland) AG, Schlieren
Switzerland

1. Introduction

Fossil fuel repository is decreasing worldwide very quickly and finding new sources becomes more and more difficult. Experts are expecting that the fossil fuel will end in a few decades. This is the reason for researchers to find alternatives. Many technical improvements have already been made for car engines and also many developments have been made in the area of fuel. FAME (fatty acid methyl esters) was found as an equivalent fuel to diesel. It is also known as "Biodiesel". In Europe, it is mostly prepared from rape, palm or soy oil. In the process of biodiesel production, the glyceride bondages are broken and methyl esters of the long chained fatty acids are formed (FAME = fatty acid methyl ester). In recent years, car engines have been developed, which run with both fossil diesel and FAME.

At a time of growing globalisation and increasing financial pressure on logistics and transport companies, cross contamination is an increasing issue. It needs extensive actions to clean a tank or a truck after having loaded FAME. Very often, traces of FAME can be found in other fuels. This was the reason, why a limit for FAME in Jet A-1 fuel needed to be defined and was set at 5 ppm (mg/kg) for aircrafts (Ministry of Defence (2008). Defence Standard 91-91 and Joint Inspection Group (2011). Aviation Fuel Quality Requirements for Jointly operated System (AFQRJOS) Bulletin No. 45).

As diesel and FAME are used in one and the same engine, one would think that cross contamination is not critical. This is correct for car drivers. However, it is well known that FAME cannot be stored for more than a couple of years. The reason for this is it's hydroscopic properties and it is also a very good alimentary for fungi.

Pure fossil diesel can be stored for decades without any problems. However, when fossil diesel is stored over several years, containing small quantities of FAME, fungi growth starts quickly and the characteristics of the diesel can change drastically. First, the odour of such contaminated diesel changes, second, FAME causes sticky deposits with water on the bottom of the containers and tanks, and third, fungi which grow in the fuel cause filter clogging.

A method was developed for sample preparation and quantification of FAME in diesel.

There is a difficulty when diesel or heating oil is analysed using a gas chromatograph connected to a mass spectrometer (GC-MS). A diesel sample contains compounds, which evaporate at high temperature. The temperature limit for the analysis using GC-MS is given by the chromatographic column. As it was found that HP-Innowax[1] shows the best

---

[1] HP-Innowax 50m, I.D. 0.200mm, Film 0.40 µm (by Agilent J&W); as an alternative column the following can be used: TBR-WAX 50m, I.D. 0.200mm, Film 0.40 µm (by Teknokroma)

separation for FAME and the temperature limit of this column is 260°C, a solution to separate the high volatile compounds from the diesel and heating oil sample needed to be found. The highly volatile compounds, as they are found in diesel, would contaminate a GC-MS injector in standard application rapidly, and cleaning would be needed too frequently. A solid phase extraction was found to be a solution for extracting FAME from diesel or heating oil samples.

2. Preparation of standards and samples

2.1 Preparation of standards

6 fatty acid methyl esters (FAME) were used to prepare the standards. The selection of these 6 FAME was already published earlier (Institute of Petroleum (2009). Norm draft document IP PM-DY/09). These are: methyl palmiate (C16:0), methyl margarate (C17:0), methyl stearate (C18:0), methyl oleate (C18:1), methyl linoleate (C18:2), and methyl linolenate (C18:3). A stock solution was prepared of approximately 50 mg of each FAME dissolved in 50 g Jet A-1[2]. From this stock solution, standard dilutions were prepared at the following concentration levels: 0.1, 0.5, 1.2, 3.0, 5.0, 12, 50, and 100 mg/kg (ppm) of each fatty acid methyl ester (FAME).

2.2 Preparation of samples

FAME free diesel and heating oil samples were used for the preparation of the samples. For the method development, they were fortified by the same stock solution as used for the preparation of standards as described above. The fortified samples were prepared at the following levels: 0.2, 2.0, 10, and 100 mg/kg of each FAME.
Later, natural mixture of FAME was used for fortification. The levels of total FAME were 1.20, 7.55, and 115 mg/kg.

3. Sample treatment

Highly volatile compounds, as they are found in diesel, contaminate a GC-MS injector when used with a HP-Innowax[3] column due to temperature limits.

3.1 Solid phase extraction

The solid phase extraction cartridge (SPE) which was found to fit the best, is a Strata SI-1 Silica (55 µm, 70A)[4]. A 12-port vacuum manifold by Supelco connected to a small vacuum pump was used for the SPE sample preparation.

3.1.1 SPE column washing and conditioning

The SPE cartridges were pre-washed with approximately 10 mL diethyl ether at a speed of approximately 2 drops per second. Right after all the diethyl ether had passed the column, it was conditioned with 10 mL n-hexane at the same flow speed. Thereafter the

---

[2] When using Jet A-1 as a solvent, it needs to be checked to be free of FAME. Other solvents such as octane or dodecane can be used as well. It is essential, that the same solvent is used for the preparation of standards as used for the sample dilution as described in section 3.1.3.

[3] See footnote 1

[4] Strata SI-1 Silica (55 µm, 70A), 1000 mg/6 mL Part Number 8B-S012-JHC by Phenomenex.

SPE cartridge was dried by vacuum for approximately 30 to 60 seconds. Then, the vacuum was stopped and the sample was applied. Both solvents, diethyl ether and n-hexane, were discarded.

3.1.2 Application of the sample

1 mL of the diesel sample or heating oil sample was passed through the cartridge at a speed of 1 drop per second. Thereafter, the diesel residue of the sample on the SPE cartridge was washed using 10 mL n-hexane. Also here, the n-Hexane from washing was discarded as well as the diesel sample which passed the column.

3.1.3 Elution and further treatment of the sample

After the n-hexane passed the SPE cartridge, it was dried for approximately 1 minute by vacuum. Thereafter, the vacuum was stopped and the adsorbed FAME were eluted with 10 mL of diethyl ether at a speed of 1 drop per second into a test tube.
The diethyl ether was evaporated by a gentle stream of nitrogen blown via a glass pipette into the test tube. Thereafter, the sample was diluted in 1 mL of FAME free Jet A-1 fuel[5]. The walls of the test tube were washed with a pipette and all of the solution was transferred into a sample vial as quantitatively as possible, closed with a crimped lid and analysed using GC-MS.

4. Analytical method

The analytical method is very similar to the one described in Literature (Institute of Petroleum (2009). Norm draft document IP PM-DY/09 and IP 585/10). However, the measuring range was extended down to 0.1 mg/kg for each FAME as the lowest standard. The preparation of standards was thus modified in terms of solvent and calibration levels. For maximum precision, the calibration curve was split into two segments as described in section 5 of this chapter.

4.1 Instrumentation

A gas chromatograph (Trace GC Ultra) connected to a mass spectrometer (DSQ II) by Thermo Scientific was used as GC-MS System.

4.1.1 GC method

Injector:	PTV
Injection:	Split
Split Flow:	20 mL/minute
Injection volume:	1.0 µL
Injector temperature:	260°C
Carrier gas:	Helium
Analytical column:	HP-Innowax 50m, I.D. 0.200mm, Film 0.40 µm (by Agilent J&W)

[5] When using Jet A-1 as a solvent, it needs to be checked whether the solvent is really free of FAME. Other solvents such as octane or dodecane can be used as well. It is essential, that the same solvent is used for the sample dilution as used for the preparation of standards as described in section 2.1.

Oven temperature:

Start temperature:	150°C (for 5 minutes)
Heating rate:	17°C/minute up to 200°C, hold time for 17 minutes, thereafter with 3°C/minute up to 252°C
End temperature:	252°C (isotherm for 3 minute)

4.1.2 MS method

Measuring mode:	Selected Ion Monitoring (SIM)
Measuring ranges:	20.00 - 27.69 minutes: SIM of 227, 239, 270, 271 Da 27.70 - 33.49 minutes: SIM of 241, 253, 284 Da 33.50 - 35.99 minutes: SIM of 255, 267, 298 Da 36.00 - 37.29 minutes: SIM of 264, 265, 296 Da 37.30 - 39.49 minutes: SIM of 262, 263, 264, 295 Da 39.50 minutes to end of run: SIM of 236, 263, 292, 293 Da
Polarity:	positive
Detector voltage:	1518V
Software used:	Xcalibur Version 2.0.7, QuanBrowser Version 2.0.7, and QualBrowser Version 2.0.7

5. Results

The standard measurement showed that it is not possible to calculate one calibration curve over the entire concentration range. Therefore, two calibration curves were created: one for the high concentration range, approximately 5 - 100 mg/kg of each FAME and a second for the range of 0.1 to 5.0 mg/kg of each FAME. An example of the high range calibration curve is shown in Figure 1 and the low concentration range is depicted in Figure 2.

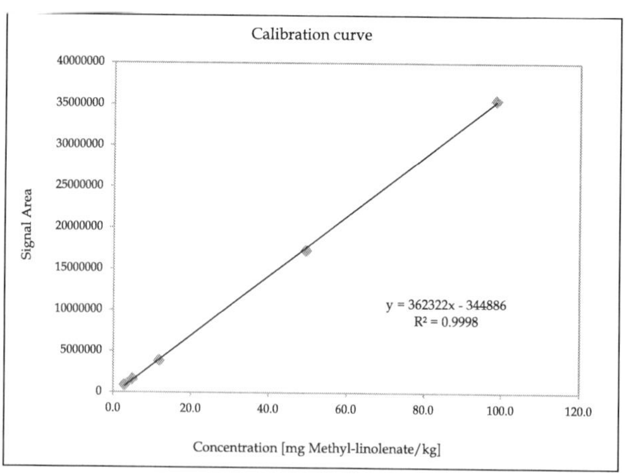

Fig. 1. Calibration curve for Methyl linolenate in high concentration range.

For each of the 6 FAME, a set of two calibration curves were calculated. Figure 3 shows the main section of the chromatograms of the standards. The depicted concentrations are 0.1,

0.5, 1.2, and 3.5 mg/kg for each FAME. The signal at approximately 26.6 minutes corresponds to methyl palmiate (C16:0), at 31.4 minutes to methyl margarate (C17:0), at 35.7 minutes to methyl stearate (C18:0), at 36.7 minutes to methyl oleate (C18:1), at 38.6 minutes to methyl linoleate (C18:2), and at 41.1 minutes to methyl linolenate (C18:3).

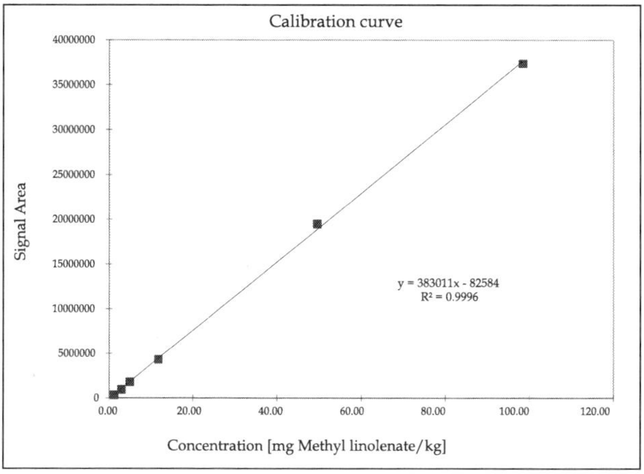

Fig. 2. Calibration curve for Methyl linolenate in low concentration range.

The expected retention time ranges are shown in Table 1 as they were also listed in the literature (Institute of Petroleum (2009). Norm draft document IP PM-DY/09 and in Purghart V. & Jaeckle H (2010). What Damage Can Biodiesel Cause to Jet Fuel? *Chimia, Volume 64, No 3*, Highlights of Analytical Chemistry in Switzerland). In the present study, slightly longer retention times were observed.

Species to be detected	Significant SIM masses [Da]	Expected retention time [minutes]
Methyl-palmitate C16:0	227, 239, 270, 271	24.9 - 26.4
Methyl-margarate C17:0	241, 253, 284	30.1 - 31.4
Methyl-stearate C18:0	255, 267, 298	34.7 - 35.5
Methyl-oleate C18:1	264, 265, 296	35.5 - 36.5
Methyl-linoleate C18:2	262, 263, 264, 294, 295	37.7 - 38.6
Methyl-linolenate C18:3	236, 263, 292, 293	40.3 - 41.1

Table 1. List of fatty acid methyl esters used as standards with the masses used for SIM detection and the approximately expected retention time ranges.

An example chromatogram of a fortified heating oil sample at a level of 2.0 mg/kg of each FAME is shown in Figure 4, the chromatogram of the one fortified at a level of 100 mg/kg of each FAME is shown in Figure 5.

A quantification of all signals is summarized in Table 2. The fortification levels were chosen to show the robustness of the method and also to cover both calibration curves with two

samples each. The fortification levels were defined as concentration of each of the 6 FAME, e.g. a fortification level of 100 mg/kg results in a total FAME concentration of 600 mg/kg as 6 FAME are considered. In later examples, fortification using natural FAME will be described. The concentration there will be given as total FAME, where the sum of 6 components is the number of interest.
As it was shown that reasonable recovery was found for each level of the fortified heating oil, samples of fortified diesel were prepared. However, if a cross contamination in a storage container or a truck occurs, then the detected signals of each fame would correspond to the FAME mixture as it comes from soy oil, rape oil, palm oil or similar. Therefore, diesel samples were prepared with natural fatty acid methyl ester mixture as commercially available. The fortification levels of total FAME were 1.20, 7.55, and 114.5 mg/kg.
An example chromatogram of a fortified diesel sample at a level of 7.55 mg/kg of total FAME is shown in Figure 6, the chromatogram of one fortified at a level of 115 mg/kg of total FAME is shown in Figure 7.

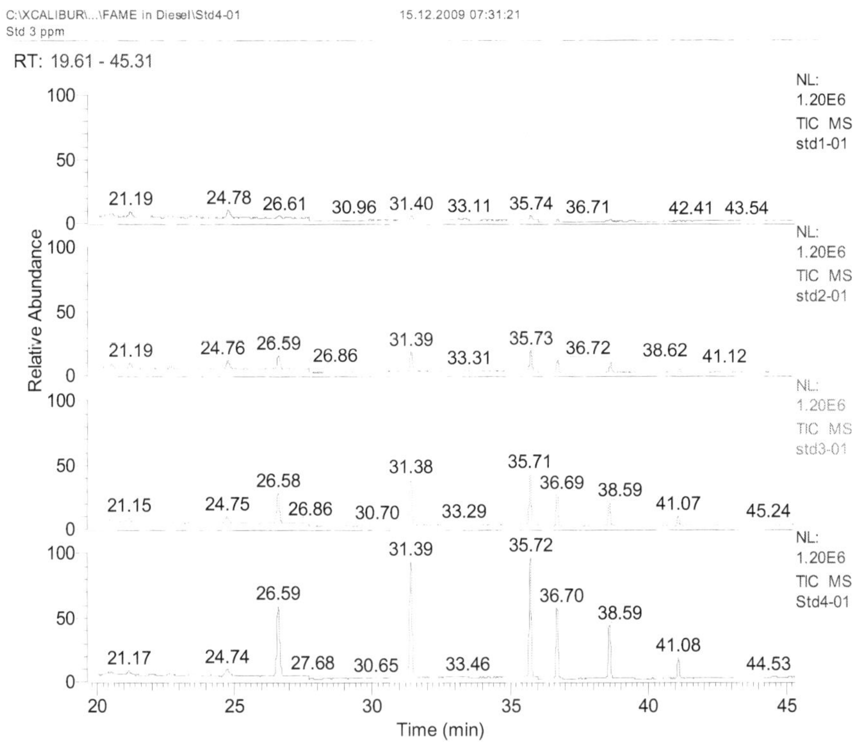

Fig. 3. Chromatograms of the standards at low concentrations i.e. 0.1, 0.5, 1.2, and 3.5 mg/kg for each FAME.

The signal at 26.56 minutes corresponds to methyl palmiate (C16:0), the signal at 35.70 minutes to methyl stearate (C18:0), the signal at 36.72 minutes to methyl oleate (C18:1), the signal at 38.59 minutes to methyl linoleate (C18:2), and the signal at 41.06 minutes corresponds to methyl linolenate (C18:3). There is no signal at approximately 31.4 minutes, which would correspond to methyl margarate (C17:0). Generally, methyl margarate is not or only very rarely at very low concentrations present in FAME prepared from rape, palm or soy oil.

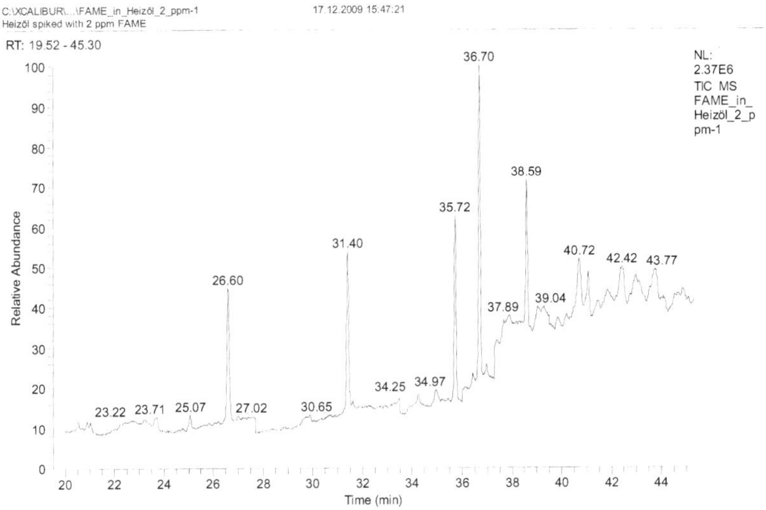

Fig. 4. Chromatogram of a fortified heating oil sample at a level of 2.0 mg/kg of each FAME.

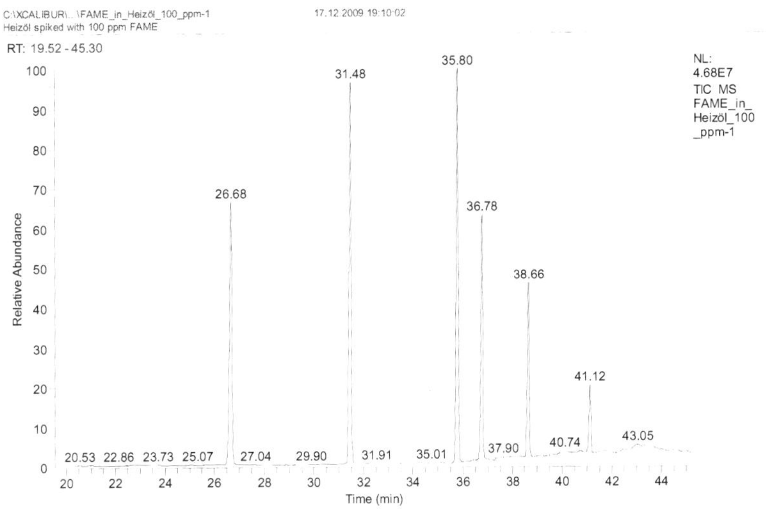

Fig. 5. Chromatogram of a fortified heating oil sample at a level of 100 mg/kg of each FAME.

Fortified level [mg/kg]	Methyl palmiate [mg/kg]	Methyl marganate [mg/kg]	Methyl stearate [mg/kg]	Methyl oleate [mg/kg]	Methyl linoleate [mg/kg]	Methyl linolenate [mg/kg]	Sum [mg/kg]
0.0	0.01	0.00	-0.02	-0.03	0.02	0.08	0.05
0.0	0.01	0.00	-0.02	0.01	0.00	0.03	0.03
0.2	0.19	0.16	0.19	0.24	0.21	0.20	1.19
0.2	0.18	0.17	0.19	0.23	0.24	0.22	1.22
2.0	1.98	2.10	1.95	1.84	2.09	2.02	11.99
2.0	1.99	2.10	1.90	2.06	2.07	1.83	11.94
10.0	9.62	11.16	9.24	9.62	9.90	10.60	60.13
10.0	9.22	10.38	9.76	9.73	9.25	10.39	58.74
100.0	104.73	103.53	94.62	99.78	102.70	105.42	610.78
100.0	104.72	102.92	94.60	99.15	100.97	101.48	603.84

Table 2. Summary of fortified heating oil samples at various levels. Each fortification level contains approximately the same amount of each FAME.

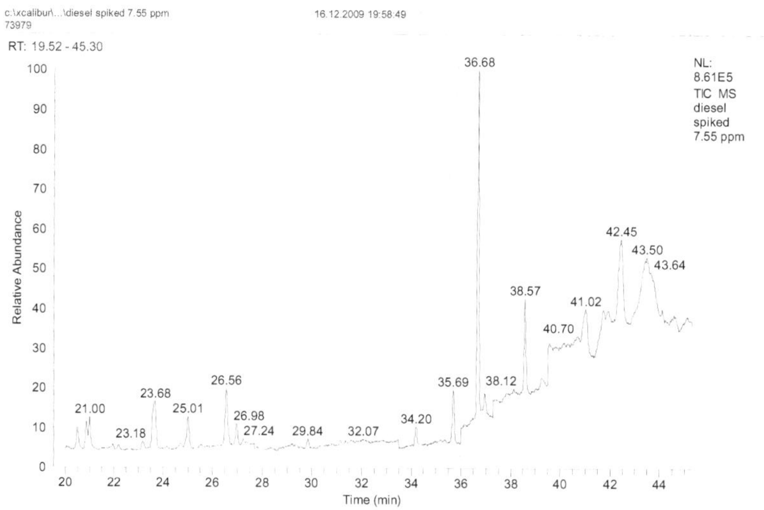

Fig. 6. Chromatogram of a fortified Diesel sample at a level of 7.55 mg/kg of total FAME.

A quantification of all signals of the fortified diesel samples is summarized in the following Table (Table 3).

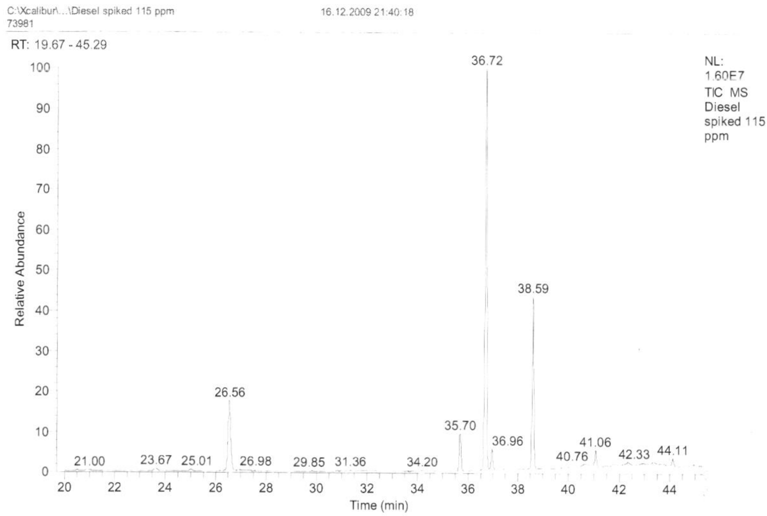

Fig. 7. Chromatogram of a fortified Diesel sample at a level of 115 mg/kg of total FAME.

Fortified level [mg/kg]	Methyl palmiate [mg/kg]	Methyl marganate [mg/kg]	Methyl stearate [mg/kg]	Methyl oleate [mg/kg]	Methyl linoleate [mg/kg]	Methyl linolenate [mg/kg]	Sum [mg/kg]
1.20	0.11	0.02	0.15	0.30	0.21	0.12	1.01
1.20	0.11	0.04	0.14	0.25	0.23	0.18	1.06
7.55	0.57	0.10	0.42	2.88	1.17	2.43	7.57
7.55	0.57	0.09	0.42	2.91	1.19	2.25	7.43
115	10.88	0.22	3.20	59.66	32.27	7.94	114.19
115	11.14	0.22	3.23	59.09	32.69	8.46	114.83

Table 3. Summary of fortified diesel samples at various levels. Each fortification level contains the sum of FAME listed in the table.

6. Conclusion

The presented analytical method for low concentration of FAME in diesel and heating oil was shown to be robust and sensitive down to low ppm level. The range of quantification was extended down to 0.1 mg/kg of each FAME. The robustness of the solid phase extraction was shown in the range of 1.2 to 600 mg/kg FAME in total. This results in a maximum total load of 600 μg FAME on the SPE cartridge.

7. References

Institute of Petroleum (2009). Norm draft document IP PM-DY/09
Institute of Petroleum (2010). Norm IP585/10

Joint Inspection Group (2011). Aviation Fuel Quality Requirements for Jointly operated System (AFQRJOS). Bulletin No. 45

Ministry of Defence (2008). Defence Standard 91-91

Purghart V. & Jaeckle H (March 2010). What Damage Can Biodiesel Cause to Jet Fuel? *Chimia, Volume 64, No 3,* Highlights of Analytical Chemistry in Switzerland

7

Analytical Methodology for the Determination of Trace Metals in Biodiesel

Fabiana A. Lobo[1], Danielle Goveia[2],
Leonardo F. Fraceto[2] and André H. Rosa[2]
[1]UFOP - Universidade Federal de Ouro Preto
[2]UNESP - Universidade Estadual Paulista
Brazil

1. Introduction

The demand for energy resources by various systems such as production and transportation, as well as for physical comfort continues to grow apace, intensifying global dependence on fossil fuels and their derivatives. For this reason, numerous private and public programs in several countries have established feasible alternatives for the substitution of petroleum derivatives (Sahin, 2011; Saint´Pierre et al., 2003). These alternatives are aimed at reducing dependence on imported and non-renewable energy, mitigating some of the environmental impacts caused by petroleum derivatives, and developing alternative technologies in the area of energy (Oliveira et al., 2002).

Biodiesel has emerged as a promising alternative to petroleum, firstly because it promotes a qualitative and quantitative reduction of the emission of various air pollutants (Agarwai, 2005; López et al., 2005; Ilkilic & Behcet, 2010; Silva, 2010;) and secondly, as a strategic source of renewable energy to substitute diesel oil and other petroleum derivatives (Chaves et al., 2008; Jesus et al., 2008).

Biodiesel, also known as vegetable diesel, is a fuel obtained from renewable sources, such as vegetable oils and animal fats, by means of chemical processes such as transesterification, esterification and thermal cracking (Chaves et al., 2010, Oliveira et al., 2009, Jesus et al., 2010; Arzamendi et al., 2008; Canakci et al., 1999; Meher et al., 2006).

In chemical terms, biodiesel is defined as a mono-alkyl ester of long-chain fatty acids with physicochemical characteristics similar to those of mineral diesel. Because it is perfectly miscible and physicochemically similar to mineral diesel oil, biodiesel can be used pure or mixed in any proportions with other solvents in diesel cycle engines without the need for substantial or expensive adaptations (Ma & Hanna, 1999; Woods & Fryer, 2007). The literature highlights several important characteristics of biodiesel: (a) its market price is still relatively high when compared with that of conventional diesel fuel; (b) its content of sulfur and aromatic compounds is lower; (c) its average oxygen content is approximately 11%; (d) its viscosity and flashpoint are higher than those of conventional diesel; (e) it has a specific market niche directly associated with agricultural activities; and lastly, (f) in the case of biodiesel from used frying oil, it has strong environmental appeal (Nigam et al., 2011). The qualitative and quantitative reduction in the emissions of various air pollutants such as sulfur, particulate material, and particularly carbon, point to biodiesel as a promising alternative to reduce the deleterious effects of petroleum and its derivatives. However, some studies are contradictory about the concentrations of NO_x emissions, with some of them reporting a reduction in emissions, while others report marginally higher emissions than those of mineral diesel (Coronado, 2010; Costa-Neto et al., 2000; Ferrari et al., 2005; López et al., 2005; Ramadhas et al., 2004).

However, it is nigh impossible for any chemical reaction to be complete, including transesterification, and therefore the products of a reaction (alky esters) are usually contaminated with other compounds. Among metals, the ones most strongly controlled are Na and K because their hydroxides are used as catalysts. These elements, which may be present as solid abrasives or as soluble soaps, can clog various mechanical parts of a vehicle (Pohl, 2010; Chaves et al., 2008; Jesus et al., 2008). In addition, other inorganic contaminants (such as Cu, Pb, Cd, Zn, Ni, etc.) may be present in biodiesel samples due to the plant's (raw material) absorption of metals from soil, and/or be incorporated during the production and storage process (Lobo et al., 2009; Lobo et al., 2011; Tagliabue et al., 2006).

The quantitative monitoring of metallic elements in fuel samples is of supreme importance in economic terms, not only for the fuel industry but also in various other sectors of industry and services. One of the most important applications is the determination of the total concentration or the monitoring of variations in concentration over time of certain metallic and semi-metallic elements. This type of analysis is crucial for maintaining quality control (Chaves et al., 2010; Jesus et al., 2010; Garcia et al., 1999). One of the most relevant aspects to consider is the phenomenon of corrosion in the combustion chamber of automotive engines, which is caused by high temperatures and by the fuels themselves (Amorim et al., 2007; Jesus et al., 2008; Haseeb et al., 2010; Saint´Pierre et al., 2006). The deactivation of catalysts through poisoning, incrustation or solid-state transformations, which lead to reduced selectivity and loss of catalytic activity, may also result in economic losses and environmental impacts (Figueiredo & Ribeiro, 1987; Meeravali & Kumar, 2001; Saint´Pierre et al., 2004).

The quality of fuels supplied to the consumer, from their production to their distribution points, can be managed by means of the efficient analytical control of incidental or accidental inorganic additives (Oliveira et al., 2002).

The metal content in fuels, which is usually low, requires the use of adequate sample preparation procedures and sensitive analytical techniques (Lobo et al., 2011). Atomic absorption spectrometry (AAS) can be employed for the quantitative determination of many elements (metals and semi-metals) in a variety of foods and in biological, environmental, geological and other types of samples. The AAA technique is widely applied for the determination of different elements (about 70) due not only to its robustness but also its sensitivity to detect trace elements in the order of $\mu g\ L^{-1}$ or even $ng\ L^{-1}$ with high accuracy and precision. The principle of the technique is based on the absorption of electromagnetic radiation from a radiation source by gaseous atoms in the fundamental state. The process of formation of gaseous atoms in the fundamental state, called atomization, can be obtained via flame, electrothermal heating, or by a specific chemical reaction such as Hg cold-vapor generation.

Graphite furnace atomic absorption spectrometry (GFAAS) with electrothermal atomization is widely used in routine analyses due to several factors. It requires small volumes of sample, the atomizer acts as a chemical reactor, excellent limits of detection are attained after separation of the analyte and matrix in the reactor, it requires no previous decomposition of the sample (direct analysis), it is multielemental, fast, relatively inexpensive, simple spectrum, and provides chemical and thermal pretreatment of the sample, among other advantages (Welz et al., 1992; Jackson, 1999).

The optimization of experimental conditions in GFAAS (chemical modifiers, sample preparation, pyrolysis temperature and atomization) normally requires numerous time-consuming and expensive experiments. However, using factorial planning enables one to extract the maximum possible amount of useful information from a given system with a minimum number of experiments. When univariate optimization is employed, it is often impossible to detect interactions among the variables under study. In addition to this problem, the number of experiments performed is usually higher when compared with those obtained through factorial design (Pereira-Filho et al., 2002; Amorim et al., 2006).

The objective of this work was to use 2^4 factorial design for the optimization of experiments for the determination of metals (Cu, Cd, Zn, Ni and Pb) by GFAAS in biodiesel samples, using different sample preparation procedures and different chemical modifiers. The chemical modifiers used were a mixture of Pd + Mg and the W permanent modifier, and the samples were prepared by microemulsion and focused microwave digestion. The advantages of applying factorial design to carry out the experiments and to determine the optimal conditions of pyrolysis and atomization temperatures are discussed.

2. Experimental procedures

2.1 Instruments and accessories

The instruments used in this study were a Zeeman electrothermal atomic absorption spectrometer (Varian, model Spectra AA240Z) equipped with an autosampler (Varian, model PSD 120) coupled to a Dell PC. The spectrometer's graphite tubes were designed with an integrated platform; Varian hollow cathode lamps (λ= 283.3 nm, bandwidth 0.5 nm for Pb; λ= 327.4 nm, bandwidth 0.5 nm for Cu; λ= 232.0 nm, bandwidth 0.2 nm for Ni λ= 228.8 nm and bandwidth 0.5 nm for Cd). The experimental setup included a Milli-Q Plus water deionizer system (Millipore®); automatic micropipettes of different volumes (fixed and variable); a Sartorius 2432 analytical balance with maximum capacity of 200 g; disposable polyethylene tubes (Corning); focused microwaves - Rapid Digestion system - SPEX. The purge gas was argon 99.9% (White Martins, Brazil). All the measurements were based on integrated absorbance.

2.2 Reagents, solutions and samples

All the solutions were prepared with high purity deionized water (18.2 MΩ.cm) purified in a deionizer system (Milli-Q Plus, Millipore®). Nitric acid (Synth-65% v/v) was used after sub-boiling of the reference analytical solutions and samples. Fresh analytical solutions of the analytes were prepared each day using 1000 mg L^{-1} of stock solutions (Normex®, Carlo Erba) in 1.0% (v/v) distilled HNO_3, Triton X-100 (Tedia), Hydrogen Peroxide (Synth - PA, 29-30% (v/v)), and Vanadium Pentoxide (Riedel-99.5% m/m). The samples biodiesel were obtained from biodiesel research laboratories in Curitiba (state of Paraná) and Cuiabá (state of Mato Grosso) and from a commercial gas station in Sorocaba (state of São Paulo). According to the biodiesel suppliers, all the samples were obtained by transesterification with ethanol, using NaOH as catalyst. Unfortunately, no additional information was supplied for sample B10 (10% v/v of biodiesel in diesel), which was obtained from the gas station in Sorocaba. Table 1 lists the characteristics of the samples and their respective designations.

Samples	Origin
A_1	Soybean
A_2	Soybean
A_3	Pure animal fat (bovine grease)
A_4	Unwashed animal fat (grease mixtures)
A_5	Washed animal fat
A_6	Sunflower
A_7	Cotton
A_8	10% biodiesel in diesel (B10)

Table 1. Analyzed biodiesel samples and their origins

2.3 Preparation of the chemical modifier

A solution of 1000 mg L^{-1} of the Pd $(NO_3)_2$ chemical modifier was prepared using a 10000 mg L^{-1} stock solution of $Pd(NO_3)_2$ in 15% HNO_3 (Perkin-Elmer, Part Nº BO190635). A solution of 1000 mg L^{-1} of the $Mg(NO_3)_2$ chemical modifier was prepared using a 10000 mg L^{-1} stock solution of $Mg(NO_3)_2$ (Perkin-Elmer, Part Nº BO190634). The Pd + Mg mixture used as chemical modifier was prepared with 5 μL of a 1000 mg L^{-1} $Pd(NO_3)_2$ solution + 3 μL of a 1000 mg L^{-1} $Mg(NO_3)_2$ solution.
The solution of 1.0 g L^{-1} of W was prepared by dissolving 0.1794 g of $NaWO_4.2H_2O$ (Merck) in 100 mL of deionized water. The atomizer was coated with the W permanent modifier in two steps: i) tungsten deposition, ii) thermal treatment of the tungsten deposited in the tube. Table 2 describes the heating program for this coating procedure.

Steps	Actions and Parameters
Deposition of W	
1	Introduction of 50 μL of a 1.0 g L^{-1} W solution into the atomizer
2	Heating program (ramp, hold) for drying and pyrolysis: 120ºC (5, 25s); 150ºC (10, 60s); 600ºC (20, 15s) and 1000ºC (10, 15s)
3	Steps 1 and 2 were repeated three times
4	Step 1 was repeated, followed by the heating program (ramp, hold): 120ºC (5, 25s); 150ºC (10, 60s); 600ºC (20, 15s), 1000ºC (10, 15s), 1400ºC (10, 5s), 2000ºC (3, 2s) and 2100ºC (1, 1s)
Thermal treatment of W	
5	The heating program was repeated four times to condition the W carbide to the average temperature (ramp, hold): 150ºC (1, 10s), 600ºC (10, 15s), 1100ºC (10, 5s), and 1400 C (10, 10s)
6	The heating program was repeated four times to condition the W carbide to high temperatures: 150ºC (1, 10s), 600ºC (10, 15s), 1100ºC (10, 5s), 1400ºC (10, 10s), 1500ºC (3, 5s), 1600ºC (1, 1s), 1700ºC (1, 1s), 1800ºC (1, 1s), 1900ºC (1, 1s) and 2000ºC (1, 1s)

Table 2. Sequence of the coating program of the atomizer with the W permanent chemical modifier (Oliveira et al., 2002)

2.4 Preparation of samples

The samples were prepared by two procedures:

a. Microemulsion: Prepared in 50.0 mL volumetric flasks by mixing 0.5 g of biodiesel and 5 g of surfactant (Triton X-100) and completing the volume with HNO_3 dist. 1% (v/v) under stirring for 20 min (Lobo et al., 2009).

b. Wet digestion: The procedure was carried out as described by Bettinelli et al. (1996), by weighing approximately 0.5 g of biodiesel, 18 mL of concentrated HNO_3 dist, and 12 mL of H_2O_2 and V_2O_5 catalyst. The mixture was allowed to rest for 24 h, after which it was processed in a focused microwave system (Rapid digestion System SPEX) for 1 hour (Liu et al., 1995). The procedure resulted in efficient and complete digestion, yielding a clear transparent solution. After this step, the volume was adjusted to 50.0 mL volume with distilled HNO_3 1.0% (v/v).

2.5 Study of the electrothermal behavior

To evaluate the thermal behavior of the elements, pyrolysis and atomization temperature curves were built and the transient absorption signals were also analyzed. Table 3 describes the heating program applied here.

Steps	Temperature (°C)	Time (s) (Ramp; hold)	Gas Flow (mL min^{-1})
1	85	5.0	300 (Ar)
2	95	40.0	300(Ar)
3	120	10.0	300(Ar)
4	*	5.0, 3.0	300(Ar)
5	**	1.0, 4.0	0 (read)
6	2250	1.0, 5.0	300(Ar)

* Pyrolysis temperature (Tp); ** Atomization temperature (Ta)

Table 3. Heating program of the graphite tube atomizer

2.5.1 Study of the electrothermal behavior to evaluate atomization and pyrolysis temperatures based on a univariate procedure

The optimal temperature was established, and only the pyrolysis temperature was varied. After the optimal pyrolysis temperature was defined, the atomization temperature was varied. The pyrolysis and atomization curves were obtained with 25 µg L^{-1} (because, in the optimization of the experiments, this concentration fell within the linear range of the calibration curve) of Cu, Pb, Ni and Cd in the presence of two modifiers (W and Pd + Mg) for the two sample preparation procedures (microemulsion and wet digestion in a focused microwave system).

2.5.2 Use of experimental design to optimize pyrolysis and atomization curves

In the process of pyrolysis and atomization temperatures, a 2^4 factorial design was used for Cu, Pb, Ni and Cd, which involved 16 assays for each analyte, as described in Tables 4 and 5. This factorial design involved two levels, one for the lowest (-1) and the other for the highest (+1) temperatures, for two variables (pyrolysis and atomization temperatures), two levels for the type of sample pretreatment procedure, digestion (-1) and microemulsion (+1), and two levels for the type of modifier, Pd + Mg (-1) and W (+1). It should be noted that the experiments were carried out randomly to avoid systemic errors.

Factors	Levels	
	Low (-1)	High (+1)
Pyrolysis temperatures, °C	*	*
Atomization temperatures, °C	*	*
Sample preparation	Wet digestion	Microemulsion
Modifiers	Pd + Mg	W

* The pyrolysis (Tp) and atomization (Ta) temperatures at the lowest level (-1) were 100°C below the values recommended by the manufacturer, while the temperatures at the highest level (1) were 100°C above the values recommended by the manufacturer. The values recommended by the manufacturer were: Pb- Tp-600°C; Ta- 2400 °C; Cu- Tp-900 °C; Ta-2300 °C ; Ni- Tp-900°C; Ta-2400°C and Cd- Tp-300°C; Ta- 1800 °C.

Table 4. Factors and levels used in the experimental factorial design

Pyrolysis and atomization curves were obtained for 25 μg L^{-1} of Cu, Pb, Ni and Cd in the presence of two modifiers, W and Pd + Mg, for both sample preparation procedures. A new factorial procedure was developed for Cd. A 2^2 factorial design involving 4 assays was developed, as described in Table 4. This factorial design involved two levels relating to the lowest (-1) and highest (+1) temperatures for the two variables (pyrolysis and atomization temperatures). The samples were prepared by microwave digestion using the Pd + Mg chemical modifier.

Experiments	Sample preparation		Pyrolysis Temperature	Atomization Temperature	Modifier	
1	Wet Digestion	-1	-1	-1	Pd + Mg	-1
2	Microemulsion	1	-1	-1	Pd + Mg	-1
3	Wet Digestion	-1	1	-1	Pd + Mg	-1
4	Microemulsion	1	1	-1	Pd + Mg	-1
5	Wet Digestion	-1	-1	1	Pd + Mg	-1
6	Microemulsion	1	-1	1	Pd + Mg	-1
7	Wet Digestion	-1	1	1	Pd + Mg	-1
8	Microemulsion	1	1	1	Pd + Mg	-1
9	Wet Digestion	-1	-1	-1	W	1
10	Microemulsion	1	-1	-1	W	1
11	Wet Digestion	-1	1	-1	W	1
12	Microemulsion	1	1	-1	W	1
13	Wet Digestion	-1	-1	1	W	1
14	Microemulsion	1	-1	1	W	1
15	Wet Digestion	-1	1	1	W	1
16	Microemulsion	1	1	1	W	1

Table 5. First factorial design for the optimization of pyrolysis and atomization

2.6 Determination of metals by the calibration procedure using the analyte addition method

Based on the results obtained with the experimental design, the methods were optimized using the two sample preparation procedures (Lobo et al., 2009, 2011).
The analyte addition method consists of adding volumes of solutions with known concentrations of analyte to the sample. This method is especially suitable when the composition of the sample is unknown or complex, as is the case of biodiesel samples. Thus, aiming to minimize interferences (since the standards and the sample had the same composition and physical properties), the analyte was quantified even when present in low concentrations, to ensure that the measures of the analytic signals would fall within a suitable interval for the technique (Harris, 2001) in the two different sample preparation procedures.
For each microemulsified sample, four standards were prepared containing different concentrations of added Cd and Ni, as shown in Table 6 (the concentration interval was used based on the manufacturer's handbook).

Flask	Va* (mL)	[Cd] added µg L^{-1}	[Ni] added µg L^{-1}	Final volume (mL)
1	1.0	0	0	10.0
2	1.0	0.5	5.0	10.0
3	1.0	1.0	10.0	10.0
4	1.0	1.5	15.0	10.0

*Va - volume of sample added in the form of microemulsion.

Table 6. Preparation of the monoelemental standards using the analyte addition method for the microemulsified samples

For each digested sample, four standards were prepared containing different concentrations of added Cu, Pb, Cd and Ni, as shown in Table 7 (the concentration interval was used based on the manufacturer's handbook).

Flask	Va* (mL)	[Cu] added µg L^{-1}	[Pb] added µg L^{-1}	[Ni] added µg L^{-1}	[Cd] added µg L^{-1}	Final volume (mL)
1	1.0	0	0	0	0	10.0
2	1.0	5.0	15.0	5.0	0.5	10.0
3	1.0	10.0	30.0	10.0	1.0	10.0
4	1.0	15.0	45.0	15.0	1.5	10.0

*Va - volume of focused microwave digested sample

Table 7. Preparation of the monoelemental standards using the analyte addition method for the digested samples

2.7 Analytical characteristics

The analytical characteristics of the procedures developed with the limit of detection (LOD), limit of quantification (LOQ), estimated standard deviation (SD) and relative standard deviation (RSD) were calculated as described by Harris (2001).

The variances were compared by the F test (Baccan et al., 1985; Vogel, 1992) and the concurrency between the means was verified by Student's *t*-test (Harris, 2001).

3. Results and discussion

3.1 Evaluation of the electrothermal behavior of the analytes by the univariate method

The elements Pb and Cd are relatively volatile, which makes the use of chemical modifiers indispensable. Therefore, Pd + Mg and W were used as chemical modifiers to thermally stabilize the analytes, enabling the satisfactory elimination of most of the matrix, and thus reducing possible interferences.

3.1.1 Determination of pyrolysis and atomization temperatures of the microemulsified samples by the univariate method

Pyrolysis and atomization curves were obtained for Cu, Pb, Ni and Cd in the presence of two modifiers, W and Pd + Mg (Lobo et al., 2009). An analysis of the curves revealed a slight difference in sensitivity when using the Pd + Mg or W modifier. Hence, analyzing solely the pyrolysis and atomization temperature curves did not allow for conclusions to be drawn about the optimal chemical modifier for the analytes in question. Table 8 lists the Tp and Ta chosen for the analytes (according to the curves).

Elements	Tp (°C)*		Ta (°C)*	
	Pd + Mg	W	Pd + Mg	W
Cu	900	900	2700	2500
Pb	900	700	2300	1600
Ni	900	1200	3000	3000
Cd	400	300	1700	2000

* Pyrolysis temperatures (Tp) and atomization temperatures (Ta)

Table 8. Values of pyrolysis (Tp) and atomization (Ta) temperatures of the microemulsified samples for the two modifiers used.

3.1.2 Pyrolysis and atomization temperature curves of focused microwave digested samples using the univariate method

Pyrolysis and atomization temperature curves were obtained for Cu, Pb, Ni and Cd in the presence of two modifiers, W and Pd + Mg (LOBO et al., 2009). The results revealed a certain similarity with the procedure using microemulsion, i.e., there was a minor difference in sensitivity when using the Pd + Mg or W modifier. Table 9 lists the Tp and Ta chosen for the analytes.

Elements	Tp (°C)*		Ta (°C)*	
	Pd + Mg	W	Pd + Mg	W
Cu	1000	1000	2600	2600
Pb	900	800	2000	1800
Ni	1500	1400	2600	2400
Cd	550	550	2000	1800

* Pyrolysis temperatures (Tp) and atomization temperatures (Ta)

Table 9. Values of pyrolysis (Tp) and atomization (Ta) temperatures of the digested samples for the two modifiers used

3.2 Evaluation of the electrothermal behavior using experimental design

3.2.1 Results of the optimization of pyrolysis and atomization temperatures using experimental design

The optimization method using factorial design evaluates the interactions between the important variables, as well as their simultaneous combination. Table 10 illustrates the results of absorbance for the elements evaluated using 2^4 factorial design.

Experiments	Cu	Pb	Ni	Cd
1	0.0974	0.1562	0.0658	0.3810
2	0.0651	0.0483	0.0912	0.3520
3	0.0988	0.1506	0.0651	0.5010
4	0.0656	0.0555	0.0824	0.5390
5	0.0991	0.1538	0.0744	0.3840
6	0.0701	0.0452	0.0968	0.3520
7	0.0945	0.1406	0.0621	0.4170
8	0.0694	0.0431	0.0908	0.3130
9	0.0971	0.1599	0.632	0.3120
10	0.0636	0.0488	0.0947	0.3310
11	0.1008	0.1597	0.0894	0.4560
12	0.0630	0.0430	0.0626	0.4600
13	0.1033	0.1673	0.0921	0.2720
14	0.0759	0.0356	0.0735	0.3040
15	0.1044	0.1455	0.0905	0.3530
16	0.0706	0.0397	0.0769	0.3950

Note: This factorial design has two levels, (-) for the lower level and (+) for the upper level, related to the sample preparation (Variable 1), i.e., digestion (-) and microemulsion (+), levels related to the lowest (-) and highest (+) temperatures for the two variables of pyrolysis temperature (Variable 2) and atomization temperature (Variable 3), as well as levels related to the type of modifier used (Variable 4) Pd + Mg (-) and W (+).

Table 10. Analyte absorbance values using factorial design

Pareto charts, graphics showing the values of effects and graphics showing special values are the ones most commonly used to evaluate the interactions between variables and to pinpoint the significant variables. In this work, values vs. probability graphics were prepared for the four analytes. In this type of graphic, the effects are calculated as a function of absorbance values, considering the levels (+ and -) presented in thematrix of the sample. The graphics of values of the effects vs. the scale of probability are presented in Figure 1 (adapted from Lobo et al., 2009).

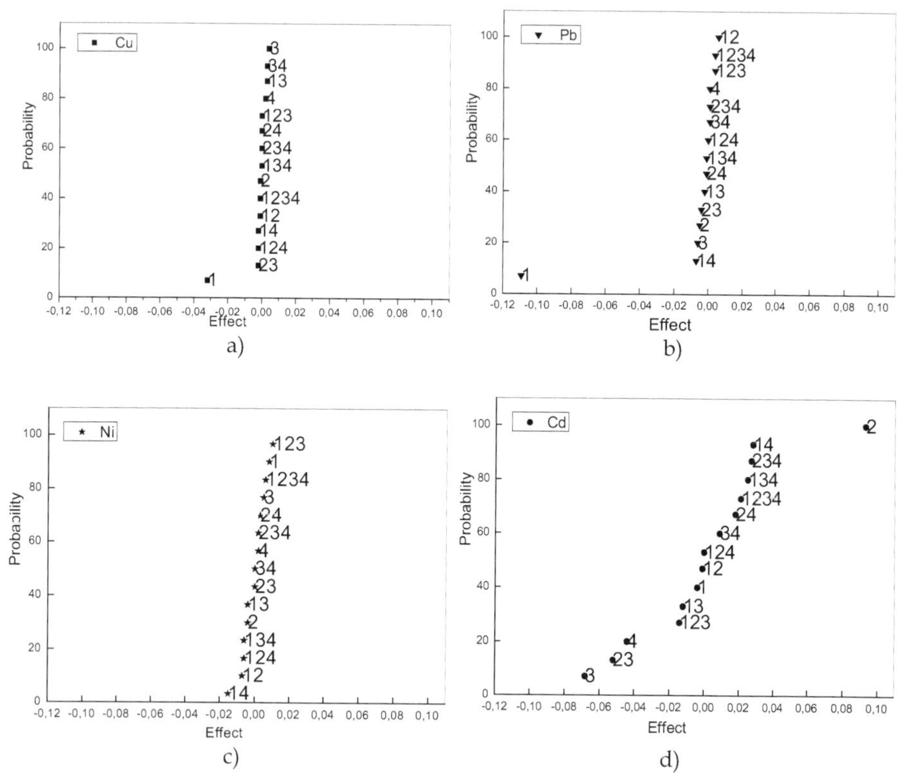

Fig. 1. Graphic of the value of the effects of the variables on the GFAAS optimization of the pyrolysis and atomization temperatures of biodiesel samples: a) Cu, b) Pb, c) Ni, d) Cd

On the plot of a graphic of scale of probability vs. value of the effect, the smaller points that tend to form a straight line are considered insignificant, while the larger (in value of the effect) and the further apart the points are, the more significant the effect.

In Figures 1a and 1b, the optimization graphic for Cu and Pb shows that variable 1 (sample preparation) in focused microwave digestion (defined as (-), Table 5) is more important for the two analytes (higher effect of absorbance), since this point was further away from the others, while the other variables were indifferent, indicating that any modifier could be used. The ideal pyrolysis and atomization temperatures were chosen according to the highest sensitivity, taking into account the lowest temperatures that least degrade the graphite tube. Thus, the pyrolysis and atomization temperatures chosen were 1000°C and 2200°C for Cu and 500°C and 2000°C for Pb, respectively.

Figure 1c shows that none of the evaluated variables were important for Ni. The pyrolysis and atomization temperatures chosen for Ni were 800°C and 2300°C. As can be seen in the optimization graph for Cd in Figure 1d, the variables that were most distant were variables 2 and 3, indicating that the temperature of pyrolysis must be increased and that of atomization decreased to ensure greater efficiency in the process. This finding points to the need to use another factorial design for this element. The variables medium (1) and modifier (4) were not important, indicating that any medium or modifier can be used.

3.2.1.1 Results of the evaluation of pyrolysis and atomization temperatures using a second experimental design for Cd

Table 11 describes the optimization method using factorial design for Cd.

Experiments	Sample Preparation	Pyrolysis Temperature*	Atomization Temperature*	Modifier	Cd
1	Digestion	-	-	Pd + Mg	0.7622
2	Digestion	+	-	Pd + Mg	0.8427
3	Digestion	-	+	Pd + Mg	0.7274
4	Digestion	+	+	Pd + Mg	0.7707

Table 11. Values of absorbance for the factorial design used for Cd

The values of the effects were too small for the design employed. Therefore, lower temperatures were chosen, Tp-500°C and Ta-1400°C, thereby increasing the service life of the graphite tube.

The formation of metal carbide (or oxycarbide) deposits on the platform of the graphite tube or directly on the wall of the tube is an interesting modification strategy to prolong the service life of the graphite tube when working with aggressive solvents or matrices. In addition to acting as a coating on the tube and/or as a modifier, these carbides can promote the decomposition of the matrix through catalytic effects, acting as catalysts and destroying the concomitants of the matrix prior to atomization of the analyte, thereby minimizing or eliminating possible interferences (Freschi et al., 2005). The results obtained for all the metals using the experimental design did not differ when working with either Pd + Mg or W. Therefore, we decided to work with W since, besides the advantages mentioned earlier, the heating programs are faster (and thus the determinations are faster), there is less contamination because purification is done *in situ*, there are few background-related problems, and the modifier's volatile impurities can be eliminated, thus improving the limits of detection (Saint´Pierre et al., 2002).

Table 12 summarizes and compares the defined temperatures using the univariate and multivariate methods.

Elements	Modifier	Microwave digestion		Microemulsion	
		Conventional Method	*Multivariate Method*	*Conventional Method*	*Multivariate Method*
		Tp / Ta*	**Tp / Ta**	**Tp / Ta**	**Tp / Ta**
Cu	Pd + Mg	1000 / 2600	1000 / 2200	900 / 2700	-
	W	1000 / 2600		900 / 2500	
Pb	Pd + Mg	900 / 2000	500 / 2000	900 / 2300	-
	W	800 / 1800		700 / 1600	
Ni	Pd + Mg	1500 / 2600	800 / 2300	900 / 3000	800 / 2300
	W	1400 / 2400		1200 / 3000	
Cd	Pd + Mg	550 / 2000	500 / 1400	400 / 1700	500 / 1400
	W	550 /1 800		300 / 2000	

* Tp - Pyrolysis temperature, °C / Ta - Atomization temperature, °C

Table 12. Comparison of the optimal values of pyrolysis and atomization temperatures using univariate and multivariate methods

An analysis of Table 12 indicates that the values of the temperatures using factorial design are usually conditions of milder temperatures. Despite a certain difference in sensitivity, the use of experimental design offers several advantages, such as the definition of suitable experimental parameters (sample preparation, chemical modifiers), help in the definition of pyrolysis and atomization temperatures and evaluation of the interactions between the variables under study.
This emphasizes the importance of the simultaneous optimization of all the factors involved in the system, with fewer experiments, greater speed, and particularly higher efficiency. The chosen temperatures were therefore those listed in Table 18 for the multivariate method, for each specific analyte.

3.3 Evaluation of the influence of standards in the determination of metals

One of the major advantages of systems that use microemulsions is precisely the fact that they use inorganic standards for the determinations (Aucélio et al., 2004). Conversely, organic standards have several disadvantages, such as high cost, instability, they are very dangerous, carcinogenic, and often require the use of special apparatuses and equipment. However, numerous studies have used emulsified standards and not aqueous standards (Chaves et al., 2008a, 2008b; Jesus et al., 2008; Saint´Pierre et al., 2004; Giacomelli et al., 2004), justifying this choice based on the fact that in most cases emulsified standards are more stable, as well as the difference in sensitivity compared to aqueous standards.
Figure 2 illustrates the results of the equations adjusted for the seven biodiesel samples evaluated by the analyte addition method, aiming at a comparison of the behavior of Cd in the different matrices (emulsified standards) and in aqueous standards. The range of concentrations used was based on the interval recommended by the manufacturer.

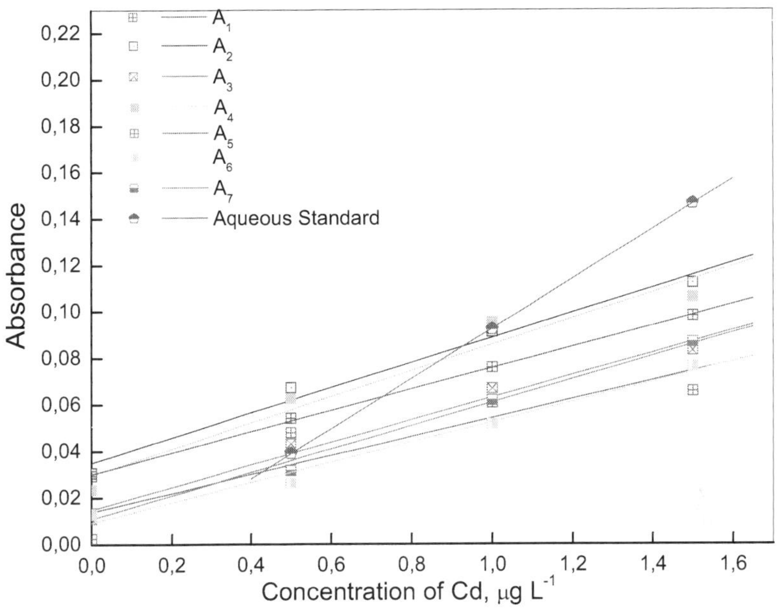

Fig. 2. Equations adjusted to the analytical curves for the determination of Cd in the 8 samples. S_1: A = 0.03017 + 0.0456[Cd], r = 0.99961; S_2: A = 0.03503 + 0.05386[Cd], r = 0.9995; S_3: A = 0.01486 + 0.04822[Cd], r = 0.98956; S_4: A = 0.02971+ 0.05622[Cd], r = 0.9737; S_5: A =- 0.01386+ 0.04042[Cd], r = 0.98695; S_6: A = 0.00931 + 0.04322[Cd], r = 0.99141; S_7: A = 0.01088 + 0.05006[Cd], r = 0.99595; r = 0.97394; Aqueous standard: A = 0.01493 + 0.1076[Cd], r = 0.99999.

As can be seen in Figure 2, there is a significant difference between the analyzed standards. The sensitivities of the analytical curves for pure biodiesel (samples S_1 to S_7) are consistently lower for the aqueous standard, confirming the reports of some studies in the literature that the sensitivities of emulsified standards differ from those of aqueous standards (Cassella et al., 2004; Reyes & Campos, 2005; Saint´Pierre et al., 2002, 2003, 2006).

It was also found that the analytical curves of the biodiesel samples did not exhibit the same sensitivity, although they were similar. This distinct behavior of the samples' analytical curves can be explained by the differences in the composition of the analyzed samples, i.e., fat content, viscosity, source, origins, production processes, etc. Thus, each sample was determined using emulsified standards.

A curve was also built for sample B10, but because most of this sample consisted of diesel (90%), it did not show good stabilization and the resulting values were inadequate for the determination of the analytes in this sample.

3.4 Determination of metals by the calibration procedure using the analyte addition method

3.4.1 Determination of Ni and Cd in microemulsified samples of biodiesel

The best way to calculate the limits of detection (LOD) and of quantification (LOQ) when using the analyte addition method is to pass standards of low concentrations until the lowest signals corresponding to the lowest detectable concentration are found. The estimated standard deviation (SD) is then calculated for 10 measures of the standard with the lowest concentration, after which the values of LOD and LOQ can be calculated according to the literature, i.e., LOD = (3sd) / K and LOQ = (10sd) / K, where k is the slope of the straight line (Baccan et al., 1985; Vogel, 1992). However, due to the experimental difficulties resulting from the need to conclude this work, the LOD and LOQ were calculated using the traditional procedure, considering the blank of the samples instead of the standard with the lowest concentration. Table 13 lists the values of the limits of detection and quantification, as well as the concentrations of Ni and Cd obtained in the samples.

Sample	LOD found, µg L^{-1}		LOQ found, µg L^{-1}		[C]* obtained in the sample, µg g^{-1} ± sd	
	Ni	*Cd*	*Ni*	*Cd*	*Ni*	*Cd*
S_1	0.78	0.10	2.67	0.36	≤LOD	0.66
S_2	0.81	0.10	2.71	0.31	≤LOD	0.61
S_3	0.85	0.10	2.85	0.34	≤LOD	0.19
S_4	0.44	0.07	1.47	0.23	≤LOD	0.33
S_5	0.41	0.12	1.37	0.40	≤LOD	≤LOD
S_6	0.29	0.12	0.97	0.39	≤LOD	0.21
S_7	0.46	0.10	1.53	0.33	≤LOD	0.21

*Concentration of the analyte

Table 13. Concentrations of Ni and Cd, and LOD and LOQ found for the samples

Table 13 indicates that low limits of detection (LOD ≤ 0.85 µg L^{-1}) were obtained for Ni, but Ni was undetectable in the samples using this method. Many explanations can be considered, since Ni is usually present in fuels as volatile Ni and low molecular mass Ni complex (Vale et al., 2004) and may have been lost during the analyses.

Low limits of detection (LOD ≤ 0.12 µg L^{-1}) were obtained for Cd by the analyte addition method. The concentrations of cadmium found in the samples were low, but were above the respective limits of detection obtained statistically.

As can be seen in Table 13, sample S_5 (washed animal fat) presented no concentration of Cd, probably because it was eliminated during the washing step of the biodiesel production process, rendering it undetectable in the sample analyzed by the proposed method.

Cd values obtained in samples of animal fat are always lower than those found in samples of vegetable biodiesel. This is probably due to the fact that vegetables can absorb metals from the soil, which is not the case in samples of animal origin. Moreover, cadystins may also be present, which are synthesized by the majority of higher plants and are also present in algae and fungi. The presence of Cd in biodiesel can be attributed to the high mobility of this element in soil (Costa-Neto et al., 2007; Oliveira, et al., 2001). Another important factor that deserves mention is that Cd has been used in the form of $CdCl_2$ as a fungicide, and can be considered a source of soil contamination (Hernández-Caraballo et al., 2004; Campos et al., 2005).

3.4.2 Analyte addition and recovery test

Due to the lack of certifies reference material for biodiesel, the analyte addition and recovery test was used to evaluate the accuracy of the method. Table 14 lists the results obtained for each sample.

Sample	Recovery rates, %	
	Ni ± RSD*	Cd ± RSD
S_1	108 ± 1.5	100 ± 4.3
S_2	105 ± 0.7	104 ± 3.3
S_3	103 ± 2.9	108 ± 0.6
S_4	102 ± 1.8	99 ± 3.7
S_5	101 ± 3.0	116 ± 3.7
S_6	93 ± 8.2	98 ± 4.7
S_7	97 ± 2.7	102 ± 4.6
S_8	93 ± 0.8	95 ± 1.8

*RSD - relative standard deviation

Table 14. Recovery rates (n=3) and relative standard deviations of the biodiesel samples, using W as modifier, prepared with 10 μg L^{-1} of Ni and 1.0 μg L^{-1} of Cd

The accuracy of an analysis depends on the matrix, on sample processing, and on analyte concentration. Data in the AOAC handbook ("Peer Verified Methods Program") indicate that, for concentrations of ≤ 10 μg L^{-1}, recovery rates of 40% to 120% are acceptable (National Health Surveillance Agency - ANVISA, 2009).

Table 14 shows that the recovery rates varied from 93% to 108% for Ni and from 95% to 116% for Cd, indicating that the method used here is suitable for the determination of Ni and Cd in matrices of biodiesel from different sources and origins. Satisfactory RSD values were obtained, i.e., ≤ 8.2% (sample S_7) and ≤ 4.7% (sample S_8) for 10.0 μg L^{-1} of Ni and 1.0 μg L^{-1} of Cd, respectively.

3.4.3 Determination of Cu, Pb, Ni and Cd in digested biodiesel samples

Due to the difference in sensitivity obtained by the emulsified and aqueous standards, the analytes were determined here using standards in the presence of digested samples by the analyte addition method.

After a certain period of storage, the sample of unwashed animal fat (sample S_4) presented nodules (like fatty stones), so this sample could not be homogenized and was therefore not subjected to the digestion procedure.

The sample of cotton origin (S_7) could not be digested because it presented fatty residues at the end of the digestion procedure. Although all the samples were similar, samples of cotton origin should show higher organic loads than other samples, and the amount of acid used was insufficient for the complete digestion of this sample.

Table 15 lists the concentrations of Cu, Pb, Ni and Cd, and the limits of detection and quantification found for the digested samples.

Sample	LOD, μg L-1				LOQ, μg L-1				[C] *, μg g-1			
	Cu	*Pb*	*Ni*	*Cd*	*Cu*	*Pb*	*Ni*	*Cd*	*Cu*	*Pb*	*Ni*	*Cd*
S_1	2.26	6.61	3.25	0.27	7.50	22	11	0.90	≤LOD	7.00	≤LOD	0.43
S_2	1.62	3.96	2.55	0.30	5.40	13	8.5	1.00	≤LOD	≤LOD	≤LOD	0.17
S_3	2.00	5.18	2.55	0.28	6.67	18	8.5	0.93	≤LOD	5.1	≤LOD	≤LOD
S_5	1.64	3.88	2.59	0.84	5.47	13	8.6	2.80	≤LOD	≤LOD	≤LOD	≤LOD
S_6	2.97	6.17	2.16	0.25	9.90	21	7.2	0.83	≤LOD	≤LOD	≤LOD	0.19
S_8	2.54	4.02	2.90	0.20	8.47	13	9.6	0.67	≤LOD	≤LOD	≤LOD	≤LOD

[C]* - Metal concentrations found in the samples

Table 15. Metal concentrations found in the samples, LOD and LOQ found for the analytical curves of the samples and for the aqueous standard

As can be seen in Table 15, all the samples presented concentrations of Cu below the LOD. As for Pb, only the concentrations found in samples S_1 and S_3 were above the LOD, but were below the LOQ. The concentrations of Ni found in the samples also fell below the limits of detection (LOD ≤ 3.25 μg L^{-1}). Low limits of detection (LOD ≤ 0.84 μg L^{-1}) were obtained for Cd by the analyte addition method, although they were higher than those obtained for the microemulsion. Cadmium was not quantified in sample A_5 by this method.
Table 18 indicates that low limits of detection (LOD ≤ 0.85 μg L^{-1}) were obtained for Ni, but Ni in the samples was also undetectable by this method. The probable reasons for this are the same as those mentioned in item 3.4.1.

3.4.4 Analyte addition and recovery test

Table 16 shows the addition and recovery results for each sample.

Sample	Recovery rates, % ± RSD			
	Cu ± RSD*	Pb ± RSD	Ni ± RSD	Cd ± RSD
S_1	99 ± 2.18	111 ± 3.30	97 ± 6.80	103 ± 0.90
S_2	98 ± 1.40	101 ± 4.70	100 ± 5.03	100 ± 0.29
S_3	101 ± 1.64	124 ± 8.50	101 ± 5.53	105 ± 1.68
S_5	106 ± 5.80	114 ± 4.90	102 ± 2.53	101 ± 2.41
S_6	91 ± 1.64	106 ± 1.38	98 ± 3.15	100 ± 1.92
S_8	103 ± 2.70	100 ± 5.30	95 ± 5.30	95 ± 1.39

*RSD - relative standard deviation

Table 16. Recovery rates (n=3) and relative standard deviations (in parentheses) of biodiesel samples prepared with 10 μgL^{-1} of Cu, 15 μg L^{-1} of Pb, 10 μg L^{-1} of Ni and 1.0 μg L^{-1} of Cd, using W as modifier

Table 16 indicates that the recovery rates varied from 91% to 106% for Cu, from 100% to 124% for Pb, 95% to 102% for Ni, and 95% to 105% for Cd. Hence, despite the low concentrations found in the samples (Table 15), the method employed here is suitable for the determination of these analytes in biodiesel matrices from different sources and origins.

Satisfactory RSD values were obtained, i.e., ≤ 5.80% (sample S_5) for 10 µg L^{-1} of Cu; ≤ 8.50% (sample S_3) for 15 µg L^{-1} of Pb; ≤ 6.80% (sample S_1) for 10 µg L^{-1} of Ni, and ≤ 2.41% (S_5) for 1.0 µg L^{-1} of Cd.

3.5 Comparison of the microemulsion and focused microwave digestion procedures

Table 17 lists the values of Cd in samples S_1, S_2, S_3, S_5, S_6 and S_8 determined by the two methods, i.e., using samples in the microemulsified and digested forms.

Sample	LOD found, µg L^{-1}		LOQ found, µg L^{-1}		[C] obtained in the sample, µg g^{-1}	
	Cd_{ME}*	Cd_D**	Cd_{ME}*	Cd_D**	Cd_{ME}*	Cd_D**
S_1	0.10	0.27	0.36	0.90	0.66	0.43
S_2	0.093	0.30	0.31	1.00	0.61	0.17
S_3	0.07	0.28	0.34	0.93	0.33	≤LOD
S_5	0.12	0.84	0.40	2.80	≤LOD	≤LOD
S_6	0.12	0.25	0.39	0.83	0.21	0.19

ME* - Microemulsified; D** - Digested

Table 17. Concentrations of Cd obtained in the samples, and LOD and LOQ of the samples using the different sample preparation procedures

As can be seen in Table 17, the analyte addition method resulted in low limits of detection (LOD ≤ 0.84 µg L^{-1}) of Cd for the two methods of sample preparation, but the LODs obtained by the digestion method were higher. The concentrations found in the digested samples were consistently lower than those found in the microemulsified samples due to a possible loss of analyte during digestion. This is because the microwave used here has a semi-open configuration, and despite the reflux, the analyte may have undergone particle evaporation (Meeravali & Kumar, 2001).

Sample S_5 (washed animal fat) did not show the same quantifiable concentration of Cd in the two procedures (Tables 13 and 15). Sample S_8 presented different results, because B10 is a sample of biodiesel mixed with diesel. The only samples that presented a consistent concentration of Cd by the two methods were S_1, S_2 and S_6.

The F-test is a hypothesis test used to ascertain if the variances of two given determinations are different, or to verify which of the two determinations shows greater variability. The F-test was also applied to verify if the variances were the same or different, and the $F_{calculated}$ values were found to be consistently lower than the $F_{tabulated}$ value at a 95% level of confidence. Thus, it can be concluded that there are no significant differences between the two accuracies at the 95% level of confidence.

The *t*-test is a statistical tool widely employed to verify the concurrency between averages. Student's *t*-test was performed to evaluate the samples by comparing individual differences, since each sample was measured by the microemulsion and digestion methods, which do not yield exactly the same results. The $t_{calculated}$ value was lower than the $t_{tabulated}$ value at a 95% level of confidence. Hence, the two methods are not significantly different at the 95% level of confidence.

Table 18 summarizes the analytical characteristics of the analytes in the two methods developed.

PARAMETERS	MICROEMULSION				MICROWAVE DIGESTION			
	Cu	Pb	Ni	Cd	Cu	Pb	Ni	Cd
Pyrolysis temperature, °C	8	500	800	500	1000	500	800	500
Atomization temperature, °C	2200	2000	2300	1400	2200	2000	2300	1400
Volume of sample, µL	20							
Linear calibration interval used, µg L^{-1}	5 – 15	15 - 45	5 – 15	0.5 – 1.5	5 - 15	15 - 45	5 - 15	0.5 – 1.5
Characteristic mass, pg	nd*	nd*	≤ 11	≤ 2	≤ 41	≤ 54	≤ 25	≤ 2
Recovery rates, %	nd*	nd*	93 – 108	95 - 116	91 -106	100 - 124	95 - 102	95 – 105
Modifier mass (µg)	200							
Graphite tube service life (avg. of the no. of firings)	520				450			
Analytical rate (determinations per hour)	40							
Relative standard deviation RSD, n=12), mL	nd*	nd*	≤ 8.20%	≤ 4.71%	≤ 5.80%	≤ 8.50%	≤ 6.80	≤ 2.41
LOD, µg L^{-1}	nd*	nd*	≤1	≤ 0.12	≤3	≤7	≤4	≤ 0.84
LOQ, µg L^{-1}	nd*	nd*	≤3	≤ 3	≤10	≤22	≤11	≤ 3

nd*- not determined

Table 18. Analytical characteristics of the proposed methods for the determination of Cu, Pb, Ni and Cd in biodiesel using W as modifier and two sample preparation procedures

4. Conclusions

Multivariate optimization techniques are currently applied preferentially in analytical chemistry because, among other advantages, they allow for the simultaneous optimization of all the factors involved in the system with fewer experiments, greater speed, and particularly higher efficiency. Despite these multiple advantages, however, multivariate techniques have only been effectively and increasingly employed in the optimization of analytical methods in the last few decades. Factorial design was employed in this work, confirming its importance in evaluating the significance of several variables, as well as in indicating optimal conditions to obtain the best results. Another aspect to be highlighted is the fewer experiments required with factorial design when compared to the traditional method (univariate). A maximum of 16 experiments were performed to optimize the pyrolysis and atomization temperatures for each element, instead of the 17 to 25 experiments the literature reports for the traditional method.

The pyrolysis and atomization temperatures for the determination of Cu, Cd, Ni and Pb were determined based on the graphics of value of the effects. Using these graphics, it was found that for the analytes Cu and Pb, preparation of the sample in digested form was the only significant variable; hence, these elements were analyzed only in focused microwave-digested samples. None of the evaluated variables were important for Ni. The optimal pyrolysis (Tp) and atomization (Ta) temperatures found were, respectively: Cu 1000°C and 2200°C, Pb 500°C and 2000°C, and Ni 800°C and 2300°C. For Cd, the pyrolysis temperature had to be increased and the atomization temperature decreased to ensure the highest efficiency of the process. A 2^2 factorial design was created with four experiments. This factorial design has two levels corresponding to the lowest (-1) and highest (+1) temperatures for two variables (temperatures of pyrolysis and of atomization). The results indicate that the values of the effects were very slight for the design used here, since the lowest temperatures were chosen, i.e., Tp-500°C and Ta-1400°C. The other variables were unimportant. It was decided to work with W because the analyses are faster, there is less contamination, few problems involving background and incompatibility among solutions, and because W is a permanent modifier, which may increase the service life of the atomizer.

The analytical procedures developed here using microemulsion can be considered satisfactory, for they exhibited good recovery rates and low RSD values.

The main advantage of the procedures employed here is that they enable the use of inorganic standards for the determinations, instead of organic solvents, which have some drawbacks such as the need for suitable equipment, connections and apparatuses in view of to their toxicity and their chemical instability.

Although some of the elements were not determined in the samples analyzed by these methods, the LOD and LOQ were low. Therefore, they are interesting since, if the respective analytes are present in the biodiesel samples analyzed here, their concentrations are lower than 3 $\mu g\ L^{-1}$ (which is the highest LOQ found). These values are much lower than those reported in the literature for these elements in fossil fuels. From the environmental standpoint, this can be considered a positive aspect of biodiesel, since some elements, for example Ni, are natural constituents of petroleum and are usually found in high concentrations in its derivatives.

This work contributes towards the establishment or proposal of a suitable standard, which is still absent from the literature and/or current legislation, in terms of the quality control of these metals in biodiesel samples. Moreover, it enables the prediction of possible environmental impacts resulting from the production, transportation and use of fuels such as biodiesel.

5. Acknowledgments

The authors thank the Brazilian research funding agencies FAPESP, CNPq and FUNDUNESP for their financial support and grants. The authors are also indebted to UFMT - Universidade Federal de Mato Grosso for providing the biodiesel samples used in this work and to Professor Edenir Pereira Filho for his assistance in the initial part of the experiments. We also thank the anonymous reviewers for their comments, which were helpful in improving the manuscript.

6. References

Agarwal, A.K. (2005). Experimental investigations of the effect of biodiesel utilization on lubricating oil tribology in diesel engines. *Proc. Inst. Mech. Eng. Transp. Eng.*, Vol. 219, pp. (703-713).

AGÊNCIA NACIONAL DO PETRÓLEO, GÁS NATURAL E BIOCOMBUSTÍVEIS. Portaria nº 311, de 27 de dezembro de 2001. Estabelece os procedimentos de controle de qualidade na importação de petróleo, seus derivados, álcool etílico combustível, biodiesel e misturas óleo diesel/biodiesel. Diário Oficial da União, Brasília, DF, 28 de dezembro de 2001. Available at: http://nxt.anp.gov.br/NXT/gateway.dll/leg/folder_portarias_anp/portarias_anp_tec/2001/dezembro/panp%20311%20-%202001.xml?f=templates$fn=default.htm&sync=1&vid=anp:10.1048/enu. Consulted on: 10 May 2009.

Amorim, F.R.; Bof, C.; Franco, M.B.; Silva J.B.; Nascentes, C.C. (2006). Comparative study of conventional and multivariate methods for aluminum determination in soft drinks by graphite furnace atomic absorption spectrometry. *Microchem. J.*, Vol. 82, pp. (168-173).

Arzamendi, G.; Arguinãrena, E.; Campo, I.; Zabala, S.; Gandiá, L.M. (2008). Alkaline and alkaline-earth metals compounds as catalysts for the methanolysis of sunflower oil *Catal. Today*, Vol. 133-135, pp. (305-313).

Aucélio, R.Q.; Doyle, A.; Pizzorno, B.S.; Tristão, M.L.B.; Campos, R.C. (2004). Electrothermal atomic absorption spectrometric method for the determination of vanadium in diesel and asphaltene prepared as detergentless microemulsions. *Microchem. J.*, Vol. 78, pp. (21-26).

Baccan, N.; Andrade, J.C.; Godinho, O.E.S.; Barone, J.S. (1985) *Química analítica quantitativa elementar*. 2. ed. Campinas: Ed. UNICAMP, Campinas.

Bettinelli, M.; Spezia, S.; Baroni, U.; Bizzarri, G. (1996).The use of reference materials in fossil fuel quality control. *Microchim. Acta,* Vol. 123, pp. (217-230).

Campos, M.L.; Silva, F.N.; Furtini Neto, A E.F.; Guilherme, L.R.G.; Marques, J.J.; Antunes, A.S. (2005). Determinação de cádmio, cobre, cromo, níquel, chumbo e zinco em fosfatos de rocha. *Pesq. Agropec. Bras.*, Vol. 40, pp. (361-367).

Canakci, M.; Monyem, A.; Van Gerpen, J. (1999). Accelerated oxidation processes in biodiesel. *Trans. of the Asae*, Vol. 42, pp. (1565-1572).

Cassella, R.J.; Barbosa, B.A.R.S.; Santelli, R.E.; Rangel, A.T. (2004). Direct determination of arsenic and antimony in naphtha by electrothermal atomic absorption spectrometry with microemulsion sample introduction and iridium permanent modifier. *Anal. Bioanal. Chem.*, Vol. 379, pp. (66-71).

Chaves, E.S.; dos Santos, E.J.; Araujo, R.G.O.; Oliveira, J.V.; Frescura, V.L.A.; Curtius, A.J. (2010). Metals and phosphorus determination in vegetable seeds used in the production of biodiesel by ICP OES and ICP-MS. *Microchem. J.*, Vol. 96, No. 1, pp. (71-76).

Chaves, E.S.; Saint´Pierre, T.D.; Santos, E.J.; Tormen, L.V.; Bascunana, L.A.F.; Curtius, A. (2008a). Determination of Na and K in biodiesel by flame atomic emission spectrometry and microemulsion sample preparation. *J. Braz. Chem. Soc.*, Vol. 19, No. 5, pp. (856-861).

Chaves, E.S.; Saint´Pierre, T.D.; Santos, E.J.; Tormen, L.V.; Bascunana, L.A.F.; Curtius, A. (2008b). Determination of Na and K in biodiesel by flame atomic emission spectrometry and microemulsion sample preparation. *J. Braz. Chem. Soc.*, Vol. 19, No. 5, pp. (856-861).

Coronado, C.R.; Villela, A.D.; Silveira, J.L. (2010). Ecological efficiency in CHP: Biodiesel case. Applied Thermal Engineering. Vol. 30, No. 5, pp. (458-463).

Costa Neto, P.R.; Rossi, L.F.S.; Zagonel, G.F.; Ramos, L.P. (2000). Produção de biocombustível alternativo ao óleo diesel através da transesterificação de óleo de soja usado em frituras. *Quim. Nova*, Vol. 23, No. 4, pp. (531-537).

Ferrari, R.A.; Oliveira, V.S.; Scabio, A. (2005). Biodiesel de soja - taxa de conversão em ésteres etílicos, caracterização físico-química e consumo em gerador de energia, *Quim. Nova,* Vol. 28, No. 1, pp. (19-23).

Figueiredo, J.L.; Ribeiro, F.R. (1987). Desativação de catalisadores. *Catálise heterogênea.* Lisboa: Fundação Calouste Gulbekian, pp. (219-252), Lisboa.

Freschi, C.S.D.; Freschi, G.P.G.; Gomes Neto, J.A.; Nobrega, J.A.; Oliveira, P.V. (2005). Arsenic as internal standard to correct for interferences in determination of antimony by hydride generation *in situ* trapping graphite furnace atomic absorption spectrometry. *Spectrochim. Acta Part B*, Vol. 60, No. 5, pp. (759-763).

Giacomelli, M.B.O.; Silva, J.B.B.; Saint'Pierre, T.D.; Curtius, A.J. (2004). Use of iridium plus rhodium as permanent modifier to determine As, Cd and Pb in acids and ethanol by electrothermal atomic absorption spectrometry. *Microchem. J.*, Vol. 77, pp. (151-156).

Harris, D.C. (2001). *Análise química quantitativa.* 5. ed. Rio de Janeiro.

Haseeb, A.S.M.A.; Masjuki, H.H.; Ann, L.J.; Fazal, M.A. (2010). Corrosion characteristics of copper and leaded bronze in palm biodiesel. *Fuel Processing Technology*. Vol. 91, No. 3, pp. (329-334).

Hernández-Caraballo, E.A.; Burguera, M.; Burguera, J.L. (2004). Determination of cadmium in urine specimens by graphite furnace atomic absorption spectrometry using a fast atomization program. *Talanta*, Vol. 63, pp. (419-424).

Ilkilic, C.; Behcet, R. (2010). Energy sources - Part A - Recovery utilization and environmental effects. The Reduction of Exhaust Emissions from a Diesel Engine by Using Biodiesel Blend. Vol. 32, No. 9, pp. (839-850).

Jackson, K. W. (1999). *Electrothermal atomization for analytical atomic spectrometry*. Chichester: John Wiley, England.

Jesus, A.; Silva, M.M.; Vale, M.G.R. (2008). The use of microemulsion for determination of sodium and potassium in biodiesel by flame atomic absorption spectrometry. *Talanta*, Vol. 74, pp. (1378-1384).

Jesus, A.; Zmozinski, A.V.; Barbara, J.A.; Vale, M.G.R.; Silva, M.M. (2010). Determination of Calcium and Magnesium in Biodiesel by Flame Atomic Absorption Spectrometry Using Microemulsions as Sample Preparation. *Energy Fuels*, Vol. 24, pp. (2109-2112).

Liu, J.; Sturgeon, R.E.; Willie, S.N. (1995). Open-focused microwave-assisted digestion for the preparation of large mass organic samples. *Analyst. Vol.* 120, pp. (1905-1909).

Lobo, F.A.; Gouveia, D.; Oliveira A.P.; Romão, L.P.C.; Fraceto, L.F.; Dias, N.L.; Rosa, A.H. (2011). Development of a method to determine Ni and Cd in biodiesel by graphite furnace atomic absorption spectrometry. *Fuel*, Vol. 90, No. 1, pp. (142-146).

López, D.E.; Goodwin Jr., J.G.; Bruce, D.A.; Lotero, E. (2005). Transesterification of triacetin with methanol on solid acid and base catalysts. *Appl. Catal.*, Vol. 295, pp. (97-105).

Ma, F.; Hanna, M.A. (1999). Biodiesel production: a review. *Bioresour. Technol.*, Vol. 70, pp. (1-15).

Meeravali, N.N.; Kumar, S.J. (2001). The utility of a W-Ir permanent chemical modifier for the determination of Ni and V in emulsified fuel oils and naphtha by transverse heated electrothermal atomic absorption spectrometry. *J. Anal. Atom. Spectrom.*, Vol.. 16, No. 5, pp. (527-532).

Meher, L.C.; Dharmagadda, V.S.S.; Naik, S.N. (2006). Optimization of alkali-catalyzed transesterification of *Pongamia pinnata* oil for production of biodiesel. *Bioresour. Technol.*, Vol. 97, pp. (1392-1397).

Nigam, P.S.; Singh, A. (2011). Production of liquid biofuels from renewable resources. Progress in Energy and Combustion Science, Vol. 37, No. 1, pp. (52-68).

Oliveira, A.P.; Gomes Neto, J.A.; Moraes, M.; Lima, E.C. (2002). Direct determination of Al, As, Cu, Fe, Mn and Ni in fuel ethanol by simultaneous GFAAS using integrated platforms pretreated with W-Rh permanent modifier together with Pd+Mg modifier. *Atomic Spectroscopy*, Vol. 23, No. 6, pp. (190-195).

Oliveira, A.P.; Villa, R.D.; Antunes, K.C.P.; Magalhães, A.; Silva, E.C. (2009). Determination of sodium in biodiesel by flame atomic emission spectrometry using dry decomposition for the sample preparation. *Fuel*, Vol. 88, pp. (764-766).

Oliveira, J.A.; Cambraia, J.; Cano, M.A.O.; Jordão, C.P. (2001). Absorção e acúmulo de cádmio e seus efeitos sobre o crescimento relativo de plantas de aguapé e de salvínia. *R. Bras. Fisiol. Veg.*, Vol. 13, pp. (329-341).

Pereira Filho, E.R.; Poppi, R.J.; Arruda, M.A.Z. (2002). Emprego de planejamento fatorial para a otimização das temperaturas de pirólise e atomização de Al, Cd, Mo e Pb por ETAAS. *Quim. Nova*, v. 25, n. 2, p. (246-253).

Pohl, P.; Vorapalawut, N.; Bouyssiere, B.; Carrier, H.; Lobinski, R. (2010). Direct multi-element analysis of crude oils and gas condensates by double-focusing sector field inductively coupled plasma mass spectrometry (ICP MS). Journal of Analytical Atomic Spectrometry. Vol. 25, No. 5, pp. (704-709).

Ramadhas, A.S.; Jayaraj, S.; Muraleedharan, C. (2004). Use of vegetable oils as I.C. engine fuels: a review. *Renewable Energy, Vol.* 29, pp. (727-742).

Reyes, M.N.M.; Campos, R.C. (2005). Graphite furnace absorption spectrometric determination of Ni and Pb in diesel and gasoline samples stabilized as microemulsion using conventional and permanent modifiers. *Spectrochim. Acta Part B*, Vol. 60, pp. (615-624).

Sahin, Y. (2011). Energy education science and technology, Part A - Energy science and research. *Environ. impacts of biofuels*, Vol. 26, No. 2, pp. (129-142).

Saint´Pierre, T.D.; Dias, L.F.; Maia, S.M.; Curtius, A.J. (2004). Determination of Cd, Cu, Fe, Pb and Tl in gasoline as emulsion by electrothermal vaporization inductively coupled plasma mass spectrometry with analyte addition and isotope dilution calibration techniques. *Spectrochim. Acta Part B*, Vol. 59, pp. (551-558).

Saint´Pierre, T.D.; Dias, L.F.; Pozebon, D.; Aucélio R.Q.; Curtius A.J.; Welz, B. (2002). Determination of Cu, Mn, Ni and Sn in gasoline by electrothermal vaporization inductively coupled plasma mass spectrometry, and emulsion sample introduction. *Spectrochim. Acta Part B*, Vol. 57, No. 12, pp. (1991-2001).

Saint´Pierre, T.D.; Frescura, V.L.A.; Aucélio, R.Q. (2006).The development of a method for the determination of trace elements in fuel alcohol by ETV-ICP-MS using isotope dilution calibration. *Talanta*, Vol. 68, pp. (957-962).

Saint´Pierre, T.D.; Aucélio, R.Q.; Curtius A.J. (2003). Trace elemental determination in alcohol automotive fuel by electrothermal atomic absorption spectrometry. *Microchem. J.*, Vol. 75, No. 1, pp. (59-67).

Silva, J.S.A.; Chaves, E.S.; Santos, E.J.; Saint´Pierre, T.D.; Frescura, V.L.; Curtius, A.J. (2010). Calibration Techniques and Modifiers for the Determination of Cd, Pb and Tl in Biodiesel as Microemulsion by Graphite Furnace Atomic Absorption Spectrometry. *J Braz Chem Soc*, Vol. 21, pp. (620-626).

Vale, M.G.R.; Damin, I.C.F.; Klassen, A.; Silva, M.M.; Welz, B.; Silva, A.F.; Lepri, F.G. (2004). Method development for the determination of nickel in petroleum using line-source and high-resolution continuum-source graphite furnace atomic absorption spectrometry. *Microchem. J.*, Vol. 77, pp. (131-140).

Vogel, A. (1992). *Química analítica quantitativa.* (5. ed.), LCD, Rio de Janeiro.

Welz, B.; Schlemmer, G.; Mudakavi, J.R. (1992). Palladium nitrate-magnesium nitrate modifier for electrothermal atomic absorption spectrometry. Part 5. Performance for the determination of 21 elements. J. Anal. At. Spectrom., v. 7, p. (1257-1271).

Woods, G.D.; Fryer, F.I. (2007). Direct elemental analysis of biodiesel by inductively coupled plasma–mass spectrometry. *Anal. Bioanal. Chem.*, Vol. 389, pp. (753-761).

Part 2

Biodiesel: Development, Performance, and Combustion Emissions

8

Analysis of the Effect of Biodiesel Energy Policy on Markets, Trade and Food Safety in the International Context for Sustainable Development

Rodríguez Estelvina[1], Amaya Chávez Araceli[1], Romero Rubí[1], Colín Cruz Arturo[1] and Carreras Pedro[2]
[1]Universidad Autónoma del Estado de México- México
[2]Universidad Americana
[1]Mexico
[2]Paraguay

1. Introduction

According by national objectives in each country to achieve energy alternatives, the reduction of gases which cause the greenhouse effect and new strategies for rural development, the production of biodiesel have increased in the last few years and a higher number of countries are adopting new policies. Nevertheless, in the annual report entitled The State of Agriculture and Food Supply presented by the FAO (Food and Agricultural Organization) (FAO, 2008b), the increase of biofuel production is presented as worrisome since the massive use of biofuels would generate more pressure on the food supply and could bring negative social and environmental consequences. However, there is no clear consensus on the level of connection between food and biofuel since high prices can also offer potential long term opportunities for agriculture and rural development. The demand for raw materials to produce biofuels could constitute a structural variation in the tendency for prices of agricultural products to decrease, creating opportunities as well as risks. The perspectives of growth in bioenergy for developing countries as well as the demand from countries of the OECE (Organization for Economic Cooperation and Development) can bring new opportunities for commerce in biodiesel and the securing of raw materials. In this way, the applied policies seem to play an important role in sustainability for this type of bioenergy. This chapter analyzes the tendencies in the market, the impact on raw materials as well as the repercussions in the food supply and in the policies of the sector, within a context of sustainable development.

The method used is an analytical approach by using data and statistics of international organizations to develop baseline scenarios and forecasts on the factors of sustainability, international policy and market and food security. The paper brings together the available knowledge and processes of the sustainability framework to support debate about the potential of biodiesel systems. Among the reflections, it is considered that the impact of biofuels depends upon the scale and type of system under consideration, and the policies, regulations and subsidies that accompany them. The discussion is extended to include energy efficiency, impact assessment and research of biodiesel technology, to contribute to sustainable development from the use of this fuel.

2. Sustainability factors for a biodiesel fuel perspective

In recent years the protection and conservation the environment has become a priority on the global agenda, considering the natural environment is the most important capital humanity has and, knowing this it is best to preserve and regenerate. The condition and quality of life for all people will be guaranteed. However, it was not until 1987, when the United Nations World Environment and Development Commission unanimously approved the Brundtland Report, known as Our Common Future, where sustainable development is defined as "that which meets the essential needs of the present without compromising the ability to meet the essential needs of future generations." That is, sustainable development was established as a Model of Wise Production, whose central objective is the preservation of natural resources, based on three concepts:

a. Human welfare, whose lines of action were established in health, education, housing, safety and protection of children´s rights,
b. The ecological well-being through actions for the care and protection of air, water and soil, and
c. The interactions established through public policies on population, equity, distribution of wealth, economic development , production and consumption and exercising government.

Sustainability, in any production process, is achieved by harmonizing three fundamental principles: cost-effectiveness, social benefit and ecological balance. Based on this foundation, biodiesel should be a part of new energy policies. Within the literature on the topic, many definitions are offered. It should be noted, however, that the concept of biodiesel needs a dual approach: from the area of environmental science (sustainability criteria) and from a multidisciplinary standpoint. Biodiesel sustainability factors are mainly related to:

a. **Raw material:** The varieties of plants used as feedstock for biodiesel include sunflowers, soya and rapeseed, among others. It is best if the source of the biodiesel is second generation, of high yields and low cost in order to avoid putting pressure on the soil, competing with food demands and increasing availability (IEA, 2004). A positive energy balance depends upon the raw materials and the technology used.
b. **Technology used:** Different technologies are used in the production of biodiesel depending upon the raw materials used and the costs involved. In the case of biodiesel, transesterification processes are used continuously or with an acid or base as a catalyst. In the case of bioethanol is generally obtained through fermentation. It is best if the technology applied does not require a large quantity of energy to operate in order to avoid the possible generation of effluent contaminants.
c. **Waste generation:** Biofuels have several advantages: they reduce CO_2 emissions and other gases which cause the greenhouse effect by 80%; reduce the sulfur emission, which is the main cause of acid rain; it is biodegradable and doubles motor life because of the optimal lubrication (Stenblik, 2007). Nevertheless, the majority of studies compare biodiesel to conventional diesel, leaving out the life cycle of the product. It is important to understand the process from the conception of the product and verify the

residual outflow, either as atmospheric emissions or effluents in the industrial process, agricultural residuals and the waste of pesticides.

d. **Development Policies and Standards:** Policies should encourage the development of biodiesel by coordinating efforts and avoiding the overlap of public resources (Mitchell, 2008). In order to operationalize the concept of sustainable development, it is necessary to use principles, criteria and indicators which cover social, environmental and economic issues for the management of resources in the production of biodiesel.

Sustainable development is a comprehensive process that requires different actors of society commitments and responsibilities in the application of the economic, political, environmental and social as well as in consumption patterns that determine quality of life.

3. Overview of international policy and markets

To date, the production of biofuels in industrialized countries has been developed under the protection of high tariffs and, at the same time, paying out large subsidies to producers. These policies hurt developing countries which are, or could become, efficient and profitable producers of biofuels in new export markets (Torero, 2010). For the most part, the recent increase in the production of biofuels has taken place in countries involved in the Organization for Economic Cooperation and Development (OCDE), mostly in the United States and in countries of the European Union (EU). It is expected that global production of biodiesel will increase, as shown in **Figure 1**, under the mandates and tax concessions arising from policies. However, trends in consumption for biodiesel haves remained stable **(Figure 2)** in relation to the percentage of total energy demand for transport.

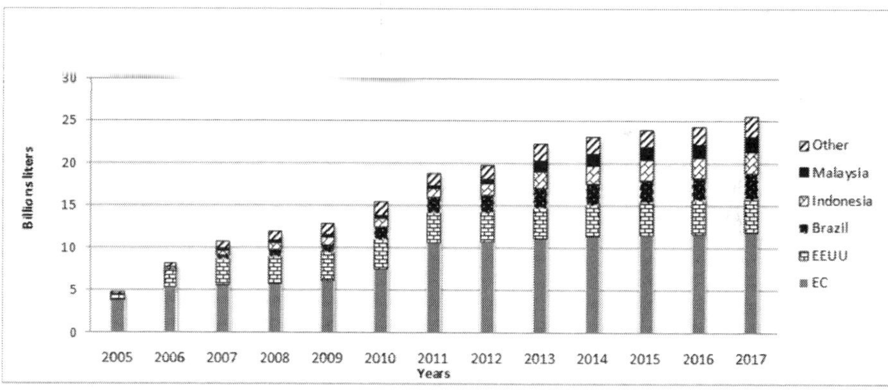

Source: Analysis based on reference data from FAO (2008).

Fig. 1. World production of biodiesel and current projections to 2017, in billion liters.

Biofuels, including biodiesel, have been promoted by policies which support and subsidize their production and consumption. At present, these policies are applied equally in various developing countries. The driving forces of these policies have been the need to ensure the supply of energy and climate change mitigation by reducing emissions of greenhouse gases in conjunction with the desire to support agriculture and promote rural development (World Bank 2007a). These worries have even more relevance in an international context.

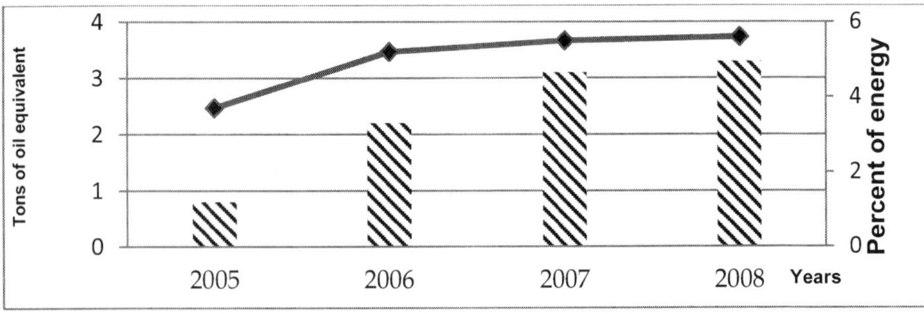

Fig. 2. Percentage of total energy demand for transport CEPAL-FAO (2007).

However, the role of biofuels in the solution for these problems with adequate policies for their application, are subject to more rigorous examination. Because the current policies are costly, their coherence and foundations are being questioned. Current subsidies for biofuels are high and have a limited role in the world supply of energy. The estimates made by the Global Subsidies Initiative for the United States and other countries of the OCDE and a large part of South America suggest the maximum level of support for biodiesel and ethanol in 2006 was between 11,00 and 12,000 million USD (Steenblik, 2007). In dollars per liter, the support fluctuates between 0.20 and 1,000 USD. With the increase in production levels, costs could also increase. It is possible to argue that the subsidies are only temporary, depending upon the long-term economic viability of biofuels. This will also depend upon the cost of other sources of energy, like fossil fuels, or, in the long-term, alternative sources of renewable energy. If we take into account the recent increase in the price of oil, of the larger producers, only the sugar cane ethanol of Brazil appears to be able to compete against fossil fuels without subsidies. Direct subsidies, nevertheless, represent only the most evident costs. Other costs are the result of a disproportionate allocation of funds, a consequence of select support for biofuels and the use of quantitative instruments for mixing.

It is difficult to identify the pertinent policies and quantify their effects in specific cases given the variety of normative instruments and the way they are applied. Nevertheless, policies can influence the economic attractiveness of its production, commerce and use. Subsidies can affect this sector at different stages. **Table 1**, adapted from the Global Subsidy Initiative (Steenblik, 2007), shows the different ways direct and indirect measures can help along the chain of biofuels production. At the same we can see that the policies cover the entire biodiesel chain, from raw material production to distribution and end use. Some of these factors are interrelated so applying policies to one category or another can be risky without considering the international context.

The policies applied, as previously mentioned, are based on quantitative and qualitative instruments **(Table 1)** which are a combination of mandates, direct subsidies, tax exemptions and technical specifications. They span the entire chain of production and commercialization of the biomass of biofuels, final use and international commerce. However, while these policies are interrelated in practice they are i confusing and inadequately implemented.

It is believed that the policies and help directed towards the levels of production and consumption are distorting the market most significantly, while help for research and development most likely distort the market less. In **Figure 3** the repercussions of eliminating biofuels policies on production and consumption of biodiesel are summarized, which distort commerce in several countries.

	Biomass production	Biofuel production	Biofuels use	Biofuels market
Quantitative requirements			Mixing duties.	Import quota.
Qualitative requirements	Obligations of land for biofuel production.	Quality standards.		Fuels estandards.
Financial incentives	Payment for energy crops. General measures of agricultural support.	Grant loans. Investment support. Public research in to the conversion process.	Tax concessions. Tax concessions for the sale of biofuel-compatible vehicles. Public research in development	Import tariffs.

Table 1. Operations and activities directly affected by the policies applied from production to market for biodiesel. Adapted from Steenblik, 2007.

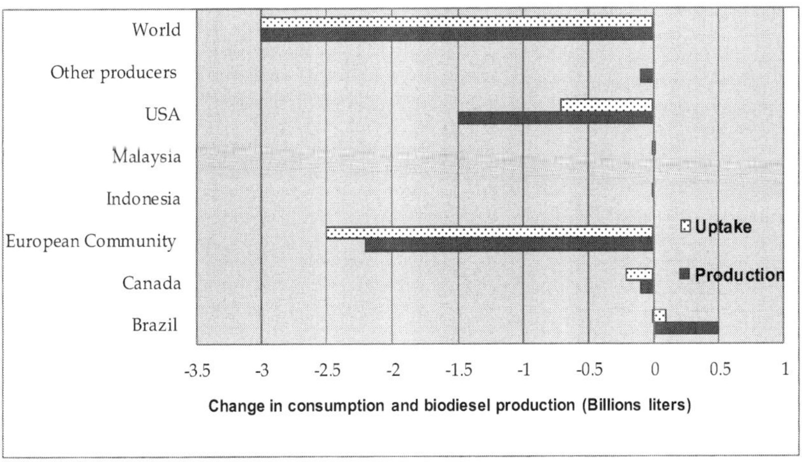

Fig. 3. Total impact of the elimination of policies that distort trade in biodiesel.

The elimination of tariffs and subsidies entail a decrease in the world production and consumption by 12% approximately. This would actually make it more competitive in the market, contributing to economic sustainability (Von Braun, J. *et al.* 2008). The European Community would be the most affected by this change. In contrast, Brazil would maintain a stable level of production and consumption since the biofuels market in that country is competitive. These data are consistent with other studies on the issue raised.

Decisions such as increasing export tariffs and withholding inventory, even when they increase the supply in a given country or region, can have a negative impact on the international offer, depending on the country's involvement as producer and exporter and the scale of fees or deductions. The barriers to biodiesel trade are summarized in **Table 2.**

	Tariff barriers	Non -tariff barriers
Trade Barriers	Stepping rate	Domestic support
	Contributions	Technical Standards

Table 2. Trade barriers of biodiesel. Dufey, 2006.

There is currently no specific customs classification for biodiesel, this biofuel in the form of esters fatty acid methyl (FAME) is internationally classified under HS code 3824 9099. 72 73 However, in neither case is it possible to establish whether the imported FAME is used as biofuel or for any other purpose. The evidence also shows that an application fee is common practice in many countries. In the U.S. a fee of 6.5 percent for biodiesel is classified under HS code 3824 909 976, the EU (European Union) applies a tariff of 5.1 percent for biodiesel from the U.S. Moreover, there are substantial tariffs on imports of raw materials for biodiesel production, including energy crops, especially on other materials with added value such as oils and molasses, and the use of tariff escalation and the use of quotas to regulate trade.
Another important trade barrier is domestic subsidies, which hinders the competitiveness of biodiesel, and the existence of divergent technical regulations in different countries. These can cause conflicts and costs for producers who wish to enter multiple markets, each with different standards.
The higher production costs of biofuels compared to conventional fuels, together with the existence of positive externalities associated with biofuel policies suggest that support could be justified to assist the development of industry in its early stages. However, the way that these policies should take and the time in which they should be implemented are issues that require further analysis.

4. Food safety

In addition to the environmental advantages of biodiesel (Marchetti *et al.*, 2007), there is a debate about the quantity of land available to cultivate biomass, in a world market with mostly first generation biodiesel. This product could compete with the availability of food, but at the same time, give farmers new and growing opportunities.
According to the definition from the Food and Agriculture Organization (FAO), "Food safety exists when all people have physical and economic access to sufficient innocuous and nutritious food to satisfy their nutritional needs". Food safety implies compliance with the following conditions:

- An adequate supply and availability of food.
- The stability of the supply without fluctuations or shortages because of the season.
- Access to food or the ability to acquire it.
- Good quality and innocuous food.

Food safety is studied in the following way (Rodriguez, 2007):
Food use: including the social value, nutritional value of the food in each region and harmlessness of the food.

Availability of food: local production, distribution and exchange (import/export). There is safety in terms of food availability nationally, when food resources are sufficient to provide an adequate diet every person in this country, regardless of the origin of the food.
Access to food: Ability to purchase, preferences and mechanisms of allocation.
The relationship between the production of biofuels and food safety is a complex topic. One of the main worries is the possible conflict of land and water use, which could have negative repercussions, since more than 50% of the impoverished population of Latin America and the Caribbean live and depend upon the rural sector (Robles & Torero 2010).
The fact that the demand for grains has increased in the last few years, while supply has decreased, has many countries worried (Heady & Fan 2008). One measure of vulnerability is the number of countries which need food assistance (FAO, 2011). In 2008, 36 countries (FAO, 2010) required external assistance because of an exceptional debt, food production/supply, general lack of access or a focused food danger. The large scale production of biodiesel in these regions without an adequate policy means more pressure. However, it should be noted that most of these countries are not exporting grains nor are they biodiesel producers, so there is not causal effect of the deflection of grain into the fuel market in this countries. With this situation and the high price of food **(Figure 4),** countries have taken measures to reduce tariffs and subsidize food.
The observed measures have weak points, above all the subsidies which are dependent upon the economy of each region and for that reason are ambiguous according to production, per capita income, etc. (CEPAL, 2008). Subsidies are not a solid foundation since it is probable that, with time, they will be discontinued.
Current technologies for liquid biofuel production, such as biodiesel and ethanol, are used as raw material in basic agricultural products. Biodiesel is based on various oleaginous crops, whose large scale production entails considerable land, given the volume of raw materials and the related needs for production.
If the price of combustibles is high enough, agricultural products can be excluded from other uses. Given that the energy markets are larger than the agricultural markets, a small change in energy demand can mean an obvious variation in the demand for raw agricultural materials, and as a consequence, the prices of crude drive the price of biofuels and, at the same time, influence the price of agricultural products (Schiff, 2008). The relationship between the price of food and the price of oil is more obvious than the relationship between

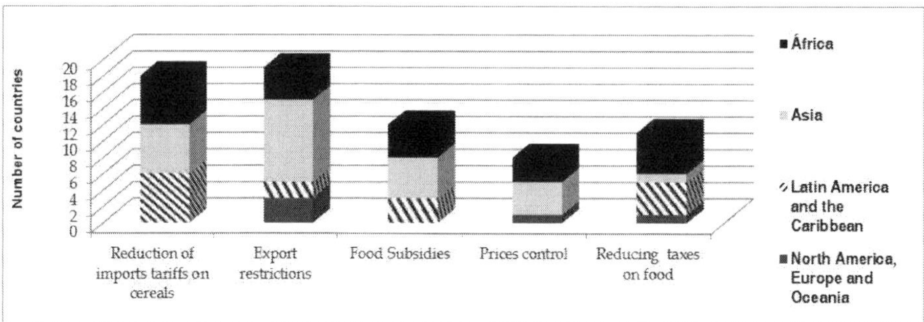

Fig. 4. Measures in response to high food prices by region. Data adapted from the World Food Program, United Nations (2009).

biofuel and agricultural products, leaving biofuels between the two. The narrow link between the price of crude and the price of agricultural products, through the demand for biofuels, establishes minimum and maximum prices for agricultural products determined by the prices of crude (FAO, 2006a). When the prices of combustible fossils reach or surpass the cost of production of substitutive biofuels, the energy market creates a demand for agricultural products. If the demand for energy is high in relation to the agricultural product markets and raw agricultural materials for the production of biodiesel are competitive in the energy market, there will be a minimum price effect for agricultural products, determined by the prices of fossil fuels. However, agricultural prices cannot increase simultaneously faster than the price of energy, since that would raise prices in the energy market.

The situational factor that has played a leading role in the sharp increase in food prices between 2007 and 2008 was financial speculation, which has injected millions of dollars in the futures markets for basic grains as a safe bet in these times of economic uncertainty, private investors and pension funds have drawn wealthy financial market investments, real estate funds in U.S. dollars and developing economies, and have gotten into commodity funds, investments and agricultural futures market (Von Braun, J. *et al.*, 2008). Investment in agricultural futures markets has had a very prominent speculative, even through this market only represents 10% of the grain traded in the world (Per Pinstrup, 2000).

Factors influencing the agricultural market and the determination of the price of food, which also depend upon supply and demand, are listed below:

a. **Climate variability**: The most recurrent source of price variability in agriculture has historically been the supply shocks caused by extreme weather events. According to OFDA / CRED International Disaster Database (EM-DAT), the frequency of floods and droughts have increased dramatically between the first half of last century and this decade. These climatic factors have led to crop losses worldwide, prompting not only fluctuations in the prices of agricultural product prices, but also famine in the most vulnerable regions.

b. **Public Policy:** Decisions, such as increasing export tariffs and withholding inventory, even when they increase the supply in a given country or region can have a negative impact on the international offer, depending on the country's involvement as producer and exporter and the magnitude of tariffs or withholding.

c. **Changes in income**: Decreases in income can occur abruptly, either because of economic crisis, the reduction of social programs or both. The effect on price volatility in these cases will be different depending on the type of product, since the income elasticity of demand for agricultural products varies considerably between products.

d. **New uses for agricultural products**: The discovery of new uses for agricultural products, driven by technological developments (such as biotechnology applied to agriculture) and social or ideological changes are other factors that can, at least in theory, lead to pressure on demand in the short term (Trostle, 2008). Although these changes are gradual, they often have incentives (such as law, investment decisions of large companies or public policy) that ultimately define their economic viability and make their effective introduction into the market. These incentives can generate volatility in the markets.

e. **Effects of foreign exchange markets and oil:** Exchange rate and international prices of agricultural products also have an effect, commonly given in U.S. dollars, they are subject to the appreciation or depreciation of that currency. In that sense, Shaun

(2010), analyzing the factors that determine changes in the longer term (over one cycle) in the volatility of international food prices, found positive and statistically significant changes.

The effects on the world prices of wheat, rice, corn, vegetable oils and sugar, in relation to the consistent reference in the maintenance of raw materials for biofuels in reported 2007 levels are reflected in these **figures 5 and 6.**

With a 14% reduction in the use of raw materials for biofuels from 2010-2011, world prices would be lower by 5% for corn, 3% for vegetable oils and 10% for sugar. By contrast, an increase in the use of raw materials for biofuels of 30% would result in an increase, but on a small scale. The sugar price would increase by 5% and between 2% and 6% for maize and vegetable oil. Since the biodiesel market represents not even 1% of global energy market. Therefore the actual impact is low. The results show little variation compared with the data

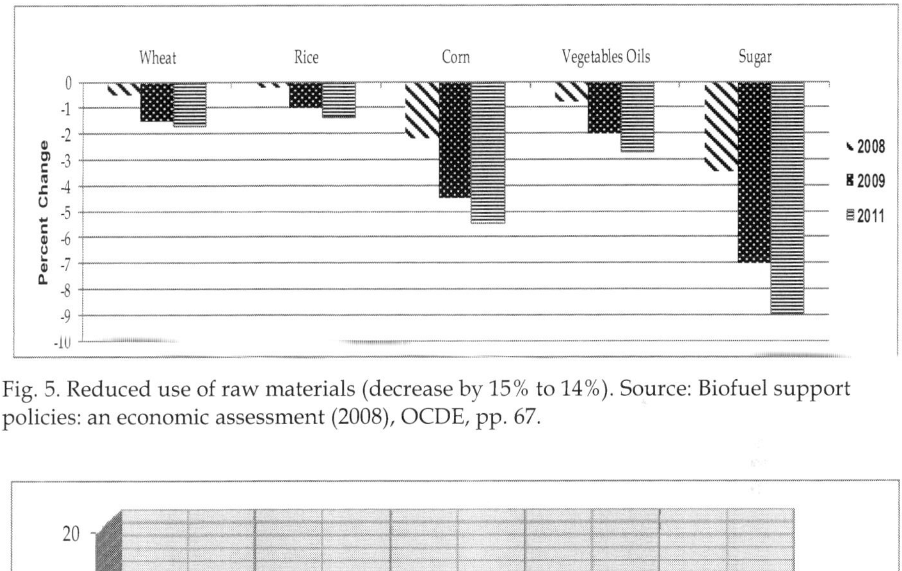

Fig. 5. Reduced use of raw materials (decrease by 15% to 14%). Source: Biofuel support policies: an economic assessment (2008), OCDE, pp. 67.

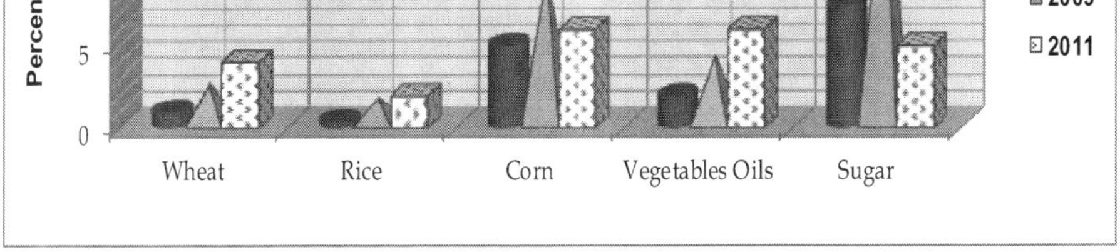

Fig. 6. Increased use of raw materials (increase of 30% in biofuels by 2010).

published by FAO 2008. The same may be due to methodological differences in the estimates this paper; the proper levels of uncertainty have been taken into account, such as agricultural markets, weather conditions, tariffs, etc.

The agricultural market has some characteristics that can be seen as an additional risk in the potential increase in food prices in the coming years or the retention of current levels of volatility for a longer period than those applied in past episodes. These market characteristics not directly attributable to biodiesel (Cotula *et al.*, 2008).

In practice, it is possible that the link between the prices of agricultural products and energy may not be so close at least until biofuel markets are sufficiently developed. And although an increase in biodiesel production worldwide is expected there is no significant increase in trading and cost, according to the outlook through 2017 **(Figure 7).** That is, it tends to remain stable so that the influence on the food market may be representative but not influential on a massive scale. Studies in recent years have analyzed the causes of the crisis in agricultural commodity prices in 2007-2008 (Heady & Fan, 2008, Mitchell, 2008, World Bank, 2008, Robles et al., 2009; Baffes & Haniotis, 2010, Sinnott et al., 2010, Shaun, 2010). Since many of the cases analyzed are structural in nature, different studies focus on analyzing more or less homogeneous factors, seen as potential causes of that crisis. It is important to note, however, that most studies make a qualitative, not quantitative diagnosis, these factors, and even present empirical results should be consulted with caution. On one side are supply-side factors such as climate variability and public policy and on the other hand, those related to demand such as changes in revenues and new uses for agricultural products in relation to the currency market (oil prices).

In the short term, the ability for response from the biofuel sector to the changes in prices relative to fossil fuels and agricultural products can be limited to a group of obstacles. Some examples are: dysfunction in distribution, technical problems during transport and mixture systems or the inability of factories to transform raw materials. The more flexible the capacity for response to demand and the signs of changing prices, the closer the link will be between the price of energy and agricultural markets

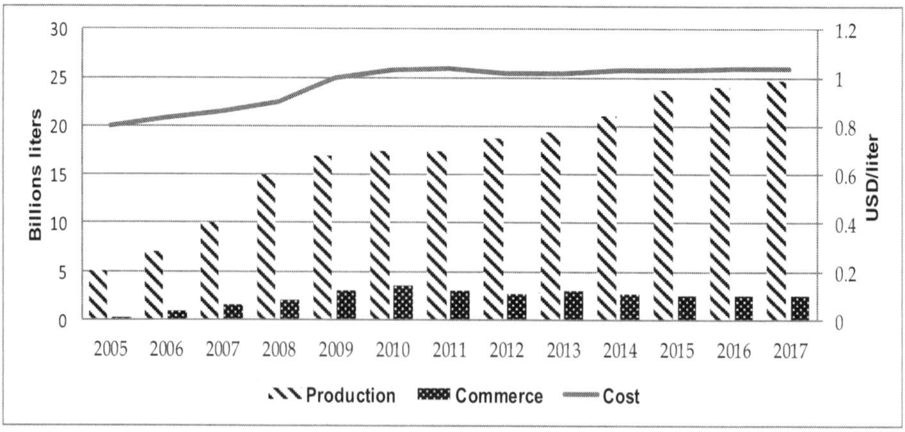

Fig. 7. Production, trade and world prices for biodiesel.

Joachim Von Braun details various factors which have influenced the increase in the price of agricultural products, together with a reaffirmation of the importance of launching biofuel. He adds additional factors like the high rate of growth experimented in Southern Asia which was close to 9% between 2004 and 2006, as well as an important growth rate in Africa which was 6%. Of the 34 countries with the most food safety problems, 22 of them had even more problems during those same years at a rate of 5 and 16%. This growth represents strong pressure on the demand for food and in countries with less income. The growth in these numbers translates to a higher demand for food. Also consider the volatility in agricultural commodity markets which has important economic implications for countries that specialize in export. Using price data from the eighteenth century, Jacks, *et al.*, (2009) concluded that the volatility of commodity prices has been higher than the prices of manufactured products. Thus, the dependence on few export commodities is a fundamental cause of instability in terms of the trade of countries that specialize in production and consequently greater economic vulnerability to which they are exposed to this excluding biodiesel.

At the same time, Manuel Chiriboga emphasizes the strong increases in the demand for food in China and India. In fact, he states that in China the average incomes increased 8 times in last 25 years. A strong change towards urbanization and the expansion of the middle class provoked changes in consumption patterns. At the same time, the per capita consumption of food grew by 30% in the last few years making China the third largest importer of food in the world, after the United States and Japan. The world production of cereals also decreased by 2.4% in 2005 because of, climate problems and a decreased of area production in countries who are main exporters of cereals. The strong presence of investors speculating on these products has also influenced the increase in the price of cereals, as well as commodities in general.

It is also important to mention that the increase in the price of food has underlying causes like: the increase in the price of oil, speculations about the market and the growing demand for grains. According to the United Nations, the cause of hunger is inequity, not the lack of food.

5. Conclusions

The criteria for economic, environmental and social sustainability should be a fundamental part of any analysis of biofuel policies. Exhaustive research is needed to identify practices for sustainable management, technological options and the environmental and social impacts at various levels of biofuels production. Guaranteeing energy sources without comprising food sources means raising rents agricultural, while at the same time reducing financial aid and subsidies. Although there are special tariffs, barriers and subsidies in several countries, the international trade of biofuels benefits from preferential schemes through trade agreements, mainly from two major importers, such as the U.S. and the European Union (EU).

While the political pressure to produce biofuels has been considerable, there are no incentives or norms which guarantee the use of new and innovative technologies to avoid the substitution of food crops.

Energy prices have been influenced for a long time by the prices of agricultural products due to the importance of fertilizers and machinery as inputs in production processes. The trend of rising food prices is positively correlated with the increase in oil prices, not increasing production of biofuels directly, because biofuels represent only 0.3% of total world energy supply.

Biofuels should be considered within a larger context. Biofuels are only part of the solution to the energy problem and should remain in that role. The development and production of biofuels should be accompanied by other alternative energy measures like the reduction of consumption and the improvement of technology. Sustainable production of biodiesel can be an opportunity for rural development and a form of clean energy when considering the appropriate economic and environmental policies. The existence of trade barriers, both tariff and non tariff is a key issue. The further liberalization of trade in biofuels is threatened by the lack of a comprehensive multilateral trade regime applicable to biofuels, which means that business conditions vary from country to country. This scenario is further complicated by the vast number of products involved in the trade - from the different types of raw materials (energy crops) to the final product (biofuels) - passed by a wide range of semi-processed products.

Available data show that economic policies are not the most appropriate and create distortions in the market. It highlights the need for sustainability criteria in each country at the same time, international trade regulation to ensure social acceptability, economic viability and environmental quality.

6. Acknowledgment

The Organization of American States (OAS) for supporting our research. Ricardo Duarte, research economist, for his contribution to the study data. And organizations mentioned in the chapter allowed free access to their databases.

7. References

Baffes, J. & Haniotis, T. (2010). Placing the 2006/08 Commodity Price Boom into Perspective. Policy Research Working Paper 5371. Washington, DC, The World Bank.

Bello, O. Cantú, F. & Heresi, R. (2010). Variabilidad y persistencia de los precios de productos básicos. Serie Macroeconomía del Desarrollo No. 105. Santiago, Chile, CEPAL.

Block, D., Thompson, M.; Euken, J.; Liquori, T.; Fear, F. & Baldwin, S. (2008). *Engagement for transformation: Value webs for local food system development.* Agric Hum Values, Vol. 25, pp. 379-388, (Jan., 2008). Doi: 10.1007/s10460-008-9113-5.

CEPAL-FAO (2007). Oportunidades y riesgos del uso de la bioenergía para la seguridad alimentaria en América Latina y el Caribe. CEPAL y FAO, Santiago, Chile.

CEPAL (2008). La volatilidad de los precios internacionales y los retos de política económica en América Latina y el Caribe. Santiago de Chile, CEPAL.

CEPAL (2010). Panorama de la inserción internacional de América Latina y Caribe, 2009-2010. Santiago de Chile, CEPAL.

CME Group (january 2011). Monthly Agricultural. December 2010. Chicago, Illinois, CME Group.

CEPAL-FAO-IICA (2009). Perspectivas de la Agricultura y del Desarrollo Rural en las Américas: una Mirada hacia América Latina y el Caribe. Accessed on March 18, 2011. Available in: *http://www.agriruralc.org*

Cotula, L., Dyer, N., & Vermeulen, S. (2008). Fuelling exclusión? The biofuels boom and poor people´s access to land. IIED and FAO, London.

Clover, J. & Eriksen, S. (2009). The effects of land tenure change on sustainability: human security and environmental change in southern African savannas. Journal Environmental Science & Policy, Vol. 12 (2009). Pp. 53-70. Doi:10.1016/ j.envsci.2008.10.012.

FAO (2011). Guide for Policy and Programmatic Actions at Country Level to Address High Food Prices. Accessed on March 24, 2011. Available in: *http://www.fao.org/fileadmin/user_upload/ISFP/ISFP_guide_web.pdf*

FAO (2010). Agricultural futures: Strengthening market signals for global price discovery. Extraordinary joint intersessional meeting of the intergovernmental group (IGG) on grains and the intergovernmental group on rice; Committee on commodity problems. Rome, Italy, September.

Heady, D. & Fan, S. (2008). Anatomy of a Crisis, The Causes and Consequences of Surging Food Prices. Discussion Paper 00831. Washington DC., International Food Policy Research Institute.

Hyman, G., Larrea, C. & Farrow, A. (2005). *Methods, results and policy implications of poverty and food security mapping assessments*. Journal Food Policy, Vol. 30 (2005). Pp. 453-460. Doi:10.1016/j.foodpol.2005.10.003.

IEA – International Energy Agency. (2004). *Biofuels for Transport: An International Perspective*, OECD Publications for the International Energy Agency, Paris, 2004.

Jakobsen, J., Rasmussen, K., Leisz, S., Folving, R., & Vinh Quang, N. (2007). *The effects of land tenure policy on rural livelihoods and food sufficiency in the upland village of Que, North Central Vietnam*. Journal of Agricultural Systems, Vol. 94 (2007). Pp. 309-319. Doi:10.1016/j.agsy.2006.09.007

Mitchell, J. (2008). A note on rising food prices. Policy Research Working Paper n° 4682. World Bank, Washington DC.

OECD and FAO, Organization for Economic Development and Cooperation (2007). *Agricultural Outlook 2007-2016*, Organization for Economic Development and Cooperation, Paris, and the U.N. Food and Agricultural Organization of the United Nations, Rome, 2007.

Per Pinstrup, A. (2000). *Food policy research for developing countries: emerging issues and unfinished business*. Food Policy, Vol. 25, Issue 2, (Abr., 2000), Pp. 125-141.Doi:10.1016/S0306-9192 (99)00088-3

Renzaho, A. & Mellor, D. (2010). *Food security measurement in cultural pluralism: Missing the point or conceptual misunderstanding*?. Journal l of Hunger & Environmental Nutrition, Vol. 26, (2010), pp. 1-9. Doi:10.1016/j.nut.2009.05.001.

Robles, M & Torero, M. (2010). Understanding the Impact of High Food Prices in Latin America. Economia, Revista de la Latin American and Caribbean Economic Association (LACEA). Vol. 2. Brookings Institution Press.

Rodríguez, A. (2008). Análisis de los mercados de materias primas agrícolas y de los precios de los alimentos. Unidad de Desarrollo Agrícola, CEPAL, en base al Banco Mundial, Commodity Price Data.

Schiff, R. (2008). *The Role of Food Policy Councils in Developing Sustainable Food Systems*. in: Journal l of Hunger & Environmental Nutrition, Vol. 3, No. 2 & 3 (August., 2008), pp. 206 - 228. Doi: 10.1080/19320240802244017.

Sinnott, E.; Nash, J. & de la Torre, A. (2010). Los recursos naturales en América Latina y el Caribe ¿Más allá de bonanzas y crisis? Estudios del Banco Mundial sobre América Latina y el Caribe.

Steenblik R, (2007). Biofuels – At What Cost? Government support for ethanol and biodiesel in selected OECD countries. The Global Subsidies Initiative (GSI) of the International Institute for Sustainable Development (IISD) ISBN 978-1-894784-03-0. Geneva, Switzerland. Accessed on May 5, 2011. Available online: in http://www.globalsubsidies.org/files/assets/oecdbiofuels.pdf

Trostle, R. (2008). Global agricultural supply and demand: factors contributing to the recent increases in food commodity prices. WRS-0801. Economic Research Services, USDA, Washington DC.

Torero, M. (2010). Agricultural price volatility: prospects, challenges and possible solutions. Presentation in Seminary "Agricultural price volatility: prospects, challenges and possible solutions", may 26 - 27, 2010, Barcelona, España. Accessed on April 25, 2011. Available in: *http://www.agritrade.org/events/2010Spring_Seminar_AgPriceVolatility.html*

UNCTAD. (2009). The global economic crisis: Systemic failures and multilateral remedies. Report by the UNCTAD secretariat task force on systemic issues and economic cooperation, New York and Geneva, 2009.

Von Braun, J. (2008). High food prices: the what, who, and how of proposed policy actions. IFPRI Policy Brief, May. Washington, DC.

World Bank (2011). Global Economic Prospects, January 2011. Washington, DC, The World Bank.

World Bank (2008). Global Economic Outlook, Chapter 5: Globalization, Commodity Prices and Developing Countries. Washington, DC, World Bank.

World Bank (2007). World Development Report 2008: Agriculture for development. Washington, DC, World Bank.

World Bank (2007). Agriculture for Development. World Development Report. Washington, DC,The World Bank. Accessed in March 30, 2011. Available in: *http://siteresources.worldbank.org/INTWDR2008/Resour ces/WDR_00_book.pdf*

9

Current Status of Biodiesel Production in Baja California, Mexico

Gisela Montero[1], Margarita Stoytcheva[1], Conrado García[1], Marcos Coronado[1], Lydia Toscano[1,3], Héctor Campbell[1], Armando Pérez[1] and Ana Vázquez[1,2]

[1]*Institute of Engineering, UABC*
[2]*School of Engineering and Business, Guadalupe Victoria, UABC*
[3]*Technological Institute of Mexicali*
Mexico

1. Introduction

As a result of declining oil reserves in the world, the rise in fossil fuel prices and growing interest in the environment, there is considerable demand for alternative fuels. Biodiesel is recognized like the "green fuel" and has several advantages compared to diesel. It is safe, renewable, nontoxic, and biodegradable (98% biodegradable in a few weeks). Contains less sulfur compounds, and has a high-flash point (> 130°C). Biodiesel could replace diesel and can be used in any compression ignition engine without modification techniques (Leung et al., 2010). It is an alternative biofuel which has a positive energy balance in their life cycle. In terms of effective use of fossil energy resources, biodiesel yields around 3.2 units of fuel product energy for every unit of fossil energy consumed in the life cycle. By contrast, petroleum diesel's life cycle yields only 0.83 units of fuel product energy per unit of fossil energy consumed (Kiss et al., 2006).

Chemically, biodiesel is a mixture of methyl esters of long chain fatty acids and is formed from vegetable oils, animal fats or waste oils and fats through transesterification in the presence of a catalyst (Ma & Hanna, 1999). A general equation for the transesterification (where R is the remainder of the molecule of triglyceride, fatty acid R_1 and R_2 is the length of acyl acceptor) is:

$$RCOOR_1 + R_2OH \xleftrightarrow{\text{Catalyst}} RCOOR_2 + R_1OH$$

2. Regulations on biofuels in Mexico

The government of Mexico initiated a series of measures to create an internal market for biofuels in order to increase efficiency levels in end-use energy and to reduce greenhouse emissions gases. On August 22nd, 2005 was published the Law of sugarcane sustainable development, which contain guidelines for the use of sugarcane as energetic.

In early 2007, the Mexican Congress promulgated the Law of Promotion and Development of Bioenergetics, which came into force on February 1st, 2008. Its purpose was the promotion and development of bioenergetics in the Mexican agriculture without jeopardizing food security and sovereignty of the country and to ensure the reduction of pollutant emissions to the atmosphere and greenhouse gases, considering international instruments contained in the treaties that Mexico has signed.

The biofuels development in Mexico according to the law and studies of Secretaría de Energía (Secretariat of Energy) starts from two raw materials with high levels of production in the country (sugarcane and corn yellow). In Article 11 of this Law Section VIII, it is stated the granting of permits for the production of biofuels from corn by the Secretariat of Agriculture, Livestock, Rural Development, Fisheries and Food, as long as there is overproduction.

Along with the development of legislation, Mexico undertook a project to determine the feasibility of liquid biofuels called "Potential and Feasibility of using Biodiesel and Bioethanol in Mexico Transport Sector" where the test result indicates that economic production of ethanol from sugarcane or corn is suitable as long as the ethanol price is between 0.55 and 0.65 U.S. dollars. The inputs considered in this study were sugarcane, maize, cassava, sorghum and sugar beet. In the case of sugarcane, it was analyzed the production of ethanol from sugarcane bagasse.

In this project, it was assessed the production of biodiesel from rapeseed, soya, jatropha, sunflower and safflower oils, and the use of animal fat and waste vegetable oil. The results suggest that farm input costs represent between 59% and 91% of biodiesel production costs and, as a result, animal tallow and waste vegetable oil are an opportunity for biofuels production (SENER, 2006b).

As for biofuels commercialization in Mexico the first steps were taken in 2009 when Secretaría de Energía (Secretariat of Energy) gave the first 12 permits of anhydrous ethanol commercialization to participate in the tender that Petróleos Mexicanos (Mexican Petroleum) issued for the supply of anhydrous ethanol in the metropolitan area of Guadalajara (SENER, 2009).

3. Energy situation in Mexico

The primary energy production in Mexico relies mainly on oil and natural gas with a share of 61.5% and 28.2% in 2009 respectively. Renewable energy sources are next in importance, with a contribution of 6.2%, wherein the biomass stands out more than half of that value. The biomass considered by the Secretariat of Energy in the national balance sheet only includes wood and sugarcane bagasse. The remaining 4.1% is made up of coal, nuclear and condensed (see Fig 1).

The entities involved and empowered by the federal government, to ensure and guarantee the energy supply in Mexico are Petróleos Mexicanos (Mexican Petroleum) and Comisión Federal de Electricidad (Federal Electricity Commission).

3.1 Energy situation in Baja California

Baja California is located in the northwestern region of Mexico on a peninsula that bears his name, bordered on the north by the State of California, USA, on the east by the Gulf of California and the west by the Pacific Ocean. It presents dry and warm weather. Its land area is 71,576 km^2 (3.6% of the country) and has a population of 3.3 million inhabitants (3% of the total population of Mexico). Baja California is made up of 5 cities: the capital is Mexicali, Tijuana, Tecate, Ensenada and Playas de Rosarito. The GDP is above the national average. From the economic point of view, it is characterized by a high industrial growth,

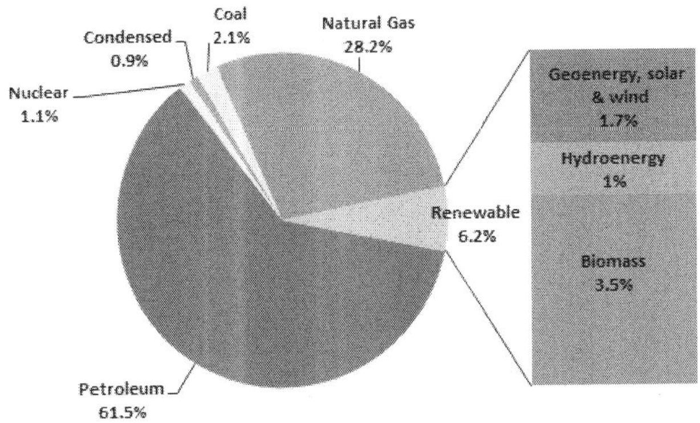

Fig. 1. Structure of primary energy production in México, 2009.

agriculture and livestock. Agriculture and livestock are intensive activities in the region, leading to the generation of large amounts of waste biomass such as animal fats and agricultural residues (wheat straw and cotton waste). Some of them are open burned *in situ*, while others are used in the production of food for livestock.

In the energy situation has primary energy sources for electricity generation, such as geothermal located in Cerro Prieto, with an installed capacity of 720 MW and wind in La Rumorosa located in the municipality of Tecate with an installed capacity of 10 MW. Besides power plants and turbo gas types exist in cities across the state and run with fuel oil and/or natural gas, its total capacity are 1,305 MW and 316 MW respectively. It is appropriate to mention that currently the Baja California's electrical system is isolated from the national grid and interconnected with the United States of America. On the other hand, it has no particular oil resources, so the fuels come the region from southern Mexico and arrive by tanker to Baja California to the Rosarito Beach, situated on the Pacific coast.

In particular may be noted that throughout the year in Mexicali weather conditions are extreme, with temperatures ranging approximately from 0°C to 50°C as shown in Fig. 2, which involve high-energy requirements to ensure physical comfort of its inhabitants.

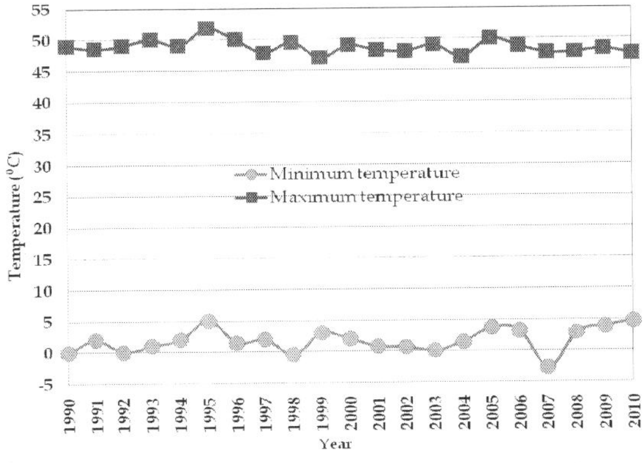

Fig. 2. Minimum and maximum temperatures of Mexicali, 1990-2010.

In 2006, the average per capita consumption of electricity in Mexico was 75% of the world average of 2,659 kWh, while Baja California and Mexicali exceeded it that by 21% and 117% respectively. As shown in Fig. 3, (adapted from Campbell et al., 2011) the annual per capita electricity consumption of Mexicali was comparable to that of Italy and was ranked ahead of countries like Brazil and Chile.

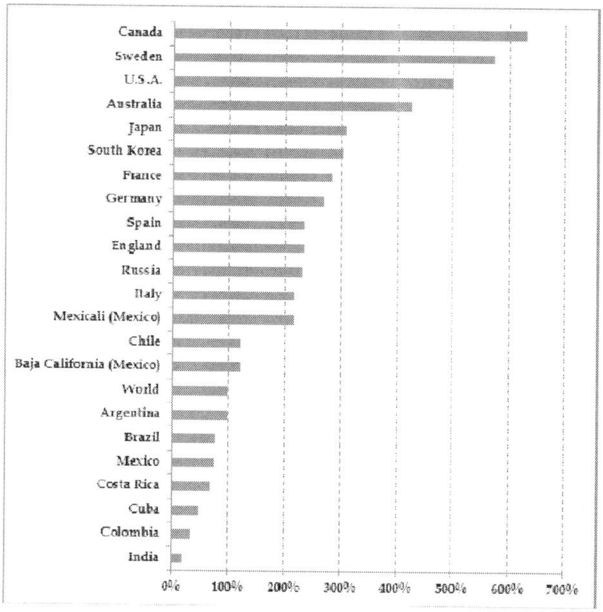

Fig. 3. Relative average annual consumption per capita 2006, World: 2,659 kWh, 100%.

Besides the high-electricity consumption, the fuel requirements of different services sectors should be meet which are supplied with fuel from Southern Mexico, as is the case of diesel.

3.2 Consumption of diesel in Baja California

Baja California has 3 diesel outlets in the cities of Mexicali, Ensenada and Rosarito Beach catering to other locations in the region. The average sales volume of diesel was 717,211 m^3 in the period 2000-2010, as presented in Fig. 4.

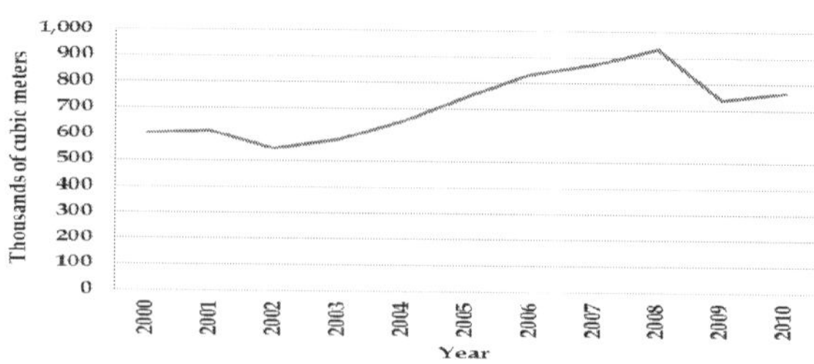

Fig. 4. Diesel sales volume of Baja California, period 2000-2010.

It can be seen that the volumes of diesel are high, and for that reason, it is proposed its replacement with biodiesel obtained from different raw materials available in the state.
With respect to biofuels such as biodiesel, currently its production is inexistent in commercial scale, in North Mexico and incipient in the rest of the country, as shown in the map of Fig 5, which presents the current state of production facilities of biodiesel in Mexico. In Baja California are only reported biogas projects. (Adapted from REMBIO, 2011).

Fig. 5. Bioenergy projects in México.

4. Feedstocks availability in Baja California

In order to evaluate the resources available in the state of Baja California for processing and use as biofuels, Biofuels Group of the Institute of Engineering of the Universidad Autónoma de Baja California, started from 2008, a series studies to identify the biomass resources and determine their potential use as feedstock for the production of biofuels. The materials considered were animal fats, waste vegetable oil (WVO), castor oil, *jatropha curcas* and agricultural waste.

4.1 Waste vegetable oil

The Waste Vegetable Oil (WVO) is highly available resource in the state, as well as in the rest of the country, and its amount varies depending on the demand for edible vegetable oil. Its generation is closely linked to the food preparation processes in various sectors: a) the restaurateur, b) food industrial and c) domestic. Traditionally, the WVO has an inadequate disposal and it is directly discharged to the sewage system and illegal dumping. This creates problems of clogging of the sewage system, soil and water pollution and increased maintenance costs and wastewater treatment.
The WVO is that oil that has been altered physical-chemical properties due to its use in a batch or continuous processes of food preparation. Mexican standard NOM-052-SEMARNAT-2005

establishes the characteristics, the procedure of identification, classification and listing of hazardous waste and does not include the WVO as hazardous waste.

The WVO for its high energy content, about 30 MJ/kg (Talens et al., 2006), is likely to be reused as raw material for bioenergy production or other manufacturing processes such as production of soap. Currently a fraction of the WVO generated in Baja California is collected by companies certified from the appropriate authorities, and sell the residues to their end use as food for beef cattle. Other companies choose to export the WVO to the United States, where it is used as feedstock in biodiesel production and eventually that biodiesel is acquired by Mexico for use in vehicles, machinery and equipment of the Comisión Estatal de Servicios Públicos de Tijuana (State Public Services Commission of Tijuana). So then, it is presented a scenario where a valuable resource from the energy standpoint, it is not processed locally, and instead is exported as raw material and imported as a finished product, missing the economic, environmental and social development in the region.

There are several reasons why it is appropriate to promote the development of biodiesel from WVO, among which it can be mention, the following (Canacki & Gerpen, 2001; Gerpen et al., 2006):

- Represents a sustainable method where is revalued and reused a resource with a high-energy content, to produce a cleaner fuel. This will no longer discard a valuable resource from the energy standpoint, and at the same time is greatly benefited the environment and society in general.
- Avoid the use of edible oil crops in the production of biodiesel, so it does not risk food security because it is the reuse of a waste.
- It is an opportunity to mitigate the environmental impact caused by emissions of greenhouse gases responsible of global warming.
- It is an opportunity to diversify the energy matrix, traditionally based on fossil fuels.
- Reduces dependence on fossil fuels.
- Emissions of carbon dioxide are integrated to the carbon cycle of plants from which oils are extracted.
- It has excellent lubricating properties for the diesel engines motors.
- They come from a renewable resource.
- They are biodegradable.

In 2008, was estimated a generation of 2.1 million liters of WVO in the restaurant sector of Mexicali city, capital of Baja California (Coronado, 2010). The results indicate that the types of foods that have greater participation in this generation are: fast food, international food, mexican and china food.

The Fig. 6 shows that 100% of oil used in food preparation 41% is consumed in food or disposed of in cookware, while the remaining 59% becomes WVO. Also, shows that 59% of WVO has different destinations: 33% are collected by companies engaged in such activity, private collectors 16%, 8% were discharged to municipal sewage system and 2% are donated to be reused in food preparation.

On the other hand, it was realized the spatial distribution of restaurants in Mexicali by using a satellite Geo positioning obtaining the geographic coordinates in UTM Zone 11 N of each food preparation facility and place them on a satellite image that is illustrated in Fig. 7.

Usually there is a greater density of restaurants in commercial areas and main avenues. It was confirmed the existence of a relationship between the density of restaurants with a higher incidence of the problem of clogging in the sewage system of the city, due to the discharge of the WVO.

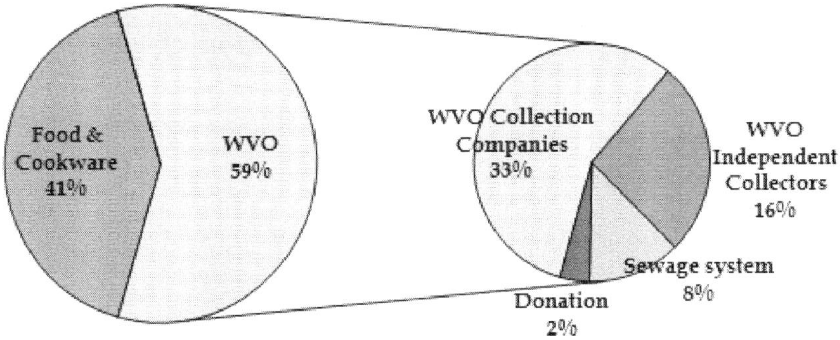

Fig. 6. Waste vegetable oil disposal.

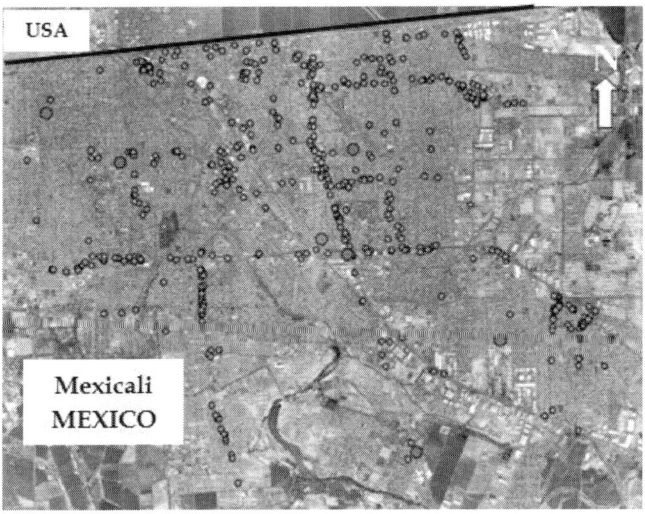

Fig. 7. Spatial distribution of Mexicali restaurants.

Finally, it was performed a dynamic model that helped to predict the WVO generation for a period of 10 years, from 2008 to 2017. The results showed an average generation rate of 3 million liters of WVO, or 3.4 L per capita/year.

In order to determine the WVO availability of the restaurant industry in Baja California, the results were extrapolated from the survey conducted in Mexicali. The volume of WVO estimated was 8 million liters per year. From this value, it would be feasible to produce 6.4 million liters of biodiesel annually, assuming a conversion efficiency of 80%, obtained experimentally.

4.2 Animal fat

According to the report of the Mexican Service of Information, Food, Agriculture, and Fishery (SIAP, 2009), in 2008 Baja California holds the 6th national rank in beef carcass production. In that year, 695,000 fowls and 259,000 heads of cattle (beef, pigs, goats, and sheep) were processed to produce meat. The number of the standing head of cattle in Baja California in 2008 is shown in Fig. 8.

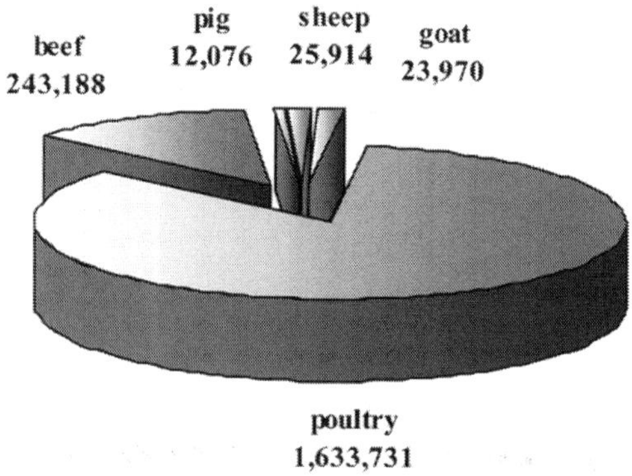

Fig. 8. Standing cattle in Baja California in 2008.

Some portions of the cow mass (49%), of the pig (44%), and of the fowl ((37%) are not suitable for human consumption (Clottey, 1985), so a considerable amount of organic residues are generated by the slaughter processes. Before the rendering process, their average composition is 60% water, 20% proteins/minerals, and 20% fat. These residues could be used to produce biodiesel, due to the fat content. These organic materials with microorganisms potentially pathogens for humans and animals are processed by rendering, which fulfills all of the basic requirements for environmental quality and disease control (Meeker & Hamilton, 2006).
The basic rendering of the materials generated in the beef processing systems is presented in Fig. 9 (Toscano et al., 2011).

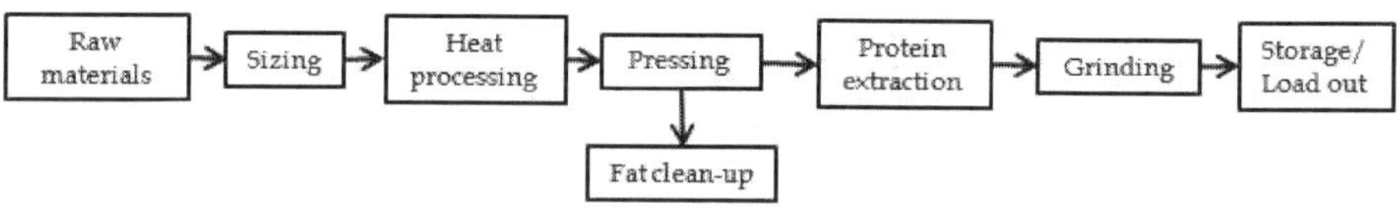

Fig. 9. Scheme of the basic rendering of the materials generated in the beef processing system.

In order to determine the potential of producing biodiesel from yellow grease, it was considered the fat fraction generated in the beef rendering process "in Baja California". The result for 2008 was 1,380 t of fat (Toscano et al., 2011). Assuming conversion efficiency of fat into biodiesel it is stated in 95% (Bhatti et al., 2008), it was estimated that the potential for producing biodiesel in 2008 was 1,311,000 L.
In addition to the residues presented previously, Baja California has oil crops, which represent potential raw material for biodiesel production. Such crops include castor oil plant (*Ricinus communis*) and *jatropha curcas* that are characterized by their high oil content and because it is not edible, so do not compete with food. In the case of the *jatropha curcas*,

experimental plantings were conducted in Mexicali Valley, to determine their adaptation to soil conditions and climate. The results highlighted that the growth of the plant was not successful. However, in the coast of Baja California, south of Ensenada with Mediterranean climate, this plant grows successfully.
The castor oil plant (*Ricinus communis*) is a Baja California endemic plant and has made significant progress in the modernization of such cultivation for mass production in order to obtain oil for its transformation in biodiesel.
Given the raw materials constituted of waste biomass as well as those derived from oilseed plants endemic from Baja California; Institute of Engineering of the Universidad Autónoma de Baja California (Autonomous University of Baja California) is developing research to adapt processes and technologies to achieve the highest conversion yields in the process of biodiesel production. This has been used in different catalytic pathways are described below.

5. Transesterification process

The transesterification reaction can be catalyzed by alkalis, acids (Canacki & Gerpen, 1999) or enzymes (Vyas et al., 2010). Several studies have been performed using different oils as raw material, alcohols, as well catalysts, including homogeneous catalysts such as sodium hydroxide, potassium hydroxide and sulfuric acid and heterogeneous catalysts such as lipase (Nielsen et al., 2008), CaO (Lim et al., 2009) and MgO (Refaat, 2010).

5.1 Transesterification by acid catalyst

The transesterification process is catalyzed by Bronsted acids, preferably sulfuric and hydrochloric acids. These catalysts show high yields, but the reactions are slow. The molar ratio alcohol/vegetable oil is one of the main factors affecting transesterification. An excess of alcohol promotes the formation of alkyl esters. On the other hand, an excessive amount of alcohol impairs the recovery of glycerol, so the ideal ratio of alcohol/oil must be established empirically, considering each individual process (Demirbas, 2009). It has been observed that the use of an acid catalyst is more effective than an alkaline catalyst when the concentration of free fatty acids is high, above 1%. These reactions require washing, because the acids involved a large amount of salts produced during the reaction which can be corrosive. The mechanism of acid catalyzed transesterification of vegetable oil is shown in Fig. 10, for a monoglyceride. However, this can be extended to di- and triglycerides. Protonation of the ester carbonyl group produces one carbon cation II which after a nucleophilic attack of alcohol causes the tetrahedral intermediate III and removes the glycerol to form the new ester IV and regenerate the catalyst H^+.
According to the mechanism, carboxylic acids can be formed by the reaction of the carbon cation II when water is present in the reaction mixture. This suggests that transesterification by acid catalyst should be done in the absence of water to avoid the formation of carboxylic acids, which reduces the yield of alkyl esters (Schuchardt et al., 1998).

5.2 Transesterification by alkali catalyst

Transesterification of vegetable oils by alkali catalyst proceeds faster than the reaction by acid catalyst. The first step is the reaction of the base with the alcohol, producing an alkoxide and a catalyst protonated. The nucleophilic attack of the alkoxide to the carbonyl group of the triglyceride generates a tetrahedral intermediate from which form the alkyl

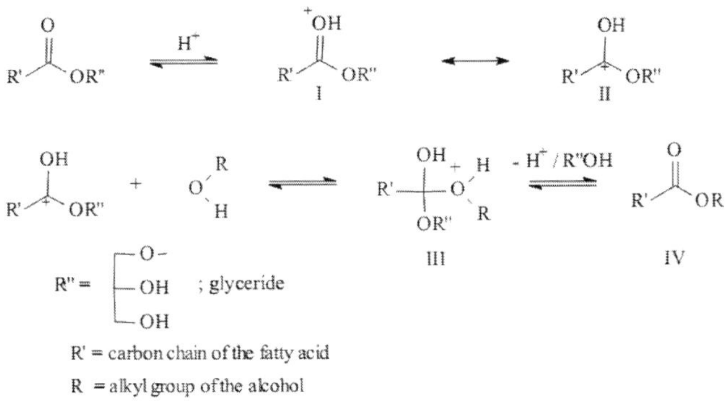

Fig. 10. Mechanism of the acid-catalyzed transesterification of vegetable oils.

ester and the corresponding anion of the diglyceride. Diglycerides and monoglycerides are converted by the same mechanism. The alkali metal alkoxides (CH_3ONa for methanolysis) are the most active catalysts which offer high yields in short reaction times (30 min). However, the reaction requires no water (Demirbas, 2009). The reaction mechanism for catalysis of transesterification is displayed in Fig.11.

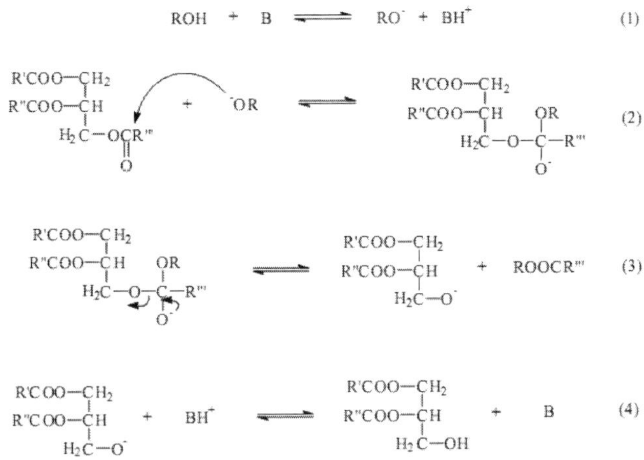

Fig. 11. Mechanism of the base-catalyzed transesterification of vegetable oils.

The transesterification process by alkali catalyst is 100% in the commercial sector, because the chemicals used have proved to be the cheapest for their high level of conversion to esters at low temperature and atmospheric pressure. The main inconvenient of this technology is the sensitivity of alkaline catalysts with the purity of the raw material. The presence of free fatty acids and water in the raw materials has a significant impact on the transesterification reaction (Marchetti et al., 2008). Besides the complex purification of the final products of reaction, this method requires treatment of waste water produced during the process. The amount of wastewater produced is about 200 kg/t of biodiesel produced which increases the costs of this technology and makes it unfriendly to the environment (Ghaly et al., 2010). Fig. 12 depicts the process of transesterification by alkali catalyst (Bacovsky et al., 2007).

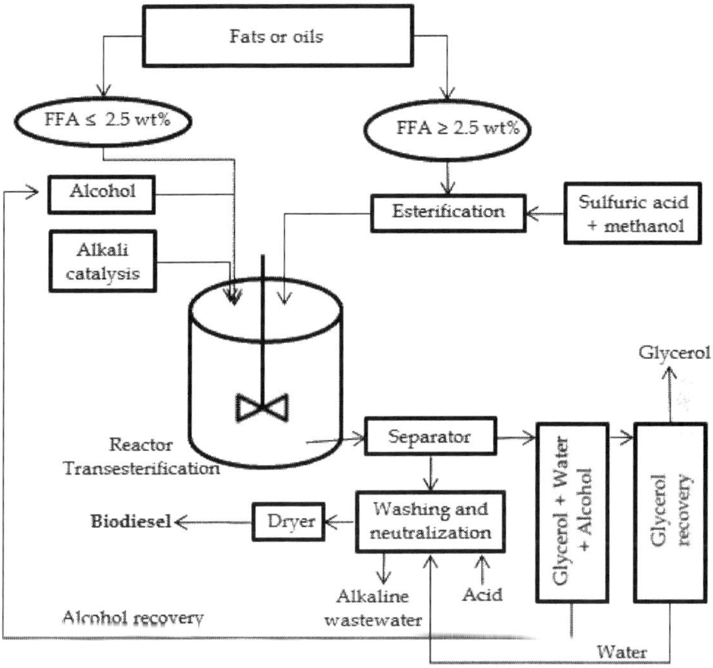

Fig. 12. Process diagram of biodiesel production by alkali catalyst.

5.3 Transesterification by enzymatic catalyst

Biological enzymes, including lipases, can be used as catalysts instead of acids or bases. The use of such enzymes in the production of biodiesel has several advantages (Hass et al., 2002):

- Requires little or no heating.
- The presence of free fatty acids (FFA) in the raw material increases performance, without the production of soap, creating a source of opportunities for the use of low-quality raw materials and low cost.
- It works even in the presence of water.
- Required less alcohol and do not produce salts.

A large number of lipases have been explored as catalysts in biodiesel production. Researchers in China have tested a wide variety of lipases, obtaining yields of up to 94%. Italian researchers tested different lipases and found that the lipase produced by *Pseudomonas cepacia* had a yield of 100% in six hours.

Similar studies were performed in the United States (Jin & Bierma, 2010). Unfortunately, lipases and other enzymes tend to be expensive due to the purification process, which increases the costs of biodiesel production (Gerpen et al., 2004).

In order to increase the duration of the activity of lipase, thereby reducing costs, research has been developed in the immobilization of lipases using physical structures to stabilize the enzyme and allow its reuse. Lipases have been covalently attached to activated polyvinyl chloride, nylon or silica gel. They have also been immobilized by entrapment in alginate gels. Adsorption on hydrophobic or hydrophilic media is some of the techniques used to lipase immobilization (Minovska et al., 2005). Although the enzymatic process is not commercially developed, a large number of publications have shown that enzymes are promising catalysts. These studies consist mainly on the optimization of reaction conditions (temperature, ratio alcohol/oil, type of microorganisms that produce lipase, lipase amount, time, etc.) to establish the characteristics of industrial application.
Fig. 13 shows the enzymatic production of biodiesel using immobilized lipase (Du et al., 2008).

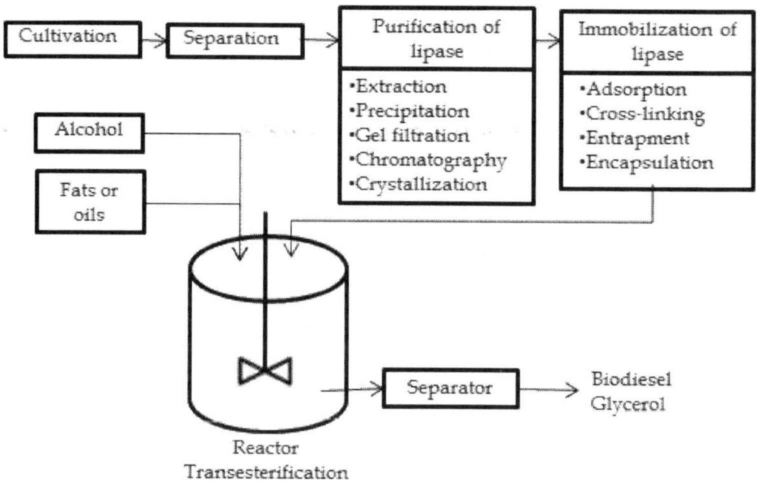

Fig. 13. Enzymatic biodiesel production by immobilized lipase.

Currently in the Institute of Engineering is being developed a project that aims to identify a microorganisms that produces a high yield of lipase, with the necessary characteristics to be used in the synthesis of biodiesel. So it was analyzed the production of lipase from 6 different fungi (*Aspergillus niger, Aspergillus oryzae, Aspergillus Awamory, Trichoderma sp, Trichoderma reesei* and *Penicillium chrysogenum*).
The advantage of this methodology is based on the fact that extracellular lipases can be produced in large quantities under standard laboratory conditions. This could be successful by using the appropriate media and the optimal process parameters.

6. Economic analysis of biodiesel production from WVO and yellow fat in Mexicali, Baja California

Mexicali has a motor vehicle fleet of diesel estimated at 14,000 units and cargo transport. The transport cargo sector with 11,861 units consumes about 169 million liters of diesel. The diesel used in Baja California comes from southern Mexico and is one of the causes of CO_2 emissions that affect air quality in Mexicali, it is therefore necessary to explore options for replacing it with biodiesel, which produces less CO_2 and can be obtained from waste material. Thus, in this analysis, was considered the use of waste vegetable oil from the Mexicali restaurant industry as a raw material for the production of 4.78 million liters of biodiesel energy equivalent to 4.45 million liters of diesel.

The environmental benefit involving the replacement of such a volume of diesel with biodiesel is to reduce emissions by about 9,700 t of CO_2, 22 t of SO_x and 11 t of PM_{10}.

To determine the economic feasibility of producing biodiesel, were applied the methodologies of net present value and internal rate of return. The results indicate that the production of biodiesel is profitable. However, the recovery time of investment, coupled with the uncertainty presented by the biofuels market, make necessary a policy that implements local tax resources to support the promotion, production and use of biodiesel for the transport sector. Therefore, under the circumstances considered in this analysis, the production of biodiesel is feasible if it is developed synergy among the productive sectors, education and government.

The profitability indicators are set at the discretion of the financial analysis methodology. The final report of economical evaluation of this project is supported with the following results:

a. The net present value with a bank interest rate of 17%, meets the acceptance criteria to generate 423,747 USD, however, the magnitude of the indicator does not provide the certainty to accept conditions of project implementation.
b. The internal rate of return is calculated based on cash flow net present value, resulting in the profitability of 23.5%; therefore the project is considered financially viable, however, an acceptance criterion is to get 10 points above the discount rate.
c. The Benefit/Cost Ratio result is 1.05, therefore, is slightly positive, meets the criteria of acceptance, but does not provide the necessary clearance to run the project within the evaluation period.
d. The Profitability Index of the project is 0.227, which does not meet the acceptance criteria for the project.

Based on the evidence derived from cost-benefit analysis it may be concluded that carrying out the project to produce biodiesel from WVO in Mexicali is profitable. However, the return time of investment and the uncertainty presented by the biofuels market, make necessary a policy that implements local tax resources to support the promotion, production and use of biodiesel for the transport sector (Vazquez et al., 2011).

7. Conclusion

Baja California has significant potential for the development of biodiesel production projects, taking into account residual material such as yellow fat or others that are not currently used as vegetable oils, which are discarded mostly. It has also been encouraged by the government of Mexico the planting of bioenergy crops such as castor and *jatropha curcas*.

The promotion for these projects, in areas without oil resources such as Baja California, will slightly shift the use of fossil fuels, and thereby avoid the emission of sulfur compounds.

The current state of biofuel development in Baja California largely reflects the current situation of production, operation and sales of biofuels, including biodiesel, in Mexico.

From an economic standpoint, the production of biodiesel in Baja California will be successful as long as the support from the productive sectors, education and government.

8. Acknowledgment

The authors thank to Institute of Engineering of Universidad Autónoma de Baja California, Consejo Nacional de Ciencia y Tecnología, Secretaría de Educación Pública for their support in the development of the present work.

9. References

Bacovsky, D.; Körbitz, W.; Mittelbach, M. &Wörgetter, M. (2007). Biodiesel production: *Technologies and European providers, IEA Editor*, (2.6.2011), Available from http://www.biofuels-dubrovnik.org/downloads/Process%20Developers%20 Catalogue-2007.07.06.pdf

Bhatti, H.; Hanif, M.; Faruq, U. & Sheikh, M. (2008). Acid and base catalyzed transesterification of animal fats to biodiesel. *Iran J Chem Chem Eng*, Vol.27, No.4, (2008).

Boey, P.; Maniam, G. & Hamid, S. (2009). Biodiesel from adsorbed waste oil on spent bleaching clay using CaO as a heterogeneous catalyst. *European Journal of Scientific Research*, Vol.33, No.2, (July 2009), pp. 347-357 10, ISSN 1450-216X

Canacki, M. & Gerpen, J. (1999). Biodiesel Production Via Acid Catalysis. *American Society of Agricultural Engineers*, (2.2.2011), Available from www.asabe.org

Canacki, M. & Gerpen, J., (2001). Biodiesel Production from oils and fats with high free fatty acids. *American Society of Agricultural Engineers 1429*, Vol.44(6), pp. 1429-1436 7, (September 2001), ISSN 0001-2351

Campbell, H.; Montero, G.; Pérez, C. & Lambert, A. (2011). Efficient energy utilization and environmental issues applied to power planning. *Energy Policy*, Vol.39, No.6, (June 2011), pp. 3630–3637 7, ISSN 0301- 4215

Clottey S. (1985). *Manual for the slaughter of small ruminants in developing countries*. (30.3.2011), (1985) Available from http://www.fao.org/docrep/003/X6552E/X6552E00.HTM

Coronado, M. (2010). *Estudio de Factibilidad de Producción de Energía a partir de Aceite Vegetal Residual, Caso: Sector Restaurantero,*. Master Thesis. Institute of Engineering, Universidad Autónoma de Baja California, Mexicali, B.C. México.

Demirbas, A. (2008). Progress and recent trends in biodiesel fuels. *Energy Conversion and Management*, Vol.50, No.1, (January 2009), pp. 14-34 20, ISSN 0196-8904

Du, W.; Li, W.; Sun, T.; Chen, X.; Liu, D. (2008). Perspectives for biotechnological production of biodiesel and impacts. *Appl Microbiol Biotechnol*, Vol.79, No.3, (June 2008), pp. 331-337 6, ISSN 1432-0614

Gerpen, J.; Pruszko, R.; Clements, D.; Shanks, B. & Knothe, G. (2006). *Building a succesful biodiesel business*, Biodiesel Basics, Second Edition, Biodiesel Basics, ISBN 097863490X

Gerpen, J.; Shanks, B. & Pruzco, R. (July 2004). *Biodiesel Production Technology*, in NREL/SR-510-36244, M.R.E. Laboratory, Editor. 2004: Boulder, CO, (13.4.2011), Available from http://www.nrel.gov/docs/fy04osti/36244.pdf

Hass, M.; Piazza, G. & Foglia, T. (2002). Enzymatic Approaches to the Production of Biodiesel Fuels, in Lipid Biotechnology, *T.M.K.a.H.W. Gardner*, (2002) Marcel Dekker, Inc: New York, pp. 587-598 11, ISBN 9780824706197

Jin, G. & Bierma, T. (2010). *Whole-cell Biocatalyst for producing Biodiesel from Waste Greases, in ISTC Reports.* Illinois State University: Normal, Illinois. (23.4.2011) Available from http://www.istc.illinois.edu/info/library_docs/RR/RR-117.pdf

Kiss, A.; Dimian, A. & Rothenberg, G. (2006). Solid Acid Catalysts for Biodiesel Production – Towards Sustainable Energy. *Advanced Synthesis & Catalysis*, Vol.348, No.1-2, (January 2006) pp. 75-81 6. ISSN 1615 4169

Leung, D.; Wu, X. & Leung, M. (2010). A review on biodiesel production using catalyzed transesterification. *Applied Energy*, (7.11.2009), pp. (1083-1095 87). ISSN 0306-2619

Ma, F.; & Hanna, M. (1999). Biodiesel production: a review. *Bioresource Technology*. Vol.70, No.1, (October 1999), pp. 1-15. ISSN 09608524

Marchetti, J.; Miguel, V. & Errazu, A. (2008). Techno-economic study of different alternatives for biodiesel production. *Fuel Processing Technology*, Vol.89, No.8, (August 2008), pp. 740-748 8, ISSN 0378-3820

Meeker, D. & Hamilton, C. (2006). *An overview of the rendering industry. In: Meeker, D., editor. Essential rendering: all about the animal by-products industry. Arlington*: National Renderers Association, Available from http://assets.nationalrenderers.org/essential_rendering_book.pdf

Minovska, V.; Winkelhausen, E. & Kuzmanova, S. (2005). Lipase immobilizaed by different techniques on various support materials applied in oil hydrolysis. *J. Serb. Chem. Soc.* Vol.70, No.4, (April 2005), pp. 609-624 15, ISSN 0352-5139

Nielsen, P.; Brask, J. & Fjerbaek, L. (2008). Enzymatic biodiesel production: Technical and economical considerations. Eur. J. *Lipid Science and Technology*, Vol.110, No.8, (July 2008), pp. 692-7008, ISSN 1438-9312

Refaat, A. (2010). Biodiesel production using solid metal oxide catalysts. *Int. J. Environ. Sci. Tech*, Vol.8, No.1, (January 2011), pp. 203-221 18, ISSN 1735-1472

REMBIO, Red Mexicana de Bioenergía. (2011). *Mapa Proyectos Bioenergia*, http://www.rembio.org.mx/MapaProyectosBioenergia.

SENER, Secretaría de Energía. (2009). *Se otorgan los primeros 12 permisos de comercialización de bioenergéticos, Boletín de prensa no. 51*, (18.3.2011) Available from http://www.sener.gob.mx/webSener/portal/Default.aspx?id=1071

SENER, Secretaría de Energía. (2006b). *Potencialidades y viabilidad del uso del bioetanol y biodiesel para el transporte en México (SENER-BID-GTZ)*, Proyectos ME-T1007 – ATN/DO-9375-ME y PN 04.2148.7-001.00, (2.3.2011), Available from http://www.sener.mx

SIAP, Servicio de Información Agroalimentaria y Pesquera Resumen Nacional de la Producción Pecuaria. (2009). (24.3.2011), Available from http://infosiap.siap.gob.mx/ventana.php

Talens, L.; Villalba, G. & Gabarrell, X. (2006). Exergy analysis applied to b iodiesel production. *Resources Conservation & Recycling*, Vol.51, (August 2007), No.2, pp. 397- 407 10, ISSN 0921- 3449

Toscano, L.; Montero, G.; Stoytcheva, M.; Campbell, H. & Lambert, A. (2011). Preliminary assessment of biodiesel generation from meat industry residues in Baja California, Mexico, *Biomass and Bioenergy*, Vol.35, No.1, (January 2011), pp. 26-31 5, ISSN 0961- 9534

Vázquez, A; Montero, G.; Sosa, J.; García, C. & Coronado, M. (2011). Economic Analysis of Biodiesel Production from Waste Vegetable Oil in Mexicali, Baja California, *Energy science and technology*, Vol.1, No.1, (2011), pp. 87-93 6, ISSN 1923-8460

Vyas, A.; Verma, J. & Subrahmanyam, N. (2010). A review on FAME production processes. *Fuel,* Vol. 89, No. 1, (January 2010), pp. 1-9 8, ISSN 0016-236

10

Development of Multifunctional Detergent-Dispersant Additives Based on Fatty Acid Methyl Ester for Diesel and Biodiesel Fuel

Ádám Beck, Márk Bubálik and Jenő Hancsók
University of Pannonia, MOL Hydrocarbon and Coal Processin Department
Hungary

1. Introduction

Nowadays fuel blending components produced from renewable sources (biodiesel, mixture of iso- and normal-paraffins produced from triglycerides, gasoline and diesel produced from synthesis gas etc.) are an important part of the blending pool (Hancsók et al., 2007; Krár et al., 2010a, 2010b). The use of fuels produced from renewable resources is supported by several EU directives (2003/30/EC (Biofuels), 2009/28/EC (Renewable Energy Directive) and 2009/30/EC (Fuel Quality Directive).

The biocomponents of diesel fuel are mainly fatty acid methyl ester (biodiesel), produced from the catalytic transesterification of vegetable oils. Their blending is allowed up to 7% by the EN 590:2009 diesel fuel standard. The application of biodiesels causes several problems due to their properties which are different from that of the fossil diesel fuel: higher cold filter plugging point (CFPP), higher viscosity, hydrolysis (corrosion), storage stability problems, lower energy content etc. As a result new challenges rose to ensure the high quality of diesel fuel and the proper function of the engine by applying high performance additives in the diesel fuel and engine oil (Beck et al., 2010; Bubálik et al., 2005).

2. Modern diesel fuel additives

Modern diesel fuels are blended from high quality blending components and high performance additives. The additives are usually synthetic materials applied in low concentration which improve the properties of diesel fuel or provide them new, advantageous ones (Hancsók et al., 1999a; Haycock & Thatcher, 2004).

By the introduction of ultra low sulfur diesel and biofuels (e.g.: fatty acid methyl ester) into the market the importance of fuel additives increased. Among the fuel additives detergent-dispersants, lubricity improvers and corrosion inhibitors there are long chained hydrocarbon molecules with a polar head. All these three types of additives provide their effect by linking to the metal surface, as a result these additives compete for the metal surface and not all of them can reach it and provide its effect. Therefore, the development of a multifunctional additive providing two or three of the previously mentioned effects is an important research field. In the current publication the mechanism of detergent-dispersants, lubricity improvers and corrosion inhibitor additives are detailed, followed by the results of the research work.

2.1 Detergent-dispersant additives

Among the different types of additives the detergent-dispersants (hereinafter: DD) have high importance. Their share in the total additive market is about 40-50%. Their role is to clean and keep clean the fuel supply system and the combustion chamber: remove deposits and prevent their formation (Figure 1 and 2) in order to ensure the proper function of the engine (Hancsók, 1999a; Haycock & Thatcher, 2004).

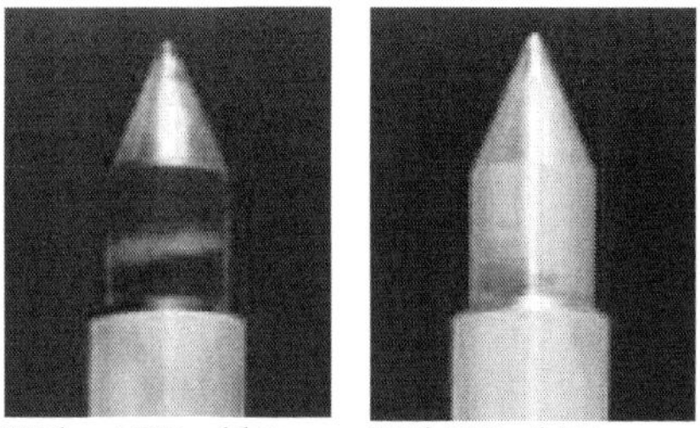

Without DD additive With DD additive

Fig. 1. Deposits on the injector

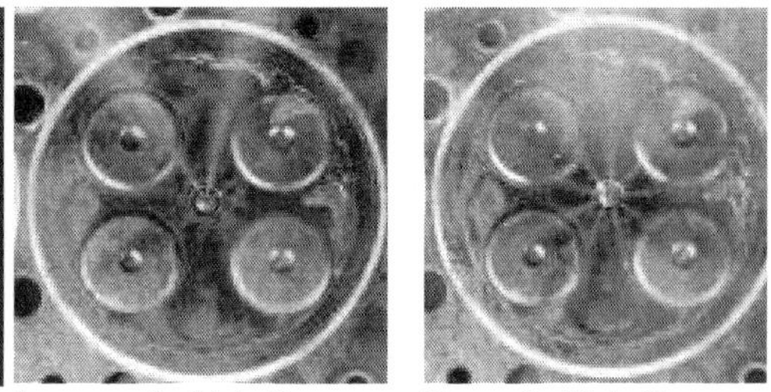

Improper fuel injection caused by deposits Proper fuel injection (without deposits)

Fig. 2. Effect of DD additives on the fuel injection

Several additive types were developed for the above mentioned purpose. These additives usually have different efficiency and they have only the detergent-dispersant function. The development and application timeline of DD additives are shown in Figure 3 (Hancsók, 1999a).

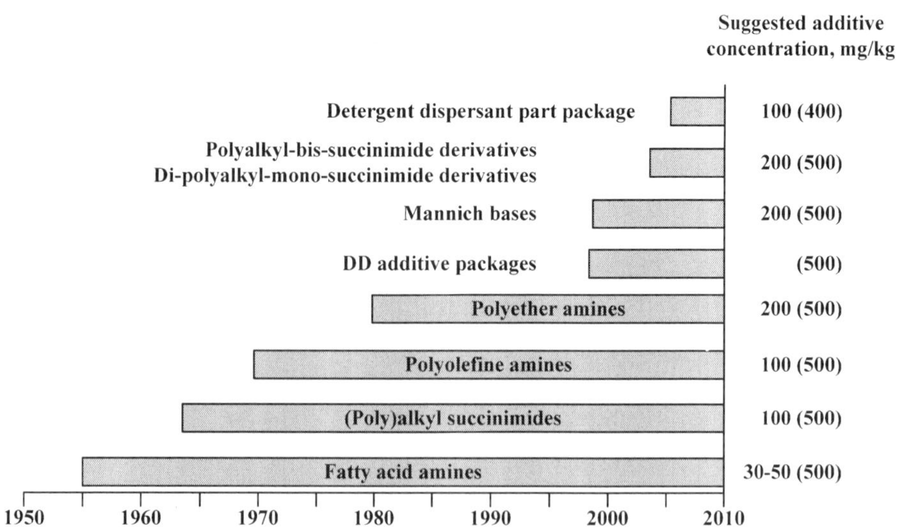

Fig. 3. Timeline of the development of detergent additives

Alkenyl-succinimides were applied as dispersants in lubricants already at the end of the '50s, in fuels in the early '60s. These were the second generation of the deposit control additives. In the '70s fuel consumption increased world-wide, as a result olefins were blended in a higher concentration. The fuel quality change and the higher olefin content resulted in lower fuel stability, and as a consequence there was a need for higher performance additives. In this period the application of polyolefin-amines started as new, high performance deposit preventing additives, followed by the application of polyether-amines, then Mannich-bases. In the latest decades the environmental and quality prescriptions towards fuels have become stricter. Nowadays, different types of polyisobutylene-mono- and bis-succinimide additives (Figure 4), polyisobutylene-amines, Mannich bases, polyether amines and their mixtures are applied in fuels (Hancsók, 1999a; Haycock & Thatcher, 2004). The application of the so-called detergent-dispersant packages containing the mixture of different, unique additives is becoming more and more wide-spread, too (Hancsók, 1999a).

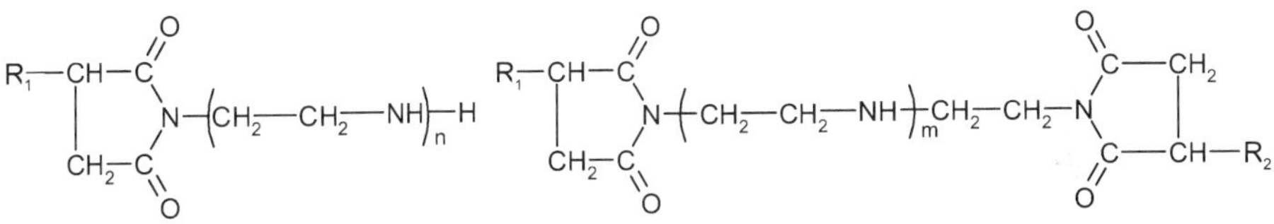

PIB-mono-succinimide PIB-bis-succinimide

Where R: polyolefin having Mn = 500-6000 average molecular weight, advantageously polyisobutylene chain
m, n: 1-5 whole number

Fig. 4. General structure of polyisobutylene mono- and bis-succinimides

The mechanism of detergent-dispersant additives is summarized in Figure 5. The additive molecules bond to the metal surface by chemisorption and prevent the formation of deposits by covering the surface. They remove deposits by their detergent action and keep in

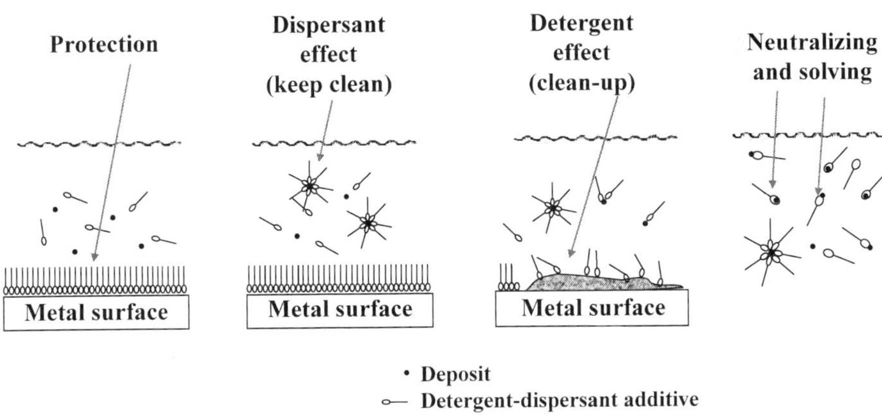

Fig. 5. Mechanism of detergent-dispersant additives

dispersion the insolubiles by absorbing to the impurities with their polar head. In such a way they prevent the formation of bigger agglomerates by steric hinder. They form a micellar colloid structure with the impurities, into this structure further impurities can enter by electrostatic or hydrogen bond, in this way the size of the micelle increases. As a result the additive prevents the deposits of polar compounds. Their other important function is the acid neutralization with their base group (Beck et al., 2010).

The application of detergent-dispersant additives provides several advantages for the end-users, such as (Caprotti et al., 2007; Hancsók et al., 1997; Haycock & Thatcher, 2004; Kocsis et al., 2001; Rang & Kann, 2003; Ullmann et al., 2009):

- Smooth fuel injection (preventing deposit formation)
- Smoother pressure increase during injection resulting in quieter motor function
- Better combustion
- Higher performance
- Better driveability
- Lower fuel consumption and lower maintenance costs
- Lower emission

The latest environmental prescriptions can only be satisfied by applying high performance detergent-dispersant additives. These additives are not only responsible for cleaning and keeping clean the fuel supply system and the combustion chamber, but for ensuring lower fuel consumption as well (Beck et al., 2009a; Caprotti et al., 2007; Rang & Kann, 2003; Ullmann et al., 2009).

The most widely applied DD additives are the succinimide type additives, which are mostly produced in two steps by thermal technology (Mach & Rath; 2006). This production method has several disadvantages: the intermediate has to be filtered due to the formation of gum-like byproducts, high energy need etc. In order to eliminate these disadvantages at the Department of Hydrocarbon and Coal Processing of University of Pannonia (thereafter Department) a new, radically initiated method was developed. The additives synthesized with this method had better performance compared to the commercial ones (Caprotti et al., 2007; Kocsis et al., 2003; Rang & Kann, 2003). During the research work the polyisobutylene of 800-1300 number average molecular weight was found to be the most advantageous for the production of such an additive structure, which is soluble in modern engine fuels due to its long hydrocarbon chain; and has such a polar functional group which is able to disperse the insoluble impurities (Beck et al., 2009a, 2009b, 2010; Bubálik et al., 2005; Kocsis et al., 2001).

One of the most important methods to measure the efficiency of detergent-dispersant additives is the engine test. Currently the Peugeot XUD-9 and the Peugeot DW-10 engine tests are the most common (Breakspear & Caprotti, 2007). The XUD-9 is a 1.9 litres indirect injection engine, while the Peugeot DW-10 is a 2.0 litres common rail diesel engine equipped with Siemens Euro 5 injectors (maximum injection pressure: 1600 bar). In 2008 March the CEC (Co-ordinating European Council for the development of performance tests for transportation fuels, lubricants and other fluids) accepted the F-98-08 injector test. According to it, the DW-10 engine is applied for the testing of diesel, biodiesel-diesel blends and deposit control additives. The DW-10 engine test does not substitute the XUD-9 engine test, because the latter is more suitable for measuring the efficiency of deposit control additives. The DW-10, however, is more sensitive for the efficiency of other additives, for example the metal deactivators.

2.2 Lubricity improver additives

The ultra low sulfur diesel - due to the strict hydrotreating - does not contain those compounds which provide good lubricity, and are originally present in the crude oil. In order to avoid wear, the application of synthetic lubricity improvers is necessary (Hancsók et al., 1999a, 2008a; Haycock & Thatcher, 2004; Kajdas & Majzer, 2003; Wei & Spikes, 1986). These additives are usually molecules with a long hydrocarbon chain and polar functional group. They form a layer on the metal surface by adsorption and decrease the friction coefficient.

Wear occurs in the fuel system, where different parts are moving on each other. As a consequence the most important thing is to form a protective layer on the surfaces by adsorption and/or chemisorption in order to reduce the metal-metal contact points (Figure 6 and 7) (Spikes & Wei, 1997; Wei & Spikes, 1986). The protective layer can be formed by the application of polar compounds.

Several compounds can be applied as lubricity improvers, such as different mixtures or esters of unsaturated hydrocarbon acids; alkenyl-succinimides; alcohols, acids, or esters having long hydrocarbon chain (Batt et al., 1996; Denecker, 2002; Hancsók et al., 1997; Haycock & Thatcher, 2004; Kajdas & Majzer, 1999; Spikes & Wei, 1997; Wei & Spikes).

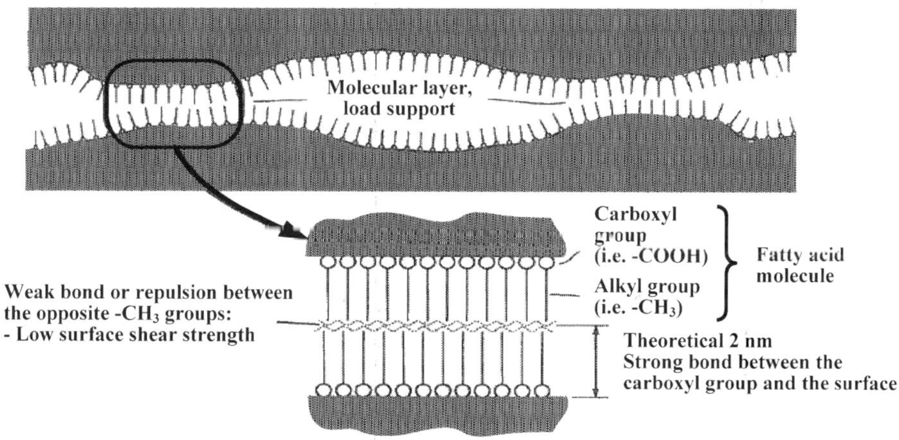

Fig. 6. Mechanism of the formation of the film by adsorption

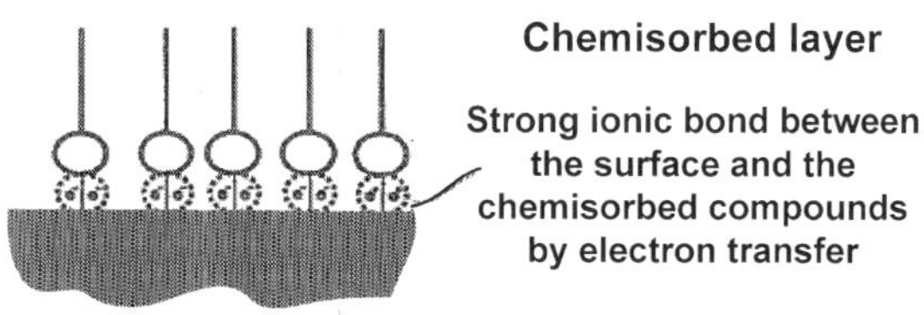

Fig. 7. Mechanism of the formation of the film by chemisorption

The latter mentioned esters can be produced from alcohols, vegetable oils or they can be bis-alkenyl-succinic esters, etc.

The most widely applied and standardized method for the measurement of the lubricity of diesel fuel is the HFRR test (High Frequency Reciprocating Rig, ISO 12156, ASTM D6079; VI. category): two metal test pieces are rubbed in continuous fuel flow and the friction and wear is measured. The wear of the upper test piece is measured in μm, according to the diesel standard (EN 590:2009) the wear has to be lower than 460 μm to assure enough lubrication - provided by the fuel - to protect the engine.

SLBOCLE (Scuffing Load Ball On Cylinder Lubricity Evaluator, ASTM D 6078; VI. category): two metal test pieces are rubbed in the presence of fuel, the test is carried out until the total wear of the test pieces, the load is measured.

Injector pump test (ISO/DIS 12156-1; III. category): wear on the test piece is measured in μm, and evaluated in the range of 1-10. The minimum requirement during the test is 6. The test load correlates to the load on the injector pump during 300,000 km normal operation. The test method has a high cost and is time consuming, therefore, it is not applied as a first selecting method, but only applied before the introduction of a new product. The results of the injector pump test correlates to the previously described two test results.

2.3 Corrosion inhibitors

Pure hydrocarbons of diesel fuel are not corrosive themselves. However, the sulfur or acidic compounds which can also be found in diesel fuel have a high corrosion effect on copper and its alloys. Storage conditions of fuel also have a significant effect on the corrosion properties of diesel fuel: during temperature change water can condensates on the roof or the wall of the storage tank and can enter into the fuel causing corrosion. These effects can be decreased by applying proper corrosion inhibitors. These additives, like the detergent-dispersants or the lubricity improvers, have polar functional group and long apolar hydrocarbon chain. They bond to the metal surface by chemisorption and form a protective layer. Corrosion inhibitors can be different alkyl- or polyalkyl-succinimides, their esters, dimer acids, amine salts (Hancsók et al., 1997; Haycock & Thatcher, 2004).

Among the different methods applied for measuring the performance of diesel fuel corrosion inhibitors, the copper strip corrosion and the steel drift tests are the most wide-spread. During the copper strip corrosion the cleaned and polished copper strips are kept in diesel fuel under controlled conditions (3h, 50°C) and after the test the colour change of the copper strips are evaluated visually according to the scale of the standard (EN ISO 216, ASTM 130). During the steel drift test water is added to the fuel to increase corrosion and after the controlled test (12h, 100°C) the corrosion degree of the steel drift is evaluated visually (ASTM 665). Both methods are based on increasing the corrosion properties of diesel fuel by elevated temperature and in case of the latter test also by the addition of water.

3. Development of multifunctional detergent-dispersant additives

In the recent years in the Department the aim of the research was to develop a multifunctional additive by applying the radically initiated synthesis method. The scope was to modify the structure of polyisobutylene succinimides in order to achieve other advantageous properties, but keeping their high detergent-dispersant efficiency (Russel,

1990; Hancsók et al., 1999b). Fatty acid methyl esters (biodiesel) have a reactive double bond, polar functional group and they have very good lubricity (Knothe, 2005a). Therefore, during our research the aim was to incorporate the fatty acid molecule into the succinimide structure.

One major issue of biodiesel blending into diesel fuel is the incompatibility of biodiesel with metals. Numerous publications report on corrosion of copper and its alloys due to contact with biodiesel, and as a result gum formation and acid number increase was also noticed. Due to the fact that long chain hydrocarbons having a polar functional group can form a protective layer on the metal surface, the corrosion inhibiting effect of the developed additive was also investigated. Another important effect of biodiesel on fossil fuels is that the fatty acid methyl esters enhance the biodegradability by co-metabolism (Pasqualino et al., 2006).

The reaction of polyisobutylene, maleic anhydride and fatty acid methyl ester was performed by radical initiation due to the fact that by the thermal reaction of maleic anhydride and fatty acid methyl ester gum-like byproducts formation was reported by other publications (Candy, 2005; Quesada, 2003).

3.1 Materials

Polyisobutylene succinimides are synthesised in two steps. In the first step commercial polyisobutylene (hereinafter: PIB) of 1000 number average molecular weight, fatty acid methyl ester (hereinafter: FAME), maleic anhydride (hereinafter: MA), a radical initiator and aromatic solvent was applied. The main properties of PIB and FAME are summarized in Table 1 and 2.

Properties	FAME
Mono-ester content, %	97.2
Density, g/cm^3	0.88
KV at 40 °C, mm^2/s	4.5
Flash point, °C	>110
Water content, %	0.02
Acid number, mg KOH/g	0.3
Methanol content, %	0.04
Iodine number, g/100g	112

Table 1. Main properties of the fatty acid methyl ester

Properties	PIB
Number average molecular weight	1050
Polydispersity, α	1.52
α-Olefin content, %	88
KV at 100°C, mm^2/s	192
Appearance	Transparent, bright
Flash Point (Cleveland), °C	204
Iodine number, g/100g	17

Table 2. Main properties of the polyisobutylene

For the second step of the synthesis the intermediate was diluted with base oil (SN-150) then the following amines were acylated: diethylene-triamine (hereinafter: DETA), triethylene-tertaamine (hereinafter: TETA), tetraethylene-pentaamine (hereinafter: TEPA), pentaethylene-hexaamine (hereinafter: PEHA), monoethanol-amine (hereinafter: MEA), diethanol-amine (hereinafter: DEA), piperazine, dibutyl-amine.

3.2 Methods

For the investigation of the properties of the intermediates and additives standard and in-house methods were applied, which are summarized in Table 3.

Properties	Methods
Kinematical viscosity	EN ISO 3104
Nitrogen content	Kjehldal method
Total Base Number (TBN)	ISO 3771
Total Acid Number	ISO 6618
Maleic-anhydride content	proprietary (titrimetic)
Active material content	local standard (column chromatography)
Molecular weight and distribution	GPC (PIB standards)
Washing Efficiency	proprietary (thin layer chromatography)
Detergent Index	proprietary (photometric)
Potential DD Efficiency (PDDE)	proprietary
Copper strip test	ISO 2160:2000
Steel drift test	ASTM D 665
Peugeot XUD9 engine test	CEC-PF-023
Lubricity improving effect (4ball test)	Modified ASTM D 2783-88
HFRR	ENISO12156

Table 3. Methods for measuring analytical properties and performance

3.2.1 Potential detergent-dispersant efficiency (thereafter PDDE)

According to our knowledge currently there is no standard method available for measuring the performance of the detergent additives of diesel fuels. Therefore, the potential detergent-dispersant efficiency (hereinafter: PDDE) method was applied, which was originally developed for motor oil additive testing.

The detergent-dispersant efficiency of the additives was measured by two methods: washing efficiency and detergent index.

The washing efficiency is measured by thin layer chromatography method. Its aim is to evaluate how effective the additive is for removing the impurities from the surface. For the test as the first step 1.5% additive is dissolved in SN-150 base oil. Then a suspension of 9.8 g of the mixture and 0.2 g carbon black is prepared by ultrasonic equipment. 10 μl of the suspension is placed on the chromatography paper and after letting it dry the paper is placed in a vertical position over heptane in such a way that only the lower ca. 0.5 cm of the paper is in contact with heptane. In this way the additive oil mixture - with the heptane - brings the suspension of carbon black upwards. The different additives bring the carbon black in different height of the paper based on the washing efficiency of the additive. The washing efficiency is measured in millimeters between the point where the suspension was put and the height where the oil brings the suspension with the heptane. Reproducibility of the measurement is ±5%.

The detergent index characterizes the dispersion stabilizing efficiency of the additive, thus, how they keep the impurities in a dispersed phase. The test is based on centrifugation and it is a modification of the original method developed at the Department. Additives with high dispersant efficiency do not let the suspended impurities accumulate even under centrifugation force. During the test the suspension prepared for the washing efficiency test is used. The suspension is diluted with petroleum in 1:5 ratio and centrifuged for 30 minutes at 500 1/min. After centrifugation the intensity of light through the upper part of the solution is measured at 530 nm. The detergent index is calculated from the intensity of light before and after the centrifugation according to the following:

$$DI=-(I_1/I_0)*100$$

Where DI: detergent index in %,
I_1: intensity of light in % transmitted through the blend containing carbon black,
I_0: intensity of light in % transmitted through the blend free of carbon black,
Reproducibility: ±1.
Based on the two methods potential detergent-dispersant efficiency of the additives is defined as follows:

$$PDDE=(DI+M)/225*100$$

Where DI: detergent index in %; its maximum value is 100%
M: washing efficiency in mm, maximum value is 125 mm
Reproducibility: ± 4%.

3.2.2 Measuring lubricity with four-ball machine

The lubricity improving effect of the additives was measured with Stanhope SETA four-ball machine. During the test the sample is put into a cup where 3 balls are in steady state and a fourth ball is pushed - with adjustable load - from above to the standing ones. The evaluation of the additive performance is carried out based on the wear scar diameter of the three standing balls and the friction coefficient. In order to improve the method a thermometer and a computer weres connected to the standard four-ball machine in order to register data and control the test. Figure 6 shows a simplified scheme of the apparatus. The lubricity of the diesel fuel additives is measured in 300 mg/kg concentration in diesel fuel under 300 N load during 1 hour. The average of the wear scar diameter of the three standing balls was evaluated (Bubalik et al., 2004, 2005).

3.3 Synthesis of additives having fatty acid methyl ester in their molecular structure

The production of the additives having fatty acid methyl ester in their molecular structure was performed in two steps. In the fist step intermediate was synthesized from PIB, MA, FAME, radical initiator and aromatic solvent in a four-neck flask equipped with a stirrer, thermometer, flow-back cooler and feeder (Hancsók et al., 2008b). The reactions were carried out at atmospheric pressure and at different temperatures (130-150°C) by applying different PIB:FAME:MA molar ratio, different solvent and initiator concentrate. The reactants were added in more portions, the reaction time was between 4 and 7 hours. The solvent and the unreacted maleic anhydrid were removed at 200°C under vacuum.
The intermediates were diluted with base oil in order to reduce their viscosity, then different amines were acylated by applying different amine:intermediate molar ratios. These reactions were carried out at 165-185°C, in 4-7 hours, by applying nitrogen atmosphere and slight vacuum. Unreacted amines and the formed water were removed at about 200°C under vacuum.

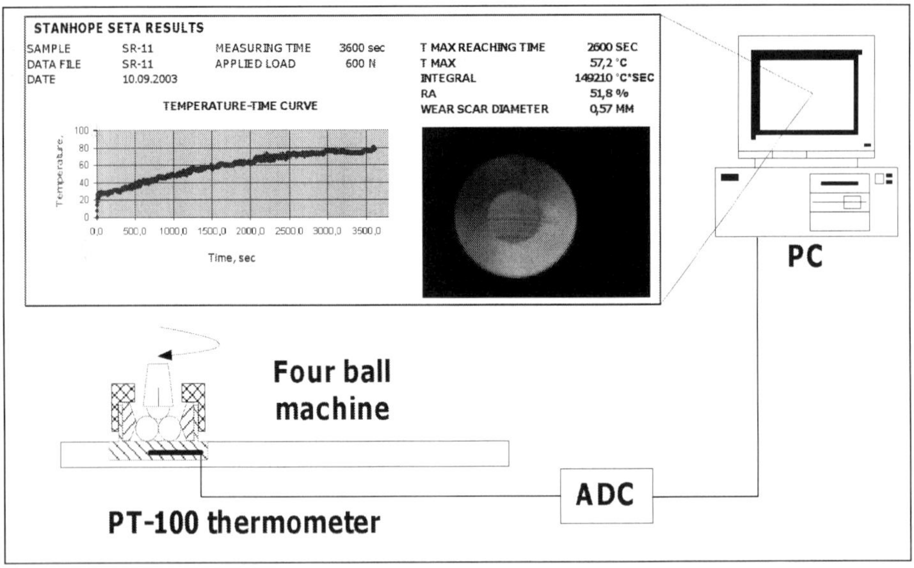

Fig. 6. Simplified scheme of the modified four-ball machine

The main properties of some intermediates are summarized in Table 4. Number average molecular weight of the PIB was about 1000, and that of the FAME about 300, while that of the intermediate was between 1300 and 1700. Based on these data we suppose that during the reaction the succinic structure is formed by one PIB, one MA and one or two FAME molecules.

Properties	KT-1	KT-2	KT-3	KT-4
Main parameters of the synthesis				
PIB:FAME:MSA molar ratio	1.0:1.1:1.0	1.0:1.1:1.1	1.0:1.1:1.3	1.0:1.1:1.4
Reaction temperature, °C	140	140	140	140
Properties of the intermediate				
Appearance	bright	Bright	Bright	Slightly opal
Active material content, %	64.5	63.3	69.3	64.2
Kinematical viscosity at 100°C, mm^2/s	122.4	136.6	186.3	168.2
Acid number, mg KOH/g	62.3	67.9	70.6	84.5
MSA content, mg/g	0.9	2.3	1.2	1.3
Number average molecular weight	1300	1340	1500	1430

Table 4. Main properties of some intermediates

The additives showed in Table 5 were produced from the intermediate KT-3 by acylating different amines. All additives synthesised by applying polyethylene polyamines had high detergent-dispersant efficiency and high total base number (hereinafter: TBN). The other amines resulted in additives with lower PDDE and TBN, except for piperazine, which had high detergent-dispersant efficiency and at the same time low total base number (see Table 5 and Figure 7). The low total base number improves the compatibility of the additive with fluoroelastomers, while the high total base number increases the acid neutralising property of the additive. The polyethylene polyamines and piperazine let us produce additives with high detergent-dispersant efficiency and an appropriate total base number, depending on what the application field requires.

Properties	FP-1	FP-2	FP-3	FP-4	FP-5	FP-6	FP-7	FP-8
Acylating agent	TEPA	DETA	PEHA	TETA	MEA	DEA	Pipe-razine	Dibutyl -amine
Average molecular weight of the amine, g/mol	189	103	232	146				
Molar ratio	1,0:1,0	1,0:1,0	1,0:1,0	1,0:1,0	1,0:1,0	1,0:1,0	1,0:1,0	1,0:1,0
Appearance	Bright	Bright	Bright	Bright	Bright	Bright	Bright	Bright
TBN, mg KOH/g	54.9	29.0	60.5	43.7	2.1	6.5	7.0	2.4
Nitrogen content, %	2.5	1.58	2.43	1.98	0.21	0.56	0.6	0.38
Average molecular weight	1500	1750	1350	1725	1350	1450	2350	5400
3 % (based on active material content) additive in SN 150 base oil								
V.I.E	112	108	107	106	105	106	107	106
Detergent Index, % (max. 100)	100	100	100	100	0	100	100	43
Washing Efficiency, mm (max. 125)	94	96	99	93	13	62	95	11
Potential Detergent-Dispersant Efficiency, % (max. 100)	86	87	88	85	6	72	87	24

Table 5. Main parameters of the additives having FAME in their molecular structure

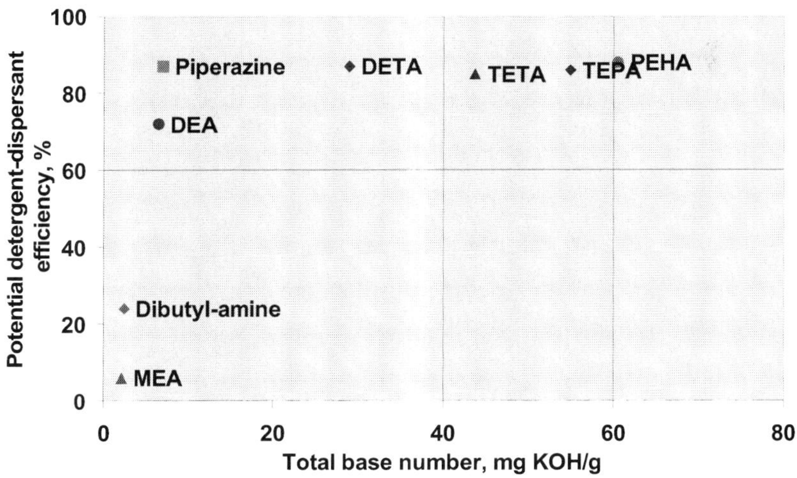

Fig. 7. PDDE of the additives in function of their TBN

The additives synthesized with polyethylene polyamines had such a strong apolar functional group that they were able to form a protective layer on the metal surface, meanwhile their long apolar hydrocarbon chain provided excellent solubility in hydrocarbons. Therefore, their corrosion inhibiting effect was investigated by applying two test methods: copper strip corrosion and steel drift test. The tests were performed in diesel fuel without biodiesel and diesel fuel with 7% biodiesel content. The additive concentration was 20 mg/kg in all cases (Table 6). Copper strip corrosion classification of the diesel fuel without additive was 1B, with the additives it became 1A. During the steel drift test after 6 hours the corrosion degree of the base diesel fuel was quite high, after 12 hours it was over the limits of the measurement. The additives decreased the corrosion during the 12 hours test; however they could not inhibit it completely.

The diesel fuel having 7% biodiesel had disadvantageous corrosion properties both during the copper strip and the steel drift test. Based on the results it was established that all tested additives decreased the corrosion degree of the 7% biodiesel containing diesel fuel. The corrosion inhibiting effect of the additives can be explained by the presence of both polar and apolar function groups in the molecular structure. The polar group enables the additives to bond to the metal surface by chemisorption, while the apolar functional group enables the solubility in hydrocarbons.

	Addtive applied in 20 mg/kg	Copper strip test, classification	Steel drift test, classification 6h	Steel drift test, classification 12h
Diesel fuel	-	1B	3	n/a
	S-1	1A	0	1
	S-2	1A	0	1
	S-3	1A	0	1
	S-4	1A	0	1
B7 (7% biodiesel in diesel fuel)	-	3A	3	n/a
	S-1	1A	0	1
	S-2	1A	0	1
	S-3	1A	0	1
	S-4	1A	0	1

Table 6. Corrosion inhibiting effect of the additives

3.3.1 Molecular structure of the additives having fatty acid methyl ester in their molecular structure

For determining the molecular structure GPC, IR and NMR spectroscopy tests were carried out.

The number average molecular weight of intermediates was in the range of 1400-1700 and the polydispersity was in the range of 1.66-1.84. It suggests that the polyisobutylene and also one or two fatty acid methyl ester compounds linked into the molecular structure. The highly reactive allyl and bis-allyl position carbon atoms (Knothe, 2005b) of the fatty acid methyl ester can react with maleic anhydride to form a succinic anhydride molecule. The results published by Candy (Candy et al., 2005) approved that there is an ene-reaction between succinic anhydride and fatty acid methyl ester.

On the basis of IR spectroscopy data (Hancsók et al., 2006, 2008) and the ^{13}C and ^{1}H NMR tests it was concluded that during the reaction of polyisobutylene, maleic anhydride and fatty acid methyl ester two alkyl chains can substitute a maleic anhydride. Based on the GPC, IR and NMR tests (see Figure 8), the most possible structure is that a polyisobutylene and a fatty acid methyl ester molecule are substituted a maleic anhydride as shown in Figure 9.

3.4 Synthesis of additives having fatty acid methyl ester and styrene comonomer in their molecular structure

The additives having fatty acid methyl ester in their structure were produced in two steps, as mentioned in the chapter before. In order to incorporate the styrene comonomer to the additive, further reaction steps were necessary. Polyisobutylene, maleic anhydrid and styrene were reacted in the presence of a radical initiator and aromatic solvent at the same reaction parameters that were applied for the production of the intermediate with FAME in

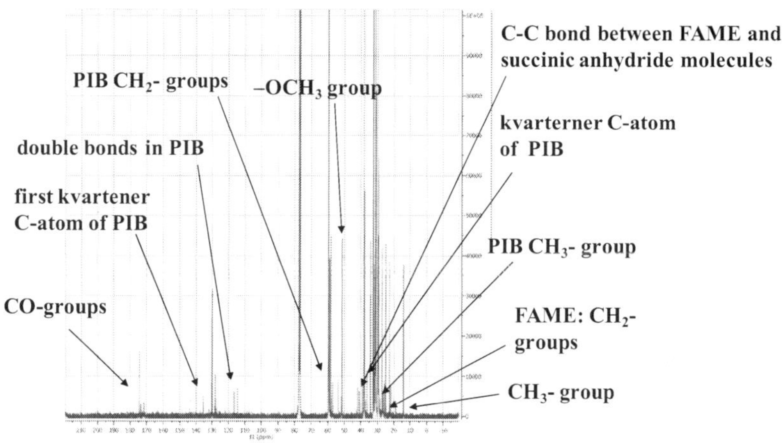

Fig. 8. 13C NMR spectra of the intermediate

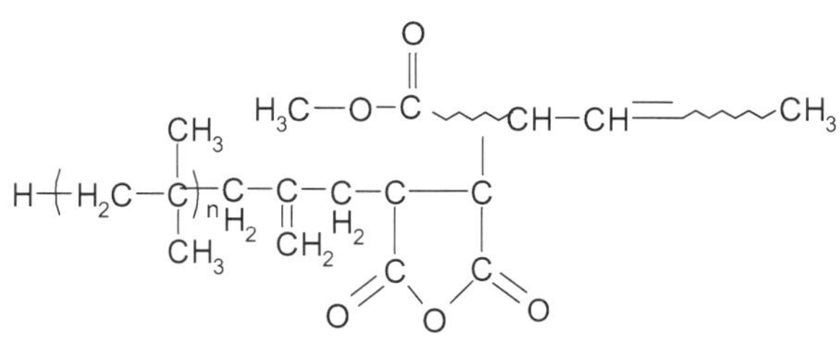

Fig. 9. The most possible structure of the intermediate

its structure. The intermediate containing the styrene was diluted with base oil, and then reacted with the previous additives (FP). In such a way a bis-succinic structure was obtained (see Figure 4).
The quality of the additive highly depends on the process parameters of the synthesis. The main properties of some intermediates (synthesized with styrene, without FAME) are summarized in table 7. The reactions were performed at the same temperature, but with different feedstock molar ratio, as a result the properties of the intermediates were significantly different. Intermediate TS-2 was found to be the most advantageous with its higher active material content, acid number and number average molecular weight. It was found that the increase of MA ratio in the feedstock composition leads to significant change in the acid number of the intermediate.

Properties	TS-1	TS-2	TS-3	TS-4
PIB:MA:styrene molar ratio	1.0:1.9:1.1	1.0:2.0:1.2	1.0:1.8:1.1	1.0:1.9:1.2
Reaction temperature, °C	140	140	140	140
Properties of the intermediate product				
Appearance	bright	bright	bright	cloudy
Active material content, %	55.8	72.3	59.6	63
Kinematical viscosity at 100°C, mm^2/s	94.0	141.2	195.8	163.4
Acid number, mg KOH/g	145	158	152	149
MA content, mg/g	1.3	1.5	1.2	1.2
(number average molecular weight)	1570	1950	1860	1700

Table 7. Main properties of some intermediates with styrene

The intermediates having styrene comonomer were reacted with the additives FP-1 and FP-3 which have fatty acid methyl ester in their structure. The result of the reaction was a bis-succinic type additive (see Figure 4) with relatively high detergent-dispersant efficiency, but with a relatively lower base number compared to the reference commercial mono-bis-succinic type additive.
The total base number of the FPS-1 - FPS-4 additives was significantly lower than that of the FP-1 and FP-3 additives (~30-60 mg KOH/g) synthesized with polyethylene polyamines, while their detergent-dispersant efficiency was quite similar.
The viscosity index improving effect of all the additives was tested in 1.5% active material concentration in SN-150 base oil. The styrene containing additives had a significant viscosity index improving effect, as a result their application as engine oil additives can be considered in the future.
The lubricity of the FAME (S-1 - S-4); and the FAME and styrene containing additives (PSS-1 - PSS-4) was tested in 300 mg/kg concentration in diesel fuel by the four-ball machine. During the four-ball test the applied load was 300 N for 1 hour. The average wear scar diameter measured on the three standing balls and the results of the HFRR test are summarized in Figure 10. Among the additives synthesized with FAME (without styrene) FP-1 was found to have the best performance. The additives synthesized with both FAME and styrene resulted to be the most advantageous in increasing the lubricity of the base diesel fuel (GO). FPS-2 and FPS-4 additives had the best performance, these additives were synthesized with high styrene:maleic anhydride molar ratio. FPS-2 was produced from the FP-1 additive which was acylated with TEPA, while FPS-4 was produced from FP-3 additive which was acylated with PEHA. Both TEPA and PEHA are long chained amines among the tested ones, with a high total base umber.

Properties	FPS-1	FPS -2	FPS -3	FPS -4	Reference
Intermediate	TS-1	TS-2	TS-1	TS-2	-
Acylating agent	FP-1	FP-1	FP-3	FP-3	-
Molar ratio of intermediate and acylating agent	1.0:1.0	1.0:1.0	1.0:1.0	1.0:1.0	-
Appearance	Bright	Cloudy	Bright	Bright	Bright
TBN, mg KOH/g	11.8	10.3	12.6	13.5	40
Nitrogen content, %	1.98	1.58	2.43	2.50	3,22
1,5 % (based on active material content) additive in SN 150 base oil					
V.I.E	108	110	124	119	104
Detergent Index, % (max. 100)	100	100	100	100	100
Washing Efficiency, mm (max. 125)	92	95	94	92	90
Potential Detergent-Dispersant Efficiency, % (max. 100)	85	87	86	85	85

Table 8. Succinic additives containing FAME and styrene comonomer

The lubricity improving effect of the additives was also tested with HFRR machine. The results showed the same tendency among the efficiency of the additives as in case of the four-ball test.

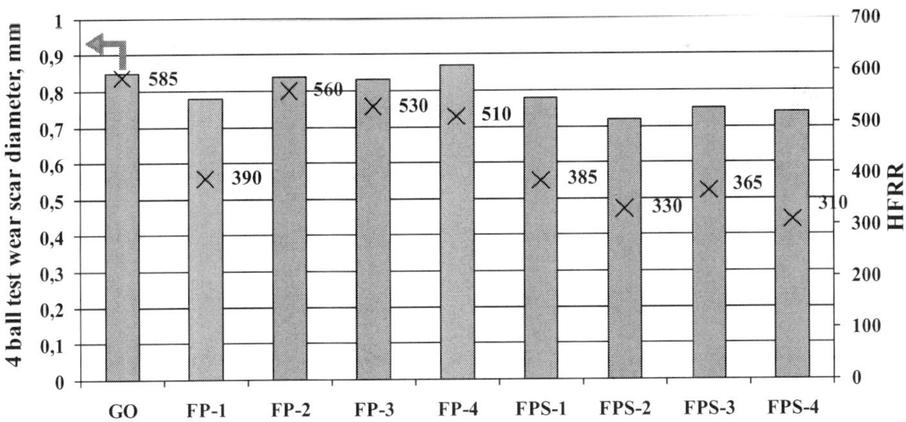

Fig. 10. Lubricity improving effect of the additives in 300 mg/kg concentration in base diesel fuel (GO)

4. Summary

By incorporating fatty acid methyl ester to the structure of the polyisobutylene succinimide a new additive was developed which had high detergent-dispersant efficiency and an additional lubricity improving and corrosion inhibiting effect.

The lubricity improving effect of the abovementioned additives could be further increased by incorporating not only fatty acid methyl ester, but also styrene comonomer to the additive. The detergent-dispersant efficiency of the additives remained relatively high, while their viscosity improving effect increased significantly. Their relatively low total base number ensures good compatibility with fluoroelastomers, thus in the near future the application of these additives in engine oil compositions can be promising research field.

5. Acknowledgement

We acknowledge the financial support of this work by the Hungarian State and the European Union under the TAMOP-4.2.1/B-09/1/KONV-2010-0003 project.

6. References

Batt, R.J., McMillan, J.A. & Bradbury, I.P. (1996). Lubricity additives - performance and no – harm effects in low sulphur fuels. SAE Paper 961943

Beck, Á., Bubálik, M., & Hancsók, J. (2009a) Development of a novel multifunctional succinic-type detergent-dispersant additive for diesel fuel. The 8th International Conference on Chemical and Process Engineering, pp. 1747-1752, ISBN 978-88-95608-01-3, ISSN 1974-9791, Rome, Italy, May 10-13., 2009; Chemical Engineering Transactions, 2009, 11, pp. 893-898.

Beck, Á., Krár, M., Pölczmann, Gy. & Hancsók, J. (2009b). Development of multifunctional additives for new generation bio-fuels. 4th International Bioenergy Conference, pp. 605-612, ISBN 978-952-5135-44-2, Finland, Jyväskylä, August 31. - September 4., 2009

Beck, Á., Pölczmann, Gy. & Hancsók, J. (2010). Improving the compatibility of multifunctional detergent-dispersant additives. 17th International Colloquium Tribology, 4 pp In Book of Synopsis 53., Germany, Stuttgart/Ostfildern, January 19-21, 2010

Breakspear, A. & Caprotti, R. (2007). Additives to provide injector detergency for Euro V. O03 Proceedings of Additives 2007 Conference, UK, London, April 17-19, 2007

Bubálik, M. & Hancsók, J. (2004). Characterization of the AF/AW Properties of Diesel Fuel. 6th International Symposium Motor Fuels 2004, MF-2218, pp 12, .Vyhne, ISBN 80-968011-3-9, 14-17 June, 2004

Bubálik, M., Hancsók, J., Molnár, I. & Holló, A. (2005). Characterization of the AF/AW properties of diesel fuel. 5th International Colloquium on Fuels 2005, pp. 279-286, ISBN 3-924813-59-0, In W.J. Bartz (Ed.), Germany, Stuttgart/Ostfildern, January 12-13, 2005

Candy, L., Vaca-Garcia, C. & Borredon, E. (2005). Synthesis of alkenyl succinic anhydrides from methyl esters of high oleic sunflower oil, Eur. J. Lipid Sci. Technol., Vol. 107, pp. 3-11

Caprotti, R., Breakspear, A. & Graupner, O. (2007). Beyond 2008: The Challenges for Diesel Detergency. 6th International Colloquium on Fuels 2007, pp. 273-276. ISBN 3-924813-51-5, Germany, Stuttgart/Ostfildern, January 15-16, 2007

Denecker, V. (2002). Diesel Fuel Sulphur Reduction and Lubricity Additive Use. Euroforum seminar - Compatibilité et évolutiondu couple moteurs et carburants, France, Paris, June, 2002

Hancsók, J. (1999). Modern engine and jet fuels II. Diesel Fuels, pp. 363, Veszprém University Press, ISBN 963 9220 27 2, Veszprém, Hungary (in Hungarian)

Hancsók, J., Bartha, L., Baladincz, J. & Kocsis, Z. (1999). Relationships Between the Properties of PIB-Succinic Anhydrides and Their Additive Derivatives, Lubrication Science, Vol.11, No.3, pp. 297-310

Hancsók, J., Bartha, L., Baladincz, J., Auer, J. & Kocsis, Z. (1997). Use of Succinic Anhydride Derivatives in Engine Oils and Fuels, Petroleum and Coal, Vol.39, No.1, 21-24

Hancsók, J., Bubálik, M., Beck, Á. & Baladincz, J. (2008b). Development of multifunctional additives based on vegetable oils for high quality diesel and biodiesel. Chemical Engineering Research & Design, Vol.86, pp. 793-799

Hancsók, J., Bubálik, M., Törő, M. & Baladincz, J. (2006). Synthesis of fuel additives on vegetable oil basis at laboratory scale, Eur. J. Lipid Sci. Technol., Vol.108, pp. 644-651

Hancsók, J., Krár, M., Magyar, Sz., Boda, L., Holló, A. & Kalló, D. (2007). Investigation of the production of high cetane number biogasoil from pre-hydrogenated vegetable oils over Pt/HZSM-22/Al_2O_3, Microporous and Mesoporous Materials, Vol.101, No.1-2, pp. 148-152

Hancsók, J.,Baladincz, J. & Magyar, J. (2008a). Mobility and Environment, pp 240, ISBN: 978-963-9696-50-1, Pannon University Press, Veszprém, Hungary, (in Hungarian)

Haycock, R.F. & Thatcher, R.G.F. (2004). Fuel Additives and Environment, Technical Committee of Petroleum Additive Manufacturers in Europe, 2004.

Kajdas, C. & Majzner, M. (1999). Boundary Lubrication of Low-Sulphur Diesel Fuel in the Presence of Fatty Acids, 2nd International Colloquium on Fuels, pp. 219-238, Germany, Stuttgart/Ostfildern, January 20-21, 1999

Kajdas, C. & Majzner, M. (2003). Diesel Fuel Lubricity - A Review, 4th International Colloquium on Fuels, pp. 369-384, Germany, Stuttgart/Ostfildern, January 15-16, 2003

Knothe G. (2005a). Lubricity of Components of Biodiesel and Petrodiesel. The Origins of Biodiesel Lubricity, Energy & Fuel, Vol.19, No.3, pp. 1192-1200.

Knothe, G. (2005b). Dependence of biodiesel properties on the structure of fatty acid alkyl esters, Fuel Processing Technology, Vol.86, pp. 1059-1070

Kocsis, Z., Baladincz, J., Bartha, L. & Hancsók, J. (2001). Possibilities of application of polyisobutenyl succinic anhydride derivatives of various molecular structures, Hungarian Journal of Industrial Chemistry, Vol.29, No.2, pp. 139-141

Kocsis, Z., Varga, G., Szirmai, L., Resofszki, G., Holló, A. & Hancsók, J. (2003). Detergents for Diesel Fuels to Improve Air Quality and Fuel Economy at Lower Operating Costs, 4th International Colloquium on Fuels 2003, pp. 273-276, ISBN 3 924813 51 5, in W.J. Bartz, (Ed.), Germany, Stuttgart/Ostfildern, January 15-16, 2003

Krár, M., Kovács, S., Kalló, D. & Hancsók J. (2010b). Fuel purpose hydrotreating of sunflower oil on CoMo/Al_2O_3 catalyst, Bioresources Technology, Vol.101, No.23, pp. 9287-9293

Krár, M., Thernesz, A., Tóth, Cs., Kasza, T. & Hancsók, J. (2010a). Investigation of catalytic conversion of vegetable oil/gas oil mixtures, Silica and Silicates in Modern Catalysis, In I. Halász, (Ed.), 435-455, Transworld Research Network, ISBN 978-81-7895-455-4, Kerala, India

Mach, H. & Rath P. (2006). Highly reactive polyisobutene as a component of a new generation of lubricant and fuel additives, Lubrication Science, Vol.11, No.2, pp. 175-185

Pasqualino, J.C., Montané, D. & Salvado, J. (2006). Synergic effects of biodiesel in the biodegradability of fossil-derived fuels, Biomass and Bioenergy, Vol.30, pp. 874-879

Quesada, J. (2003). Produciton of alkenyl succinic anhydrides from low-erucic and low linolenic rapeseed oil methyl esters, MCB University Press, 0036-8792, Eur. J. Lipid Sci. Technol, Vol.105, pp. 281-287

Rang, H. & Kann, J. (2003). Advances in petrol additives research, Proc. Estonian Acad. Sci. Chem, Vol. 52, No.3, pp. 130-142

Russel T. (1990). Diesel Fuel Additives, Diesel Fuel Quality Trends - the Growing Role of Additives, The College of Petroleum Studies, Course RF 6, April, 1990

Spikes H. A. & Wei, D.P. (1997). Fuel Lubricity - Fundamentals and Review, 1st International Colloquium on Fuels, pp. 249-258, Germany, Stuttgart/Ostfildern, January 16-17, 1997

Ullmann, J., Stutzenberger, H., Caprotti, R. & Hess, D. (2009). Effects of Fuel Impurities and Additive Interactions on the Formation of Internal Diesel Injector Deposits, 7th International Colloquium Fuels, pp. 377-388., ISBN 3-924813-75-2, Germany, Stuttgart/Ostfildern, January 14 - 15, 2009

Wei, D. & Spikes, H.A. (1986). The lubricity of diesel fuels, Wear, Vol.111, pp. 217-235

11

Research on Hydrogenation of FAME to Fatty Alcohols at Supercritical Conditions

Yao Zhilong
Beijing Institute of Petrochemical Technology, Beijing, PRC

1. Introduction

It is hard to develop biodiesel industry currently as a consequence of the rapid increase in the prices of animal and vegetable oil in recent years. Production of high-value bulk chemicals from biodiesel (fatty acid methyl esters) and its by-product of glycerol is an effective way to overcome the difficulties and to promote the steady development of biodiesel industry. However, the choice of target products and technologic routes for producing them should follow three principles: 1) biomass feedstocks are cheaper than petroleum products when they are used as raw materials, 2) process being developed is simple and environmentally-friendly, 3) the target products are in line with the market demands [1]. Production of fatty alcohols (FA) from fatty acid methyl esters (FAME) by hydrogenation under supercritical conditions is in line with these principles. The study on hydrogenation of FAME to FA using propane and carbon dioxide as solvents under supercritical conditions has been reported previously. But the reaction is carried out above 15.0 MPa [2-5], and such high operation pressure will seriously increase the capital and operating costs. It is thus clear that a low-pressure hydrogenation process for FAME conversion to FA under supercritical conditions can lead to reduction in the capital and operating costs of the process while the advantages of supercritical reactions are maintained, which can also lead to great social and economic benefits.

2. Experimental

2.1 Materials

The composition of the FAME of palm oil (supplied by Shijiazhuang petrochemical Co.) was shown in table 1.

A commercial copper-chromium oxide (supplied by Nanjing catalysis Factory) was used as the catalyst in this study, in the form of granule of 0.5~0.8mm average diameter, with a surface area of $73m^2*g^{-1}$ and a pore volume of $0.17cm^3*g^{-1}$.

Solvent A: Butane, chemical pure.

Solvent B: n-pentane, chemical pure.

Solvent C: Hexyl hydride, chemical pure.

2.2 Experimental set up

The hydrogenation reaction of FAME was carried out in 316 stainless steel downflow fixed beds, with an internal diameter of 17mm and 0.6m length, placed in oven and packed with 15g of catalyst. The temperature of reaction was measured and controlled with thermocouples at axial positions. The production was analyzed by HP5890 gas chromatography (HP Co.), using a InoWax capillary column (HP Co.,).

Contents	FAME in feedstock, Wt%
C_{12}FAME	0.301
C_{14}FAME	1.166
C_{16}FAME (0)	41.214
C_{16}FAME (1)	0.330
C_{18}FAME (0)	3.655
C_{18}FAME (1)	38.914
C_{18}FAME (2)	12.176
Others	2.244

Table 1. The composition of the FAME of palm oil

3. Results and discussion

3.1 Critical parameter measuring and solvent choosing

The compositions of FAME and solvent system have been measured at bubble point conditions. Peng-Roinson (PR) EOS and gas-liquid equation are used to calculate the binary interaction parameters. The critical parameter of the ternary system in which the ratio of FAME to solvent in weight is 90:10 has been predicted by PR model. The prediction reveals that the critical pressure of the ternary system is higher than 10.0 MPa as solvent A is used, and the critical temperature of the ternary system is higher than 300°C as solvent C is used. So, solvent B is chosen as the supercritical solvent in this work. It can be concluded from the study on the phase equilibrium that the critical temperature of this system increases as the ratio of FAME to solvent increases; the critical pressure of this system increases as the ratio of hydrogen to solvent increases.

3.2 Effect of the molar ratio of hydrogen to FAME

These data in figure1 were obtained under the operating conditions of the ratio of FAME to solvent B in weight is 90:10, 240℃, 9.5MPa and the reaction space velocity of 2.5-4.0h^{-1}.

These data in figure 1 show that the conversion of FAME increased with the molar ratio of hydrogen to FAME increasing when the molar ratio of hydrogen to FAME is below to 7.2 at different reaction space velocity. But, if the molar ratio of hydrogen to FAME is above 7.2, the conversion of FAME would decrease while the molar ratio of hydrogen to FAME increased. The result comparing with the conventional process of liquid-phase hydrogenolysis of fatty ester was different. The reason would be the reaction system was on supercritical conditions while the molar ratio of hydrogen to FAME was blow to 7.2 according the result of phase equilibrium. However, while the molar ratio of hydrogen to FAME was above 7.2, the reaction system was gas-liquid phase in the same way as the conditions process. So, the result shows that, to increase the reaction rate of hydrogenolysis of fatty ester, the reaction should be completed on the supercritical conditions.

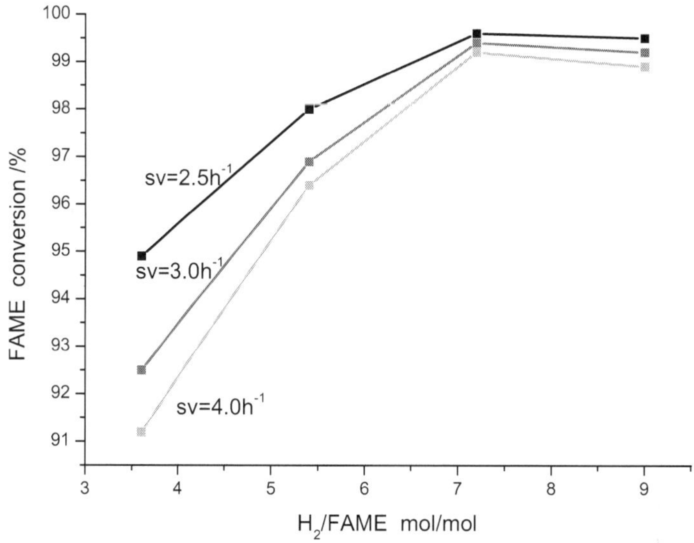

Fig. 1. The effect of the molar ratio of hydrogen to FAME

In the tradition process of hydrogenolysis of fatty acid ester, the selectivity of fatty acid ester generated lower alcohol with the reaction of hydrogen molar ratio. However, in the supercritical reaction system, the selectivity of alcohol generated with hydrogenated palm oil methyl ester increased, accompanied with molar ratio of fatty acid ester. The extreme value turned out to be around 7.2 mole ratio of hydrogen and palm oil fatty acid methyl ester, according to the date in Table 2 and Figure 2 When the mole ratio of hydrogen and palm oil fatty acid methyl ester was 7.2, the critical pressure was 9.05MPa. When the mole ratio went up to 9.0, the critical pressure went to 10.52MPa. In fact, the experimental pressure was 9.5MPa, which was in the rage of radio of hydrogen and acetate. Around 7.2 was the turning point of the change of reaction system. When the mole ratio of hydrogen and palm oil fatty acid methyl ester was less than 7.2, the reaction system was in supercritical state. Due to the polarity differences among supercritical fluids, there was a strong clustering effect between solute and solvent, solvent and solvent molecules, which made a large negative value of partial molar volume between produce and reactant near the critical point. The increase of reaction rate acted as an accumulation role to remove products promptly from the catalyst surface in order to prevent the secondary reaction with a further attempt, which means to improve the selectivity of reaction. This was also the kinetic and thermodynamic interpretation on the role of supercritical fluids to enhance the chemical reaction rate and selectivity. Among the reaction system of hydrogenated palm oil fatty acid methyl ester, the polarities between the products of alcohol and methanol were strong. However, the polarities the such reactants as palm oil fatty acid methyl ester, alkane solvents, and the reaction by products of six-alkanes and octadecane were weak. An associated effect might be caused by the products of reaction between fatty alcohol and methanol, which could remove quickly from the surface of catalyst and inhibit the further reaction of fatty alcohol to generate the side reactions of alkane. It could lead to the abnormal phenomena from the increase of the mole ratio of hydrogen and palm oil fatty acid methyl ester, to the selectivity of palm oil fatty acid methyl ester conversion and fatty alcohol generation. When the mole ratio of hydrogen and palm oil fatty acid methyl ester was 9.0, the phase state of reaction system was deviated from supercritical state. With the decrease of associated effect, the selectivity of generated palm oil fatty acid methyl ester declined in associated with the increase of adverse reaction.

SW/h^{-1}	H_2/FAME mol/mol	X_{FAME}/%	S_{FA}/%	Phase state
2.5	3.6	94.9	97.7	supercritical
	5.4	98.0	98.1	supercritical
	7.2	99.6	99.1	supercritical
	9.0	99.5	98.4	Nearly critical
3.0	3.6	92.5	98.6	supercritical
	5.4	96.9	99.1	supercritical
	7.2	99.4	99.3	supercritical
	9.0	99.2	98.7	Nearly critical
4.0	3.6	91.2	98.8	supercritical
	5.4	96.4	99.2	supercritical
	7.2	99.2	99.4	supercritical
	9.0	98.9	99.0	Nearly critical

Table 2. The effect of the molar ratio of hydrogen to palm oil FAME to the selectivity of fatty acid ester

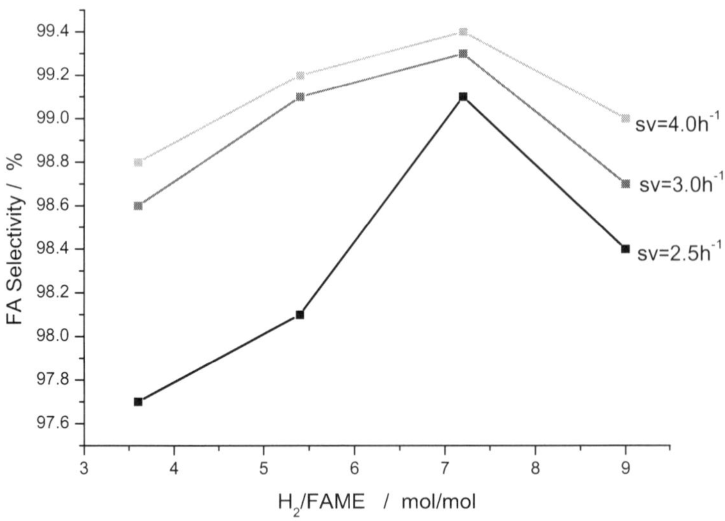

Reaction Temperature: 240°C; Reaction pressure: 9.5MPa

Fig. 2. The effect of the molar ratio of hydrogen to palm oil FAME to the selectivity of fatty acid ester

3.3 The impact of reaction temperature

As what has been discussed above, the thermodynamic analysis that fatty alcohol was generated by hydrogenolysis of fatty acid methyl esters has shown that the equilibrium constant of this reaction decreased by the increase of reaction temperature. The side reaction of fatty alcohol with further hydrogenation was an endothermic reaction. The equilibrium constant increased with increasing temperature. As a matter of fact, from the point of thermodynamic view, the reaction should be done in a lower temperature in order to obtain higher product selectivity purpose. From the reaction kinetics point of view, the reaction rate increased as reaction temperature raised. Therefore, it was necessary to compromise between reaction rate and purpose product selectivity, and to select appropriate conditions of supercritical hydrogenation process on fatty acid methyl ester. According to catalyst pretreatment and catalytic properties results and the results of reaction system equilibrium, in the condition of the test materials which the mass ratio of fatty acid methyl ester and solvent was 10:90, the response pressure was 9.5MPa, the molar ratio of hydrogen to fatty acid methyl ester of 7.2 , the results have been shown on Table 3, Figure 3 and Figure 4 in the range of 230~270°C.

Table 3, Figure 3 and Figure 4 shown the conversion rate of fatty acid methyl ester increased slightly with the raise of reaction temperature in the supercritical state. Although from the thermodynamic perspective, the reaction equilibrium constant of hydrogenolysis of fatty acid methyl esters decreased with temperature rising, from reaction kinetics perspective, the reaction rate was accelerated with increasing reaction temperature. Due to that fact, when reaction deviated from chemical equilibrium, the overall performed that the conversion rate of fatty acid methyl esters increased with the reaction temperature rising. The side reaction of hydrogenated alkane is an endothermic reaction. The purpose product selectivity was unfavorable whether from the perspective of thermodynamics or kinetics. The experiential results also showed that the purpose product selectivity of fatty acid methyl ester decreased significantly as the reaction temperature raised.

SW/h^{-1}	Temperature /°C	X_{FAME}/ %	S_{FA}/ %	Phase state
	230	98.9	99.4	supercritical
2.5	240	99.5	99.1	supercritical
	260	99.8	94.4	supercritical
	230	98.6	99.5	supercritical
	240	99.4	99.3	supercritical
3.0	250	99.6	98.8	supercritical
	260	99.7	96.9	supercritical
	230	97.9	99.6	supercritical
	240	99.1	99.4	supercritical
4.0	250	99.2	98.3	supercritical
	260	99.4	97.1	supercritical

Table 3. The Impact of Reaction Temperature (Reaction pressure 9.5MPa、 the molar ratio of hydrogen to FAME 7.2)

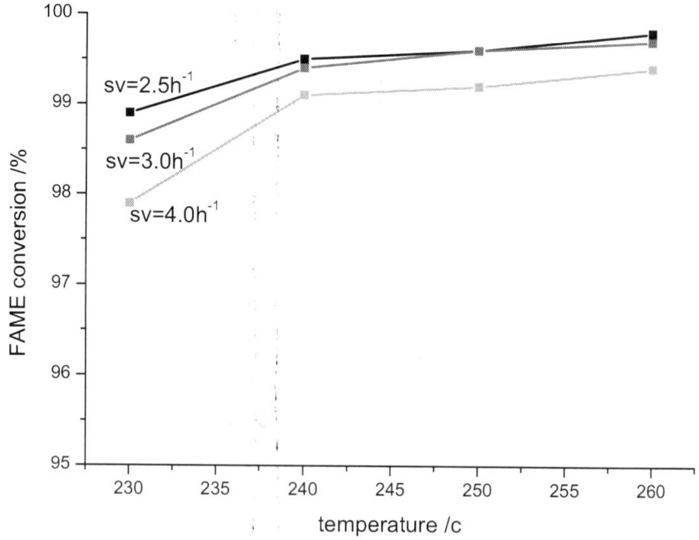

Reaction conditions: Reaction pressure: 9.5Mpa, the molar ratio of hydrogen to FAME:7.2

Fig. 3. The effect of the Reaction temperature to the conversion of the palm oil FAME

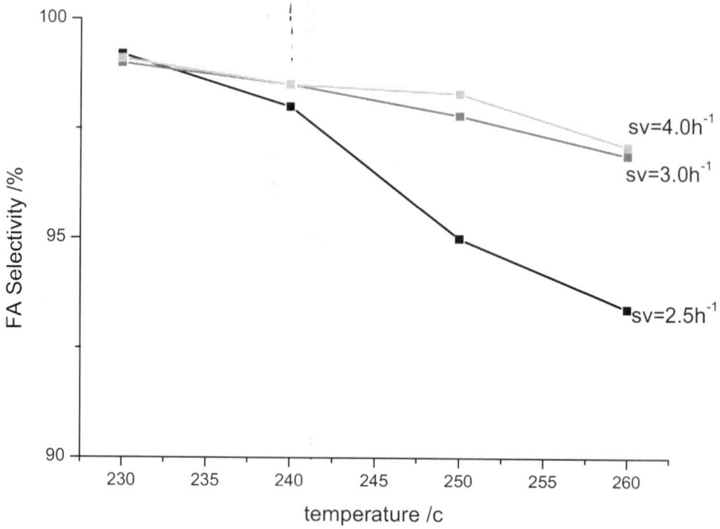

Reaction conditions: Reaction pressure: 9.5Mpa, the molar ratio of hydrogen to FAME:7.2

Fig. 4. The effect of the Reaction temperature to the selectivity

3.4 The impact of space velocity

In the process of heterogeneous catalysis, the space velocity of reactant, which is one of the important parameters in the process of heterogeneous, reflected the reaction materials and the catalyst contact time directly. The experimental results which contained reaction materials made of palm oil fatty acid methyl ester and solvent in the ratio of 10:90 have shown in Table 4and Figure 5 in the condition of reaction temperature 240 °C, molar ratio of hydrogen esters 7.2, reaction pressure 9.5MPa, in order to research how the space velocity (including mixed material with solvent contained) affected the reaction results.

According to the data from Table 4 and Figure 5, the conversion rate of fatty acid methyl ester increased with the space velocity decreased. However, the selectivity of product fatty alcohol generated by fatty acid methyl ester increased slightly with the space velocity increased. This was due to the increasing space velocity, short time contacted by reaction mixture and catalyst.

Reaction Space Velocity /h-1	2.5	3.0	3.7	5.0	7.5
Methyl ester conversion / %	99.6	99.35	99.1	86.6	71.2
The selectivity of alcohol / %	99.0	99.2	99.3	99.3	99.6
System state	Supercritical	Supercritical	Supercritical	Supercritical	Supercritical

Reaction conditions: Reaction temperature 240°C, the molar ratio of hydrogen to FAME 7.2, Reaction pressure 9.5MPa

Table 4. The Impact of Space Velocity

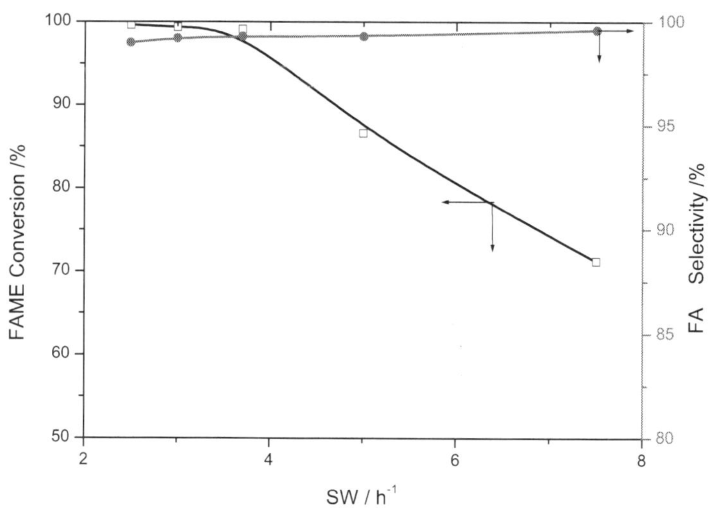

Reaction conditions: Reaction temperature 240°C, the molar ratio of hydrogen to FAME 7.2, Reaction pressure 9.5MPa

Fig. 5. The effect of Space Velocity to the conversion and selectivity

This also indicated that reaction was controlled by kinetics in the condition of the space velocity. Besides, the date in Table 4 and Figure 5 shows that the conversion rate of fatty acid methyl ester was above 99% with the condition of less than $4.0h^{-1}$ space velocity. While in terms of purpose products, it was more than 90%, and increased slightly with space velocity increased. Compared with $0.15\sim0.4h^{-1}$ space velocity which calculated according to tradition process of hydrogenolysis of fatty acid methyl ester, supercritical reaction technology was taken, but there was 90% solvent remaining in reaction materials. When the space velocity of mixture made of solvent and fatty acid methyl ester was $4.0h^{-1}$, the space velocity of fatty acid methyl ester was around $0.4h^{-1}$. There was no significant difference between them. That means compared with tradition technology, taking new supercritical reaction technology to deal with the same amount of fatty acid methyl ester did not change the size of reactor by increasing supercritical solvent. In other words, it was able to enhance the production strength of fatty alcohol.

3.5 The impact of operating pressure

According to the situation of hydrogenation of fatty acid methyl ester, when the composition of fatty acid methyl ester, solvent and hydrogen was constant, the reaction system could change into different phase state by transforming the system operating pressure in a certain degree. Therefore, operating pressure could not only affect the reaction result, but also affect the system operating pressure. The experimental materials on the pressure effect research is the mixture solution with mass ratio of 10:90 on fatty acid methyl ester and solvent. Experimental results by different operating pressures are shown in Table 5 and Figure 6 on the condition of reaction temperature 240°C, the molar ratio of hydrogen fatty acid methyl ester 7.2, and the space velocity of mixed solution weight $4.0h^{-1}$.

According to Table 5 and Figure 6, as the pressure increased, the conversion rate of fatty acid methyl ester increased. When the pressure rose to 8.0MPa, the conversion rate increased rapidly. The purposed product selectivity of fatty alcohol rose with the reaction pressure. The minimum was occurred with around 8.0MPa of the operating pressure. The research results on the reaction equilibrium shows that the critical pressure of reaction system was 9.0MPa due to the composition of experimental materials. It means that when the operating pressure reached 8.0MPa, the phase reaction was close to the critical state. When the operating pressure was up to 9.5MPa, the reaction system was in the supercritical state. That also shows that when the operating pressure was lower than 8.0MPa, the hydrogen solubility increased in the reaction medium as the increasing operating pressure. It reduced the mass transfer resistance, and led to the fatty alcohol generated by hydrogenation of fatty acid methyl ester and the reaction rate with hydrogen added to generate alkanes increased. The conversion rate of fatty acid methyl ester increased as the pressure increased, and the purposed product selectivity decreased as the pressure increased. The reaction system was up to the supercritical state in the same situation. The reaction system was changed from gas-liquid phase to supercritical fluid phase. The mass transfer resistance of hydrogen in the reaction medium was eliminated. The conversion rate of fatty acid methyl ester increased significantly, when the operating pressure was over 8.5MPa. Due to the fact that when the reaction pressure was close or up to the reaction system critical point, the product of fatty alcohol, the association of the product of fatty alcohol and methanol improved gradually. The speed of removing from fatty alcohol to catalyst surface in time was increased. It could benefit the local composition which was good to purpose response. It also reduced the further alkane generation of adverse reaction from fatty alcohol. Therefore when the reaction pressure was close or up to the reaction system critical point, the purposed product selectivity of fatty alcohol rose with the reaction pressure. That led to the minimum of selectivity of fatty alcohol was occurred by the change of the operating pressure. The results of phase equilibrium and calculated reaction system stimulation which were reasonable were proved by the regularity that the operating pressure affected the reaction results.

Reaction pressure /MPa	6.5	7.5	8.0	8.5	9.5
Methyl ester conversion / %	65.2	72.3	80.5	98.4	99.1
The selectivity of fatty alcohols / %	99.2	98.5	97.8	98.8	99.4
System state	Gas - liquid	Gas - liquid	Subcritical	Nearly critical	supercritical

Reaction conditions: Reaction temperature 240°C, the molar ratio of hydrogen to FAME7.2, Weight Space Velocity of the mixed solution $4.0h^{-1}$

Table 5. The Impact of Operating Pressure

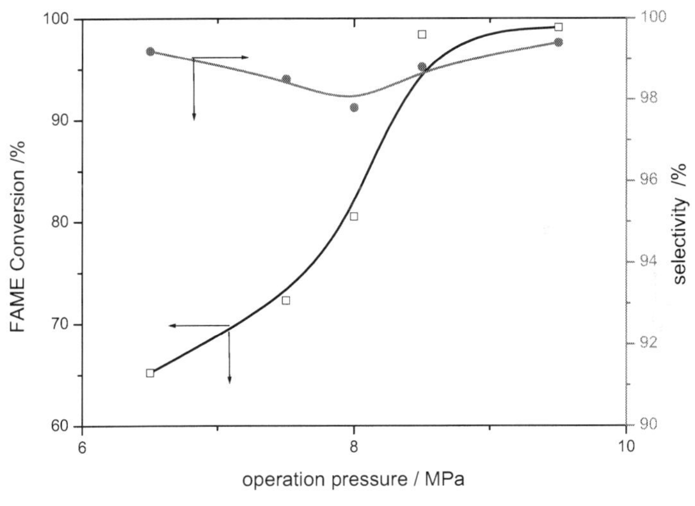

Reaction conditions: Reaction temperature 240°C, the molar ratio of hydrogen to FAME 7.2, Weight Space Velocity of the mixed solution $4.0h^{-1}$

Fig. 6. The Impact of Operating Pressure

3.6 Analysis of reaction products

The fatty alcohols produced from this process have been evaluated, as shown in table 6. Comparing the date in table 6 and table 7, the results show that the properties of these products meet the GB/T 16451-1996 specifications.

Compose	Content /%
FAME	0.73
FA	98.77
Alkane	0.50

Table 6. The compose of reaction products

Type		Appearance	MP/°C	Content of alkane/%	Content of FA/%
C_{16}~$_{18}$FA	Superior quality	White flaky	48-58	≤0.5	≥95
	First quality			≤1.0	≥90
	Qualified			≤2.0	≥85

Table 7. GB/T 16451-1996 specifications of nature fatty alcohols

4. Conclusion

For FAME of palm oil, the proper supercritical conditions are as follows: molar ratio of hydrogen to FAME of 5.4-7.2, 240-250°C, 9.0-9.5MPa, and the reaction space velocity of 2.5-4.0h^{-1}, respectively. It has been shown that conversion of more than 99% and about 99% yield of fatty alcohol can be achieved at the optimized reaction conditions. A comparison between the novel hydrogenation technology under supercritical conditions in this work and the conventional process of liquid-phase hydrogenolysis of fatty ester shows that the ratio of hydrogen to FAME in reaction system is reduced by 90-98%, and the temperature and pressure of reaction is decreased significantly also in this work. It can be concluded from the reaction conditions that the pressure of this novel process here is lower by about 5.0 MPa than that for the process reported by other researcher[4].

5. References

[1] Min Enze, Yao Zhilong. The Development of Biodiesel Industry in Recent Years -- Peculiarity, Predicament and Countermeasures. Progress In Chemistry (Chinese), Vol 19, No 7(2007):1050-1059

[2] Brand D.S., Poels E.K. Solvent-based fatty alcohol synthesis using supercritical butane: thermodynamic analysis. JAOCS, Vol 79, No 1 (2002): 75~83

[3] Brand D.S., Poels E.K. JAOCS, Vol 79, No 1 (2002): 85~91

[4] Sander van den Hark, Magnus Harrod. Ind. Eng. Chem. Res., 2001(40): 5052~5057

[5] Sander van den Hark, Magnus Harrod. Applied catalysis A: General 210(2001): 207~215

12

The Use of Biodiesel in Diesel Engines

S. Chuepeng
Kasetsart University
Thailand

1. Introduction

Biodiesel has been increasingly used in diesel engines as a neat or partial substitute with diesel within the past few decades. It is mainly due to its comparable properties to those of diesel, environmental concerns, and energy security. This chapter describes impacts of the use of biodiesel as a fuel for diesel engines, collected from previous research work recently published in journals, proceedings, or other references involved.

1.1 Advantages of biodiesel use in diesel-powered vehicles

Promotions to use alternative bio-fuels in transportation and environmental concerns on carbon dioxide (CO_2) emissions are the main reasons for instigating the use of biodiesel as an alternative fuel for compression ignition (CI) engines (usually known as diesel engines). Presently, vehicles currently circulated in Europe and other countries are fuelling with low percentage of biodiesel without problem, due to a consequence of technological advances. In Europe, there is a European Union (EU) Directive to promote the use of bio-fuels for transportation (Directive 2003/30/EC, 2003) with an objective of increasing use of bio-fuels towards CO_2 emission reduction in transportation.

As one among other bio-fuels, biodiesel is considered to be CO_2 neutral in terms of the global carbon cycle (Quirin et al., 2004). In the production aspect, the cost for bio-fuel for transportation is normally higher than those of conventional fossil fuels. However, there are benefits of biodiesel in the view of environment, not just conserving fossil fuel resources. Other distinctive advantages comprise near-zero sulphur content in the fuel and its combustion emissions, superior capability of biological degradation in aquatic environment, and a reduction in greenhouse effect gas due to a more favourable energy and CO_2 balance over the full life cycle (Camobreco et al., 2000). The latter revealed that overall energy used with soybean-based biodiesel production (feedstock production, feedstock transportation, conversion, fuel transportation) and use (combustion in a diesel engine) will drop by 74% compared to fossil diesel. Though, this report makes no account between CO_2 fixation by the soybean crop and the use of land for farming.

In addition, the Commission Green Paper (CEC, 2000) described an ambitious EU programme that has set a target of 20% alternative fuel substitution in conventional fuel in the road transport sector by the year 2020. However, for compliance to the relevant legislation on emission standards, the EU Directive suggests that high proportion blends (>5% v/v) of biodiesel used in non-adapted vehicles should be monitored.

1.2 Emission regulation and controls

Diesel engines are normally encounter with combustion noise, engine vibration, and the problem of nitrogen oxides (NO_x)- particulate matter (PM) trade-off emissions. The latter is considered to impact in global part while world emission legislations are increasingly stringent. Table 1 shows an example of the EU emission standards for heavy-duty diesel engines since EURO I which came into force in 1992.

Tier	Year	CO	HC	NO_x	PM	Smoke	Test Method
		$g·kW^{-1}·h^{-1}$	$g·kW^{-1}·h^{-1}$	$g·kW^{-1}·h^{-1}$	$g·kW^{-1}·h^{-1}$	m^{-1}	
EURO I	1992 (<85 kW)	4.5	1.1	8.0	0.612		ECE R-49
	1992 (>85 kW)	4.5	1.1	8.0	0.36		ECE R-49
EURO II	1996	4.0	1.1	7.0	0.25		ECE R-49
	1998	4.0	1.1	7.0	0.15		ECE R-49
EURO III	2000	2.1	0.66	5.0	0.1	0.8	ESC and ELR
EURO IV	2005	1.5	0.46	3.5	0.02	0.5	ESC and ELR
EURO V	2008	1.5	0.46	2.0	0.02	0.5	ESC and ELR
EURO VI	2013	1.5	0.13	0.4	0.01		ESC and ELR

Table 1. EU emission standards for heavy-duty diesel engines (Source: www.dieselnet.com)

Researchers have made efforts to reduce pollutant and greenhouse gases emitted from engines. A number of approaches have been conducted and developed since internal combustion (IC) engines were invented. Nowadays, clean diesel engine technologies have been introduced and widely used such as (1) fuel and additives, (2) in-cylinder technology, (3) lubricant oil, and (4) exhaust gas after-treatment devices.

In the view of fuel technology, biodiesel fuels in forms of ethyl or methyl esters have been proven to lower hydrocarbon (HC), carbon monoxide (CO), and PM but generating higher NO_x emissions (Graboski & McCormick, 1998; Lapuerta et al., 2008) when combusted in diesel engines. A synthetic gas-to-liquid (GTL) (as well as X-to-liquid) fuel derived by the Fischer-Tropsch method has been introduced to increasing numbers of countries. The synthetic diesel fuel properties are comparable to those of fossil diesel but higher cetane number, lower sulphur, and lower aromatic hydrocarbons (Oguma et al., 2002). The combustion of synthetic diesel improves fuel consumption and emissions, i.e. NO_x, PM, CO, and HC, compared to fossil diesel. Such reported problems e.g. sliding part lubricity, seal material compatibility, and low temperature flowability can be improved with additives (McMormick et al., 2002). In the past decades, hydrogen as a gas has been tested and substituted diesel in the IC engines. It contains no carbon and therefore does not produce CO_2. The addition of hydrogen to the main fossil diesel was favorably reported in terms of brake power, thermal efficiency, and reduction of HC, CO, CO_2, and PM emissions (Kumar et al., 2003).

In-cylinder fuel injection system is one of the effective strategies in reducing emissions from diesel engines (Mahr, 2002). For the engine induction system, both fresh air and exhaust gas recirculation (EGR) was proven to reduce NO_x emissions. Yet another charging system, variable geometry turbocharger (VGT) can recently provide acceleration for a wide range of load and speed (Filipi et al., 2001). Furthermore, new combustion concepts, i.e. multiple stage diesel combustion (MULDIC) (Hashizume et al., 1998), late fuel injection strategies (Kimura et al., 2001), premixed diesel combustion (PREDIC) (Klingbeil et al., 2003), homogeneous charge compression ignition (HCCI) (Ibara et al., 2006), and partially

premixed compression ignition engines (PPCI) (Weall & Collings, 2007) have been tested and some of them are in markets today.
Auxiliary emission control devices makes possible an optimisation between fuel consumption (in term of thermal efficiency) and NO_x-PM trade-off emissions, thank to the advent of new control technologies. Examples of after-treatment techniques are diesel oxidation catalysts (DOC), diesel particulate filters (DPF), NO_x adsorber catalyst (NAC), and selective catalytic reduction (SCR). These come into common use nowadays.

1.3 Diesel engine operation

Diesel engine operates at high compression ratios as only air is inducted into the cylinder and compressed. Fuel and air are therefore mixed internally (Ferguson, 1986). The injection process of a high pressurised fuel takes place under high temperature compressed air condition in the cylinder near the end of the compression stroke. This fuel jet atomises into droplets, evaporates, and entrains in the compressed air to form a combustible charge. At that time, the air temperature and pressure are beyond the fuel's ignition point, and after a short delay, auto-ignition of the fuel-air mixer spontaneously initiates the combustion process. This concomitantly occurs in all over the combustion chamber unlike propagated flame in the spark ignition (SI) engine (gasoline or petrol engine).
The overall diesel combustion process described in Heywood (1988) can be summarised here by identifying in a typical heat-release-rate diagram of a direct injection engine with one injection per engine cycle as shown in Fig. 1. This may differ from that of multiple injection engines such today's engines with common rail fuel injection system. Ignition delay is the period between the start of fuel injection (SOI) into the combustion chamber and the start of combustion (SOC). The phase of rapid combustion of the premixed fuel with air under the flammability limit during the ignition delay period is called premixed combustion, resulting in the high heat-release rate characteristics of this phase. Subsequently, mixing-controlled combustion phase occurs once the fuel-air pre-mixture during the ignition delay has been consumed. The burning rate is controlled in this phase primarily by the fuel vapour-air mixing process and the heat release rate is controlled by the mixture becoming available for burning. Late combustion is the phase well into the expansion stroke that heat release continues in low rate, due to a small fraction of the fuel yet has not been burnt, promoting more complete combustion.

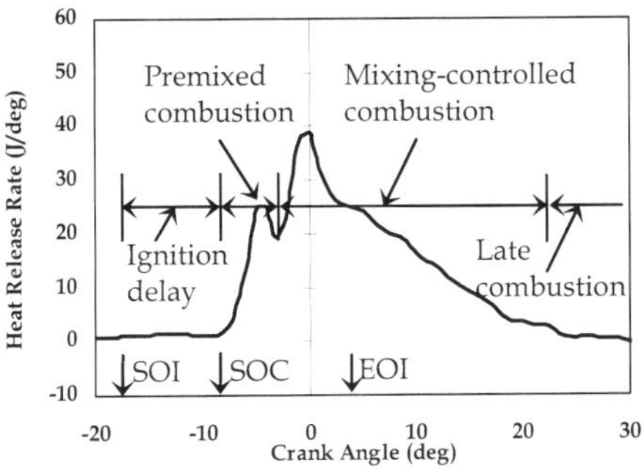

Fig. 1. Typical heat release rate diagram of direct injection engine identifying diesel combustion phase (Heywood, 1988, with modification)

In the expansion stroke, the exhaust valves start to open about two-third of the way. At this time, the blow-down process takes place as the cylinder pressure is higher than the exhaust manifold pressure. The piston simultaneously pushes the burned gases out of the cylinder during exhaust stroke through the valves, into exhaust port and manifold. Just before top dead centre (TDC), the intake valves open while the exhaust valves close just after TDC; this is called valve overlapping. The next cycle starts again.

1.4 Diesel fuel injection

Just about to reach TDC in the compression stroke, the fuel is injected into the cylinder of a diesel engine by high pressure pump through a nozzle orifice. High injection pressures ranging from 200 to 2,000 bar, depending on specific combustion strategies, are required. This is as the injected liquid fuel jet enters the combustion chamber at high velocity to atomise the fuel into droplets for rapid evaporation and to traverse the combustion chamber in a short time for fully utilising the air charge.

It is necessary to develop the fuel injection pump to serve increasing demands for fuel injection systems (Bosch, 2005) as well as the tightening exhaust gas emission standards. The followings are common types of fuel injection pump systems.

a. Distributor injection pumps with mechanical and electronic governors producing injection pressures up to 700 bar, especially popular in high-speed diesel engines for passenger cars and light-duty trucks.
b. In-line injection pumps with mechanical governors or electronic actuators timing devices producing injection pressure up to 1,150 bar, generally used for commercial vehicles and stationary engines.
c. Single-plunger injection pumps, directly actuated by the engine's camshaft with injection pressure up to 1,500 bar, usually used with large marine engines, construction machinery and low displacement engines.
d. Unit injector-pump system, commonly employed in commercial vehicles and passenger cars with injection pressures up to 1,500 bar.
e. Common rail injection system, fully equipped with sensors and actuators.

For the common rail fuel injection system, the injection pressure and timing are independent (Flaig et al., 1999). The injection timing is controlled by an engine electronic control unit (ECU) which can communicate with a fast control area network (CAN). This can be applied to both naturally aspirated and turbocharged engines. Additionally, hydraulic actuation of conventional pump-line-injector fuel systems can be eliminated. By this manner, multiple injections within an engine cycle are enabling. Therefore, engine torque and noise levels can be potentially improved.

2. Biodiesel production

Biodiesel production from oil-bearing crops, animal fats, and waste cooking oils is literary investigated. These include several operating parameters (e.g. feedstock, catalyst, techniques, etc.) which impact on biodiesel production processes, i.e. transesterification and esterification reactions. Selected fuel standards for biodiesel are gathered and presented.

2.1 Biodiesel production techniques

Biodiesel is oxygenated compounds, defined as the mono alkyl esters of long chain fatty acids derived from lipid feedstock for example, vegetable oils, animal fats, or even waste cooking oils. Biodiesel can be used in diesel engines as some of its key properties are similar to those of fossil diesel. However, pure oils are unsuitable for diesel diesel due to being a cause of carbon deposit and pour point problems (Graboski & McCormick, 1998). Additionally, they can also lead to engine problems, e.g. long-term engine deposit, injector plugging, or lube oil gelling (Kalam & Masjuki, 2005).

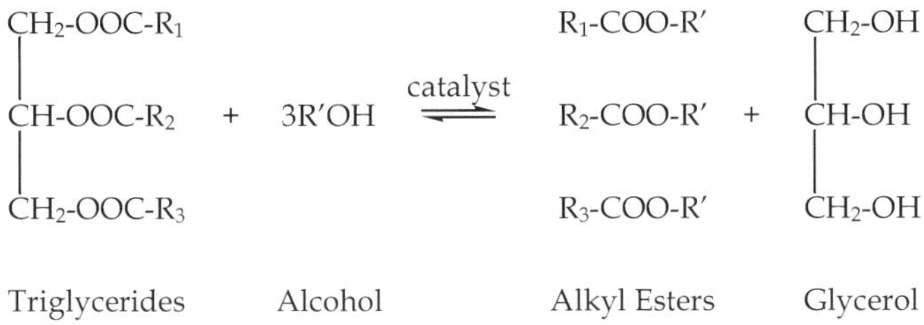

Fig. 2. Transesterification of triglyceride with alkyl alcohol (Komintarachat & Chuepeng, 2009)

To prepare biodiesel, the most commonly used process is the base catalyst (e.g. sodium hydroxide, NaOH) reaction (Graboski & McCormick, 1998), due to its cost effectiveness and reaction stability. Biodiesel is produced through a transesterification from pure oils (Van Gerpen et al., 2004). In the transesterification simply depicted in Fig. 2, feedstock in forms of triglycerides reacts with methanol in the presence of a catalyst to yield fatty acid methyl ester (FAME) and by-products (Kinast, 2003). The by-products generally are glycerol, water, methanol and catalyst traces, and un-reacted triglycerides (Babu & Devaradjane, 2003).

In the biodiesel production from waste used cooking oil (WCO), the methanol based transesterification over the synthesized solid acid catalyst at high free fatty acid (FFA) of 15% w/w was extensively studied by Komintarachat & Chuepeng (2009). The type of porous support of the catalyst affected the amount of the FAME yield. Under the reaction conditions of 383 K temperature, 0.3 methanol/WCO weight ratio, 1.0% w/w catalyst to WCO ratio within 2-hour reaction time, it was found that the WO_x/Al_2O_3 support yielded the maximum FAME of 97.5% and the rest were in the following order: silicon oxide (SiO_2) > tin oxide (SnO_2) > zinc oxide (ZnO). This is due to higher surface area and greater volume of the porous aluminum oxide support compared to the others.

The catalytic activities of conventional catalysts, i.e. potassium hydroxide (KOH), potassium carbonate (K_2CO_3), sulfuric acid (H_2SO_4) with the WO_x/Al_2O_3 catalyst under the optimum condition, previously mentioned were also compared and studied by Komintarachat & Chuepeng (2009). The WO_x/Al_2O_3 and KOH catalysts gave the highest activity by yielding the maximum FAME. However, the latter promotes soap formation which may be a problem on separation (Jitputti et al., 2006).

The conversion of WCO at 15.0% w/w FFA to biodiesel over potassium hydroxide (KOH) catalyst through transesterification reactions was reported in Komintarachat & Chuepeng (2010). The effects of alcohol and catalyst quantity, reaction time, and temperature on the FFA conversion and biodiesel production were studied. The optimum use of 5% w/w KOH catalyst at 70°C for 2 h yielded 88.20% FFA conversion and 50% biodiesel recovery of WCO. It was observed that the produced biodiesel has exhibited the same functional group as of the biodiesel blend sold in local gas station. Summarily, the produced biodiesel may be used in diesel engines if other properties are tested for compatibility. This provides one more choice for alternative energy.

2.2 Biodiesel standard

The quality of biodiesel in Europe is described in the European Standard EN 14214 "Automotive fuels - Fatty acid methyl esters (FAME) for diesel engines - Requirements and test methods". Both neat biodiesel and its blend component are required to conform to this standard while requirements and test methods are shown in Table 2.
Neat biodiesel is named B100 and may be blended with fossil diesel. In case of the blend, it is designated as BXX, where XX represents the volumetric percentage of neat biodiesel contained in the blend. In Directive 2003/30/EC (2003), biodiesel used for vehicles in pure form or as a blend should comply with the quality standard to ensure optimum engine performance.

Property	Unit	Minimum limit	Maximum limit	Test method
Viscosity @ 40°C	$mm^2 \cdot s^{-1}$	3.50	5.00	ISO 3104
Flash point	°C	120	-	ISO 3679
Sulphate ash	% wt	-	0.02	ISO 3987
Cetane number		51.0	-	ISO 5165
Carbon residue	% wt	-	0.30	ISO 10370
Acid value	$mg\ KOH \cdot g^{-1}$	-	0.50	EN 14104
Total glycerol	% wt	-	0.25	EN 14105
Oxidation stability @ 110°C	h	6.0	-	EN 14112
Sulphur	$mg \cdot kg^{-1}$	-	10.0	ISO 20846/84

Table 2. Requirements for fatty acid methyl ester

3. Biodiesel properties

Biodiesel fuels in the form of methyl or ethyl esters are oxygenated organic compounds that can be used in diesel engines as some of their properties are comparable to those of diesel. Table 3 shows the key properties of biodiesel derived from rapeseed oil (rapeseed methyl ester, RME) and ultra low sulphur diesel (ULSD) in comparison.
Biodiesel feedstock does not inherently contain sulphur but however, it may be present in biodiesel because of prior contamination during the transesterification process and in storage (EMA, 2003). Some other physical properties of biodiesel affect characteristics of the combustion in diesel engine such as density and viscosity (Rakopoulos & Hountalas, 1996) and bulk modulus of compressibility (Tat & Van Gerpen, 2002; Boehman et al., 2004). The bulk modulus of compressibility of biodiesel (property not shown in Table 3) is higher than that of fossil diesel. This yields better fuel atomisation by increasing the number and shifting the fuel droplets to smaller sizes. Generally, the bulk modulus of compressibility is a function of injection pressure. This suggests that the pressure in the pump-line-injector fuel system with biodiesel fuelling can be built-up and distributed faster even at the same pump timing (Szybist & Boehman, 2003).

Fuel analysis	Unit	Test method	Ultra low sulphur diesel	Rapeseed methyl ester
Viscosity at 40°C	cSt	ASTM D445	2.467	4.478
Density at 15°C	$kg \cdot m^{-3}$	ASTM D4052	827.1	883.7
Cetane number		ASTM D613	53.9	54.7
Lower heating value	$MJ \cdot kg^{-1}$		42.7	39.0
Sulphur	$mg \cdot kg^{-1}$	ASTM D2622	46	5
Molecular weight			209	296
50% distillation	°C		264	335
90% distillation	°C		329	342
Carbon	% wt		86.5	77.2
Hydrogen	% wt		13.5	12.0
Oxygen	% wt		-	10.8

Table 3. Fuel properties (Chuepeng et al., 2007)

4. Biodiesel-fuelled engine performance

In general, typical heating value for biodiesel is lower than that of fossil diesel (see Table 3). A greater amount of fuel is subsequently required to maintain the same engine brake torque. Greater fuel consumption of up to 13% by the use of D-2 diesel-biodiesel mixture were reported with heavy-duty engines over the United States Federal Test Procedure (US-FTP) cycle (Sharp et al., 2000a). However, the energy efficiency is independent of fuel consumption (Graboski et al., 1996).

The engine power is dependent upon the energy density stored in the fuel (Chuepeng, 2008). Sharp et al. (2000a) revealed their findings that 8% and 2% engine power losses are measured with neat biodiesel and B20 blends, respectively. In addition, Graboski et al. (1996) found the reduction in maximum torque respective to the increase of biodiesel blend in a D-2 diesel. The brake torque from the combustion of neat biodiesel is lower by 5.4% compared to that from pure D-2 diesel which is in good agreement as expected from the energy density ratio of the two base fuels. However, Senatore et al. (2000) found that the engine torque and performance are substantially unaffected when comparing in terms of equivalence ratio.

5. Combustion characteristics of biodiesel and its blends

Both physical and chemical properties can affect combustion characteristics of biodiesel and its blends. The biodiesel blends combustion increases the average peak cylinder pressure due to the shorter ignition delay over the baseline diesel combustion (Chuepeng, 2008). The advanced injection timing and increased injection pressure (and thereby increased fuel injection rate) have been frequently reported for the use of biodiesel (Szybist & Boehman, 2003). The main reasons are due to their differences in density (Rakopoulos & Hountalas, 1996) and bulk modulus of compressibility (Boehman et al., 2004).

Chuepeng et al. (2007) studied quantitative impacts on combustion characteristics and exhaust emissions by the use of high proportion biodiesel blends. Fuel mixtures of 0%, 25% and 50% RME by volume in ULSD were experimentally investigated in a single cylinder diesel engine in terms of the effects of engine load, exhaust gas recirculation (EGR) rate, and injection timing. By keeping engine with the same load, the RME blends increased proportion of the fuel burnt in the premixed phase and the combustion is advanced to earlier crank angle positions, with shortened ignition delay and increased peak cylinder pressure. Increasing the EGR rate of up to 20% at the same load and speed appeared to reduce peak pressure slightly and increase ignition delay, for all tested fuels. Without EGR, the SOI was studied by advancing and retarding by 2 °CA from the standard injection timing (22 °CA BTDC). For the same blended fuel, the retarded SOI lowered peak pressure and shorter ignition delay for all tested fuels, and the adverse effects were observed with advanced SOI.

6. Combustion-generated emissions

Without exhaust catalyst and timing change, common trends of exhaust gas emissions from a stock engine fuelled with neat or blended biodiesel are (1) increased NO_x, (2) decreased PM, CO, and HC, and (3) decreased soot (solid carbon fraction of PM) mass emission (Lapuerta et al., 2008). Summarily, biodiesel and its blends mostly reduce engine emissions compared to fossil diesel, while the only regulated emission shown to increase consistently with biodiesel is NO_x. There are three main strategies to mitigate the increasing engine NO_x:

- Determining biodiesel properties which can be modified to lower NO_x emissions or modifying fuel properties using a proper base fuel and additives for biodiesel blending (McCormick et al., 2002).
- Improving combustion chamber design to inhibit NO_x production by lowering combustion temperatures.
- Calibrating the engine when using biodiesel fuel. NO_x can be controlled by tuning injection strategy to optimise all engine outputs specifically for biodiesel (Postrioti et al., 2003).

For other unregulated emissions from an engine fuelled with biodiesel, polycyclic aromatic hydrocarbon (PAH) and nitro PAH compounds are substantially reduced, as well as the lower levels of some toxic and reactive HC species (Sharp et al., 2000b).

The PM composition (i.e. volatile material and elemental carbon) from the combustion of RME-based biodiesel blend (B30) in a turbo-charged engine with EGR operation was studied using thermo-gravimetric analysis (TGA) (Chuepeng et al., 2008a). Generally, total PM mass from B30 combustion was lower than that for diesel in all engine operating conditions. Elemental carbon PM mass fractions were slightly lower for the B30. The volatile material portions of the B30 particulates are greater than those of diesel particulates irrespective of engine operating condition. For both fuels used in the test, volatile material was observed to be higher at idle speed and light load when exhaust gases were at low temperature. For other carbonaceous emissions, the combustion of B30 tends to reduce visible smoke, HC and CO emissions.

For particle number size characterisation, Tsolakis (2006) examined the exhaust PM from a single cylinder diesel engine equipped with pump-line-injector fuel system and fuelled with neat biodiesel. The particle size distributions were found to be affected by the use of EGR. The results previously obtained were consistent with those conducted by Chuepeng et al. (2008b) using a V6 diesel engine equipped with a common rail fuel injection system. In summary, the particle size of B30 combustion aerosol without EGR is smaller than that of diesel while giving higher number concentration. When EGR were in use, the total particle number and mass were increased along with the increase in particle size for both B30 and diesel. The total calculated particle masses of B30 combustion aerosol are lower than those of the diesel case (Chuepeng et al., 2009). This confirms the results obtained by the TGA previously mentioned.

7. Emission control technology for biodiesel-fuelled engine

Emission control technology for biodiesel-fuelled engine is composed of two main ideas, i.e. engine and after-treatment technologies. These have been tested and widely introduced to diesel engine vehicles. For the engine technology, two popular methods comprise fuel injection strategy (both fuel injection timing and pressure) and EGR. With the advent of advance technology in electro-mechanics, the common rail fuel injection system can accomplish splitting fuel injection, choosing injection event and timing, and controlling injection pressure. By this way, the rate shaping strategies of the fuel injection are controllable (Mahr, 2002). The NO_x emissions can be reduced using pre-injection with small amount of fuel; this prevents a long period of ignition delay, resulting a reduction of peak pressure occurred when the premixed fuel combusts.

Technology from research on NO_x emission reduction by the use of EGR is obviously effective. The reduction of the in-cylinder global temperature by the EGR is the main reason for the NO_x reduction. The research work by Andree & Pachernegg (1969) has shown impacts on ignition conditions as oxygen concentration is decreased due to the dilution by EGR. In addition, Ladommatos et al. (1998) also revealed that the reduction in combustion temperature is a consequence of the reduced peak rate of the premixed phase combustion due to the lower oxygen availability when EGR is applied.

8. Other automotive applications of biodiesel

Biodiesel is not only used as a fuel for automotive fuel, but also used for other automotive application: for example, exhaust gas-assisted fuel reforming. This manner is a way to produce hydrogen on-board in stead of carrying a massive hydrogen vessel in the vehicle for combusted in engine. This exhaust gas emission control concept has been originally applied to SI engines (Jamal & Wyszynski, 1994; Jamal et al., 1996). In a catalytic reformer, the exhaust gas reforming process takes place by injecting a portion of fresh fuel (reformer fuel) to react with an extracted exhaust gas stream to generate a hydrogen-rich reformed exhaust gas which is routed to mix with fresh intake charge before entering the engine combustion chamber; this method is called reformed exhaust gas recirculation (REGR).

Similarly to the gasoline reforming, in a diesel engine, hydrogen is generated using a direct catalytic interaction of hydrocarbon fuel with partial exhaust gases at sufficiently high temperatures with plenty of oxygen and steam (unlike gasoline exhaust). Tsolakis et al. (2003) firstly studied on an open-loop engine reformer system. The addition of EGR in combination with small amounts of hydrogen was found to affect the combustion and exhaust gas emissions. The added hydrogen replaced the main injected fossil diesel and maintained the same engine load, resulting in simultaneous reductions of both smoke and NO_x emissions without significant impacts on engine efficiency.

A feasibility study on producing hydrogen on-board from biodiesel by catalytic exhaust gas fuel reforming was carried out using a laboratory reforming mini reactor. Tsolakis &

Megaritis (2004b) experimentally studied the reforming of RME-based biodiesel and diesel in comparison and had found that the former produced more hydrogen (up to 17%) with higher fuel conversion efficiency. The appropriated addition of reformer fuel and water to the reformer promotes reactions, yielding more hydrogen production even in the low temperature diesel exhaust gas conditions (Tsolakis & Megaritis, 2004a). Though the reformer fuel added to produce REGR is required, the produced hydrogen-rich gas, substituting part of the main engine fuel resulted in improved fuel economy, during close-loop engine-reformer operation (Tsolakis et al., 2005).

9. Conclusion

Biodiesel is oxygenated ester compounds produced from a variety sources of feedstock such as vegetable oils, animal fats, or waste cooking oils. Biodiesel is widely use as a part substitute for fossil diesel in the present day due to its comparable properties to those of fossil diesel. The use of biodiesel blends in diesel engines has affected engine performance as well as combustion characteristics, i.e. ignition delay, injection timing, peak pressure, heat release rate, and so on. This results in different composition and amounts of both engine exhaust gaseous and non-gaseous emissions. The combustion of biodiesel in diesel engines has normally improved the most regulated emissions except nitrogen oxides emissions. However, there are techniques to mitigate this problem, e.g. exhaust gas recirculation and exhaust gas-assisted fuel reforming. One of the main serious problems in diesel engines is smoke emissions especially particulate mass which can be dramatically reduced by the use of biodiesel. Summarily, with the advent of advanced engine control technology, it is prospective in using biodiesel as an alternative not only combusted in internal combustion engines but also used in other automotive applications.

10. References

Andree, A. & Pachernegg, S.J. (1969) Ignition conditions in diesel engines. *Society of Automotive Engineering Transaction*, Vol. 78, No. 2, pp. 1082–1106

Babu, A.K. & Devaradjane, G. (2003) Vegetable oils and their derivatives as fuels for CI engine: an overview, *Society of Automotive Engineers*, Paper No. 2003-01-0767

Boehman, A.L., Morris, D. & Szybist, J. (2004) The impact of the bulk modulus of diesel fuels on fuels injection timing. *Energy & Fuels*, Vol. 18, pp. 1877–1882

Bosch. (2005) *Diesel-engine management systems and components* (4th ed.), John Wiley, ISBN 0-470-02689-8, West Sussex

Camobreco, V., Sheehan, J., Duffield, J. & Graboski, M. (2000) Understanding the lifecycle costs and environmental profile of biodiesel and petroleum diesel fuel, *Society of Automotive Engineers*, Paper No. 2000-01-1487

CEC (2000) *Green paper: towards a European strategy for the security of energy supply*, Commission of the European Communities, Brussels

Chuepeng, S., Tsolakis, A., Theinnoi, K., Xu, H.M., Wyszynski, M.L. & Qiao, J. (2007) A study of quantitative impact on emissions of high proportion RME-based biodiesel blends, *Society of Automotive Engineers*, Paper No. 2007-01-0072

Chuepeng, S. (2008) Quantitative impact on engine performance and emissions of high proportion biodiesel blends and the required engine control strategies, PhD Thesis, The University of Birmingham

Chuepeng, S., Xu, H.M., Tsolakis, A., Wyszynski, M.L., Price, P., Stone, R., Hartland, J.C. & Qiao, J. (2008a) Particulate emissions from a common rail fuel injection diesel engine with RME-based biodiesel blended fuelling using thermo-gravimetric analysis, *Society of Automotive Engineers*, Paper No. 2008-01-0074

Chuepeng, S., Theinnoi, K., Tsolakis, A., Xu, H.M., Wyszynski, M.L., York, A.P.E., Hartland, J.C., & Qiao, J. (2008b) Investigation into particulate size distributions in the exhaust gas of diesel engines fuelled with biodiesel blends. *Journal of KONES Powertrain and Transport*, Vol. 15, No. 3, pp. 75-82

Chuepeng, S., Xu, H.M., Tsolakis, A., Wyszynski, M.L., & Hartland, J.C. (2009) Nano-particle number from biodiesel blends combustion in a common rail fuel injection system diesel engine equipped with exhaust gas recirculation. *Combustion Engines*, Vol. 138, No. 3, pp. 28-36

Directive 2003/30/EC (2003) The promotion of the use of biofuels or other renewable fuels for transport. *Official Journal of the European Union*, Vol. L123, pp. 42–46

EMA (2003) Technical statement on the use of biodiesel fuel in compression ignition engines, Date of access 23 June 2011, Available from: http://www.reefuel.com/data/info/EMA_Position_on_Biodiesel_Use_Mar_2003.pdf

Ferguson, C.R. (1986) *Internal combustion engines: applied thermosciences*, John Wiley, ISBN 0-471-88129-5, Newyork

Filipi, Z., Wang, Y. & Assanis, D. (2001) Effect of variable geometry turbine (VGT) on diesel engine and vehicle system transient response, *Society of Automotive Engineers*, Paper No. 2001-01-1247

Flaig, U., Polach, W. & Ziegler, G. (1999) Common rail system (CR-system) for passenger car DI diesel engines: experiences with applications for series production projects, *Society of Automotive Engineers*, Paper No. 1999-01-0191

Graboski, M.S., Ross, J.D. & McCormick, R.L. (1996) Transient emissions from no. 2 diesel and biodiesel blends in a DDC series 60 engine, *Society of Automotive Engineers*, Paper No. 961166

Graboski, M.S. & McCormick, R.L. (1998) Combustion of fat and vegetable oil derived fuels in diesel engines. *Progress in Energy and Combustion Science*, Vol. 24, pp. 125–164

Hashizume, T., Miyamoto, T., Akagawa, H. & Tsujimura, K. (1998) Combustion and emission characteristics of multiple stage diesel combustion, *Society of Automotive Engineers*, Paper No. 980505

Heywood, J.B. (1988) *Internal combustion engine fundamentals*, McGraw-Hill, ISBN 0-07-100499-8, Singapore

Ibara, T., Lida, M. & Foster, D.E. (2006) Study on characteristics of gasoline fueled HCCI using negative valve overlap, *Society of Automotive Engineers*, Paper No. 2006-32-0047

Jamal, Y. & Wyszynski, M.L. (1994) On-board generation of hydrogen-rich gaseous fuels-A review. *International Journal of Hydrogen Energy*, Vol. 19, pp. 557–572

Jamal, Y., Wagner, T. & Wyszynski, M.L. (1996) Exhaust gas reforming of gasoline at moderate temperatures. *International Journal of Hydrogen Energy*, Vol. 21, No. 6, pp. 507–519

Jitputti, J., Kitiyanan, B., Rangsunvigit, P., Bunyakiat, K., Attanatho, L. & Jenvanitpanjakul, P. (2006) Transesterification of crude palm kernel oil and crude coconut oil by different solid catalysts. *Chemical Engineering Journal*, Vol. 116, pp. 61-66

Kalam, M.A. & Masjuki, H. (2005) Emissions and deposits characteristics of a small diesel engine when operated on preheated crude palm oil, *Society of Automotive Engineers*, Paper No. 2005-01-3697

Kimura, S., Aoki, O., Kitahara, Y. & Aiyoshizawa, E. (2001) Ultra-clean combustion technology combining a low-temperature and premixed combustion concept for meeting future emission standards. *Society of Automotive Engineers Transaction*, Vol. 110, No. 4, pp. 239–246

Kinast, M.A. (2003) Production of biodiesels from multiple feedstocks and properties of biodiesel and biodiesel/diesel blends, Date of access 23 June 2011, Available from: http://www.nrel.gov/docs/fy03osti/31460.pdf

Komintarachat, C. & Chuepeng, S. (2009) Solid acid catalyst for biodiesel production from waste used cooking oils. *Industrial & Engineering Chemistry Research*, Vol. 48, pp. 9350–9353

Komintarachat, C. & Chuepeng, S. (2010) Methanol-based transesterification optimization of waste used Ccooking oil over potassium hydroxide catalyst. *American Journal of Applied Sciences*, Vol. 7, No. 8, pp. 1073-1078

Kumar, M.S., Ramesh, A. & Nagalingam, B. (2003) Use of hydrogen to enhance the performance of a vegetable oil fuelled compression ignition engine. *International Journal of Hydrogen Energy*, Vol. 28, pp. 1143–1154

Klingbeil, A.E., Juneja, H., Ra, Y. & Reitz, R.D. (2003) Premixed diesel combustion analysis in a heavy-duty diesel engine, *Society of Automotive Engineers*, Paper No. 2003-01-0341

Ladommatos, N., Abdelhalim, S.M., Zhao, H. & Hu, Z. (1998) Effects of EGR on heat release in diesel combustion, *Society of Automotive Engineers*, Paper No. 980184

Lapuerta, M., Armas, O. & Rodíguez-Fernández, J. (2008) Effect of biodiesel fuels on diesel engine emissions. *Progress in Energy and Combustion Science*, Vol. 34, pp. 198–223

Mahr, B. (2002) Future and potential of diesel injection systems, *THIESEL 2002 Conference on Thermo- and Fluid- Dynamic Processes in Diesel Engines*, pp. 5–17

McCormick, R.L., Alvarez, J.R., Graboski, M.S., Tyson, K.S. & Vertin, K. (2002) Fuel additive and blending approaches to reducing NO_x emissions from biodiesel, *Society of Automotive Engineers*, Paper No.2002-01-1658

Oguma, M., Goto, S., Konno, M., Sugiyama, K. & Mori, M. (2002) Experimental study of direct injection diesel engine fuelled with two types of gas to liquid (GTL) , *Society of Automotive Engineers Transaction*, Vol. 111, No. 4, pp. 1214–1220

Postrioti, L., Battistoni, M., Grimaldi, C.N. & Millo, F. (2003) Injection strategies tuning for the use of bio-derived fuels in a common rail HSDI diesel engine, *Society of Automotive Engineers*, Paper No. 2003-01-0768

Quirin, M., Gärtner, S.O., Pehnt, M. & Reinhardt, G.A. (2004) *CO_2 mitigation through biofuels in the transport sector: status and perspective*, Date of access 23 June 2011, Available from: http://www.biodiesel.org/resources/reportsdatabase/reports/gen/ 2004 0801_gen-351.pdf

Rakopoulos, C.D. & Hountalas, D.T. (1996) A simulation analysis of a DI diesel engine fuel injection system fitted with a constant pressure valve. *Energy Conversion and Management*, Vol. 37, No. 2, pp. 135–150

Sharp, C.A., Howell, S.A. & Jobe, J. (2000a) The effect of biodiesel fuels on transient emissions from modern diesel engines, part I regulated emissions and performance, *Society of Automotive Engineers*, Paper No. 2000-01-1967

Sharp, C.A., Howell, S.A. & Jobe, J. (2000b) The effect of biodiesel fuels on transient emissions from modern diesel engines, part II unregulated emissions and chemical characterization. *Society of Automotive Engineers Transaction*, Vol. 109, No. 4, pp. 1784–1807

Senatore, A., Cardone, M., Rocco, V. & Prati, M.V. (2000) A comparative analysis of combustion process in DI diesel engine fuelled with biodiesel and diesel fuel, *Society of Automotive Engineers*, Paper No. 2000-01-0691

Szybist, J.P. & Boehman, A.L. (2003) Behavior of a diesel injection system with biodiesel fuel, *Society of Automotive Engineers*, Paper No. 2003-01-1039

Tat, M.E. & Van Gerpen, J.H. (2002) Physical properties and composition detection of biodiesel - diesel fuel blends, *American Society of Agricultural and Biological Engineers*, Paper No. 026084

Tsolakis, A. (2006) Effects on particulate size distribution from the diesel engine operating in RME-biodiesel with EGR. *Energy & Fuels*, Vol. 20, pp. 1418–1424

Tsolakis, A., Megaritis, A. & Wyszynski, M.L. (2003) Application of exhaust gas fuel reforming in compression ignition engines fuelled by diesel and biodiesel fuel mixtures. *Energy & Fuels*, Vol. 17, pp. 1464–1473

Tsolakis, A. & Megaritis, A. (2004a) Catalytic exhaust gas fuel reforming for diesel engines-effect of water additional on hydrogen production and fuel conversion efficiency. *International Journal of Hydrogen Energy*, Vol. 29, pp. 1409–1419

Tsolakis, A. & Megaritis, A. (2004b) Exhaust gas assisted reforming of rapeseed methyl ester for reduced exhaust emissions of CI engines. *Biomass and Bioenergy*, Vol. 27, pp. 493-505

Tsolakis, A. & Megaritis, A. (2004c) Exhaust gas fuel reforming for diesel engines- A way to reduce smoke and NO_x emissions simultaneously, *Society of Automotive Engineers*, Paper No. 2004-01-1844

Tsolakis, A., Megaritis, A., Yap, D. & Abu-Jrai, A. (2005) Combustion characteristics and exhaust gas emissions of a diesel engine supplied with reformed EGR, *Society of Automotive Engineers*, Paper No. 2005-01-2087

Weall, A. & Collings, N. (2007) Investigation into partially premixed combustion in a lightduty multi-cylinder diesel engine fuelled with a mixture of gasoline and diesel, *Society of Automotive Engineers*, Paper No. 2007-01-4058

Van Gerpen, J.H., Shanks, B., Pruszko, R., Clements, D. & Knothe, G. (2004) *Biodiesel production technology: August 2002 - January 2004*, Date of access 23 June 2011, Available from: http://www.nrel.gov/docs/fy04osti/36244.pdf

13

Toxicology of Biodiesel Combustion Products

Michael C. Madden[1], Laya Bhavaraju[2] and Urmila P. Kodavanti[1]
[1]*Environmental Public Health Division, US Environmental Protection Agency, Research Triangle Park, NC*
[2]*Curriculum in Toxicology, University of North Carolina, Chapel Hill, NC USA*

1. Introduction

The toxicology of combusted biodiesel is an emerging field. Much of the current knowledge about biological responses and health effects stems from studies of exposures to other fuel sources (typically petroleum diesel, gasoline, and wood) incompletely combusted. The ultimate aim of toxicology studies is to identify possible health effects induced by exposure of both the general population as well as sensitive or susceptible populations, including determination of the exposure threshold level needed to induce health effects. The threshold should include not only a concentration but a duration metric, which could be acute or repeated exposures. From such information on sensitive groups and pollutant concentrations needed to induce effects, strategies can be put in place if deemed needed to improve public health. Because possible health effects may take years of exposure to discern, e.g., lung cancer, fibrosis, emphysema, mitigation of the exposure and/or effects may be too late for an individual. Typically markers and biological responses believed to be an early step leading to a clinical disease are measured as a surrogate of the health effect. A biological marker, or "biomarker", indicates a homeostatic change in an organism or a part of the organism (ranging from organ systems to the biochemicals within cells), that will ultimately lead to a disease induced by exposure to a pollutant (Madden and Gallagher, 1999). So with the previous example of lung cancer, damage to lung DNA induced by an exposure would substitute as the biomarker of effect, or possibly examination of the mutagenic potential of the combustion products through an Ames assay using bacterial strains.

For brevity, this chapter will primarily examine human responses to combustion products though an extensive literature exists on nonhuman animal effects. Discussion of nonhuman animal findings will be used to present findings where human data are sparse or nonexistent, and to provide information on health effects mechanisms. Much of the nonhuman findings fill in data gaps concerning extrapulmonary effects of combustion emissions, particularly cardiac and vascular effects.

2. Combustion emissions composition

Products of incomplete fuel combustion from various sources have some similarities, including some of the same substances and induction of related biological responses. Identification of the compounds, and quantities of the compounds, of the emissions from various combustion sources may allow a prediction of the biological responses that occur in exposed people. Additionally, examination of the compounds could indicate unique markers that would serve as an indicator of exposure to that source, as well as raising unique biological responses. For example, levoglucosan is a unique marker of woodsmoke combustion and can be used to determine an individual's exposure to fireplace emissions. A fairly comprehensive list of the chemical species in onroad emissions in California, U.S. derived primarily from gasoline and petroleum diesel powered engines is given in the report by Gertler et al (2002). It is not the focus of the chapter to comprehensively list all emission species; however briefly, the types of components in the gas and particulate matter (PM) phases include single aromatic and polyaromatic hydrocarbons (PAHs) and related compounds (e.g., alkylbenzenes, oxy- and nitro- PAHs), metals, alkanes, alkenes, carbonyls, NOx, CO and CO_2, inorganic ions (e.g., sulfates, carbonates), among other chemicals. Woodsmoke particles tend to be relatively rich in certain metals, including iron, magnesium, aluminum, zinc, chromium, nickel, and copper (Ghio et al., 2011).

Biodiesel combustion produces gaseous and PM phases. Compared to other petroleum diesel fuels, biodiesel combustion in "modern" engines generally tends to produce lower concentrations of PAHs, PM, sulfur compounds, and carbon monoxide (CO) ((McDonald and Spears 1997; Sharp, Howell et al. 2000; Graboski, McCormick et al. 2003). There are conflicting reports of whether nitrogen dioxide (NO_2) levels are decreased (Swanson et al., 2007). Regarding biodiesel PM, the soluble organic fraction of the biodiesel PM is commonly a greater percentage of biodiesel exhaust emissions, but a smaller percentage of organic insoluble mass is present relative to petroleum diesel soot (Durbin, Collins et al. 1999). A decreased production of biodiesel PM but coupled with a greater concentration of soluble organic material may impact the biological effects of biodiesel exhaust PM. Combusted biodiesel PM is lower in metal content than ambient air PM. Combustion of gasoline generally tends to produce less PM but more gas phase amounts than petroleum diesel combustion.

Gas phase components of biodiesel exhaust have been studied. A U.S. Environmental Protection Agency report (EPA420-P-02-001) comparing standard petroleum diesel and biodiesel emissions of specific compounds termed Mobile Source Air Toxics (e.g., volatile substances such as acrolein, xylene, toluene, etc) concluded that while the total hydrocarbon (THC) measurement decreased from biodiesel emissions, there was a shift in the composition towards more unregulated pollutants. (U.S. EPA, 2002a). However the shift was too small to increase total air toxics compared to petroleum diesel emissions. Biodiesel fuel with a high glycerol content (indicative of poor post-transesterification refining) produces greater acrolein emissions (Graboski and McCormick 1998). Ethanol and methanol are used in biodiesel production to provide ethyl and methyl esters, respectively. These alcohols are aldehyde precursors if not removed from the biodiesel and lead to increased formaldehyde and acetaldehyde formation. Biodiesel combustion leads to fatty acid fragments of the starting material (i.e., methylated fatty acids, or FAMEs). The gas phase exhaust of 2002 Cummins heavy duty engine operated under a wide range of operating conditions was reported to produce methyl acrylate and methyl 3-butanoate (Ratcliff et al, 2010); these compounds are believed to be unique markers for biodiesel combustion. It is unclear whether intact FAMES are emitted in the exhaust due to incomplete and /or poor combustion, but the possibility has implications for toxicity. Intact FAMES from biodiesel fuel can be released into the environment via 1) spills such as in the Black Warrior River in Alabama, USA (New York Times, 2008) and 2) the introduction of the fuel into lubrication oil, with subsequent leakage from the engine (Peacock et al, 2010); however the toxicity of biodiesel fuel not being combusted is not the focus of this chapter.

Plant oils are utilized in biodiesel production on a commercial scale in the United States, though some biodiesel fuel can be produced from animal fats. At present, the main plant oil feedstocks for the United States and Europe are soybean oil and rapeseed oil, respectively (Swanson et al, 2007). Other sources globally potentially include switchgrass, jatropha, and palm oil. Algal feedstocks potentially can produce more energy per volume due to their increased fatty acid content. It is unclear if the fatty acid composition is significantly different among the feedstocks, or within feedstocks grown under different conditions.

3. Human health effects

3.1 Nonbiodiesel combustion sources

Identification of health effects observed in humans exposed either acutely or repeatedly to combustion sources other than biodiesel provides guidance for which effects, or surrogate biomarkers of the effects, to examine with combusted biodiesel exposures. Although the epidemiological studies linking biofuel exhausts and impaired human health have not yet surfaced, diesel exhausts, biomass burning, forest fires, and coal burning have been strongly associated with adverse effects and mortality. Recently increases in emergency room visits for asthma symptoms, chronic obstructive pulmonary disease, acute bronchitis, pneumonia, heart failure, and other cardiopulmonary symptoms were noted for people exposed to a peat fire in eastern North Carolina, USA (Rappold, Stone, et al., 2011). These studies are supported by the further evidence of increases in blood pressure in near-road residents (diesel exhaust can be the primary contributor of near road PM in certain locations) (Auchincloss, Diez Roux et al. 2008) and add into consistency of evidence that can be linked to emissions from biologically based and fossil fuels. A number of clinical studies have similarly shown vasoconstrictive and hypertensive effects with petroleum diesel exhaust (PDE) (Peretz, Sullivan et al. 2008) including a decrease in brachial artery diameter in humans. These human studies supporting evidence of adverse cardiovascular impairments have been concurrently proved to be true with animal toxicological studies. However, the mechanism of these apparent cardiovascular impairments without pulmonary health effects are not understood due to inherent variability in the chemical nature of exhaust PM examined and varied exposure scenarios and the variable responsiveness of animal models. Moreover, the physiological relationship between vasoconstrictive effect and change in blood pressure are not understood. PDE have been long studied for their immunological and carcinogenic effects on the lung, however more recent evidence also points to the effects on cardiovascular system.

3.1.1 Lung cancer

With PDE exposures, lung cancer is of concern. The International Agency for Research on Cancer (IARC), the U.S. EPA, the U.S. National Institute for Occupational Safety and Health (NIOSH), and the National Toxicology Program (NTP) have classified PDE as a probable carcinogen, likely carcinogen, potential occupational carcinogen, and reasonably anticipated to be a human carcinogen, respectively, regarding human exposures. There is some question of PDE as a carcinogen due to confounding variables and uncertainties related to exposure levels in some of the epidemiological studies. The increased risk for lung cancer associated with diesel exhaust exposure are derived primarily from epidemiological findings performed prior to 2000. A recently published study involved trucking industry workers regularly exposed to diesel exhaust and the development of lung cancer (Garshick, 2008). The findings showed an elevated risk for the development of lung cancers in those with greater exposure compared to workers (e.g., office workers) with a lower exposure.

3.1.2 Lung inflammation and immune system

Controlled exposures of humans to whole PDE typically results in lung inflammation as shown with neutrophils entering the lungs; these studies are generally 1-2 hr at approximately100-300 μg /m^3 with healthy adults (Holgate 2003). In these same exposures, several soluble substances which mediate inflammation, e.g., interleukin-8 (IL-8) were shown to be increased by use of lung lavage or inducing sputum production to recover airways secretions. PDE PM induced an adjuvancy effect using nasal instillations of 300 μg particles in allergic subjects as common biomarkers of allergy (e.g., increased IgE production and histamine release) increased in nasal secretions (Diaz-Sanchez et al, 1997). Neutrophil influx into the lungs of healthy volunteers exposed to nearly 500 μg/m3 woodsmoke for 2 hr was observed (Ghio et al, 2011) suggesting a common outcome from different combusted fuel sources. There are no studies of human volunteers exposed in a controlled manner to gasoline exhaust.

3.1.3 Cardiac physiology

Biomass, wood smoke and PDE have been linked to increased blood pressure in humans (Sarnat, Marmur et al. 2008). More mechanistic understanding of combustion induced effects have been derived from studies in nonhuman animal models.

Animal toxicology studies have provided some understanding of how diesel exhausts inhalation, while producing small effects in the lung, could have profound effects on the vasculature and myocardium. A few studies have considered the balance of sympathetic and parasympathetic tone, and how these may be altered by PDE. In early high concentration PM studies, classical arrhythmias were apparent, along with heart rate changes, but, when doses fell to more relevant levels, these effects became more difficult to discern (Watkinson, Campen et al. 1998). Increased arrhythmogenicity after aconitine challenge has been noted following environmentally relevant low concentrations of PDE in rats, suggesting that prior air pollution exposure increases the susceptibility to develop arrhythmia in response to severe cardiac insult (Hazari et al., 2011). This increased arrhythmogenic effect of PDE has been postulated to occur as a result of increased intracellular calcium flux. It is not known if preexistent arrhythmogenic status might result in mortality following subsequent air pollution exposure. Thus, PDE exposures, together with compromised cardiac function (especially ischemia), myocardial infarction, hypertension, or heart failure, likely cause arrhythmogenicity in susceptible humans. Biodiesel exhaust might have similar effect on cardiac performance but these studies are needed to understand the influence of compositional similarities and differences in PDE- and BDE-induced cardiac injuries.

The lack of cardiac inflammation, myocardial cell injury, or mitochondrial damage despite cardiac physiological impact in many studies (Campen et al., 2005; Cascio et al., 2007; Hansen et al., 2007; Sun et al., 2008; Toda et al., 2001), supports the findings that PDE induces physiological transcriptome response without altering pathological abnormalities in short-term exposure scenarios (Gottipolu et al., 2009).

3.1.4 Systemic thrombogenic effects

While some clinical studies provide negative evidence of systemic thrombogenic effects of PDE most clinical studies are consistent with increased systemic thrombus formation (Lucking et al 2011) in humans. Animal studies have shown fairly consistent results in regards to increased vascular thrombogenicity of PDE. Exacerbation of systemic thrombus formation in response to UV-induced vascular injury in hamsters and mice exposed to PDE has been known for few years (Nemmar, Nemery et al. 2002; Nemmar, Nemery et al. 2003). The increase in intravascular thrombosis in these earlier studies coincided with inflammation and mast cell degranulation. In hamsters, the thrombogenic effect of PDE was diminished by pretreatment with the anti-inflammatory agents dexamethasone or mast cell stabilizing sodium cromoglycate, implicating the role of inflammatory cells–specifically mast cells (Nemmar, Nemery et al. 2003; Nemmar, Hoet et al. 2004). Pulmonary injury was postulated to cause procoagulant changes and the systemic vascular response to PDE. A number of studies since then have shown prothrombotic effects of PDE exposure in the thoracic aorta of mice and rats (Kodavanti et al., 2011). The precise mechanisms of how PDE or other biodiesel particles might induce thrombogenic effects and the role of pulmonary versus systemic vasculature are now well understood. The evidence supports the role of pulmonary injury/inflammation in eliciting this vascular effect.

3.1.5 Vascular physiology and inflammation

Human clinical and animal studies have provided the evidence that inhalation of PDE and woodsmoke results in peripheral vasoconstriction and increased prothrombotic effects (Mills et al., 2007; Peretz et al., 2008; Lucking et al., 2008; Laumbach et al., 2009; Törnqvist et al., 2007; Campen et al., 2005; Knuckles et al., 2008; Barregard et al, 2006). Vasoconstrictive effects of PDE have been noted even at environmentally relevant inhalation concentrations (Peretz et al., 2008; Brook, 2007). A reproducible decrease in vasodilation in response to various agonists for about 2-24 hr after petroleum diesel exposure has been demonstrated (Mills et al, 2005). Healthy and compromised animal models show alterations in the NO-mediated vasorelaxation and endothelin-mediated vasoconstriction (Nemmar et al., 2003; Knuckles et al., 2008; Lund et al., 2009). PDE-included vasoconstrictive response has been thought to involve impairment of vasodilation due to decreased availability of NO (Mills et al., 2007). Newer studies suggest that vascular effects of PDE and gasoline exhausts might be primarily due to gaseous components such as carbon monoxide and nitrogen oxides. Numerous studies done using PDE and gasoline exhausts have used ApoE-/- mouse model of atherosclerosis and shown that PDE and gasoline exhausts exacerbate lesion development and molecular changes associated with atherogenic susceptibility of ApoE-/- mice.

An array of plasma markers, including cytokines; biomarkers of coagulation and thrombosis; antioxidants; adhesion molecules; and acute phase proteins have been evaluated in a number of studies where animals or humans are exposed to PDE. Although a number of effects have been reported, the results from systemic biomarker studies lack consistency in terms of a similar effect on a given biomarker regardless of some differences in the protocols; in one study, one marker might be increased, whereas, in the other, a different marker may be affected. For example, in one study, PDE exposure has been shown to increase IL-6 (Tamagawa, Bai et al. 2008), whereas, in another, it may show no effect (Inoue, Takano et al. 2006). This discrepancy could result from a small magnitude of effects with a limited sample size; insensitivity of the methods, difficulty in controlling human behavior variables among sequential testing; variable composition of PDE; low exposure concentrations; and, perhaps more importantly, the overwhelming variability in individual host factors. Owing to the fact that biodiesel exhaust might contain more gas-phase components, the systemic biomarkers might respond differently.

3.1.6 Other organ systems

Common symptoms of combustion emissions exposures typically reported include nausea, headache, eye and throat irritation, and dizziness (US EPA, 2002). Other possible biological responses and health effects induced by PDE have been initially investigated by use of epidemiological approaches and rodent models. These endpoints are typically difficult to be examined in controlled exposure studies with humans. For instance, rodent spermatogenesis decreased with exposure in utero (Watanabe et al, 2005), and atrial defects (odds ratio of 2.27) was observed in newborns in seven Texas (USA) counties (Gilboa et al, 2005) and were associated with PM and CO concentrations. These findings of reproductive and in utero atrial defects and the initial observations of decreased spermatogenesis need to be followed up for reproducibility of the findings.

3.2 Biodiesel combustion products

Mutagenicity of substances is typically assessed in bacterial or cellular mutagenicity assays.The vast majority of mutagens are also carcinogenic. Studies indicate that petroleum diesel is more mutagenic than biodiesel. The soluble organic faction of PDE had more mutagenic potential than biodiesel originated from rapeseed in a mutagencity assay using cultured rat hepatocytes. Similar results were found with PDE using bacterial culture in the ames assay. (Eckl et al 1997) Soluble organic fraction of PDE regardless of the various engine cycle combustion conditions still induces more bacterial mutagenesis when compared to biodiesel (Rapeseed methy ester). (Bunger et al 1998) The same organic extracts were tested for potency of mutagenesis after incubation with enzymes extracted from the S9 fraction, and produced the same results indicting PDE is more mutagenic even after liver detoxification. Comparison of PDE from high sulfur and low sulfur content fuel results in more mutagenic activity from high sulfur fuel exhaust regardless of engine mode and incubation with liver metabolic enzymes. (Kado and Kuzmicky 2003) Similar studies with combusted vegetable oils including sunflower seed, cotton seed, soybean and peanut all indicated the soluble extract was less mutagenic than PD extract. (Jacobus et al 1983) However recent regulations have shifted PD over to low sulfur diesel and some have reported biodiesel extracts to be more mutagenic than the new low sulfur PD combustion extracts.

Biodiesel exhaust extract from methylated feedstocks of soy, canola, and beef tallow were found to be more mutagenic than Philips Petroleum- certified PD. (Bunger et al 2000a AND Bunger et al 2000b) In the same study they combusted non-methylated rapeseed oil along with rapeseed methyl esters and found the non-methylated to be more mutagenic than either the methylated or PD. Additionally the gas phase components were collected by cooling and extraction into a solvent. The condensates of the gas phase showed little difference between the combusted PD and biodiesel mutagencity. The BD and PD extracts have recently been used in in vitro toxicity testing. Exposure of PD and BD (soy methyl and ethyl) soluble organic extracts to cultured human airway epithelial cells (BEAS-2B) resulted in elevated cytokine production (IL-6, IL-8) from BD after 24hr exposure. (Swanson et al 2009) An immortal lung epithelial cell line (A549) after exposure to PM from both biodiesel and PD revealed cell morphological changes. The control (unexposed cells) had baseline of

7% multinucleated cells, where as exposure to Biodiesel blend of 80% increased multinucleated cells to 16%. Biodiesel blend of 20% (80% petroleum) increased the multinucleation rate up to 52%. (Ackland et al 2007) Cultured mouse fibroblast cells also indicate BD exhaust soluble extract to be more cytotoxic relative to the PD extracts. (Bunger et al 2000b) Some speculation as to components driving this shift toward increased mutagenicity in biodiesel indicate the increased carbon and carbonyl content in the biodiesel to interfere with cells for longer lengths before the components can be metabolized. The variability of responses can be due to the contents of the soluble extract based on type of solvent and combustion conditions or to the robustness of the cell line.

Animal exposure studies eliminate some of the in vitro variability. Rats exposed to filtered air, PD, B50, and 100% BD (soy ethyl ester) for 1hr were analyzed for lung inflammation. Results indicate lung lavage to have increase in total cell count in the three treatment groups but non were statistically greater in cell count indicating one PM doesn't cause more inflammation. The lung parenchymal tissue was analyzed for inflammation and also resulted positive for inflammation but non of the PM types induced significantly elevated levels. (Brito et al 2010) A second study utilized intratracheal instillation of exhaust PM collected as water aerosol from PD, gasoline, and Biodiesel powered engines (without oxidation catalyst). The aerosols were instilled into mice and the lungs were examined 24hrs later for inflammatory response. The instillation from the gasoline and diesel engines were the most potent to induce an increased neutrophill influx into lungs (inflammatory response), relative to saline control mice. (Tzamkiozis et al 2010)

Chronic exposure with BD and PD produce similar results however the extent of the inflammation may vary. Particle laden alveolar macrophages, lung neutrophilia and fibrosis are detectable in BD exposed rats however the difference from PD an BD exposure was not statistically significant. (Finch et al 2002, Mauderly 1994, Hobbs et al 2002). Human exposure to delivery truck workers, road maintenance workers, and industrial fork lift truck drivers all exposed to BDE or PDE occupationally were asked to report their symptoms in a questionarie. The results of the questionnaire indicate dose related respiratory effects but nothing to indicated significant differences between the combustion of different fuels.

3.3 Summary

Based on the literature available at present, biodiesel exhaust can have more, less, or the same potency in inducing biological responses and health effects as PDE. This may be due to the chemical mix of exhausts and the differences between various types of exhaust emissions. Better reproducibility of design from study to study in the future would assist in the assessment of whether biodiesel exhaust induced the same biological responses. The designs should try to narrow down the fuel type utilized, minimize fuel impurities, utilize an engine commonly available and in use, standardize the run conditions (load, ambient temperature of intake air, etc), so that emissions used in biological test system are fairly similar.

4. Components found in biodiesel combustion with known health effects

Some compounds present in combusted biodiesel exhaust can induce known toxicity in exposed human populations down to cellular effects. The literature on these components may allow a research strategy to determine if these substances exist in great enough concentrations to induce health effects in humans, and if so, how to attenuate the effects through management of the emissions quantities. Additionally, examination of whether the gas and/or the PM phase is primarily responsible for the induction of any observed effects could also be utilized relative to decreasing biologically active substances.

4.1 Filtered particle exhaust studies

PDE studies provide preliminary information for predicting BDE toxicity specifically the studies can give insight on the potency of gas phase and PM.

The removal of particles from petroleum diesel exhaust can attenuate the adverse effects caused by inhalation of diesel exhaust. The exhaust can be filtered to completely remove particles or minimize the amount. Controlled human studied conducted in exhaust chambers fitted with ceramic filters (temperature maintained to eliminate PM nucleation) to capture the particulate successfully reduced the PM by 25%. In this study, the exposures without particles significant increased activated immune response cells (CD3-labeled T lymphocytes) more than particle laden exposures. (Rudell, Blomberg et al. 1999) A lung lavage sample from each exposure indicated no changes in total cell number indicating no significant inflammatory responses. However there was a noticeable decrease in the number of macrophages collected from the bronchial location of the lungs in individuals exposed to the filtered exhaust. A number of explanations for the lack of sentinel macrophages can be concluded, including the filtered exhaust was eliminating larger PM which removes interference from PM deposition and the immediate immune response resulting in two completely different immune responses. However not all studies with PDE indicate gas phase to have more potency. In a mouse exposure study with particle (3.3mg/m3) and filtered PDE (PM < 0.1mg/m3) followed by immediate challenge with pollen, results indicate similar increases in IgE and IgG2 sera titer for the mice exposed to both the filtered and non-filtered exposures. (Maejima, Tamura et al. 2001) However, there was no detectable dose dependent increase to the pollen in only the group exposed to the diesel exhaust gas components. This study proposes an allergic challenge is attenuated after exposure to filtered PDE or PDE with particles, increasing the confidence that each exposure is unique. The use of low sulfur diesel fuel has been indicted to reduce the PM by reducing the soot nucleation rate. (Karavalakis, Bakeas et al. 2010) A study using both low sulfur fuel and a particle trap to reduce the emissions was successful in reducing the toxic health effects relative to regular emissions. (McDonald, Harrod et al. 2004) In this study mice were exposed to the two exhaust types and results indicate with reduced emissions there is significant reduction in the number of potentially toxic inflammatory responses and reactive oxygen species generation. Lung toxicity measured with IL-6, interferon-γ and tumor necrosis factor-α (TNF- α) and antioxidant enzymes (heme oxygenase-1) were all reduced after exposure to reduced PM. The study measured inflammatory response in the mice after a seven day exposure. The study concluded most components of both exhausts were in the range of background air however the responses indicate particles have substantial roles in inflammation and oxidative stress. Not all endpoints of injury indicate filtered exhaust to be less harmful. In an experiment with healthy male subjects who were exposed to both filtered and unfiltered diesel exhaust exposure indicate a reduced response of vasomotor function in subjects exposed to diluted diesel exhaust. Lucking, Lundbäck et al, 2011). Specifically there was reduced vasodilatation even with agonists to promote constriction after exposure to dilute diesel exhaust but not with filtered exhaust. In this study the effects of pure carbon nanoparticles was utilized as a control for the particles however there was no significant alterations of vasoconstriction abilities inhibited by the pure nanoparticles. The particles of diesel exhaust consist of surface bound hydrocarbons or other charged components and are likely interfering with the localized cellular response.

4.2 Carbonyls

Carbonyls (aldehydes and ketones) are common components of fossil fuel combustion. Common species in combustion exhaust are short chain aldehydes such as acetaldehyde and formaldehyde. The use of catalyzed diesel particle filter and plant based fuel reduces carbonyl emissions; however with biodiesel blends there noticeable increases in formaldehyde and acetaldehyde emissions from diesel alone. (Ratcliff, Dane et al. 2010; Jayaram, Agrawal et al. 2011) Petroleum diesel combustion also releases formaldehyde and acetaldehyde with a larger percentage of total carbonyl release being acrolein. Acrolein is a highly reactive aldehyde which creates adducts leading to various degrees of toxicity. Inhalation of acrolein can lead to onset of pulmonary edema, respiratory disturbance and asthma like symptoms. New research indicates acrolein may initiate platelet activation, an event both beneficial and detrimental if induces plaque buildup. Due to the nature of the highly reactive acrolein, specific measures were taken to identify acrolein adducts were not the primary cause of platelet activation but acrolein works directly on platelets as it forms covalent adducts. (Sithu, Srivastava et al. 2010) The study conducted by Sithu et al, utilized fresh mice platelets and vaporized acrolein to conduct exposures. Removal of the blood and isolation of the platelets also found increases in activation proteins like fibrinogen and platelet derived growth factor and platelet factor 4 with exposure to acrolein alone. The observed events were not inflammatory responses because the study measured mRNA expression of pro-inflammatory cytokines and found none were increased above control. Recent studies have indicated increased release of formaldehyde from the combustion process of soy based biodiesel. (Ratcliff, Dane, et al. 2010; Karavalakis, Bakeas et al. 2010) Recently formaldehyde has been classified as a carcinogen. Many studies have addressed the mutagenic characteristics of formaldehyde. In a study of formaldehyde exposure to rat nasal epithelial cells, multiple toxic end points were increased. The study measured the frequencies of micronuclei formations, un-regulated cell proliferation, and pathological changes. Exposure doses larger than 2ppm resulted in site specific increase in cell proliferation. Additionally lesions and metaplastic changes were observed in only the formaldehyde exposed. Histopathology of the nasal regions indicted increases in leukocytes, indicating inflammatory response. Epithelial cell were sloughing off as well as abundant indications of squamous cell metaplasia and the nasopharyngeal duct displayed transitional cell metaplasia. (Speit, Schutz et al. 2011) Basic cellular observations of increased aldehydes released by biodiesel combustion needs to be better understood for any adverse health effects.

4.3 Fatty acids and derivatives

Biodiesel fuel is created with trans-esterification of fatty acids. The composition of BDE has found a number of methyl esters, cyclic fatty acids and nitro fatty acids. Fatty acids including palmitic acid, oleic acid, and stearic acids are considered pulmonary irritants. Fatty acids can be simply classified as unsaturated or saturated and the complexity increases with the types of functional groups bound. Some of the more complex components derived from fatty acids are created with enzymatic reactions others are not. A characteristic of fatty acid derived structures, specifically lipids, is their ability to have dual polarity. Phospholipids create the barriers established in all cell membranes. Normally the fatty acid tail is the hydrophobic region and the carboxyl head is hydrophilic. Tampering with the membrane structures can lead to cell death. Fatty acids play a crucial role in maintaining the pliability of surfaces. Lungs are an important location of fatty acid mediated flexibility, as the lungs fill up with air there is limited distension however when the air is exhaled the ability to expand should remain unaffected. A component of the lungs that allows for this rapid intermittent expansion and contraction is surfactant. Surfactant is a complex mixture of proteins and lipids that is largely dipalmitoyl phosphatidylcholine and its purpose can be adversely affected with intrusion by other fatty acids. Many studies have been conducted on the disruption of surfactant by an increase of a type of lipid, oleic acid. (Hall, Lu et al. 1992) One set of experiments conducted was able to measure the surface pressure changes created with the addition of oleic acid. Surface pressure and the created tension by the lipid molecules, can be inferred with correlation to the absorption of surface pressure. In this study the excised calf lung were lavaged with saline instead of air and the lungs were either treated with the oleic acid or control. Overall results indicated oleic acid disruption of the dynamic lung compression and expansion model can't be correlated directly to an absolute concentration, however inhibition occurs when oleic acid is relatively higher than surfactant concentrations. The incorporation of the oleic acid prevented the spreading of the surfactant film to occur during contraction of a simulated compression. The ability of the lungs to maintain elasticity weakens as the repetitive cycles increased. General observations of extraneous lipid incorporation include disruption of the surfactant films created for lung flexibility and can cause harm to the mechanical physiology of the lung.

4.4 Transition metals

Metals are more abundant in petroleum diesel combustion exhaust than biodiesel. The metals originate from multiple sources including the fuel. Metal particles can be emitted from engine components. Several studies indicate there is a decrease in the concentration of transition metals in biodiesel combustion exhaust. (Brito, Luciano Belotti et al. 2010) Biodiesel blends result in increases in the transition metals Cu, Fe, and Zn in soy based B50 compared to B100.(Brito, Luciano Belotti et al. 2010) Transition metals are highly oxidative species and can lead to intracellular redox cycling. Metals have the ability to generate radicals which likely lead to depletion of antioxidants and increases in DNA, and protein adducts. Both biodiesel and diesel exhaust particles analyzed for elemental metal composition, were found to have metals bound to the carbon core. Several studies have observed the decrease in DNA adduct formation with the pre-treatment of particles with a metal chelator. One study was also able to develop a method to measure the indirect products of ROS and they concluded, diesel exhaust particles treated with a diethylenetriamine pentaacetic acid (DTPA) generate fewer ROS products. The study utilized the same method to measure the amount of 2,3- and 2,5- dihydroxybenzoate (DHBA) generated in the presence of known amounts of Cu and Fe; both are toxic and highly reactive metals. (DiStefano, Eiguren-Fernandez et al. 2009) Other studies were able to study the inflammatory effects residual oil fly ash (ROFA), a dust rich in transitions metals especially V, Ni, and Fe, alone and with pre-treatment with a metal chelator. Similar results were found indicating cytokine induction and depletion of antioxidants is partly due to the metals that are bound to the various particles. (Carter JD, Ghio AJ et al. 1997) Metals are essential elements within cells however too much metals can cause harm to cellular homeostasis and induce cellular toxicity.

4.5 PAH and PAH-related compounds

Polyaromatic hydrocarbons classified by the functional group attachments most prevalent in combustion byproducts are nitro- and oxy- species. They can also vary in reactivity based on their molecular weight. Biodiesel blends up to 50% are analyzed to have large decreases in PAH emissions when compared to diesel fuel combustion. (Brito, Luciano Belotti et al. 2010) A commonly measured sample PAH released during diesel combustion is phenathraquinone (PQ). PQ can be reduced by flavin enzymes including NADPH located in the mitochondria and along energy transport membranes. The reduction leads to generations of semiquinone radicals, oxidative stress and DNA damage followed by cytotoxicity. Experiments with PQ exposure to human pulmonary epithelial cells have observed increases in toxic byproducts of ROS generation. Some increases measured are increase in protein carbonyl formation, increased levels of superoxide dismutase (Cu/Zn SOD) and heme oxygenase (HO-1). (Rika Sugimotoa, Yoshito Kumagaia et al. 2005)Protection from the damaging consequences of protein carbonyl formation originated from both the use of iron chelators and antioxidants. High emission of NO_2 lead to the nitration of the available PAH's forming nitro-PAH's. Using soy based biodiesel, species identified included few volatile nitro compounds that were more abundant in B100 as opposed to the B20, however the overall trend was a decrease in the nitro-PAH emission when biodiesel was combusted. Detailed analysis of 7 nitro PAH emission concluded in several products decreased by more than 50% with the blending of B20 into the petroleum diesel and further decreases with B100. (Ratcliff, Dane et al. 2010) Naphthalene is still a larger percent of the combustion emissions, in both biodiesel and diesel fuel engines. (Ratcliff, Dane et al. 2010; Jayaram, Agrawal et al. 2011) Naphthalene vapors are toxic and are commonly used as pesticides. There are several signaling pathways that have been identified which are initiated with the binding of PAH's to the acryl hydrocarbon receptor, however many PAH's have not been identified as ligands. Other indirect increases in cellular toxicity from PAH's involves thiol generation which inactivate proteins with sulfhydryl groups. Quinones generally are not alkylating agents but they can generate redox cycling which generates thiol oxidants including hydrogen peroxides. PAHs will cause cellular and regional increases in ROS generation and further deplete antioxidants while repair processes work to increase antioxidant defenses.

4.6 Other hydrocarbons

Toluene is a common aromatic hydrocarbon emitted with the combustion of fossil fuels. Diesel emissions contain detectable amounts of toluene in both the vapor phase and particle. Toluene is more reactive than benzene due to the methyl group and is easily nitrated in the presence of increases NO_2. Toluene has been found to interact with the aryl hydrocarbon receptor in cells. In a study conducted using Drosophila flies to study genotoxicity and apoptosis, toluene exposure produced large amounts of cell death. (Singh, Mishra et al. 2011)The study also measured the amount of apoptosis after treating the cells with a known aryl hydrocarbon receptor blocker before exposure to toluene. The results of the study with AHR blocker producing less toxicity can justify the observations indicating toluene works via the AHR. Activation of the AHR can increase transcription of antioxidants.(Singh, Mishra et al. 2011) Apoptosis increased with toluene exposure was measured with TUNEL assay. Previous research with other aromatic hydrocarbons has observed increases in inflammation and increased activation of T-lymphocytes and eosinophils. Hydrocarbons like toluene, with reactive functional groups are likely to enter into the cell and cause cellular apoptosis as they are to accumulate in tissue and cause regional inflammation.

4.7 Carbon monoxide and nitrogen dioxide

Primary concern of carbon monoxide (CO) poisoning involves the ability of CO to bind to hemoglobin in the blood and inhibit binding oxygen molecules to hemoglobin. Cardiac compromised patients, such as ones with angina, are a sensitive population to the effects of

CO. Nitrogen dioxide (NO_2) is well known to cause lung function decrements and increase airways hyperresponsiveness, especially in asthmatic individuals. Comprehensive reviews of CO and NO_2 toxicity have been published by the U.S. EPA (U.S. EPA 2000; U.S. EPA 2008).

5. Sensitive and susceptible populations

Human responses to air pollutants are heterogeneous. Certain factors can make an individual sensitive or resistance. Some factors identified that affect the type of response as well as the magnitude of a response include age, genetics (i.e., genotypes), diet, medication, body mass index (BMI), and disease status. Lung function decrements (e.g., the forced expiratory volume exhaled in 1 sec, or FEV_1) induced by ozone inhalation are dependent on age in normal healthy individuals (McDonnell et al, 2007); smaller decrements are observed in older individuals compared to adolescents. Nonsteroidal anti-inflammatory medications such as ibuprofen have been shown to attenuate ozone-induced FEV1 decrements, but not lung neutrophil influx (Hazucha et al, 1996). Women with BMI > 25 had greater lung function decrements to ozone exposure (Bennett et al, 2007). Individuals placed on high antioxidant intake had smaller FEV_1 decrements than those on placebo regimens (Samet et al, 2001) with less indication of ozone-induced oxidative stress to lung tissue (Sawyer et al, 2008). Individuals with the specific genotypes of glutathione-S-transferase, e.g., the M1 and P1 null types, had augmented nasal ragweed allergic responses such as increased histamine production when PDE PM was instilled in the nose (Gilliland et al, 2004). Diabetics had more hospital admissions for cardiopulmonary illnesses associated with ambient levels of carbon monoxide and coarse ambient PM (PM between 2.5 and 10 um) in Los Angeles (Linn et al, 2000). These are select examples of some factors associated with exaggerated responses of sensitive individuals to certain air pollutants and are not intended as a comprehensive review of susceptible populations.

It is unclear at present which human population may be more sensitive to biodiesel combustion emissions. What is known from sensitivity factors for other air pollutants will assist in designs for examining potentially susceptible groups. Potentially if fatty acids and/or fatty acid fragments are emitted from biodiesel combustion and are deposited in the lung, these substances may induce greater responses and health effects in those individuals with defects in fatty acid metabolism.

6. Future issues and challenges

6.1 Fuel additives

Biodiesel fuel has several classes of substances intentionally added to cover several purposes. Antimicrobials, cold-flow improvers, detergents, corrosion inhibitors, and fuel stabilizers are blended into fuel depending on the need, e.g., storage duration, ambient temperature, etc. Additionally, the possibility of pesticides being unintentionally present in fuel due to residues in the fuel stock being carried through the production process has not been confirmed as well as the possible health implications addressed. As previously mentioned, poor quality fuels have alcohols and/or glycerol present. Hence biodiesel is not solely FAMEs. The combustion of the substances non-FAME components likely contributes to the emissions, but it is unclear whether the combustion products contribute to the toxicity, or modify the toxicity (negatively, additively, synergistically) of the FAME combustion products.

6.2 Fuel blends

Currently biodiesel is primarily used commercially as a 20% blend with petroleum diesel fuel in the United States. It is unclear whether this ratio of biodiesel to petroleum fuel will increase and to what extent. Some vehicles will continue to operate on 100% biodiesel. A potential problem in assessment of biodiesel toxicity is that changing the proportion of biodiesel in blends can alter the amounts of some combustion products emitted in a nonlinear manner. For instance, in changing from 100% to 50% to 0% petroleum diesel fuel (make up fuel being biodiesel), metals changed in the exhaust fairly linearly and predictably, i.e., from 1.0 to 0.9 (i.e., a 10% decrease) to 0.8 (a 20% decrease) relative concentration units, respectively (Brito et al, 2010). However CO and black carbon changed in a nonlinear (concave shaped) fashion, i.e., for CO, from 1.0 to 1.6 to 0.7 relative concentration units, respectively, while volatile organic compounds (VOCs) and PAHs changed in a nonlinear (convex shaped) fashion, i.e., for VOCs, from 1.0 to 0.2 to 0.6 relative concentration units, respectively. Such nonlinear changes in emissions from blended biodiesel make prediction of the combustion product concentrations more difficult, and hence prediction of human responses or health effects harder to characterize if the products affect the toxicity. The potential shapes of the changes in an emission component are presented in Figure 1 below.

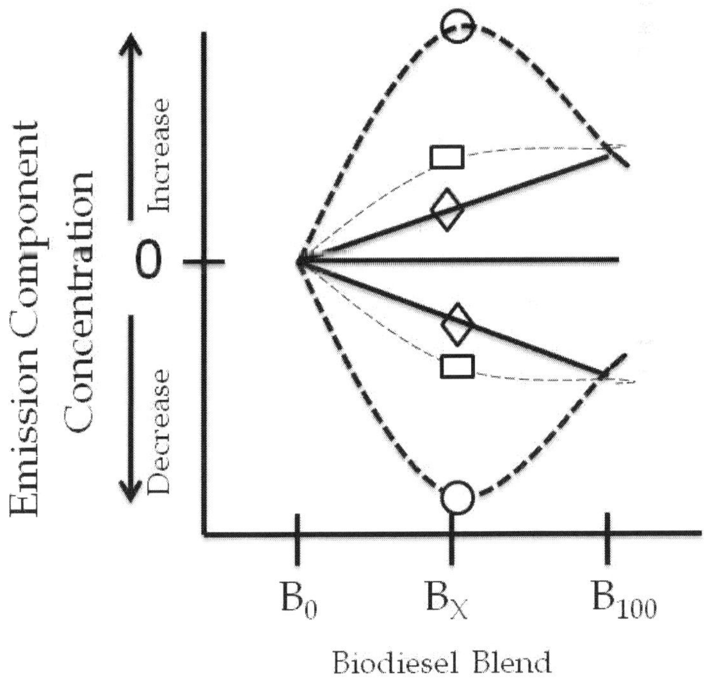

Fig. 1. Theoretical changes in an emission component as the proportion of biodiesel changes from 0 (B0) to 100 (B100) %. Linear changes are reflected by straight lines and diamonds, concave and convex changes are reflected by dashed lines and circles, and sigmoidal changes are reflected by dashed lines and rectangles.

6.3 Evolving fuel standards and engine technology

As fuels evolve, emissions will also change. For instance, petroleum diesel fuel sulfur content for onroad use has decreased in most countries, resulting in lower PM exhaust concentrations. As mentioned previously, petroleum diesel is currently blended with biodiesel in the U.S and other countries; hence changes in fuel components will likely affect emission components. This constantly changing fuel composition will be driven by requirements for meeting specific standards. The first national biodiesel specification in the USA was the ASTM standard D 6751, *"Standard Specification for Biodiesel Fuel (B100) Blend Stock for Distillate Fuels"*, adopted in 2002. Findings from toxicology studies using fuels created before current standards in affected countries will likely have different emissions, and possibly different health effects and responses. For instance, Brito et al, 2010 used petroleum diesel fuel containing ethylated (not methylated) fatty acids and relatively high sulfur content (500 ppm) in their studies. This is likely due to the fuel being produced in Brazil with abundant ethanol production and standards allowing higher sulfur content in petroleum diesel. A potential major issue is whether fuels derived from 3rd generation feedstock, such as algae, produce a different fuel than those of 1st or 2nd generation feedstocks, and if so, do the emissions change considerably along with biological responses being altered.

6.4 Risk assessment

The attractiveness of biodiesel in part stems from lower emissions of some pollutants such as PM and CO, and additionally lower mutagenic potential associated with the PM phase relative to petroleum diesel emissions. However some studies report increased inflammatory mediator release (Swanson et al, 2009), and increased cell death (Bunger et al, 2000). These health effects need to be examined in the context of the amount of pollutant emitted per mile or unit work as biodiesel replaces petroleum diesel. Health effects may need prioritization based on the degree of adversity, reversibility of the effect, and proportion of the general population and also potentially sensitive population(s) exposed to biodiesel exhaust.

7. Challenges

There are several challenges ahead for assessing the implications of increased biodiesel end use. The toxicology of what is emitted from combusted biofuels needs more establishment. This establishment would be aided if reproducible study designs could be established. In part, experiments could be fairly similar with exposures using the same atmospheres. Hence, standardized biodiesel fuels, use of engines with a large market penetration to simulate what most individuals may be exposed to, and several similar endpoints (e.g., mutagenicity, lung inflammation, vascular and cardiac changes) should be incorporated. The appropriateness of whole animal and cultured cells models to human exposures and effects will need to be established, as is currently being determined in the field of PDE toxicology. Included in the validation of nonhuman models would be extrapolation of effects are relatively high doses to low dose exposures of humans, especially if sensitive human populations to biodiesel exhaust are identified. The toxicology of PDE has been advanced to some extent with the creation of some standardized PDE particles to use as an internal control condition, such as those Standard Reference Materials (SRMs) at the National Institutes of Standards and Technology in Gaithersburg, MD (http://www.nist.gov/). The bioactivities of the biodiesel gas phase and PM phase are still largely unknown, so research effort at present must be put into both phases in order to eventually determine if adverse health effects exist, and if so, which phase to manipulate to effect fewer effects. Studies are still scant where health effects and biological responses have been measured when individuals are exposed to whole biodiesel exhaust. Only one study is currently published, though a few are underway currently, or have finished and are awaiting publication. Now would be an opportunistic time to design and implement studies, especially in an occupational setting, as biodiesel fuels replace petroleum based fuels. Data can be collected from workers in regards to possible adverse symptoms and other health effects induced by PDE, and similar endpoints at a later time after biodiesel is introduced into the workplace. A final challenge ultimately will be to incorporate the knowledge of human health effects induced by exposures to combusted biodiesel emissions into a comprehensive strategy for management of 1) issues related to increased biodiesel production (soil use, production, transport and distribution) and 2) issues related to future energy production in general, such as how well biodiesel measures up to other fuel alternatives (ethanol, butanol, wind, solar, nuclear, etc) in terms of feasibility and public health impacts.

8. Disclaimer

This manuscript has been reviewed by the National Health and Environmental Effects Research Laboratory, U.S. Environmental Protection Agency and approved for publication. Approval does not signify that the contents necessarily reflect the views and policies of the Agency, nor does mention of trade names or commercial products constitute endorsement or recommendation for use.

9. Acknowledgements

Partially funded by the EPA/UNC Toxicology Research Program, Training Agreement T829472, with the Curriculum in Toxicology, University of North Carolina at Chapel Hill.

10. References

Ackland, M.L., Zou, L., et al. (2007). Diesel exhaust particulate matter induces multinucleate cells and zinc transporter-dependent apoptosis in human airway cells. Immunol Cell Biol. 85(8):617-22.

Auchincloss, A. H., A. V. Diez Roux, et al. (2008). "Associations between recent exposure to ambient fine particulate matter and blood pressure in the Multi-ethnic Study of Atherosclerosis (MESA)." Environ Health Perspect 116(4): 486-491.

Barregard L, Sällsten G, et al. (2006). Experimental exposure to wood-smoke particles in healthy humans: effects on markers of inflammation, coagulation, and lipid peroxidation. Inhal. Toxicol. 18(11):845-53.

Bennett WD, Hazucha MJ, et al. (2007). Inhal Toxicol. 19(14):1147-54. Acute pulmonary function response to ozone in young adults as a function of body mass index.

Brook, R.D. (2007). Is air pollution a cause of cardiovascular disease? Updated review and controversies. Rev Environ Health. 22(2):115-37.

Brito, J. M., Luciano Belotti, et al. (2010). "Acute Cardiovascular and Inflammatory Toxicity Induced by Inhalation of Diesel and Biodiesel Exhaust Particles " Toxicological Sciences 116(1): 67-78.

Bünger J, Krahl J et al. (1998). Mutagenic and cytotoxic effects of exhaust particulate matter of biodiesel compared to fossil diesel fuel. Mutation Res. 8:415(1-2):13-23.

Bünger J, Müller MM, et al. (2000a). Mutagenicity of diesel exhaust particles from two fossil and two plant oil fuels. Mutagenesis. 15(5):391-7.

Bünger J, Krahl J, (2000b). Cytotoxic and mutagenic effects, particle size and concentration analysis of diesel engine emissions using biodiesel and petrol diesel as fuel. Arch Toxicol. 74(8):490-8.

Bünger J, Krahl J, et al. (2007). Strong mutagenic effects of diesel engine emissions using vegetable oil as fuel. Arch Toxicol. 81(8):599-603.

Carter JD, Ghio AJ, et al. (1997). "Cytokine production by human airway epithelial cells after exposure to an air pollution particle is metal-dependent." Toxicology and Applied Pharmacology 146(2): 180-188.

Diaz-Sanchez D, Tsien A, et al. (1997). Combined diesel exhaust particulate and ragweed allergen challenge markedly enhances human in vivo nasal ragweed-specific IgE and skews cytokine production to a T helper cell 2-type pattern. J Immunol. 158:2406-13.

DiStefano, E., A. Eiguren-Fernandez, et al. (2009). "Determination of metal-based hydroxyl radical generating capacity of ambient and diesel exhaust particles." Inhalation Toxicology 21(9): 731-738.

Durbin, T. D., J. Collins, et al. (1999). Evaluation of the effects of alternative diesel fuel formulations on exhaust emission rates and reactitivity. Final Report for South Coast Air Quality Management District Technology Advancement Office (98102). Riverside, CA, Center for Environmental Research and Technology, University of California.

Eckl P, Leikermoser P, et al. (1997). The mutagenic potential of diesel and biodiesel exhausts. In: Plant oils as fuels-Present state of science and future developments (Martini N. and Schell J., eds). Berlin: Springer, 124-140.

Finch, G. L., C. H. Hobbs, et al. (2002). "Effects of subchronic inhalation exposure of rats to emissions from a diesel engine burning soybean oil-derived biodiesel fuel." Inhal Toxicol 14(10): 1017-1048.

Ghio AJ, Soukup JM, et al. (2011). Exposure to wood smoke particles produces inflammation in healthy volunteers. Occup Environ Med. Jun 30.

Gilboa SM, Mendola P, et al. (2005). Relation between ambient air quality and selected birth defects, seven county study, Texas, 1997-2000. Am J Epidemiol.162:238-52.

Gilliland FD, Li YF, et al. (2004). Effect of glutathione-S-transferase M1 and P1 genotypes on xenobiotic enhancement of allergic responses: randomised, placebo-controlled crossover study. Lancet. 363(9403):119-25.

Graboski, M. S. and R. L. McCormick (1998). "Combustion of fat and vegetable oil derived fuels in diesel engines." Progress in Energy and Combustion Science 24: 125-164.

Graboski, M. S., R. L. McCormick, et al. (2003). The effect of biodiesel composition on engine emissions from a DDC series 60 diesel engine, Colorado Institute for Fuels and Engine Research, Colorado School of Mines, Golden, Colorado.

Hall, S., R. Z. Lu, et al. (1992). "Inhibition of pulmonary surfactant by oleic acid: mechanisms and characteristics." The American journal of Physiology: 1708-1716.

Hasford, B., M. Wimbauer, et al. (1997). Respiratory symtoms and lung function after exposure to exhaust fumes from rapeseed oil in comparison to regular diesel fuel. Proceedings of the 9th International Conference on Occupational Respiratory Diseases: Advances in Prevention of Occupational Respiratory Diseases, Kyoto, Japan, Elsevier.

Hazari MS, Haykal-Coates N, et al. (2011). TRPA1 and Sympathetic Activation Contribute to Increased Risk of Triggered Cardiac Arrhythmias in Hypertensive Rats Exposed to Diesel Exhaust. Environ Health Perspect. 119(7):951-7.

Hazucha, M.J., M.C. Madden, et al. (1996). Effects of cyclooxygenase inhibition on ozone-induced respiratory inflammation and lung function changes. Eur. J. Appl. Physiol. 73: 17-27.

Inoue, K., H. Takano, et al. (2006). "Pulmonary exposure to diesel exhaust particles enhances coagulatory disturbance with endothelial damage and systemic inflammation related to lung inflammation." Exp Biol Med (Maywood) 231(10): 1626-1632.

Jacobus, M.J., S.M. Geyer, et al. (1983). Single-Cylinder Diesel Engine Study of Four Vegetable Oils. SAE paper no. 831743.

Jayaram, V., H. Agrawal, et al. (2011). "Real time gaseous, PM, and ultrafine particle emissions from a modern marine engine operating on biodiesel." Environmental Science and Technology 45: 2286-2292.

Kado, N.Y. and Kuzmicky, P.A. Bioassay (2003). Analyses of Particulate Matter from a Diesel Bus Engine Using Various Biodiesel Feedstock Fuels. NREL Report No. SR-510-31463, National Renewable Energy Laboratory, Golden, CO.

Karavalakis, G., E. Bakeas, et al. (2010). "Influence of oxidized biodiesel blends on regulated and unregulated emission from a diesel passenger car." Environmental Science and Technology 44: 5306-5312.

Kodavanti UP, Thomas R, et al. (2011). Vascular and cardiac impairments in rats inhaling ozone and diesel exhaust particles. Environ Health Perspect. May 11.

Linn WS, Szlachcic Y, et al. (2000). Air pollution and daily hospital admissions in metropolitan Los Angeles.Environ Health Perspect.108(5):427-34.

Lucking AJ, Lundbäck M, et al. (2011). Particle traps prevent adverse vascular and prothrombotic effects of diesel engine exhaust inhalation in men. Circulation. 123(16):1721-8.

Lund AK, Lucero J, et al. (2009). Vehicular emissions induce vascular MMP-9 expression and activity associated with endothelin-1-mediated pathways. Arterioscler Thromb Vasc Biol. 29:511-7.

Madden, MC, and J.E. Gallagher. (1999). Biomarkers of Exposure. In: Air Pollution and Health. ST Holgate, HS Koren, J Samet, and R Maynard, eds. Academic Press, London. pp. 417-430.

Maejima, K., K. Tamura, et al. (2001). "Effects of the inhalation of diesel exhaust, Kanto loam dust, or diesel exhaust without particles on immune responses in mice exposed to Japanese cedar (Cryptomeria japonica) pollen." Inhalation Toxicology 13(11): 1047-1063.

Mauderly, J. L. (1994). "Toxicological and epidemiological evidence for health risks from inhaled engine emissions." Environ Health Perspect 102 Suppl 4: 165-171.

McDonald, J. and M. W. Spears (1997). Biodiesel: effects on exhaust constituents. Plant oils as fuels-Present state of science and future developments. Martini N. and Schell J. Berlin, Springer: 141-160.

McDonald, J. D., K. S. Harrod, et al. (2004). "Effects of Low Sulfur Fuel and a Catalyzed Particle Trap on the Composition and Toxicity of Diesel Emissions." Environmental Health Perspectives 112(13): 1307-1313.

McDonnell WF, Stewart PW, et al. (2007). The temporal dynamics of ozone-induced FEV1 changes in humans: an exposure-response model. Inhal Toxicol.19(6-7):483-94.

Mills NL, Törnqvist H, et al. (2007). Diesel exhaust inhalation causes vascular dysfunction and impaired endogenous fibrinolysis. Circulation. 112(25):3930-6.

Mills NL, Törnqvist H, et al. (2007). Ischemic and thrombotic effects of dilute diesel-exhaust inhalation in men with coronary heart disease. N Engl J Med. 357(11):1075-82.

Nemmar, A., P. H. Hoet, et al. (2004). "Pharmacological stabilization of mast cells abrogates late thrombotic events induced by diesel exhaust particles in hamsters." Circulation 110(12): 1670-1677.

Nemmar, A., B. Nemery, et al. (2003). "Pulmonary inflammation and thrombogenicity caused by diesel particles in hamsters: role of histamine." Am J Respir Crit Care Med 168(11): 1366-1372.

Nemmar, A., B. Nemery, et al. (2002). "Air pollution and thrombosis: an experimental approach." Pathophysiol Haemost Thromb 32(5-6): 349-350.

New York Times, (2008). "Pollution Is Called a Byproduct of a 'Clean' Fuel". http://www.nytimes.com/2008/03/11/world/americas/11iht-11biofuel.10914638.html?pagewanted=1. Brenda Goodman, author. Published 3/11/2008. Accessed July 5, 2011.

Peacock,E.E. , Arey,J.S. et al. (2010). Molecular and Isotopic Analysis of Motor Oil from a Biodiesel-Driven Vehicle. Energy Fuels 24: 1037–1042.

Peretz, A., J. H. Sullivan, et al. (2008). "Diesel exhaust inhalation elicits acute vasoconstriction in vivo." Environ Health Perspect 116(7): 937-942.

Rappold, A.G., S.L. Stone, et al. (2011). "Peat Bog Wildfire Smoke Exposure in Rural North Carolina Is Associated with Cardio-Pulmonary Emergency Department Visits Assessed Through Syndromic Surveillance." Environ. Health. Perspect. Epub. June 27. http://dx.doi.org/10.1289/ehp.1003206

Ratcliff, M., A. J. Dane, et al. (2010). "Diesel particle filter and fuel effects on heavy duty diesel engine emissions." Environmental Science and Technology 44(21): 8343-8349.

Sugimotoa, R., Y. Kumagaia, et al. (2005). "9,10-Phenanthraquinone in diesel exhaust particles downregulates Cu,Zn–SOD and HO-1 in human pulmonary epithelial cells: Intracellular iron scavenger 1,10-phenanthroline affords protection against apoptosis " Free Radical biology and Medicine 38(3): 388-395.

Rudell, B., A. Blomberg, et al. (1999). "Bronchoalveolar inflammation after exposure to diesel exhaust: comparison between unfiltered and particle trap filtered exhaust." Occupational Environmental Medicine 56: 527-534.

Samet JM, Hatch GE, et al. (2001).Effect of antioxidant supplementation on ozone-induced lung injury in human subjects.Am J Respir Crit Care Med. 164(5):819-25.

Sarnat, J. A., A. Marmur, et al. (2008). "Fine particle sources and cardiorespiratory morbidity: an application of chemical mass balance and factor analytical source-apportionment methods." Environ Health Perspect 116(4): 459-466.

Sawyer, K., Samet, J.M. et al. (2008). Responses measured in the exhaled breath of human volunteers acutely exposed to ozone and diesel exhaust. J. Breath Research, 2 037019 (9pp)

Sharp, C., S. Howell, et al. (2000). "The Effect of Biodiesel Fuels on Transient Emissions from Modern Diesel Engines, Part II Unregulated Emissions and Chemical Characterization." Technical Paper 2000-01-1968; SAE: Warrendale, PA.

Singh, M., M. Mishra, et al. (2011). "Genotoxicity and apoptosis in Drosophila melanogaster exposed to benzene, toulene, and xylene: attenuation by quercetin and curcumin." Toxicology and Applied Pharmacology 253(1): 14-30.

Sithu, S., S. Srivastava, et al. (2010). "Exposure to acrolein by inhalation causes platelet activation." Toxicology and Applied Pharmacology 248: 100-110.

Speit, G., P. Schutz, et al. (2011). "Analysis of micronuclei, histopathological changes and cell proliferation in nasal epithelium cells of rats after exposure to formaldehyde by inhalation." Mutation Reseach 721: 127-135.

Swanson, KJ, Madden, MC, et al. (2007). Biodiesel Exhaust: The Need for Health Effects Research.. Env. Hlth. Perspect. 115:496-499.

Swanson, KJ, Funk, W. et al. (2009). Release of the pro-inflammatory markers IL-8 & IL-6 by BEAS-2B cells following *in vitro* exposure to biodiesel extracts. The Open Toxicology Journal. 3:8-15.

Tamagawa, E., N. Bai, et al. (2008). "Particulate matter exposure induces persistent lung inflammation and endothelial dysfunction." Am J Physiol Lung Cell Mol Physiol 295(1): L79-85.

Toda N, Tsukue N et al. (2001). Effects of diesel exhaust particles on blood pressure in rats.J Toxicol Environ Health A. 63(6):429-35.

Tzamkiozis T, Stoeger T, et al. (2010). Monitoring the inflammatory potential of exhaust particles from passenger cars in mice. Inhal Toxicol. 22 Suppl 2:59-69.

United States Environmental Protection Agency. 2002a. Health Assessment Document for Diesel Exhaust. EPA/600/8-90/057F. Washington, DC.

U.S. Environmental Protection Agency. 2000. Air Quality Criteria Document for Carbon Monoxide. EPA/600/P-99/001F.Washington, DC: U.S. Environmental Protection Agency.

U.S. Environmental Protection Agency. 2002b. A Comprehensive Analysis of Biodiesel Impacts on Exhaust Emissions EPA420-P-02-001. Washington DC: U.S. Environmental Protection Agency.

U.S. Environmental Protection Agency. 2008. Integrated Science Assessment for Oxides of Nitrogen- Health Criteria. EPA/600/R-07/093aB. Washington DC: U.S. Environmental Protection Agency

Watanabe N. (2005). Decreased number of sperms and Sertoli cells in mature rats exposed to diesel exhaust as fetuses. Toxicol Lett. 155:51-8.

Watkinson, W. P., M. J. Campen, et al. (1998). "Cardiac arrhythmia induction after exposure to residual oil fly ash particles in a rodent model of pulmonary hypertension." Toxicol Sci 41(2): 209-216.

14

Utilization of Biodiesel-Diesel-Ethanol Blends in CI Engine

István Barabás and Ioan-Adrian Todoruț
Technical University of Cluj-Napoca
Romania

1. Introduction

The biodiesel's use can be considered as an alternative for compression ignition engines, but some of its properties (density, viscosity) present superior values compared with diesel fuel. These properties can be improved by adding bioethanol, witch on one side allows the content's increasing of the bio-fuel in mixture, and on the other side brings the reminded properties in the prescribed limits of the commercial diesel. First of all, the bioethanol is destined as an alternative for the spark ignition engines, but has applications for compression ignition engines, too.

The undertaken researches about partial replacement of the diesel fuel destined to diesel engines with mixtures biodiesel-diesel fuel-bioethanol (BDE), have as main purpose the identification and the testing of new alternative fuels for compression ignition engines, with similar properties of the commercial diesel fuel, having a high content of bio-fuel. In this case, it was started from the fact that by using BDE mixtures, some properties of the biodiesel and of the ethanol are mutually compensated, resulting mixtures with properties very similar with the ones of the diesel fuel. In the research, were used binary mixtures (BD, DE) and triple mixtures (BDE) between biodiesel (B) obtained from rapeseed oil, commercial diesel fuel (D) and bioethanol (E), in different proportions of these ones (the bio-fuel content varied from 5 % v/v to 30 % v/v, in scales of 5 % v/v, also for ethanol, and for biodiesel), having the purpose of evaluating the mixtures'(BDE) main properties and of comparing these ones with the diesel fuel.

The BDE mixtures were noted so the volumetric composition of the new fuels to be reflected. For example, the mixture B10D85E5 indicates the following volumetric composition of the component parts: 10 % biodiesel, 85 % commercial diesel fuel and 5 % ethanol.

At the established scale of researched fuels were taken into consideration the following criteria:

- the mixture's cetane number has not fall under the minimum value of the diesel fuel and of the biodiesel (51);
- the mixture's density has not be smaller than the one of the diesel fuel and has not be bigger than the one of the biodiesel;
- the mixture's caloric power has not fall with more than 5 % than the diesel fuel's caloric power;
- the three component parts has to be miscible until 0 °C temperature, and the formed mixture has to be long-term stable (at list three months from the preparation);

- the bio-fuel content has to be minimum 5 % v/v and maximum 30 % v/v;
- the mixtures' viscosity has to be near of the commercial diesel fuel's one.

The objective of this research, was focused on fitting the biodiesel-diesel fuel-bioetanol blends to compression ignition engines. This obiective carried out by:

- evaluating the use of biodiesel (rapeseed oil methyl esters) as an additive in stabilizing ethanol in diesel blends;
- blends selecting based upon mixture solubility and stability;
- determining the key fuel properties of the blends such as density, viscosity, surface tension, lubricity, flash point and cold filter plugging point;
- second mixtures selection based on phisical and chemical properties;
- engine performance and emission characteristics evaluation in laboratory condition;
- vehicle performance evaluation on chassis dynamometer;
- road test performances of biodiesel-diesel fuel-bioethanol blend.

Based on the undertaken researches regarding the miscibility, the stability, the lubrication ability and the main physicochemical properties (chemical composition, density, kinematic viscosity, limited temperature of filterability, the ignition temperature and superficial tension), from 27 mixtures BD, DE and BDE were selected three fuels (B10D85E5, B15D80E5, B25D70E5), which have similar properties as the diesel fuel.

The fuels thus selected were used for making the tests regarding the evaluation of the performances and regarding the pollution made by a Diesel engine, compared with the diesel fuel use, thus: *tests on the experimental stand for testing the compression ignition engines*, through the determination of the fuel's specific consumption, through the determination of the engine's performance and through the determination of the pollution emissions (CO, CO_2, NO_x, HC, smoke), at different tasks of its; *tests on the inertial chassis dynamometer* through the determination of the passenger car's power and torque; *road tests* through the determination of some dynamic features (vehicle elasticity, overtaking and accelerations parameters) of the tested passenger car.

2. The main properties of the biodiesel-diesel fuel-ethanol mixture component parts

The solubility, stability and properties of biodiesel-diesel fuel-ethanol ternary mixtures were investigated using commercial diesel fuel, biodiesel produced from rapeseed oil and ethanol with purity of 99.3 %. For eight selected blends viscosity, density, surface tension, lubricity, flash point and cold filter plugging point were measured and compared with those of diesel fuel to evaluate their compatibility as compression-ignition engine fuels. Standard recommended test methods were used in EN 590 to determine density at 15 °C (EN ISO 12185), flash point (EN ISO 2719), lubricity (EN ISO 12156-1), cold filter plugging point (EN 116). Viscosity at 40 °C was determined using ASTM D7042-04 and for determining surface tension the stalagmometric method was used. The main properties of the biodiesel, diesel fuel and ethanol used (Barabás & Todoruţ, 2009; Barabás & Todoruţ, 2010; Barabás et al., 2010) are shown in Table 1.

3. The miscibility and stability of the biodiesel-diesel fuel-ethanol mixtures

During the preparation of the mixtures BD, DE and BDE, it was observed their aspect before and after the homogenization. The mixtures, preserved 30 hours long at 20 °C temperature,

Properties \ Fuels	D100	B100	E100
Carbon content, % wt.	85.21	76.97	52.14
Hydrogen content, % wt.	14.79	12.24	13.13
Oxygen content, % wt.	0	10.79	34.73
Kinematic viscosity at 40 °C, mm^2/s	2.4853	5.5403	1.0697
Density at 15 °C, kg/m^3	843.3	887.4	794.85
Cetane number	52	55.5	8
Lower heating value, kJ/kg	42600	39760	26805
Flash point, °C	61	126	13
Lubricity WSD, μm	324	218	-
Surface tension at 20 °C, mN/m	29.0	38.60	19.19
Cold filter plugging point (CFPP), °C	-9	-14	-

Table 1. Main properties of the fuels (biodiesel, diesel fuel, ethanol)

were visually re-inspected (all the mixtures become homogeneous, transparent and clear), after that they were cooled at 0 °C. The experiment was repeated also for the -8 °C temperature (with one grade Celsius over the diesel fuel's cold filter plugging point - CFPP, which is the highest one).

Regarding the BDE mixtures' miscibility and stability it can be mentioned that these can be realized in different proportions, becoming homogeneous and clear after about 30 hours from the preparation. The mixtures' stability depends on their temperature, thus: at 20 °C temperature the mixture up to 15 % v/v bioethanol content remain stable; at 0 °C temperature the mixture up to 15 % v/v bioethanol content remain homogeneous (clear or diffuse), with the exception of the binary mixtures, which take place at the alcohol separation, found phenomenon also at the triple mixtures with a content over 15 % v/v bioethanol; at -8 °C temperature, the mixtures gain different aspects, thus: homogeneous and clear remain only the B30D70 and B25D70E5 mixtures; homogeneous, but diffuse become the B10D90, B5D95, D95E5 mixtures; clear with sediments (ice crystals) gain the B25D75, B20D80, B20D70E10, B20D75E5, B15D70E15, B15D75E10, B15D80E5 mixtures; separated in two levels (bioethanol + diesel fuel-biodiesel mixture) in case of the mixtures with an intermediate level of bio-fuel (B10D80E10, B10D85E5, B5D90E5) or in four levels (one level ethanol, followed by a paraffin emulsion level, diesel fuel-biodiesel mixture and emulsion formed by ice crystals and diesel fuel- biodiesel mixture) at the other mixtures.

The 27 types of studied mixtures comparative with diesel fuel have been realized respecting the presented compositions from figure 1. The results of these observations are shown in Figure 1 and are the first selection criteria of the blends. In the case of mixtures under the marking lines, the separation of the components was visible, while those located above remained stable (homogeneous).

4. The main properties of the selected biodiesel-diesel fuel-ethanol mixtures

4.1 Determining the key fuel properties of the investigated blends

After first selection of the blends we determined the mixtures key fuel properties under recommanded standard methods and calculus. In order to make the second selection, density, viscosity, surface tension, cold filter plugging point, lubricity, flash point, carbon

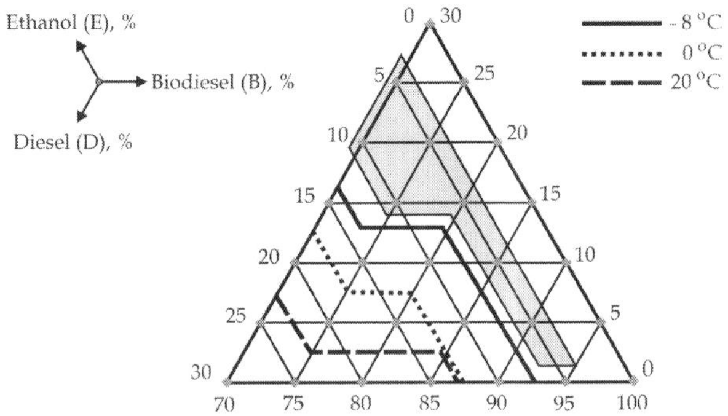

Fig. 1. Solubility and stability of biodesel-diesel fuel-ethanol blends

content, hydrogen content, oxygen content, cetane number and heating value of the blends was evaluated (measured or calculated) (Barabás & Todoruţ, 2009).

Density (ρ) is a fuel property which has direct effects on the engine performance characteristics (Sandu & Chiru, 2007). Many fuel properties such as cetane number and heating value are related to density. Fuel density influences the efficiency of fuel atomization and combustion characteristics (Sandu & Chiru, 2007). Because diesel fuel injection systems meter the fuel by volume, the change of the fuel density will influence the engine output power due to a different mass of injected fuel. Ethanol density is lower than diesel fuel density, but biodiesel density is higher.

Viscosity (ν) is one of the most important fuel properties. The viscosity has effects on the atomization quality, the size of fuel drop, the jet penetration and it influences the quality of combustion (Sandu & Chiru, 2007). Fuel viscosity has both an upper and a lower limit. It must be low enough to flow freely at its lowest operational temperature. Too low viscosity can cause leakage in the fuel system. High viscosity causes poor fuel atomization and incomplete combustion, increases the engine deposits, needs more energy to pump the fuel and causes more problems in cold weather because viscosity increases as the temperature decreases. Viscosity also affects injectors and fuel pump lubrication (Sandu & Chiru, 2007).

The surface tension (σ) of the fuel is an important parameter in the formation of droplets and fuel's combustion. A high surface tension makes the formation of droplets from the liquid fuel difficult.

The cold filter plugging point (CFPP) of a fuel is suitable for estimating the lowest temperature at which a fuel will give trouble-free flow in certain fuel systems. The CFPP is a climate-dependent requirement (between -20 °C and 5 °C for temperate climate).

Lubricity describes the ability of the fuel to reduce the friction between surfaces that are under load. This ability reduces the damage that can be caused by friction in fuel pumps and injectors. Lubricity is an important consideration when using low and ultra-low sulfur fuels. Fuel lubricity can be measured with High Frequency Reciprocating Rig (HFRR) test methods as described at ISO 12156-1. The maximum corrected wear scar diameter (WSD) for diesel fuels is 460 μm (EN 590). Reformulated diesel fuel has a lower lubricity and requires lubricity improving additives (which must be compatible with the fuel and with any additives already found in the fuel) to prevent excessive engine wear. The lubricity of biodiesel is good. Biodiesel may be used as a lubricity improver, especially unrefined biodiesel, while ethanol lubricity is very poor (Emőd et al., 2006; Zöldy et al., 2007; Rao et al., 2010).

The flash point (FP) is defined as the lowest temperature corrected to a barometric pressure of 101.3 kPa at which application of an ignition source causes the vapor above the sample to ignite under specified testing conditions. It gives an approximation of the temperature at which the vapor pressure reaches the lower flammable limit. The flash point does not affect the combustion directly; higher values make fuels safer with regard to storage, fuel handling and transportation (Rao et al., 2010). The flash point is higher than 120 °C for biodiesel (EN 14214), must be higher than 55 °C for diesel fuel (EN 590), and is below 16 °C for bioethanol.

The carbon content of the fuel determines the amount of CO_2 and CO in the burnt gas composition. Hydrogen content together with oxygen content determines the energy content of the fuel. Oxygen content contributes to the oxygen demand for combustion, providing more complete fuel combustion. The carbon, hydrogen and oxygen contents were calculated based on the composition of the constituents.

Cetane number (CN) is a measurement of the combustion quality of diesel fuel during compression ignition. It is a significant expression of diesel fuel quality among a number of other measurements that determine overall diesel fuel quality. The cetane number requirements depend on engine design, size, nature of speed and load variations, as well as starting and atmospheric conditions. Increase of cetane number over the values actually required does not materially improve engine performance. Accordingly, the cetane number specified should be as low as possible to ensure maximum fuel availability. Diesel fuels with a cetane number lower than minimum engine requirements can cause rough engine operation. They are more difficult to start, especially in cold weather or at high altitudes. They accelerate lube oil sludge formation. Many low cetane fuels increase engine deposits resulting in more smoke, increased exhaust emissions and greater engine wear. The cetane number was assessed based on the cetane numbers of the constituents and the mass composition of the blends (Bamgboye & Hansen, 2008).

The lower heating value (LHV) of the fuel determines the actual mechanical work produced by the internal combustion engine and the specific fuel consumption value. Since diesel engine fuel dosage is volumetric, the comparison of the volumetric lower heating value is more suitable. For this purpose it is useful to determine the *Fuel Energy Equivalence (FEE)*, which is the ratio of the heating value of the blend and the heating value of diesel fuel.

The main properties of the selected blends used (Barabás & Todoruţ, 2009; Barabás & Todoruţ, 2010; Barabás et al., 2010) are shown in Table 2. The densities of the biodiesel-diesel fuel-ethanol blends are in the range of 843.7...851.9 kg/m^3, very close to the diesel fuel requirement related in EN 590. In the case of the investigated blends kinematic viscosity is in the range of 2.3739…2.756 mm^2/s. The blends flash points that containing 5 % ethanol are in the range of 14…18 °C, and which containing 10 % ethanol are less than 16 °C. Measured values of surface tensions are in the range of 30.66…34.83 mN/m.

A significant decrease in the blends' flash point can be observed. The flash point of a biodiesel-diesel fuel-ethanol mixture is mainly dominated by ethanol. All of the blends containing ethanol were highly flammable with a flash point temperature that was below the ambient temperature, which constitutes a major disadvantage, especially concerning their transportation, depositing and distribution, which affects the shipping and storage classification of fuels and the precautions that should be taken in handling and transporting the fuels. As a result, the storage, handling and transportation of biodiesel-diesel fuel-ethanol mixtures must be managed in a special and proper way, in order to avoid an explosion.

Properties \ Blends	B5 D90 E5	B10 D85 E5	B15 D80 E5	B20 D75 E5	B25 D70 E5	B15 D75 E10	B20 D70 E10	EN 590
ρ, kg/m³	843.7	845	847.2	849.6	851.9	844.7	846.8	820...845
ν, mm²/s	2.4353	2.4205	2.5269	2.6447	2.756	2.3739	2.4796	2...4.5
FP, °C	17.5	14	16	17	18	15.5	16	55 (min.)
WSD, μm	305	232	276	243	252	272	264	460 (max.)
CFPP, °C	-18	-17	-13	-17	-16	-4	-7	climate-dependent
σ, mN/m	30.79	34.62	34.66	32.86	34.83	30.66	31.77	not specified
c, % wt.	83.22	82.79	82.37	81.94	81.52	80.80	80.38	not specified
h, % wt.	14.58	14.44	14.31	14.18	14.05	14.23	14.10	not specified
o, % wt.	2.20	2.76	3.32	3.88	4.43	4.96	5.52	not specified
CN	51.04	51.20	51.36	51.52	51.68	49.24	49.41	51 (min.)
LHV, kJ/kg	41707	41560	41414	41269	41124	40668	40524	not specified
LHV, kJ/L	35011	34979	34948	34916	34885	34219	34188	not specified
FEE	0.979	0.978	0.977	0.976	0.975	0.957	0.956	not specified

Table 2. Main properties of the blends

Concerning the cold filter plugging point (CFPP) it was observed that in the case of 5 % ethanol blends it decreases, but it gets higher in the case of 10 % ethanol blends because of the limited ethanol miscibility, which restricts its use at low temperatures (Barabás & Todoruţ, 2009).

Surface tension for blends containing 10 % ethanol is comparable to that of diesel fuel. Blends with high biodiesel content have a surface tension higher by up to 20 %, due to the higher surface tension of biodiesel (Barabás & Todoruţ, 2009).

Mixtures' density variation depending on temperature is depicted in Figure 2. Density of investigated mixtures varies depending on the content of biodiesel and ethanol in diesel. Increasing biodiesel content increases mixture's density, while increasing ethanol content leads to decrease its density. Comparing density of (Barabás & Todoruţ, 2009; Barabás et al., 2010) investigated fuels at 15 °C can be seen in Figure 3. It can be observed that mixtures in which the relation biodiesel content/ethanol content is less than 2 are within the imposed limits for diesel density EN 590, in terms of density.

Mixtures' viscosity variation with temperature (Barabás & Todoruţ, 2009; Barabás et al., 2010) is depicted in Figure 4. It can be observed that the ethanol reduced viscosity compensates biodiesel higher viscosity, and biodiesel-diesel fuel-ethanol blends have a closer viscosity to diesel, especially at temperatures above 40 °C. From Figure 5 it can be noted that all studied mixtures correspond in terms of kinematic viscosity to diesel imposed quality requirements EN 590 (Barabás & Todoruţ, 2009).

Surface tension of mixtures was determined at a temperature of 20 °C by an stalagmometric method (non-standard). Based on obtained results (Fig. 6) can be said that most biodiesel-diesel fuel-ethanol mixtures have a close superficial tension to diesel, ethanol successfully offsetting surface tension of a biodiesel (Barabás & Todoruţ, 2009).

The *flash point* was determined for all investigated blends using a HFP 339 type Walter Herzog Flash Point Tester, according to Pensky Martens method. Because the ethanol flash

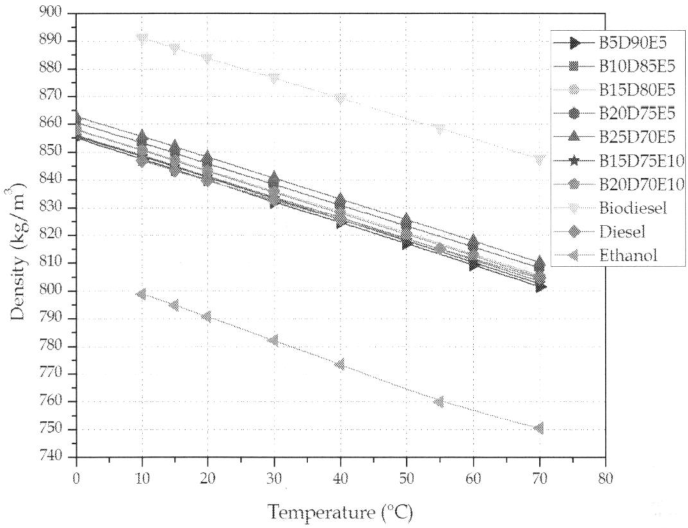

Fig. 2. Density variation with temperature

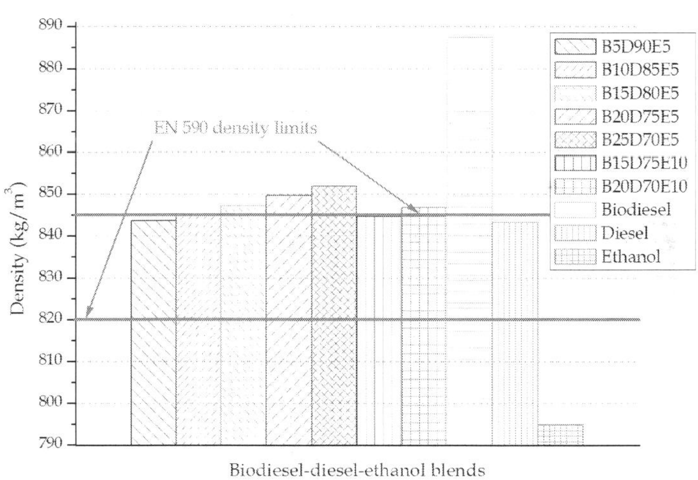

Fig. 3. Density of investigated fuels at 15 °C

point is very low, measured (Barabás & Todoruț, 2009) flash points for biodiesel-diesel fuel-bioethanol blends are very close to bioethanol flash point (Fig. 7).
The investigated blends *cold filter plugging points* were measured (Barabás & Todoruț, 2009) using an ISL FPP 5Gs type tester. CFPP is very different for each and also depends by solubility of biodiesel-diesel fuel-ethanol blends in test temperature (Fig. 8).

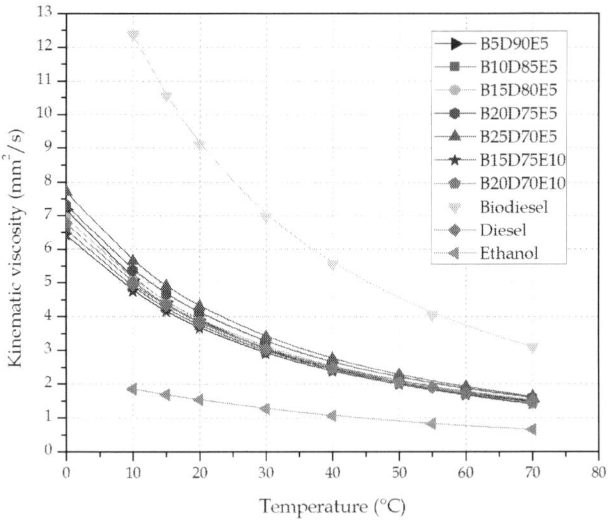

Fig. 4. Kinematic viscosity variation with temperature

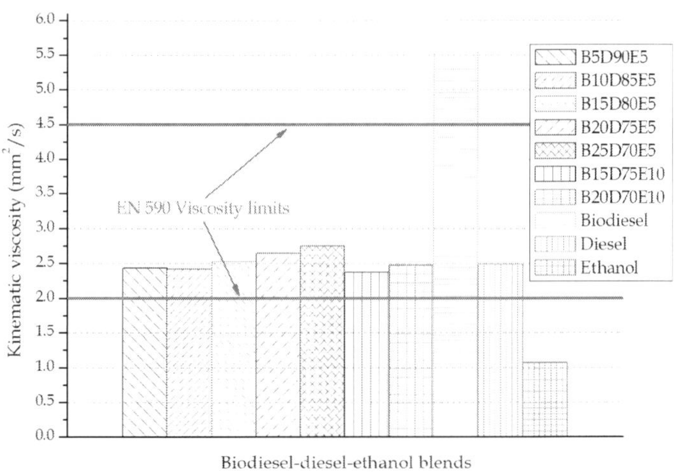

Fig. 5. Kinematic viscosity at 40 °C

4.2 Second mixtures selection based on phisical and chemical properties

For the second selection the following criteria were considered: volumetric lower heating value should not decrease with more than 3 %; cetane number should be over 51; density should not exceed the maximum limit imposed in EN 590 (845 kg/m^3) by more than 3 %, biofuel content should be above 7 % v/v (commercial diesel fuel may already contain max. 7 % v/v biodiesel) and various biodiesel/ethanol relations should be observed.

Based upon evaluated fuel properties (Table 2, Fig. 2 - Fig. 8), second mixtures selection was made. Selected blends was: B10D85E5, B15D80E5 and B25D70E5.
It can be seen that the biodiesel-diesel fuel-ethanol blends have a very close density to diesel fuel on the whole considered temperature domain.
There may be seen that the blends' viscosity is very close to that of diesel fuel, and the differences get smaller with temperature increase. Because the ethanol vaporizing temperature is quite small (approximately 78 °C), it will be in vapor state at the operating

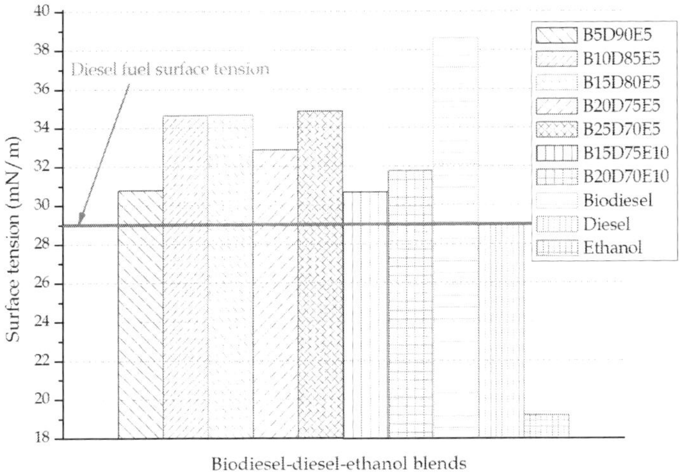

Fig. 6. Surface tension at 20 °C

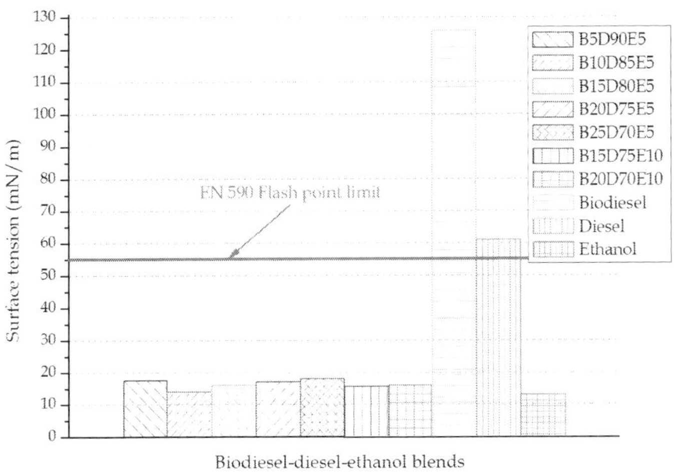

Fig. 7. Measured flash points for investigated biodiesel-diesel fuel-ethanol blends

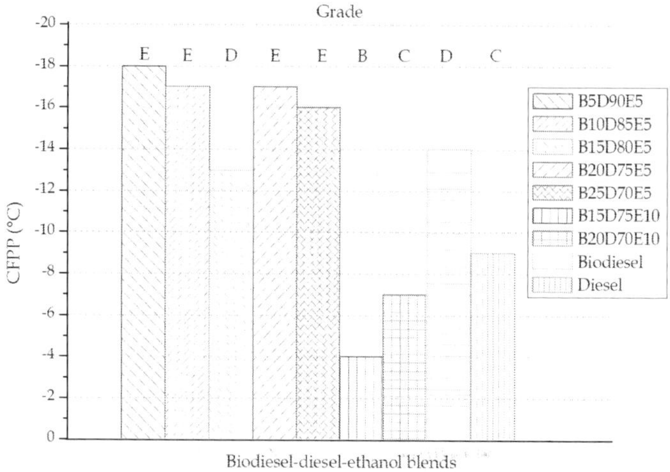

Fig. 8. Cold filter plugging point measured for different biodiesel-diesel fuel-ethanol blends

injector temperature. The compensation of biodiesel higher density and viscosity levels is important especially at low engine operating temperatures.

At the same time, a significant decrease in the blends flash point can be observed (14...18 °C) (Barabás et al., 2010). The flash point of a biodiesel-diesel fuel-ethanol mixture is mainly dominated by ethanol. All of the blends containing ethanol were highly flammable with a flash point temperature that was below the ambient temperature, which constitutes a major disadvantage, especially concerning their transportation, depositing and distribution, which affects the ship- ping and storage classification of fuels and the precautions that should be used in handling and transporting the fuel. As a result, the storage, handling and transportation of biodiesel-diesel fuel-ethanol mixtures must be managed in a special and proper way, in order to avoid an explosion.

Concerning the cold filter plugging point (CFPP) it was observed that in the case of 5 % ethanol blends it decreases (Barabás et al., 2010).

5. The performance and the emission evaluation features in the test bench

5.1 Engine performance and emission characteristics evaluation in laboratory condition

The experimental research concerning the ICE performances and pollution have been directed toward three fuel blends of biodiesel-diesel fuel-ethanol (B10D85E5, B15D80E5 and B25D70E5), for which diesel fuel has been used as reference. The experimental researches concerning the performances and the determination of pollutant emissions were developed on a test bench, equipped with a CI engine (number of cylinders - 4 in line; bore - 110 mm; stroke - 130 mm; compression ratio - 17:1; rated power - 46.5 kW at 1800 rpm; rated torque - 285 Nm at 1200 rpm; displacement volume - 4.76 l; nozzle opening pressure - 175 ± 5 bar; size of nozzle - 4 x 0.275 mm; injection system - direct, mechanical), hydraulic dynamometer and a data acquisition system for recording the operating parameters. For the evaluation of pollutant emissions the Bosch BEA 350 type gas analyzer was used (Barabás et al., 2010). The load characteristics have been drawn at 1400 rpm engine speed, this one being between the maximum torque speed and the maximum power speed. Before each test the fuel filters were replaced and the engine was brought to the nominal operating temperature. For evaluation, the obtained results were compared with those obtained in the case of diesel fuel. The results-evaluation has been made for three engine loading domains: small loads (0–40 %), medium loads (40–80 %) and high loads (>80 %).

Engine power and actual torque of the engine decreases with 5-9 % using the researched mixtures versus base diesel fuel. Also found that the engine speed corresponding to the maximal power decreases with 70-100 rpm when engine is fuelled with biodiesel-diesel fuel-ethanol blends.

Break specific fuel consumption (BSFC). The obtained results (Barabás et al., 2010) in the case of specific fuel consumption related to engine load are presented in Figure 9. The brake specific consumption is greater at smaller loads, but it decreases at medium and higher loads. The brake specific fuel consumption is greater for the blends, because their heating value is smaller. The sequence is D100, B10D85E5, B15D80E5 and B25D70E5 being the same at all engine loads, maintaining the increasing sequence of biofuels content. The increase is higher at small loads (32.4 % in the case of B25D70E5); at medium and high loads the determined values for blends are comparable with the values for diesel fuel, being between 6.2 % and 15.8 %.

Brake thermal efficiency (BTE). The engine efficiency variation with load for the studied fuels (Barabás et al., 2010) is shown in Figure 10. As it was expected, the engine efficiency decreases for fuel blends, the tendencies being similar with those of brake specific fuel consumption. The engine efficiency decrease is between 1.3 % and 21.7 %.

For *pollution evaluation* the emissions of CO, CO_2, NO_x, HC and smoke have been measured. The CO emissions (Fig. 11) vary according to the used fuel and according to the engine load (Barabás et al., 2010). Such as, at small and medium loads, the highest emissions were measured in the diesel fuel case, and the lowest ones in the B15D80E5 mixture case.

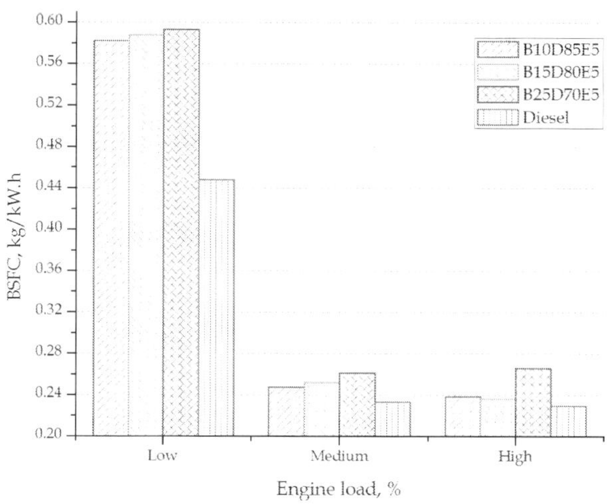

Fig. 9. Variation of brake specific fuel consumption of different fuels

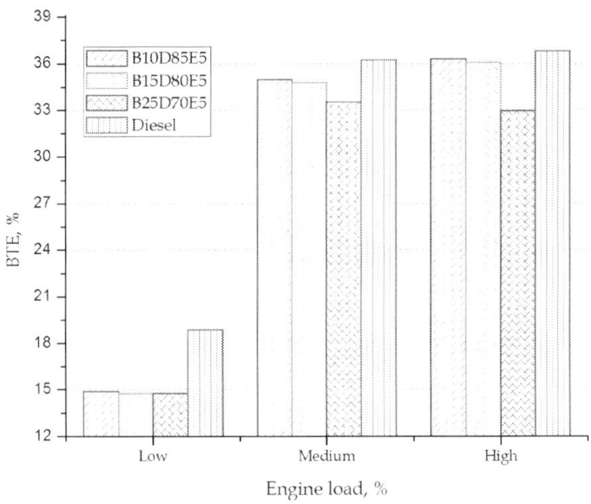

Fig. 10. Engine's efficiency variation with load for analyzed fuels

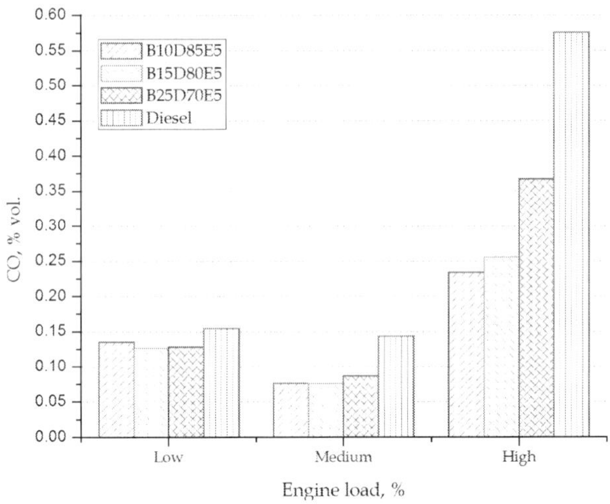

Fig. 11. Variation of CO emission with percentage of load for different fuels

As expected, at high loads increase the CO emissions, being lower in case of the researched mixtures with about 50 %. This fact is explained in (Subbaiah et al., 2010) by the high oxygen content of the biodiesel and of the ethanol witch sustained the oxidation process during the gas evacuation, too. The experimental results (Barabás et al., 2010) showed that at high engine loads, the lowest CO emission is for the B10D85E5 mixture (0.234 % vol.) which comparatively with the one seen in the diesel fuel case (0.575 % vol.) represents a 59 % reduction.

The CO_2 emissions (Fig. 12) in case of the researched mixtures are superior to those measured in case of the diesel engine function at all three regimes of loads taken into consideration (Barabás et al., 2010). The increasing level of the CO_2 emissions can be put on the decreasing CO emissions' account, which further oxidizes because of the high oxygen content of the researched mixtures providing a more complete combustion. Also, the oxygen excess made possible the CO oxidation during the evacuation process, too, including on the evacuation route of the combustion gas. This explication is also sustained by the decreasing of the CO emissions towards those seen in the diesel fuel case. The increasing of the CO_2 emissions cannot be considered as a negative consequence, because they are re-used (consumed) in the plants' photosynthesis process from which bio-fuels are fabricated.

Regarding the NO_x emissions (Fig. 13) of the Diesel engine tested with the researched fuels at different loads it was seen (Barabás et al., 2010) that the presence of the oxidized chemical component parts in the fuel at low loads has insignificant influence over the NO_x emissions levels, usually showing a slight reduction, but at medium and high loads the NO_x emissions are superior with 10-26 % to those seen in case of the diesel fuel. The increasing of the NO_x emissions at medium or high loads can be explained by the increasing of the fuel's combustion temperature, because of the oxygen content of biodiesel and ethanol, which made possible a more complete combustion and a increasing of the combustion temperature, which favors the formation of the NO_x. Also, because of the ethanol's reduced cetanic number, the mixture's cetanic number is reduced. This fact leads to the increased delay to ignition of the fuel, because of this the cumulated fuel/air mixture will burn more rapidly, creating a more rapid heat release at the beginning of the combustion process, resulting a higher temperature which favors the NO_x formation.

Regarding the HC emissions (Fig. 14) of the alternatively fueled engine with the researched BDE mixtures and diesel fuel, function by its load, it was seen (Barabás et al., 2010) that in case of the mixtures with 5 % ethanol content, the hydrocarbon emissions are reduced in significant way from diesel fuel in all three domains of the engine's load, the most significant reduction being seen in the high loads field about 50 %. The ethanol's presence in

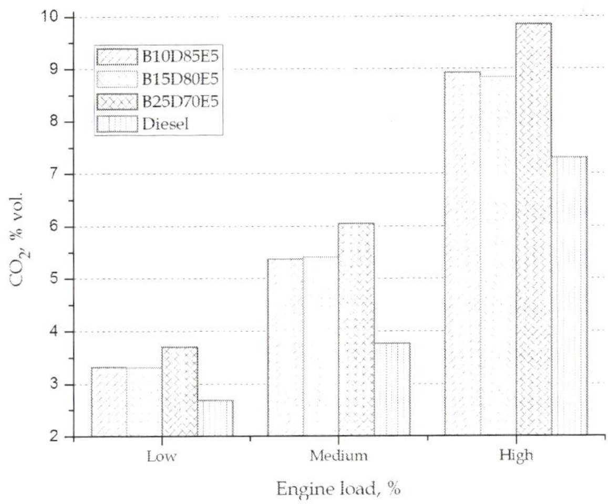

Fig. 12. Variation of CO_2 emission with percentage of load for different fuels

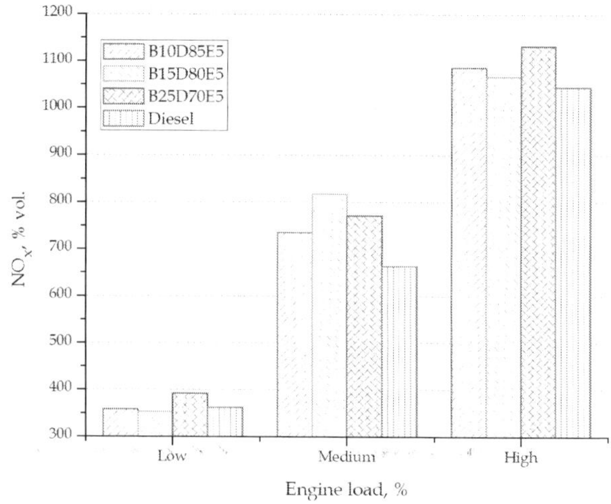

Fig. 13. Variation of NO_x emission with percentage of load for different fuels

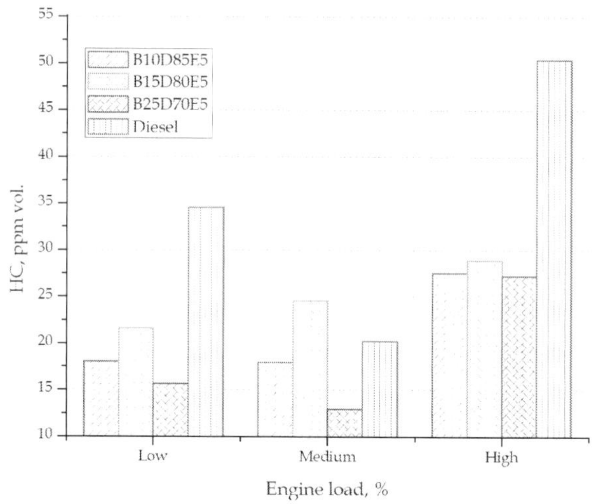

Fig. 14. Variation of HC emission with percentage of load for different fuels

mixture is an increasing factor of the HC emissions, while the biodiesel's presence leads to their reduction. An explanation it could be given through the cetanic number: the biodiesel having the cetanic number superior to the one of the diesel fuel favors easy ignition and a more complete combustion of the mixture, while the reduced cetanic number of the ethanol acts in opposite way. Because of the reduced cetanic number, the ethanol ignites later and it will burn incompletely, thus increasing the un-burnt hydrocarbons content from the evacuation gas composition.

The smoke emissions (Fig. 15) of the tested engine were evaluated by the measurement of the evacuation gas opacity, emphasized by the light's absorption coefficient (Barabás et al., 2010). The evacuation gas opacity it was significant reduced (with over 50 %) in case of the all mixtures, especially at low and medium loads. At high loads, the decreasing is between 27.6 % in the B25D70E5 mixture case and 50.3 % in the B10D85E5 mixture case. The smoke's formation takes place in the fuel reach fields of the combustion chamber, especially in the field of the injected jet's vein.

Concerning smoke opacity it has been observed that it decreases compared to the smoke opacity recorded in the case of diesel fuel, being higher for the fuel blends with high biofuel content.

Generally it may be concluded that the studied fuel blends have lower pollution levels, exceptions being CO_2 and NO_x, in which cases the recorded values are superior to those recorded for diesel fuel.

5.2 Vehicle performance evaluation on chassis dynamometer

For the comparative evaluation of the inquired fuel types, these were tested on a passenger car, equipped with a Diesel engine with a four strokes and six cylinders in line, with a maximum developed power of 86 kW at 4800 rpm and 220 Nm torque at 2400 rpm. To this end, tests for the evaluation of power and torque against engine speed were conducted on an inertial dynamometer, and road tests using GPS technology - to determine the dynamic characteristics of the test passenger car.

Tests on the dynamometer. On the dynamometer variation of power and torque measured at the wheel and engine power and torque were calculated for each fuel. Six tests were performed for each fuel and the average values of maximum power and maximum moment were calculated. The results obtained (Barabás & Todoruţ, 2010) are shown in Figure 16. When tested against diesel there was a reduction of maximum power with 3.6 % for the

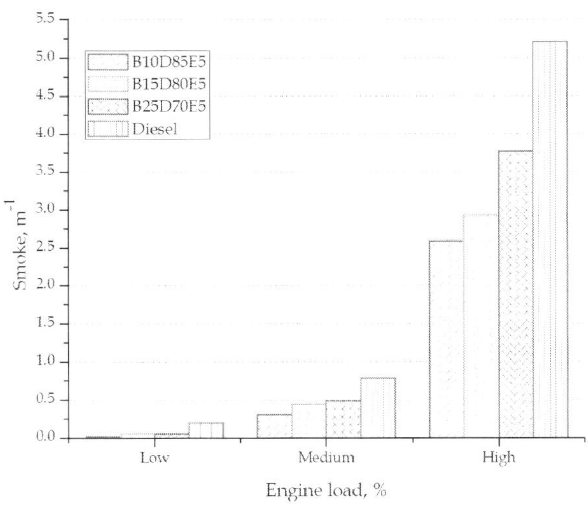

Fig. 15. Particle emissions

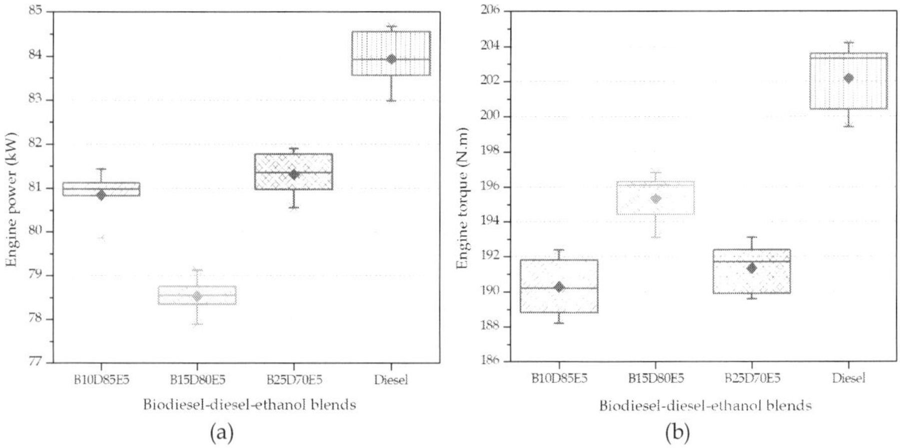

Fig. 16. Maximum engine power (a) and torque (b) for selected fuels

B10D85E5 blend, with 6.4 % for the B15D80E5 blend and with 3.1 % for the B25D70E5 blend. Engine speed changes corresponding to maximum were observed, with 4750 rpm for diesel and 5050 rpm with B10D85E5 mixture and 5000 rpm for biodiesel, such a change wasn't detected with B15D80E5 mixture. Maximum engine torque also decreased using blends, when compared to diesel fuel: 5.8 % for the B10D85E5 blend, with 3.3 % for the B15D80E5 blend and 5.3 % when using the B25D70E5 blend.

5.3 Road test performances of biodiesel-diesel fuel-bioethanol blend

For road tests the following blends have been selected: B10D85E5, B15D80E5 and B25D70E5. The performed dynamic tests were intended to determine some of the passenger car's dynamic features like (Barabás & Todoruţ, 2010): vehicle elasticity, overtaking and accelerations parameters. The configuration of the vehicle and its attitude has been as determined by the manufacturer. The vehicle was clean, the windows and air entries were closed. The tire pressures were according to the specifications of the vehicle manufacturer. The mass of the vehicle has been its kerb mass plus 180 kg. Immediately before the test, the parts of transmission and tires were warmed up during a 30 km course. The measurements have been carried out on a 5 km long, straight, with hard, smooth, good adhesion track. Longitudinal slope was max. 0.5 % and transverse slope hasn't exceeded 3 %. The corrected value of air density during the test hasn't varied by more than 7.5 % from the air density in the reference conditions (temperature: 20 °C, pressure: 1000 mbar). The average wind speed measured at a height of 1 m above the ground was less than 3 m/s; gusts were less than 5 m/s.

Vehicle performance and speed test were evaluated over acceleration ability (acceleration 0-100 km/h and 0-400 m), elasticity in 4th gear - $t_{60\text{-}100\ km/h}$, elasticity in 5th gear - $t_{80\text{-}120\ km/h}$, overtaking in 3/4th gear - $t_{60\text{-}100\ km/h}$, overtaking in 4/5th gear - $t_{80\text{-}120\ km/h}$. To determine the elasticity and overtaking capability, 12 tests were conducted with each fuel, upon which the average values were calculated (Barabás & Todoruţ, 2010). The obtained road test results are shown in Figure 17.

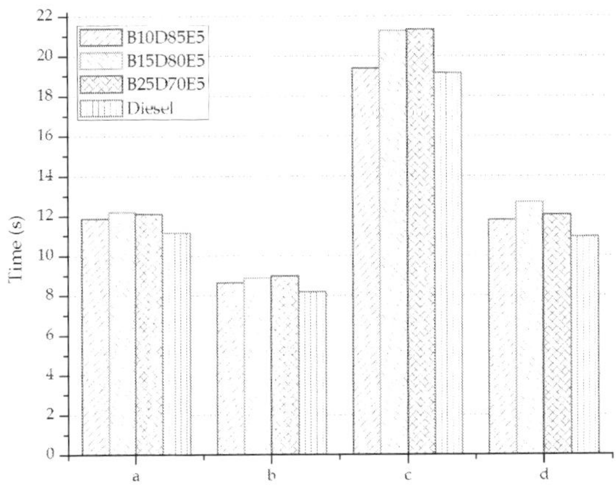

Fig. 17. Comparing dynamic parameters of the tested vehicle, when using mixtures, over the use of diesel-100 %: a - Elasticity in 4th Gear, $t_{60\text{-}100km/h}$; b - Overtaking in 3/4th Gear, $t_{60\text{-}100km/h}$; c - Elasticity in 5th Gear, $t_{80\text{-}120\ km/h}$; d - Overtaking in 4/5th Gear, $t_{80\text{-}120\ km/h}$

Characteristic \ Fuel type	D100	B10D85E5	B15D80E5	B25D70E5
Acceleration (0-100 km/h), s	17.40	17.95	22.06	19.25
Acceleration (0-400 m), s	21.15	22.48	24.53	23.42

Table 3. The acceleration parameters results of the tested passenger car

It was found that the dynamic performances were reduced for all the blends studied, the weakest performance being obtained in case of the mixture B15D80E5. Performances obtained with blends B10D85E5 and B25D70E5 are comparable, but the latter has the advantage of a higher biofuel content.

When determining (Barabás & Todoruț, 2010) the acceleration parameters, 6 tests were conducted and the best results were considered (Table 3).

6. Conclusion

This chapter presents the selection of biodiesel-diesel fuel-ethanol blends with a maximum biofuel content of 30 %, used to power compression ignition engines without their significant modification. It was found that among the original 27 mixtures only seven are suitable in terms of miscibility and stability, having an ethanol content of maximum 5 %. A comparison of the blends' main properties with those of diesel fuel has reduced the number of usable mixtures to three, having biodiesel content between 10 and 25 %.

The fuel blends used in the research for this paper have similar properties to those of commercial diesel fuel. The density of the blends is located near the maximum limit specified in EN 590. The kinematic viscosity values and the lubricity values are within the limits mentioned in the quality standard. Low flash point of mixtures requires special measures for handling and storage.

After the *performances of the Diesel engine evaluation*, by tests on the experimental stand for compression combustion engine testing, it was seen an increasing of *the specific fuel consumption* in the case of selected fuels (B10D85E5, B15D80E5, B25D70E5), on average with 15.85 %, respecting the ascending order of the bio-fuel content, and in case of the *engine's efficiency* it was seen a decrease around 10.36 % in case of the researched BDE mixtures used. From the analysis of the obtained results regarding *the pollution produced by the tested engine,* charged with the new types of fuels, in which the diesel fuel was present, it is seen that the diesel fuel can be replaced with success by the BDE mixtures taken into the study, situations in which it is provided the noticed decreasing of the environmental chemical pollution (Table 4). The international targets regarding the gas decreasing which contribute to global environmental changes and to the improvement of the air quality at local level are satisfied by the bio-fuels properties compared with those of the classic fuels. The bio-fuels obtained by energetic plants are clean fuels, biodegradable and renewable, and their obtained technology is clean.

After the inertial chassis dynamometric tests, in case of BDE mixture use, it was seen a decreasing of the tested passenger car engine's power, on average with 4.41 % (Fig. 16.a), and of the torque engine with about 4.87 % (Fig. 16.b) in spite of diesel fuel use.

The road tests spotlighted a decreasing of the tested car performances, in case of all researched fuels compared with the diesel fuel use, presenting similar tendencies with those from the inertial chassis dynamometric tests. The comparison of the obtained results after road tests, regarding the elasticity and overtaking capability of the tested passenger car, spotlights the difference between the dynamic parameters obtained in case of the researched BDE mixture use, compared with the case of diesel fuel use, thus (Figure 17 and Table 3): elasticity in 4th gear, $t_{60\text{-}100\ km/h}$ - with about 8.23 %; overtaking in 3/4th gear, $t_{60\text{-}100\ km/h}$ - with about 7.88 %; elasticity in 5th gear, $t_{80\text{-}120\ km/h}$ - with about 7.84 %; overtaking in 4/5th gear, $t_{80\text{-}120\ km/h}$ - with about 11.25 %; acceleration 0-100 km/h - with about 13.52 %; acceleration 0-400 m - with about 11 %.

Engine loading domains and blends / Pollutant	small loads (0-40 %)			medium loads (40-80 %)			high loads (> 80 %)		
	B10 D85 E5	B15 D80 E5	B25 D70 E5	B10 D85 E5	B15 D80 E5	B25 D70 E5	B10 D85 E5	B15 D80 E5	B25 D70 E5
CO	–	– – –	– –	– – –	– – –	– –	– – –	– –	–
CO_2	+	+	+ +	+	+ +	+ + +	+ +	+	+ + +
NO_x	–	–	+	+	+ + +	+ +	+ +	+	+ + +
HC	– –	–	– – –	– –	–	– – –	– – –	– –	– – –
Smoke	– – –	–	– –	– – –	– –	–	– – –	– –	–

Table 4. The synthesis of the obtained results, compared with diesel fuel, regarding the emissions of the diesel engine's chemical pollutions tested with the researched fuels (B10D85E5, B15D80E5, B25D70E5)

In general, it was seen that from the point of view of the tested passenger car's performances, the BDE mixtures can successfully replace the diesel fuel.
It was found that in terms of performance, the B10D85E5 and B25D70E5 blends can succesfully replace diesel fuel.
The researches regarding partial replacement of the diesel fuel destined to diesel engines with mixtures biodiesel-diesel fuel-bioethanol (BDE), can be continued through out the determination of the influences of research fuels on the research engine's technical condition (comparative evaluation of deposits on the engine parts; evaluation of engine parts wear; assessment of lubricating oil quality evolution).

7. Acknowledgment

This work was supported by the Romanian National University Research Council, grant number 88/01.10.2007.

8. References

Bamgboye, A. I. & Hansen, A. C. (2008). Prediction of cetane number of biodiesel fuel from the fatty acid methyl ester (FAME) composition. *International Agrophysics*. Vol. 22, No. 1, (March, 2008), pp. 21-29, ISSN: 0236-8722.

Barabás, I. & Todoruţ, A. (2009). Key Fuel Properties of Biodiesel-diesel fuel-ethanol Blends. *Proceedings of SAE 2009 International Powertrains, Fuels & Lubricants Meeting*, Session: Alternative and Advanced Fuels (Part 1 of 4), Paper Number: 2009-01-1810, ISSN 0148-7191, Florence, Italy, June 15-17, 2009.

Barabás, I. & Todoruţ, A. (2010). Chassis Dynamometer and Road Test Performances of Biodiesel-diesel Fuel-Bioethanol Blend. *Proceedings of SAE 2010 Powertrains, Fuels & Lubricants Meeting*, Session: Alternative and Advanced Fuels (Part 2 of 3), Paper Number: 2010-01-2139, ISSN 0148-7191, San Diego, California, USA, October 25-27, 2010.

Barabás, I.; Todoruţ, A. & Băldean, D. (2010). Performance and emission characteristics of an CI engine fueled with diesel-biodiesel-bioethanol blends. *Fuel - The Science and Technology of Fuel and Energy*, Vol. 89, No. 12, (December, 2010), pp. 3827-3832, Published by Elsevier Ltd., ISSN 0016-2361.

Emőd, I.; Tölgyesi, Z. & Zöldy, M. (2006). *Alernatív járműhajtások*. Maróti Könyvkereskedés és Kiadó, ISBN 963-9005-738, Budapest, Hungary.

Rao, G.L.N.; Ramadhas, A.S.; Nallusamy, N. & Sakthivel, P. (2010). Relationships among the physical properties of biodiesel and engine fuel system design requirement. *International Journal of Energy and Environment*, Vol. 1, No. 5, (2010), pp. 919-926, ISSN 2076-2895.

Sandu, V. & Chiru, A. (2007). *Automotive fuels*. Matrix Rom, ISBN: 978-973-755-188-7, Bucharest, Romania.

Subbaiah, G.V.; Gopal, K.R.; Hussain, S.A.; Prasad, B.D. & Reddy, K.T. (2010). Rice bran oil biodiesel as an additive in diesel- ethanol blends for diesel engines. *International Journal of Research and Reviews in Applied Sciences*, Vol. 3, No. 3, (June, 2010), pp. 334-342, ISSN: 2076-734X.

Zöldy, M.; Emőd, I. & Oláh, Z. (2007). Lubrication and viscosity of the bioethanol-biodiesel-bioethanol blends, presented at *11th European Automotive Congress,* ISBN 963-420-817-7, *Budapest, Hungary,* 30 May - 1 June, 2007.

15

The Key Role of the Electronic Control Technology in the Exploitation of the Alternative Renewable Fuels for Future Green, Efficient and Clean Diesel Engines

Carlo Beatrice, Silvana Di Iorio, Chiara Guido and Pierpaolo Napolitano
Istituto Motori, CNR, Naples
Italy

1. Introduction

Great concerns are growing up on environmental impact of fossil fuel and poor air quality in urban areas due to traffic-related air pollution. In the last years, special attention was paid mainly to particulate matter (PM) and NOx emissions of diesel engines since these pollutants are associated to environmental and health issues. In particular, NOx contributes to the formation of ozone and acid rains and PM could cause injuries to the pulmonary and the cardiovascular systems. Nowadays, the overall concern about the global warming determines an increased interest also for CO_2 emissions, one of the major greenhouse gas (GHG). In this respect, a significant improvement can be reached with the increased use of "clean" and renewable fuels. It is well known, in fact, that the use of biofuels can contribute to a significant well-to-wheel (WTW) reduction of GHG emissions. The most interesting biofuel is the biodiesel and the fuels synthesised from fossil or biogenic gas.

Biodiesel designates a wide range of methyl-esters blends and is generally indicated with the acronym FAME, Fatty-Acid Methyl Esters. Biodiesel is produced from vegetable oils and animal fats through the transesterification, an energy efficient process that gives a significant advantage in terms of CO_2 emission and that features both high energy conversion efficiency and fuel yield from processed oil. These two characteristics are the main responsible for the overall GHG emissions benefit of biodiesel in WTW analyses [1].

More recently, starting from the well-known Fischer-Tropsch synthesis process, another generation of alternative diesel fuel was developed. It is usually indicated with XTL, where X denotes the specific source feedstock and TL (to Liquid) highlights the final liquid state of the fuel. It has minor interferences with the human food chain, since non-edible biomasses can be employed or, in case of animal-edible biomasses, the whole plant can be processed, as for the cellulosic ethanol production.

From the engine fuelling point of view, the significant difference between the two biofuels lies in their chemical composition. The first is essentially a blend of methyl-esters and the second of paraffin and olefin hydrocarbons. Because of the growing concerns about the energy crops impact on environment and food price, an increasing number of countries and stakeholders have recently challenged FAME biofuels. On the contrary, the XTL fuels, which

show high energy yield in the production process as well as the capability to extend input feedstock to cellulosic biomasses, are considered very attractive [2]. Within this framework, biofuel producers and OEMs are jointly devoting significant efforts in optimizing benefits from first generation biofuels while making second generation technologies economically viable soon. In particular, in order to enlarge biofuel market penetration, common fuel standards need to be defined and the compatibility of the engines with biofuels improved.
Biodiesel is the most important type of alternative fuels used in compression ignition engines because of its advantages in terms of emission reduction without significant changes to engine layout [3, 4, 5] and, at the same time, for being partly bound by future European legislation [6]. Anyway, several studies showed that the impact of biodiesel on the modern diesel engines is significant also in terms of engine performance mainly because of the interaction between the biodiesel characteristics and the engine-management strategies [7, 8]. One of the main differences of FAME with respect to petro-based diesel fuel is its oxygen content. It exceeds 10% of the total mass and it is directly responsible of the Low Heating Value (LHV) reduction of the same magnitude and eventually of the engine performance loss of about 12-15% at rated power and up to 30% in the low end torque. An efficient use of alternative diesel fuels, allowing to fully exploit all their potentials, can only be achieved through an "ad hoc" calibration of engine parameters and its control strategy (injection set and EGR rate) [8].
To create a flexible engine that can work efficiently both with conventional diesel and with biodiesel, it appears extremely important to develop a system able to detect the diesel biodiesel blending ratio present in the fuel tank and, automatically, to adapt engine calibration in order to fully exploit the fuel properties. In this respect, the adoption in the modern engines of the last recent combustion control methodology, named Closed Loop Combustion Control (CLCC) and based on the engine torque control by means of the instantaneous cylinder pressure measurement of the Electronic Control Unit (ECU) [8], has opened new scenarios for the development of the actual flex-fuel diesel engines.
On the basis of the previous experiences [5, 8], specific research activities were addressed to exploit and assess the capabilities offered by the CLCC technology in the development of the flex-fuel diesel engine.
The investigation was focused on three main aspects:

- development of a biodiesel-diesel blending detection (BD) methodology;
- mitigation of the impact of alternative fuels on emissions;
- exploitment of the alternative fuel quality for engine performance improvement.

The investigation was carried out on a 2.0L Euro5 diesel engine equipped with embedded pressure sensors in the glow plugs. Various blends of biodiesel were tested, notably 20% by volume (B20), 50% (B50) and pure biodiesel (B100). Tests on the multi-cylinder engine were carried out in a wide range of engine operating points for the complete characterization of the biodiesel performance in the New European Driving Cycle (NEDC) cycle.

2. Fuels

The measurements were performed fuelling the engine both with pure fuels and blends to achieve a reliable biodiesel blending detection. The reference diesel fuel (RF) was an EU certification diesel fuel (CEC, RF-03-A-84) compliant with EN590, while the tested biodiesel was an EU-widely-available Rapeseed Methyl ester (RME) compliant with EN14112. Table 1 reports some of the most important parameters of the pure fuels.

Feature	Method		RF	RME100
A/Fst			14.54	12.44
Low Heating Value [MJ/kg]	ASTM D 4868		42.965	37.570
Carbon [%, m/m]	ASTM D 5291		85.220	77.110
Hydrogen [%, m/m]	ASTM D 5291		13.030	11.600
Nitrogen [%, m/m]	ASTM D 5291		0.040	0.030
Oxygen [%, m/m]	ASTM D 5291		1.450	11.250
Cetane Number	EN ISO 5165		51.8	52.6
Density @ 15 °C [kg/m3]	EN ISO 12185		833.1	883.1
Viscosity @ 40 °C [mm2/s]	EN ISO 3104		3.141	4.431

Feature	Method		RF	RME100
Distillation [°C]	EN ISO 3405	IBP	158.9	318.0
°C		10% vol.	194.3	331.0
°C		50% vol.	267.6	335.0
°C		90% vol.	333.4	341.0
°C		95% vol.	350.0	353.0
°C		FBP	360.9	355.0
Oxydation stability [mg/100ml]	EN ISO 12205		-	0.6
Oxydation Thermal Stability @ 110°C [h]	EN 14112		-	6.5
C.F.P.P. [°C]	EN 116		-	-14
Lubricity @ 60°C [m	EN ISO 12156-01		-	179
POV [meq O2/Kg]	NGD Fa 4			16.60
TAN [mg KOH/g]	UNI EN 14104			0.13

Table 1. Main fuel parameters

The combustion and the exhaust gas properties are mainly influenced by the lower LHV and stoichiometric air fuel ratio (A/Fst) of the RME fuel with respect the RF fuel due to the higher oxygen content of biodiesel. Moreover, RME's higher density and viscosity, coupled with its distillation curve stretched in the temperature interval corresponding to the high-temperature boiling fractions of conventional diesel, increase significantly the penetration of its sprays with respect to the reference diesel, especially in cold conditions. Spray over-penetration is actually one of the most important concerns of diesel-FAME blends because it leads to increased oil dilution, as well as risks of piston and liner scuffing. Furthermore, higher boiling curve of biodiesel leads to higher resident time of fuel in the oil. Such drawback could be even more critical in case of specific injection strategies involving late injections in the exhaust stroke, as for example the DPF/DeNOx regeneration. In such cases, fuel injection occurs in low-density charge and is targeted well above the bowl edge: experimental verifications reported in literature using pure biodiesel have assessed an oil dilution rate up to three times the baseline, depending on engine type, operating conditions and injection strategy [9].

3. Experimental apparatus and test plan

The main characteristics of the adopted four-cylinder in-line Euro5 diesel engine are reported in Table 2. The Euro 5 engine features the closed-loop combustion control (CLCC), which enables individual and real-time control of the angular position corresponding to the 50% of Burned Fuel Mass, with respect to the top dead center (MFB50) and the Indicated Mean Effective Pressure (IMEP), cycle-by-cycle and cylinder-by-cylinder. In particular, based on in-cylinder pressure traces, heat release rate analysis is performed by ECU EDC17 using proprietary algorithms. The actual values for MFB50 and IMEP are compared to the target ones. As a consequence, the deviations of these two parameters are continuously resettled by adjusting the main injection timing and quantity for the following combustion cycle [10]. Based on these operating characteristics, the CLCC technology has been employed in order to develop a new diesel-biodiesel BD methodology, to mitigate or improve the engine emissions and increase the full load engine performance.

The engine was installed on a dyno test bench fully instrumented for indicated signal measurements (cylinder pressure, injection pressure, energizing injector current). Such measurements were carried-out by means of an AVL-based indicating acquisition system with an high-accuracy pressure sensor fitted on first cylinder, that was used as reference for validating the accuracy of the ECU-based CLCC. At the engine exhaust, smoke was measured by a high-resolution (0.01 filter smoke number, FSN) smoke meter (AVL415S), while gaseous emissions were measured upstream and downstream of the diesel aftertreatment device by means of a raw emission analysis test bench (AVL-CEB-2).

Engine type	4 cylinders in-line
Certification	EURO5
Bore x Stroke [mm]	83 x 90
Compression Ratio	16.5
Valves per cylinder	4
Rated power and torque	118kW @ 4000rpm 380Nm @ 2000rpm
Injection system	Common Rail
Injector and nozzle	Solenoid CRI 2.2+, 7 holes
Turbocharger	Single stage VGT
Catalyst system	Integrated closed-coupled DOC & DPF

Table 2. Main features of the engine

For all fuel blends, the engine was tested in nine steady-state operating points (k-points). The first seven test points were selected as the most representative of the engine operation on NEDC when matched to a D-class vehicle (1590kg IW). The eighth (2500 rpm at 16 bar of BMEP) and ninth (2500 rpm full load) test points were devoted to the characterization of the engine performance in real life aggressive driving.
The selected NEDC k-points are summarized in Figure 1, where the operating area of the engine running over NEDC is also displayed.

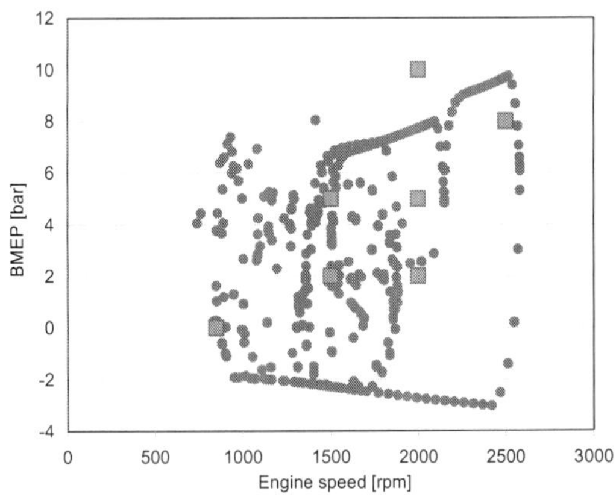

Fig. 1. Test points (k-points) together with the entire speed/load trace of the engine over NEDC for a D-class vehicle.

The measurements in the seven part-load test points enabled the NEDC-cycle vehicle performance estimation for all fuel blends by means of a well consolidated correlation procedure between the specific emissions at steady-state dyno engine testing with vehicle emissions on the chassis dynamometer. For each test point the injection strategy was composed by Pilot + Main events.

4. Results

4.1 Description of blending detection procedures

In order to be widely implemented, the blending detection (BD) strategy needs to be reliable, sufficiently accurate, robust towards biodiesel types and aging, as well as cost-effective. Taking in consideration all those key factors, the research activity was focused towards BD strategies that employ sensors already installed on the engine, whose reliability is proven. Therefore, to estimate the blending rate, the strategies combine the information given by sensors with the quantitative information derived from the diesel/biodiesel mixture properties.

In particular, the methodology described by Ciaravino et al. [11], leverages the information carried out by in-cylinder pressure transducers about the actual cycle-averaged IMEP, denoted as IMEP1. This last is obtained by integration of the high-frequency pressure signal, measured by the instrumented glow plugs of each cylinder:

$$IMEP_i = \frac{\oint p_i dV}{V} \tag{1}$$

being i the i-th cylinder and by its following averaging:

$$IMEP1 = \frac{\sum_i IMEP_i}{n} \tag{2}$$

where n is the number of cylinders.

IMEP1 can be compared, over each engine cycle, with the estimated IMEP produced on the basis of the IMEP mapping performed with pure diesel fuel, as a function of engine speed and accelerator pedal position.

Since the actual IMEP mapped for pure diesel fuel depends, for a certain engine speed and accelerator pedal position, on fuel conversion efficiency (FCE), fuel injected quantity (Qfuel) in terms of mass per cycle, LHV and friction losses (FMEP), the only quantity appreciably impacted by biofuel blending is LHV. In fact, in previous investigations, the authors verified that biodiesel does not significantly affect the engine FCE when the engine runs at the same operating point [5, 8], while FMEP, mapped as a function of the operating point and coolant temperature, is characteristic of the whole engine system architecture. Thus, the estimated IMEP from the engine mapping can be defined as IMEP2:

$$IMEP2 = \frac{FCE \cdot Qfuel \cdot LHV}{n \cdot V} + FMEP \tag{3}$$

Hence, two formulations can be leveraged for BD depending on the diesel engine operation mode, with either open-loop or closed-loop IMEP control [11]. In particular, if the engine is in IMEP open-loop control, the Blending Ratio BR is:

$$BR = 100 \cdot \frac{\left(IMEP2 / IMEP1 \right) - 1}{\left(LHV_{diesel} / LHV_{FAME} \right) - 1} \tag{4}$$

and the BR is linked to the reduction of IMEP when a biodiesel is burned. Actually, the BR calculation in open loop mode is affected by inaccuracy, due to the drift in engine operating point (i.e. different lambda, heat losses, etc), which impacts the FCE to an extent that does not allow to consider it constant in the working point with IMEP1 (using the diesel/FAME blend) and the estimated working point from engine speed and accelerator pedal position. However, as shown by the authors in two previous papers [5, 8], in open loop control mode, the differences in FCE between diesel and FAME become significant only for the medium load range (e.g. 2500 rpm and 8 bar of BMEP). Therefore, notwithstanding its inaccuracy, this method is suitable for a first rough estimation of the BR of the burned fuel.

The case with the engine closed-loop IMEP control is different; in fact, the BR calculation formula becomes:

$$BR = 100 \cdot \frac{\left(Qfuel_{FAME} / Qfuel_{diesel} \right) - 1}{\left(LHV_{diesel} / LHV_{FAME} \right) - 1} \tag{5}$$

and it is linked to the increase of fuel consumption, that is to say the Qfuel, when a biodiesel is used.

In closed loop operation mode, the variation of Qfuel is only dependent on the variation of LHV of the used fuel. Since the variation in FCE between diesel and diesel/FAME blend was estimated less than 2%, it is negligible and not affecting the accuracy of the method. For completeness' sake, as reported in other papers [5, 12], the LHV variation from B0 to B100 is about 13÷14%, with a small difference among biodiesel feedstock (≈ 2‰ of the B100 value).

Another BD methodology, the RAFR (relative air-fuel ratio) method, has been patented [14] and it is based on the comparison between:

- the relative air-fuel ratio RAFR1 estimated from the air and fuel flow rates, and the stoichiometric diesel A/F ratio, assuming the fuel was pure diesel;
- the relative air-fuel ratio RAFR2 directly evaluated through the lambda sensor installed at the engine exhaust, whose composition stems from the engine fuelling with the actual diesel-biodiesel mixture. In particular:

$$RAFR1 = \frac{Qair}{Qfuel} \cdot \frac{1}{A / F_{st,diesel}} \tag{6}$$

being Qair the measured air mass-flow by hot-film sensor and Qfuel the fuel mass-flow interpolated from the injector look-up table (stored in the ECU) as a function of actual injector energizing time and injection pressure, both already used by the ECU. Finally, the A/Fst to be employed in eq. (6) is the one for reference diesel (~14.6). On the other hand:

$$RAFR2 = f(V\_lambda\_sensor) \tag{7}$$

being the relative air-fuel ratio a monotonic function of the lambda sensor output signal, which shows a weak dependence on the biodiesel type as well as biodiesel-diesel blending ratio. In fact, the lambda sensor output signal, for the lean operation which characterizes diesel engines, is a function only of carbon/hydrogen ratio of the fuel, which is almost unchanged from diesel to biodiesel. Hence, the estimated blending ratio BR, can be calculated as:

$$BR = 100 \cdot \frac{\left({RAFR2}/{RAFR1} \right) - 1}{\left({A / F_{st,diesel}}/{A / F_{st,FAME}} \right) - 1} \quad (8)$$

In this method the BR evaluation is therefore linked to the variation of RAFR1 between diesel and FAME fuelling due to the Qfuel increase of this latter.
The combination of both the above described methodologies in a real engine can be useful in order to improve the overall accuracy and stability over time by performing cross-checks and confidentiality interval estimation.

4.2 IMEP BD method results

As claimed above, also with a reduced accuracy, the BD method is suitable for application when the engine runs in open-loop combustion control, but it is evident that its potential lies on the use of the closed-loop control. So, both to simplify the results analysis and highlight the potentiality of the closed-loop control in BD method, only the results relative to this last control mode will be shown and discussed in this section.
Before starting the tests campaign, a check of the engine hardware equipments has been done; in particular the ECU injection maps have been checked with a reference fuel flow mass meter (AVL Fuel Balance 731). The check indicated a deviation between ECU fuel flow estimation and fuel balance measurement within the 3%, in line with the normal engine to engine variation from production line.
Figure 2 reports the results of the IMEP method in the nine test points for the detection of 0% (reference diesel fuel), 20%, 50% and 100% of RME blending level. For mineral diesel, the standard deviation bars have been also reported (in orange), in order to characterize the specific variability of each engine operative point. The blending detection was calculated (in accordance with the above described algorithm), adopting as input variables the fuel consumption calculated by ECU, Qfuel; so these first results represent the current capability of the tested hardware.
The not zero value of BR burning pure diesel fuel derives from the drift between the estimated $Q_{fuelnominal}$ in the operating point (as mapped in the calibration, function of engine speed and accelerator pedal position and corrected for coolant temperature) and the corrected Q_{fuel} ($Q_{fuelactual}$) actuated by ECU by means of measured IMEP value. As it can be seen, the drift is variable point to point and depends on the differences between the laboratory engine configuration in the test cell (as air path layout, auxiliary components, deviation of injector flow characteristics with respect to the nominal values, etc.) and the reference engine configuration as installed on the vehicle.
An overall analysis of the results highlights that the method is able to detect the blending trend, showing an increase of the estimated blends with the blending level. As expected, the method is more and more precise as the fuelling is increased when high speed/load conditions are approached. The highest error was detected at minimum operating point 1000 rpm and zero load, for brevity indicated in the following 1000x0. The reason why the highest BR error occurs at 1000x0 is the very little injected quantity in this operating condition which, in turn, leads to:

- an higher injected quantity estimation error because the accuracy and the repeatability of the injection system decreases in the case of small injected quantities;
- an higher IMEP estimation error because of relatively small in-cylinder pressure variations which are more sensitive to sensor noise and accuracy.

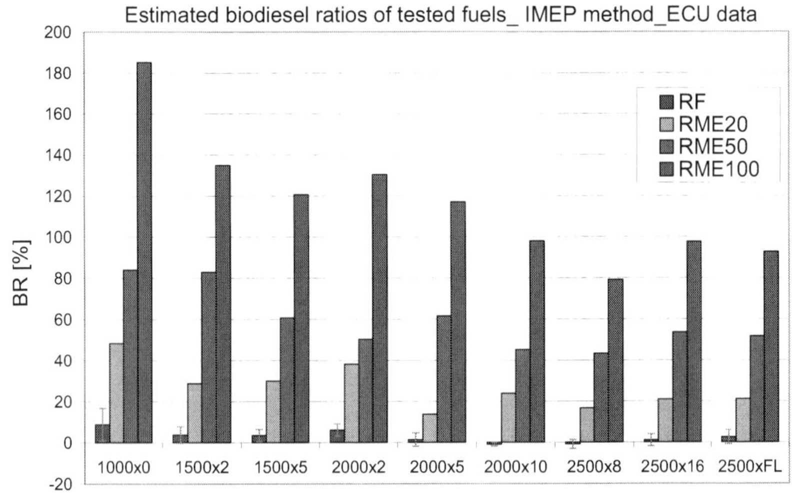

Fig. 2. Blending Ratio by means of IMEP method. ECU input data. Individual test points results.

Furthermore, in this test point, the very low fuelling condition gives very small variation in ET between RF and RME and does not exhibit acceptable results. However, without any correction/refining remedy on the $Q_{fuelnominal}$ with respect to the actual injection quantity, and except the minimum point, the potential of the method is evident also looking at the Blending Ratio values of the first row of Table 3, where the BR average values over the all tested points (except 1000x0) are reported.

Nominal Blending Ratio	B0	B20	B50	B100
Result mean value w/o injector drift correction [%]	2.1	24.2	56.2	108.8
Result mean value with injector drift correction [%]	0	22.1	54.1	106.7
Result mean value, fuel mass from the fuel flow meter [%]	0	17.2	49.2	102.5

Table 3. BR mean values with IMEP method. ECU input data and fuel flow meter data.

When a learning procedure for $Q_{fuelnominal}$ correction is implemented in the ECU, the difference between $Q_{fuelactual}$ and $Q_{fuelnominal}$ is resettled and the method shows better average results, as reported in the second row of Table 3. The averaged results show a quite good physical correspondence, so indicating that the method is certainly sensible to the different blending levels.

Looking at the method algorithm (5), taking into account the reset of the drift between nominal and actual value of Q_{fuel}, the main cause of the inaccuracy of the IMEP method lies in the ECU calculation of the fuel consumption, $Q_{fuelactual}$. In fact, the IMEP calculation relies on robust pressure signal integration over cycle, being the measured pressure signal source already accurate (max error of 2%) and the sensitivity of IMEP to pressure signal error weak. The accuracy of the ECU for the fuel consumption estimation, in each tested engine point, $Q_{fuelactual}$, has been evaluated by means of a comparison with the value of consumption measured by the fuel flow meter. This last has been assumed as the "real" value of the fuel consumption, taking into account that the precision of the flow meter instrument has been previously checked.

The results evidenced the presence of a little difference between the Qfuel values derived from the two systems, as detailed in [15]. In general, a random pattern at low and medium engine speeds and loads and an overall ECU overestimation of fuel consumption at high speeds and loads have been observed. However, the average error in all the nine test points (sum of the individual errors) is an overestimation of ECU of +0.4 %.

A further evidence of the sensibility and accuracy of the IMEP method is well illustrated in Figure 3 and in the third row of Table 3, where the results of blending detection for all the individual points and the BR mean values, referred to the IMEP algorithm calculated by using the fuel flow meter measurements are reported, respectively. The orange bars in Figure 3 represent, for each engine operating point, the uncertainty in the blending detection procedure stemming from the statistical error propagation in the calculation chain.

The adoption of the assumed "real" Q_{fuel} values leads to an evident improvement of the results; the percentages in the mean BR values of Table 3 clearly show as the BR mean values are very close to the effective level of biodiesel in the tested blends. The maximum drift between real and estimated blend was for B20 and equal to about 3%, corresponding to a measurement error of 10.5%.

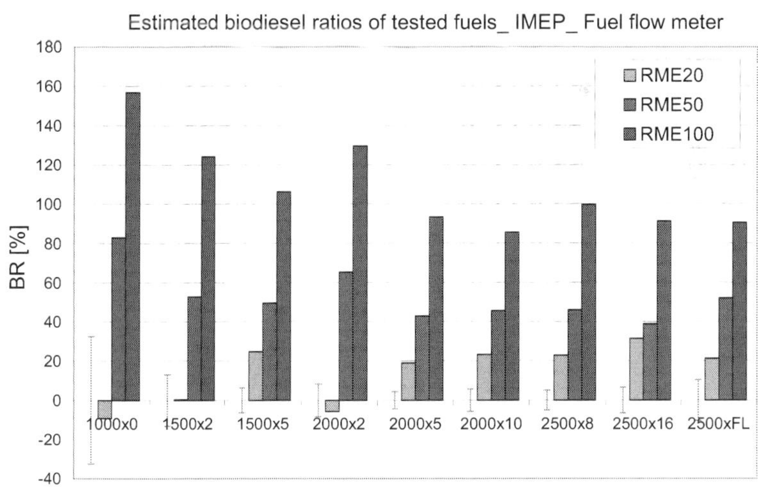

Fig. 3. Blending ratio by means of IMEP method. Fuel flow meter input data. Individual test points results.

In order to obtain a statistical value of the precision of the fuel consumption provided by ECU, the ECU Qfuel estimation has to be checked in a wide number of engine types. Moreover, to evaluate the global robustness of the IMEP BD method, the accuracy of all components of the measurement chain has to be evaluated in a statistical way. This aspect is out on the aim of the described research activity, which was more addressed to a first screening of the quality of the IMEP BD method, and it will be subject of future work. On the basis of the presented results, the accuracy of the method can be reasonably and preliminarily estimated as ±5% if a suitable engine operating is chosen to enable the BD event in the ECU (i.e. high speed, high load area). Such value has to be considered as the minimum diesel/FAME blend detectable by the method. However, BR variation within the accuracy of method gives negligible effects on engine performance and emission, as already proved by past experiences of the authors [5].

4.3 RAFR BD method results

As in the case of IMEP BD method, the RAFR BD has been at first applied adopting Qfuel, Qair and O_2 estimated by ECU as input variables (see procedures description section), so evaluating the current capability of the hardware.

The results have been summarized in the first row of Table 4, that reports the BR average values over all the test points, with the usual exception of 1000x0, not considered for the above described reason. The mean values in the first row highlight that the method is sensitive to the different blends, but the results cannot be considered satisfactory, because, due to some inaccuracies in the evaluation of the above-mentioned quantities, the BR values do not have a reasonable physical meaning in all the tested points. However, some considerations can be done to identify the positive and critical aspects and so to unlock the potential of the method.

Nominal Blending Ratio	B0	B20	B50	B100
Result mean value with ECU data [%]	39.9	55.6	84.3	159.8
Result mean value with fuel flow meter data [%]	37.3	55.4	86.1	131.4
Result mean value, corrected values [%]	0	18.1	48.7	94

Table 4. BR mean values with RAFR method. ECU input data, fuel flow meter data and "corrected" values.

For most of the test conditions the true trend of the tested blends has been grasped. An additional consideration concerns the BR values obtained for B0 (reference fuel). In many k-points the result of BR for B0, in fact, is far from zero and also the average value is not correct (about 40% instead of the expected 0%). Furthermore, the general overestimation of BR results is characterized by an increase of the numerical values increasing the blending level in the fuel, with an inaccuracy approximately proportional to the expected value of blending. It is possible that one or more quantities used in the RAFR procedure are not accurately evaluated by the tested hardware (engine sensor equipment and ECU) and so the method needs an "adjustment" process. The analysis of the inaccuracy of each quantity involved in the algorithm helps in the investigation of the difference between the obtained results and the expected ones. The evaluation of the accuracy of Qfuel estimated by ECU has been already performed. The error of ECU-estimated Qfuel, involved in calculation of both RAFR1 and RAFR2, seems responsible of the observed increase of the numerical values range among the tested blends. To confirm this consideration, the results of the method obtained using as Qfuel the values provided by fuel flow meter (instead of the ECU ones) have been reported in the second row of Table 4. An evident improvement is observable. The growing spread of BR values from B0 to B100 disappears and so a more even and physically consistent scaling among the BR values estimated for the four blends is detectable. Hence, the overestimation of the BR results is now only characterized by the offset of the B0 results, that includes the measurement errors of all the other values involved in the algorithm, except the Qfuel ones. Eventually, in order to provide an estimate of the effect of a method recalibration, an "adjustment" process has been done, subtracting the BR value obtained for the B0 from the other blends; the resulted BR values are illustrated in Figure 4 and in the third row of Table 4.

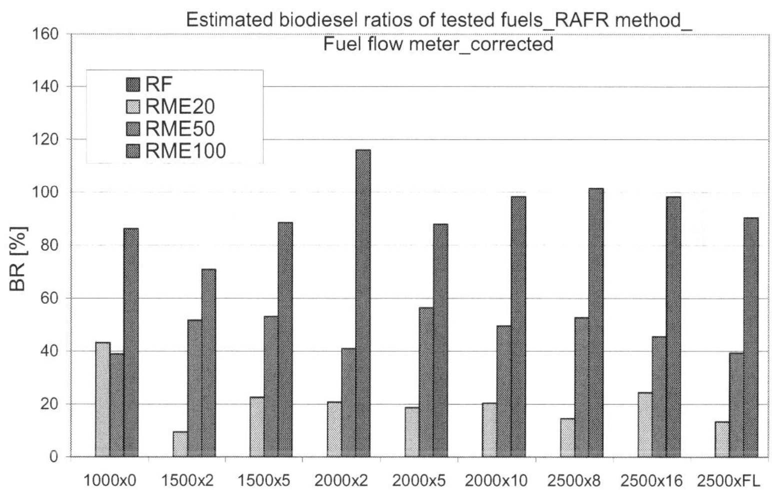

Fig. 4. Blending ratio by means of RAFR method, "corrected" values. Fuel flow meter input data. Individual test points results.

The adjustment allows to a further decisive improvement of the results, as can be observed in Figure 4: only in two test points out of nine (1000x0 and 2000x2) the results are not yet satisfactory, but the overall blending detection is now good (see third row of Table 4). The "adjusted" values represent the theoretical potential of the RAFR method, following a complete compensation of the uncertainties of the measurements involved (Qair measured by HFM, O_2 concentration by LSU, correlation between O_2 and RAFR), and could be well approached as engine re-centering strategies are enabled during on-vehicle operation.
The benefits offered by the BD methods are significant. First of all, the achievement of a reliable blending detection is a pre-requisite to exploit the application of the CLCC technology, as described in following sections.

4.4 Engine emissions improvement by using CLCC system

The potentiality offered by the CLCC to mitigate the effects of the use of biodiesels has been highlighted in some authors' works [8, 15]. In particular, it was put in evidence that, thanks to the use of CLCC technology, the drift in engine operating condition, caused by the

fuelling with a lower LHV fuels, can be avoided. Such drift is the main cause of the correspondent increment of NOx emission, as observable in the next Figure 5.
In the figure, the NEDC engine emission performance estimation without and with the employment of the CLCC technology is reported.
The RME gives slight benefits with respect to diesel fuel in HC emissions, while the CO emissions are worsened by the high emission at low speed/load conditions. The high CO emissions at low loads are mainly due to the low FCE coming out from the low pilot combustion efficiency and the delayed combustion timing [8]. Looking at the NOx and PM emissions, there are two important factors affecting them. They are the "calibration shift" and the "chemical" factor. The first one controls the NOx emission level in the NEDC test, while the second one is the main driver of the PM exhaust level even if the "calibration shift" factor is not marginal. When CLCC is not applied, both effects can lead to an increase of the level of NOx at the exhaust of about 60% and to a reduction of PM of about 90%. On the contrary, looking at NOx and PM emission charts on the right side of Figure 5, by means of the implementation of the CLCC without any engine calibration drift, the NOx emissions fall down in the RF STD bar, resetting both the calibration and fuel composition effects. For this last engine control mode, the PM emissions rise with respect to the other control modes but remain very low, about 80% lower than RF ones.

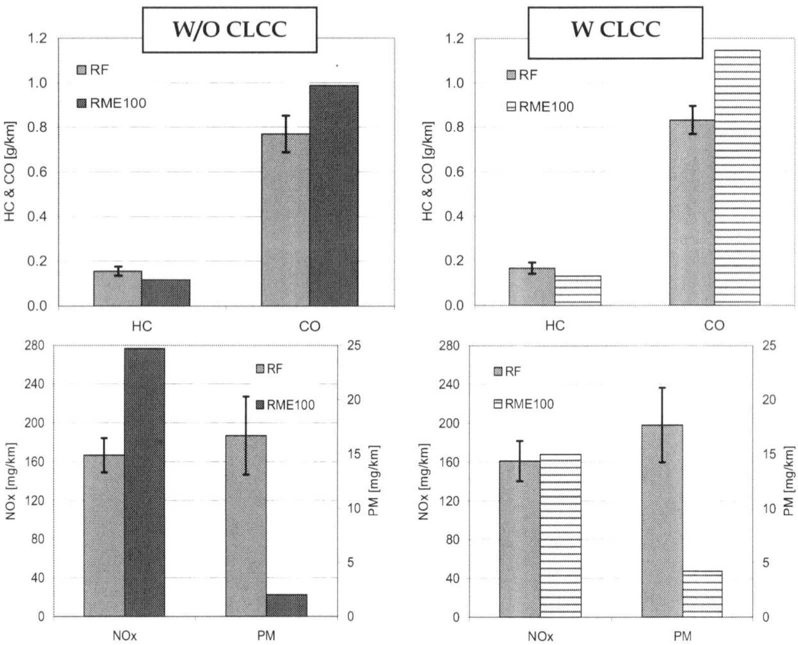

Fig. 5. HC and CO, NOx and PM over NEDC cycle for diesel reference fuel, and RME.

The above described analysis highlights the benefit of the employment of the CLCC technology in modern engines, showing that the use of oxygenated alternative fuels

characterized by lower A/Fst ratio, like the RME, gives significant PM reduction at nearly constant NOx emission level. Such effect on emission offered by the RME can be seen as a higher EGR tolerability of the fuel. So, it could be exploited by increasing the EGR rate at same exhaust PM loading and further reducing the NOx emissions.
In order to validate this calibration strategy, NOx-PM trade-offs were carried out by varying EGR rate in CLCC mode both for reference diesel fuel and RME. The diagrams in Figures 6, 7 and 8 show the obtained NOx-PM trade-offs in the three engine operating points: 1500x2, 2000x5 and 2500x8.
The PM values were converted from the smoke meter FSN values according to the well consolidated AVL procedure reported in [13]. In the first diagram (1500x2) also the CO trade-off is reported, as at low-speed low-load operating points its emission level becomes critical for the emission targets; in the other two diagrams the BSFC vs NOx trade off is also plotted.
Looking at the Figures 6 and 7, the comparison between the two fuels shows how the EGR recalibration with RME leads to a significant decrease in the exhaust NOx level, permitting to approach the estimated Euro6 NOx emission targets, also reported in the diagrams. The correspondent gap of BSFC between RME and CEC is only dependent on LHV differences between the two fuels and tends to a progressive increase with EGR level, as expected. It is possible to note that also at 2500x8 burning RME a NOx reduction of about 30% with respect to the RF at the same PM load on the DPF, was measured, while at 2000x5 the NOx decrease can reach the value of 68%. Afterwards, the estimation of the biodiesel-diesel blending level by means of the BD method could also permit the automatic recalibration of the EGR map and then a significant improvement in NOx emission without penalties in engine out PM level.

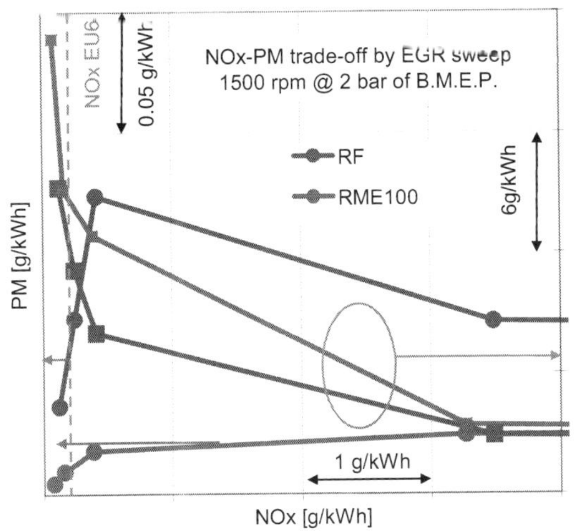

Fig. 6. PM and CO vs NOx trade-off by EGR sweep for RF and RME at 1500x2.
Solid dots markers refer to PM emissions (left axis) while solid squares to CO emissions (right axis).

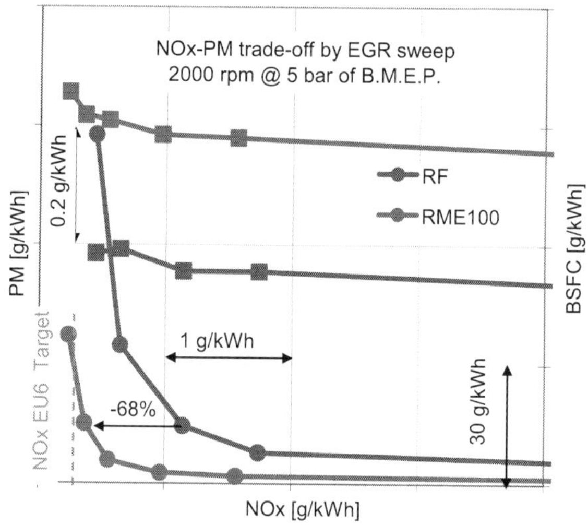

Fig. 7. PM and BSFC vs NOx trade-off by EGR sweep for RF and RME at 2000x5. Solid dots markers refer to PM emission (left axis) while solid squares to BSFC (right axis).

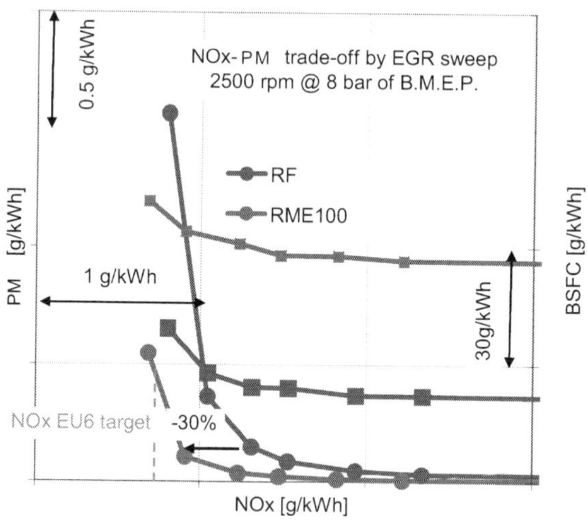

Fig. 8. PM and BSFC vs NOx trade-off by EGR sweep for RF and RME at 2500x8. Solid dots markers refer to PM emissions (left axis) while solid squares to BSFC (right axis).

4.5 Engine full load performance improvement with CLCC

The use of low LHV fuels in diesel engines reduces the full load performance [3, 9]. However, the torque reduction versus speed is dependent not only from the LHV reduction of the FAME, but also on the engine technology and the engine operating characteristics at full load. As an example, like it has been described by Millo et al. [9] about the effects of the biodiesel usage in small displacement diesel engines, the particular operating characteristics of the turbocharger in the low speed range (featuring turbo rack always totally closed and no boost margin) cause a reduction of the maximum boost pressure with a consequent full torque decrement up to 30% when pure FAME is employed, due to the turbine inlet enthalpy decrease.

The interaction between FAME characteristics and full load engine performances has been evaluated also for the engine used in the present work. The diagram in Figure 9 reports the torque values for reference fuel and RME both in OLCC and in CLCC modes. As expected, when CLCC is disabled, and so the injected fuel quantity at full load is volume limited, a torque reduction burning FAME is present in the whole engine speed range and it is higher at the highest engine speed conditions. The torque reduction is about 3-4% between 1250-1750rpm, while at medium engine speed is progressively increased, reaching 7-8% above 3000rpm. In CLCC mode such differences are completely absorbed, confirming the capability of IMEP closed-loop control to compensate any influence of the fuel properties on the engine maximum torque curve. As a matter of fact, at every engine speed, the deviations between diesel and biodiesel in CLCC mode are less than one percent, so they can be easily considered inside the test to test repeatability.

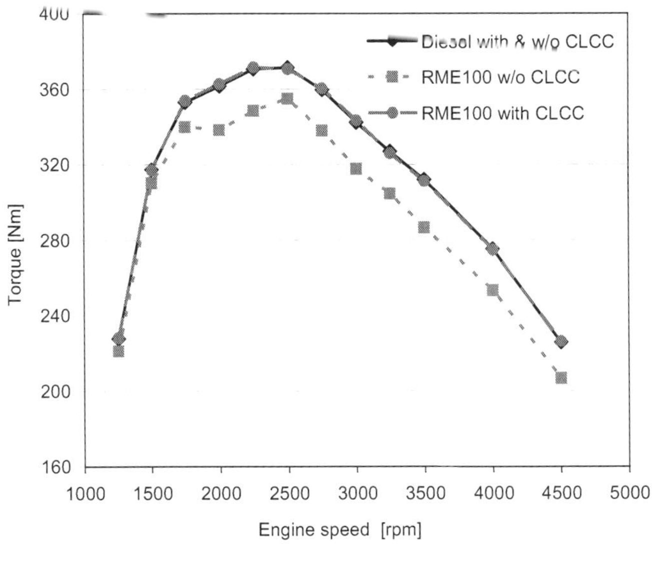

Fig. 9. Engine maximum torque curve comparison versus engine speed.

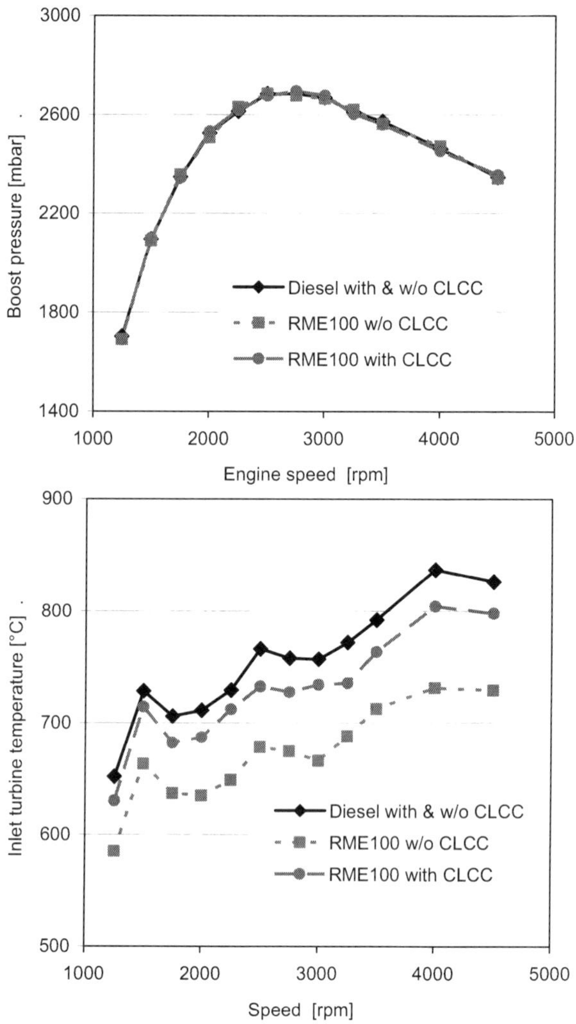

Fig. 10. Boost pressure (upper) and exhaust gas temperature (bottom) comparison versus engine speed.

Even considering the intake and exhaust manifold conditions, as represented in Figure 10 by boost pressure and exhaust temperature, very similar engine operating conditions between petrol diesel and RME are reached when CLCC technology is enabled.

There are no apparent boost pressure differences between RF and RME results in both the combustion control modes. This suggests that turbocharger is able to provide the same boosting level even employing FAME. The reduction of exhaust gas temperature when CLCC is disabled is directly linked to the reduction of chemical energy introduced with fuel (lower LHV). However, even when CLCC is enabled, a slight reduction in engine-out gas flow temperature was measured. This phenomenon is linked both to the advanced main injection timing to be used for biodiesel in order to keep constant the combustion phasing (MBF50% value), due to its extended injection duration (lower LHV) as well as to its higher thermal capacity which reduces charge temperature during fuel evaporation.

Generally, the three main limitations on the maximum attainable torque versus engine speed are the maximum cylinder peak pressure, maximum exhaust gas temperature upstream the turbine and maximum engine out smoke level. In particular, in the low engine speed range, and especially for the maximum attainable low-end torque, the smoke emission is the limiting parameter. Nevertheless, for FAME fuelling, the engine works at same maximum torque in the low speed range with significant lower smoke emission with respect to the diesel fuel because of its oxygen content and lower A/F stoichiometric ratio. Such result is clearly observable in the diagrams of Figure 11, that reports smoke emissions for both fuels and combustion control modes.

In Figure 11 it is notable that also in CLCC mode the smoke emissions of RME are always lower than diesel fuel ones although, due to the increase in power delivered, the increment is bigger than in NEDC cycle based comparison (Fig. 5). Such result suggests the possibility to increase the maximum attainable full torque where the exhaust smoke emission is the limiting parameter. Specific tests were performed with this objective, the re-calibration of the engine in order to increase the full torque until the same smoke emission level of the diesel fuel was reached. Figure 12 shows the result in the engine speed range 1250-2250 rpm. A torque increment of about 4% was attained at low engine speed (see left diagram of Figure 12) and at same engine out smoke emissions (see right diagram of Figure 12). Such result represents a good improvement in engine performance. As mentioned, this increment is obtained by calibration adaptation to a known biodiesel blending ratio in the fuel and, therefore, highlights the importance of the BD strategies.

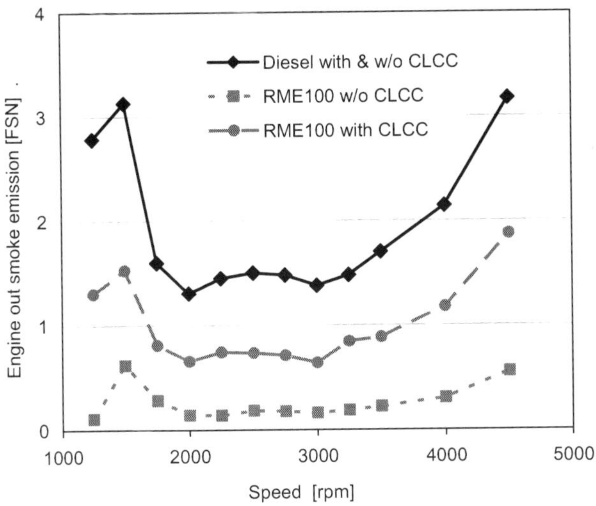

Fig. 11. Smoke emissions versus engine speed for both fuels and both combustion control modes.

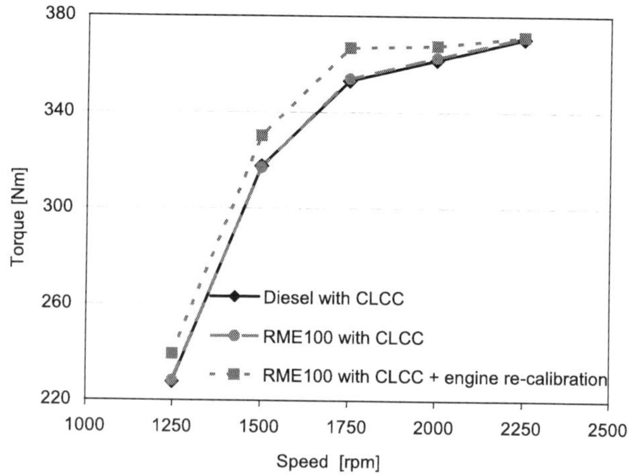

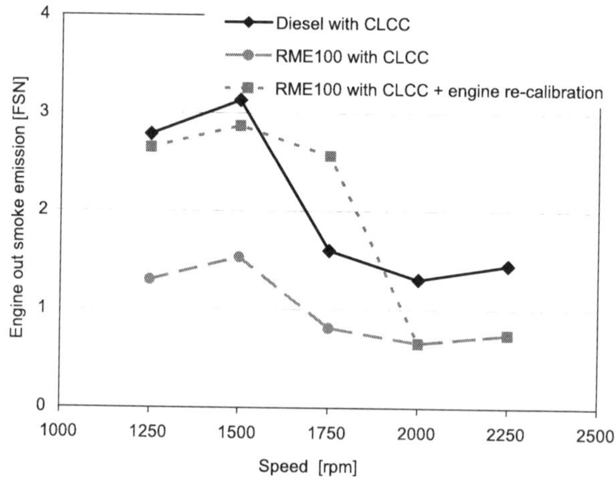

Fig. 12. Low-end torque curve (upper) and engine out smoke emissions (bottom) for reference diesel fuel, RME in CLCC mode, RME in CLCC mode and engine re-calibration.

5. Conclusions

The present chapter has been dedicated to illustrate the potentiality offered by the electronic control technology to fully exploit the biodiesel use in automotive diesel engine.
In particular, based on the employment of innovative biodiesel blending detection methodologies, the capability of closed loop combustion control to improve both pollutant emissions and full load engine performance has been investigated.
Two blending detection methods have been described and tested, assessing the potential offered by the production sensor system to detect the blending level of biodiesels. The IMEP method is certainly sensitive to the different blending levels, showing a good physical

correspondence with the effective level of biodiesel in the tested blends. By employing corrective remedy or learning procedures for fuel flow estimation, the accuracy of the method could be further improved.

The method based on lambda and air mass flow sensors (RAFR), although well reflects the actual trends, shows a more imprecise results, and needs "corrections" of some parameters involved in the algorithm and a deep investigation of the accuracy of different parts of the experimental system.

The use of a blending detection method opens new possibilities on the automatic adaptation of the engine calibration to the fuel blend characteristics. In particular, it has been showed that, thanks to the closed loop combustion control, just with a simple optimization of the EGR calibration map for low "sooting" fuels, significant improvements are possible on NOx emissions at same engine out concentration of particulate matter.

The potentiality offered by the CLCC technology to reset the torque loss burning biodiesel has been illustrated. It was also put in evidence the possibility to exploit the very low smoke emissions burning FAME fuels, increasing the low-end max torque curve with respect to a conventional diesel fuel; also in this sense the quantitative information of the BR is important.

Based on the discussed results, the authors believe that the benefits offered by the BD methods are significant. All this represents a fundamental step towards the development of flex-fuel engines that, by means of the adaptation of calibration parameters to the optimal set for the actual burned fuel, can minimize their environmental impact.

Finally, it is important to clarify that this chapter has presented only the functional assessment of blending detection methodology and adaptation strategy: this represents the first step of a new technology development. While their potential has been verified, in the next future additional work has to be performed to define how to realize robust blending detection and the related engine drift in the real driving conditions over a significant vehicle mileage and also to assess the robustness of the system when exposed to high rate of biodiesel fuelling for a long time.

6. Acknowledgment

A special thank goes to Mr. Giovanni Alovisi, Mr. Giuseppe Corcione, Mr. Augusto Piccolo and Mr. Roberto Maniscalco for their technical assistance in the engine testing.

7. References

[1] Sheehan, J.; Camobreco, V.; Duffield, J.; Graboski, M. & Shapouri, H. (1998). *An Overview of Biodiesel and Petroleum Diesel Life Cycles*, National Renewable Energy Lab, May. Available from http://devafdc.nrel.gov/pdfs/3812.pdf

[2] Report of the Commission of the European Communities, *Biofuels Progress Report*, Bruxelles, 2007. Available from: http://www2.epia.org/documents/NL_0711_037.pdf

[3] Lapuerta, M.; Armas, O. & Rodríguez-Fernández, J. (2008). *Effect of biodiesel fuels on diesel engine emissions*, Progress in Energy Combustion Science, Vol. 34:198–223

[4] Kawano, D.; Hajime, I.; Yuichi, G.; Akira, N. & Yuzo, A. (2006). *Application of Biodiesel Fuel to Modern Diesel Engine*, SAE Tech. paper 2006-01-0233

[5] Beatrice, C.; Di Iorio, S.; Guido, C.; Mancaruso, E.; Vaglieco, B.M. & Vassallo, A. (2009). *Alternative Diesel Fuels Effects on Combustion and Emissions of an Euro4 Automotive Diesel Engine*, SAE Tech. paper 2009-24-0088

[6] DIRECTIVE 2009/28/EC OF THE EUROPEAN PARLIAMENT AND OF THE COUNCIL, Bruxelles, 2009. Available from: www.energy.eu/directives/pro-re.pdf

[7] Soltic, P.; Edenhauser, D.; Thurnheer, T.; Schreiber, D. & Sankowski, A. (2009). *Experimental investigation of mineral diesel fuel, GTL fuel, RME and neat soybean and rapeseed oil combustion in a heavy duty on-road engine with exhaust gas aftertreatment*, Fuel, Vol. 88: 1–8

[8] Beatrice, C.; Guido, C. & Di Iorio, S. (2010). *Experimental analysis of alternative fuel impact on a new 'torque-controlled' light-duty diesel engine for passenger cars*, Fuel, Vol. 89: 3278–3286

[9] Millo, F.; Bianco, A.; Vezza D.; Vlachos, T.; Ciaravino, C.& Vassallo, A. (2010) *Biodiesel fuelling impact on performance and dilution of Euro5 small displacement passenger car diesel engine*, conference proceedings of SEET 2010, Jun 29th-Jul 2th, Bari, Italy,

[10] Mueller, C.; Ras Robotti, B.; Wesslau, M.; Drangel, H., Agricola, U.; Catanese, A. & Manta, E. (2009). *Opel´s new 2.0L BiTurbo Diesel engine with highly innovative CleanTech combustion concept*, MTZ Worldwide, Vol. 12

[11] British Patent Application GB 0918272.6, 2009, Inventors: Ciaravino, C., Guglielmone, F., and Vassallo, A.

[12] Rakopoulos, C. D.; Antonopoulos, K.A.; Rakopoulos, D.C.; Hountalas, D.T. & Giakoumis, E.G. (2006). *Comparative performance and emissions study of a direct injection Diesel engine using blends of Diesel fuel with vegetable oils or bio-diesels of various origins*, Energy Conversion and Management, Vol. 47: 3272–3287

[13] AVL documents, *Smoke value measurement with filter paper method*, available from: https://www.avl.com/c/document_library/get_file?uuid=2b39210c-6937-43e4-b223-b5c2c11f91ac&groupId=10138.

[14] Serra, G., and De Serra, M., "Method for Recognizing the Fuel Type in a Diesel Engine", European Patent EP 1854982.A1, 2007.

[15] Guido C., Beatrice C., Di Iorio S., Napolitano P., Di Blasio G., Vassallo A., Ciaravino C. "Assessment of Closed-Loop Combustion Control Capability for Biodiesel Blending Detection and Combustion Impact Mitigation for an Euro5 Automotive Diesel Engine" SAE Tech. paper 2011-01-1193.

Part 3

Glycerol: Properties and Applications

16

Glycerol, the Co-Product of Biodiesel: One Key for the Future Bio-Refinery

Raúl A. Comelli
Instituto de Investigaciones en Catálisis y Petroquímica - INCAPE (FIQ-UNL, CONICET)
Argentina

1. Introduction

The uncertainty of supply and prices of oil and the difficulty to establishing a sustainable model of economic and environmental development are weaknesses in the economies that depend entirely on fossil fuels, as in the most industrialized countries. In addition, the chemical industry uses about 15% of total oil consumption, 10% as raw material and 5% as fuel, being this industry still largely based on non-renewable raw materials, whose use does not control. Accordingly, from 1990 has been increasing interest in finding alternative sources of raw materials, emerging as obvious one the biomass, because it is an abundant, sustainable, and renewable energy-primary resource that can provide transport fuel, organic chemicals, and materials currently produced by fossil sources [Fulton, 2004]. In the past few years, research and technological projects related to the use of biomass as raw material to produce energy and industrial products are receiving support in countries like U.S.A., Germany, and Canada [Archambault, 2004; Werpy and Petersen, 2004; Oertel, 2007].
The use of biomass as raw material implies its transformation into chemicals and materials of commercial interest. It means moving from a petroleum-based economy to a biomass-based one; some potential advantages associated with this transition are: exploiting of unused productive capacity in agriculture and forest industry, developing new materials not available from petrochemical sources, revitalizing rural economies through local production and processing of renewable raw material sources, a more balanced development between urban and rural areas, economically sustainable development and environmental sustainability easier to achieve by using renewable raw materials, a decrease in net CO_2 emissions to the atmosphere, and a reduced outside dependence on both energy and raw material sources. In this context and by similarity with the refinery, which is the base industrial unit of petro-economy, emerges the concept of "Biorefinery", which is the production facility in which biomass is transformed into energy and bioproducts.
Bio-compounds, those ones produced from renewable sources, have a growing importance among the transport fuels. The EU imposed a 2% content of biocompounds into transport fuels by the end of 2005, increasing to 5.75% by the end of 2010 [EU Directive 2001/0265]. Transesterification of vegetable oils with methanol produces biodiesel, which is a mixture of methyl esters of fatty acid [Ali et al., 1995; Peterson et al., 1996; Vicente and Martínez, 2004]. This process generates glycerol as by-product, approximately 10 wt.% of total product. In this context, Province of Santa Fe in Argentine produces more than 2,500,000 tons of

biodiesel per year; consequently, 250,000 tons of glycerol are available. Glycerol is a chemical compound widely used in medicines, cosmetics, and sweetening agents, but its world demand is limited. In recent years, a significant increment in biodiesel production is generating an increased supply of glycerol, causing a progressive decline in its price; it encourages to research applications that allow synthesize chemicals with added value, and develop novel processes that utilize glycerol. It is highly desirable to improve the economics of biodiesel production process. Moreover, the conversion of glycerol has benefits because it comes from renewable raw materials, enabling a sustainable environmental development.

Glycerol has three hydrophilic alcoholic hydroxyl groups and it is an intermediate in the synthesis of a large number of compounds used in industry [Corma et al., 2007]. Some products and the corresponding reactions are: propyleneglycol and 1,3-propanediol by hydrogenolysis; acetol and acrolein by dehydration; dihydroxyacetone, and glyceric and hydropiruvic acids by oxidation; glycidol by epoxidation; glycerol carbonate by trans-esterification; mono- and diglycerides by selective etherification; and polyglycerol by polymerization. Another possible use of glycerol is as substrate to produce bio-hydrogen, proposed as the next-generation renewable fuel. All these uses allow considering the glycerol as a key-compound in the environment of future biorefinery.

The objective of chapter is to review reactions of glycerol such as hydrogenolysis, oxidation, and steam reforming, also including own results.

2. Glycerol reactions

2.1 Hydrogenolysis: Glycerol to propanediols

2.1.1 Characteristics, processes, catalysts, reaction conditions, and mechanisms

The hydrogenolysis of glycerol is presented as an alternative to the problem of its growing by increasing the production of biodiesel, allowing that reaction to obtain compounds with added value, such as 1,2-propanediol (1,2-PD) and 1,3-propanediol (1,3-PD). By considering the 1,2-PD, also known as propyleneglycol, in 2007, it was announced the introduction of propyleneglycol derived from renewable resources in the context of sustainable chemical technology search. 1,2-PD is a chemical with a variety of applications such as a solvent for the production of unsaturated polyester resins, medicines, cosmetics, and food, being also used as antifreeze and deicing fluid [Corma et al., 2007]. 1,3-PD can have the same applications as ethylene glycol, 1,2-PD, 1,3-butanediol, and 1,4-butanediol [Maervoet et al., 2011], and it is used as a solvent, for the production of adhesives, laminates, and paints, and in coolant formulations [Sauer et al., 2008]. However, a particular interest is its use as monomer in polycondensation reactions to produce polyesters, polyethers, and polyurethanes; 1,3-PD is copolymerized with acid to produce polytrimethylene-terephthalate (PTT) polymers, which are recognized for their excellent elastic properties, and marketed by Shell Chemical Company and DuPont with the commercial name Corterra™ and Sorona®, respectively [Biebl et al., 1999; Kurian, 2005]. The largest commercial synthesis of 1,3-PD is made by DuPont and Shell; the first hydrating acrolein to 3-hydroxypropanal followed by hydrogenation to form 1,3-PD, while the latter produces 1,3-PD by hydroformylation of ethylene oxide followed by hydrogenation [Saxena et al., 2009]. The problems in these processes are the high pressure applied in the hydroformylation and hydrogenation steps together with the high temperature, the use of expensive catalysts, and the release of toxic intermediate compounds [Saxena et al., 2009].

Propanediols can be produced by alternative routes such as the selective dehydroxylation of glycerol through either chemical hydrogenolysis or biocatalytic reduction [Zheng et al., 2008]. The selective conversion to propyleneglycol by hydrogenolysis in liquid phase has been reported using catalysts such as Raney Ni, Ru/C, Pt/C, Ni/C, and copper chromite, reaching this last material the largest production. Table 1 presents catalysts, operating conditions, and catalytic behavior to hydrogenolysis of glycerol, including results of patents and open literature on supported catalysts containing Ru, Cu, Pt, Ni, Co, Rh, and Re. Catalytic performance during hydrogenolysis on supported metal catalysts followed the order Cu ≈ Ni ≈ Ru > Pt > Pd [Roy et al., 2010]. Supported catalysts of the Group VIII metal (especially Ru/C) combined with a strong solid acid (Amberlyst 15, 70), were active in hydrogenolysis [Kusunoki et al., 2005; Miyazawa et al., 2006-2007a-2007b] Using an ion-exchange resin combined with a Ru/C supported catalyst, the first material produce acetol, an intermediate product, being the second one responsible to the conversion of acetol into propyleneglycol [Miyazawa et al. 2006]. Feng et al. [2008] investigated the effect of supports (TiO_2, SiO_2, NaY, γ-Al_2O_3, and activated carbon) in Ru catalysts, stating the type of support can influence the metal particle size and the reaction pathways. Lahr and Shanks [2005] investigated the effect of addition of sulfur on Ru/C in the hydrogenolysis at high pH and found that higher loads of sulfur increase the selectivity to 1,2-PD, without modifying the selectivity to ethyleneglycol. Maris et al. [2007] reported that Pt-Ru/C was more stable while Au-Ru/C was altered by the harsh operating conditions. Ma et al. [2008-2009-2010] studied the promoting effect of Re on Ru catalysts on several supports (SiO_2, ZrO_2, Al_2O_3, C, and ZSM5). Balaraju et al. [2009] reported Nb_2O_5 and TPA/ZrO_2 (TPA: 12-phosphotungstic acid) having moderate acid sites as the most active co-catalysts combined with Ru/C, while the amount of acidity in the acid solid influences its catalytic activity. They also studied the effect of preparation conditions of Ru/TiO_2 on the behavior during the hydrogenolysis of glycerol [Balaraju et al., 2010]. Roy et al. [2010] used a mixture of Ru/Al_2O_3 and Pt/Al_2O_3 catalysts to produce 1,2-PD from glycerol without added external H_2. Vasiliadou et al. [2009] reported the effect of both support and Ru precursor, obtaining a linear relationship between the total acidity of catalyst and the activity in hydrogenolysis. Unfortunately, Ru catalysts promote an excessive breakage of C-C bonds decreasing the selectivity to 1,2-PD. Copper catalysts having a poor activity in the rupture of C-C bonds and a high efficiency for the hydrogenating/dehydrogenating of C-O bonds, have been proposed as an alternative [Huang et al., 2009]. Commercial copper chromite [Dasari et al., 2005; Chiu et al., 2006-2008] and also prepared by impregnation and co-precipitation [Liang et al., 2009: Kim et al., 2010a-2010b], have been reported as efficient catalysts to produce 1,2-PD by hydrogenolysis. Cu/ZnO [Wang and Liu, 2007; Balaraju et al., 2008], Cu/ZnO/Al_2O_3 [Meher et al., 2009; Zhou et al., 2010], Cu/Al_2O_3 [Guo et al., 2009; Akiyama et al., 2009; Mane et al., 2010], Cu/SiO_2 [Huang et al., 2008-2009; Huang et al., 2009], and Cu/MgO [Yuan et al., 2010] were also suitable catalysts for hydrogenolysis, while Cu supported on zeolites such as HY, 13X, H-ZSM5, and Hβ did not produce 1,2-PD from glycerol [Guo et al., 2009]. By considering 1,3-PD, a selective hydroxylation involving three steps, acetalization, tosylation, and detosylation was reported as a synthesis way [Wang et al., 2003]. Kurosaka et al. [2008] also reported the selective hydrogenolysis of glycerol to 1,3-PD on Pt/WO_3/ZrO_2, using 1,3-dimethyl-2-imidazolidinone (DMI) as a solvent; using the same catalyst, Gong et al. [2009] studied the effect of both protic and aprotic solvents such as sulfolane, DMI, ethanol, and water. Other catalysts employed to produce propanediols by the hydrogenolysis of glycerol are Pt/WO_3/TiO_2/SiO_2 [Gong et al., 2010], Pt/amorphous silica-alumina [Gandarias et al., 2010], and Pt on MgO, HLT (hydrotalcite), and Al_2O_3 [Yuan et al., 2009].

Catalyst	*P* (*MPa*)	*T* (°*C*)	*Time* (*h*)	*X* (*%*)	$S_{1,2}$ (*%*)	$S_{1,3}$ (*%*)	*Reference*
$Rh(CO)_2$	31.7	200	24	n.a.	3.8[a]	3.5[a]	Che and Westiefeld, 1987
CuO/ZnO	10.0	270	2	99.9	80.8	-	Casale and Gomez, 1993
CuO/ZnO/Al_2O_3[b]	15.0	230	n.a.	93.2	94.0	-	Casale and Gomez, 1993
Ru/C	13.0	240	2	100	75.2	-	Casale and Gomez, 1994
Co/Cu/Mn/P/Mo	25.0	250	6	100	91.2	-	Schuster and Eggersdorfer, 1997
Co/Cu/Mn/P/Mo[b]	29.5	210	n.a.	100	92.0[c]		Schuster and Eggersdorfer, 1997
Pd-BCPE	6.0	140	10	n.a.	21.8	30.8	Drent and Jager, 2000
Base of Cu[b]	2.0	195	n.a.	99.5	96.5	n.a.	Tuck and Tilley, 2008
CuO/ZnO	3.0	200	12	100	97.7	-	Franke and Stankowiak, 2010
Pt/C	3.0	160-200	3	9.0-22.0	17.0-9.0	41.0	Susuki et al., 2010a
Pt-W/Al_2O_3	3.0-5.5	160	3	20.0-23.0	4.0	67.0	Susuki et al., 2010a
Cu/SiO_2[b]	2.0-20	210-230	36	94.0-93.0	94.0-98.0	-	Susuki et al., 2010b
CuO/ZnO/MnO_2	8.0	200	12	100	97.5[c]	-	Stankowiak and Franke, 2011
Ru/Al_2O_3 + Pt/Al_2O_3	1.4[d]–4.1	220	6	50.2-62.8	47.2-31.9	-	Roy et al., 2010
Ru/C	8.0	180	10	6.3	17.9	0.5	Kusunoki et al., 2005
Ru/C+Amberlyst 15	8.0	180	10	15.0	53.4	1.6	Kusunoki et al., 2005
Ru/TiO_2	3.0	170	12	66.3	47.7	-	Feng et al., 2008
Pt-Ru/C	4.0	200	5	42.0-100	24.0-18.0	-	Maris et al., 2007
Ru/Al_2O_3	8.0	160	8	18.7	34.5	3.4	Ma et al., 2008
Ru/Al_2O_3+$Re_2(CO)_{10}$	8.0	160	8	53.4	50.1	6.4	Ma et al., 2008
Ru/ZrO_2	8.0	160	8	25.4	31.9	1.8	Ma and He, 2009
Re-Ru/ZrO_2	8.0	160	8	56.9	47.2	5.5	Ma and He, 2009
Ru/SiO_2	8.0	160	8	16.8	39.0	6.4	Ma and He, 2010
Re-Ru/SiO_2	8.0	160	8	51.0	49.1	8.3	Ma and He, 2010
Ru/C + TPA/ZrO_2	6.0	180	8	44.0	64.3	-	Balaraju et al., 2009
Ru/CsPW	0.5	150	10	21.0	95.8	-	Alhanash et al., 2008
Rh/CsPW	0.5	180	10	6.3	65.4	7.1	Alhanash et al., 2008
Copper chromite	1.4	200	24	54.8	85.0	-	Dasari et al., 2005
Cu/Cr[e]	4.2	210	10	51.0	97.1	-	Liang et al., 2009
Cu/Cr	8.0	220	12	80.0	86.0	-	Kim et al., 2010
Cu/ZnO	4.2	200	12	10.4-33.9	27.9-77.5		Wang and Liu, 2007
Cu/ZnO	2.0	200	16	37.0	92.0	-	Balaraju et al., 2008
Cu/ZnO/Al_2O_3[b]	4.0	220	n.a.	81.5	93.4	-	Zhou et al., 2010
Cu/Al_2O_3	3.6	200	10	34.6	93.9	-	Guo et al., 2009
Cu/Al_2O_3[b]	0.0	200	2-5	100	78.2	-	Akiyama et al., 2009
Cu/SiO_2-P	9.0	200	12	73.4	94.3	-	Huang et al., 2008
Cu/SiO_2-P[b]	6.0	180	200	83.0	96.0	-	Huang et al., 2008

Cu-STA/SiO_2[b]	0.5	210	n.a.	83.4	22.2	32.1	Huang, et al., 2009
Cu/SBA-15[b]	4.0	250	6-7	96.0	92.4	-	Zheng et al., 2010
Cu/MgO	3.0	180	20	72.0-82.0	97.6-95.8	-	Yuan et al., 2010
Raney Cu	1.4	200	n.a.	100	91.0	<0.1	Schmidt et al., 2010
Pt/WO_3/ZrO_2	8.0	170	18	85.8	14.6	28.2	Kurosaka et al., 2008
Pt/WO_3/ZrO_2	5.5	170	12	24.7-45.6	13.6-18.9	5.3-29.3	Gong et al., 2009
Pt/WO_3/TiO_2/SiO_2	5.5	180	12	15.3	9.2	50.5	Gong et al., 2010
Pt/MgO	3.0	220	20	50.0	81.2	1.6	Yuan et al., 2009
Pt/HLT	3.0	220	20	92.1	93.0	-	Yuan et al., 2009
Pt/Al_2O_3	3.0	220	20	39.0	81.2	1.5	Yuan et al., 2009
Raney Ni	1.0	190	20	47.0-63.0	68.0-81.0	-	Perosa et al., 2005
Ni/SiO_2	6.0	200	10	56.9	44.4	-	Zhao et al., 2010
Ni/NaA	6.0	200	10	65.3	46.8	-	Zhao et al., 2010
Ni/NaX	6.0	200	10	94.5	72.1	-	Zhao et al., 2010
Ni/γ-Al_2O_3	6.0	200	10	97.1	44.2	-	Zhao et al., 2010
Ni/C	5.0	200	6	43.3	76.1	-	Yu et al., 2010a
Ni-Ce/C	5.0	200	6	90.4	65.7	-	Yu et al., 2010b
Co/MgO	2.0	200	9	44.8	42.2	-	Guo et al., 2009
Raney Co	3.0	200	4	97.0	45.4	-	Korolev et al., 2010
Rh/SiO_2	8.0	120	5	3.3	38.5	8.5	Shinmi et al., 2010
Rh-ReO_x/SiO_2	8.0	120	5	66.7	40.0	14.4	Shinmi et al., 2010
Ir-ReO_x/SiO_2	8.0	120	24	62.8	10.0	49.0	Nakagawa et al., 2010

BCPE: 1,2-bis(1,5-cyclooclylenofosfine)ethane; n.a.: non available; a): grams of PD produced from 20 g of glycerol; b): experience in gas phase; c): %wt. of products different to water; d) pressure of nitrogen; e) prepared using carbon as template.

Table 1. Catalysts, operating conditions, and catalytic behavior during the hydrogenolysis of glycerol, expressed as conversion of glycerol (X) and selectivity to 1,2-PD $(S_{1,2})$ and 1,3-PD $(S_{1,3})$, including the corresponding reference.

Different mechanisms have been proposed to understand the hydrogenolysis of glycerol. Dasari et al. [2005] detected acetol (monohydroxyacetone) between the reaction products, therefore proposing that glycerol is dehydrated to acetol which is hydrogenated to 1,2-PD. The reversible dehydrogenation of glycerol to glyceraldehyde followed by dehydration and/or retroaldolization of glyceraldehyde to 2-hydroxyacrolein and/or glycolaldehyde, and finally both precursors of glycols hydrogenated to 1,2-PD and ethylene glycol, was proposed, while the dehydration of glycerol to 3-hydroxypropianaldehyde, the latter being hydrogenated to 1,3-PD was suggested by Miyazawa et al. [2006].

2.1.2 Results and discussion of hydrogenolysis in both liquid and gas phase

In order to value glycerol, the hydrogenolysis reaction to obtain propanediols and/or acetol was studied in both liquid and gas phases, using commercial and prepared catalysts. Commercial materials were: copper chromite (43% CuO and 39% Cr_2O_3, specific surface area 49 $m^2\ g^{-1}$ and pore volume of 0.21 $cm^3\ g^{-1}$) and copper chromite stabilized with Ba (38% Cu, 31% Cr, 6% Ba, specific surface 30 $m^2\ g^{-1}$). Prepared catalysts were impregnated following the incipient-wetness technique; base materials for impregnation were $Zr(OH)_4$ (Sigma-Aldrich), ZrO_2 (from $Zr(OH)_4$, by calcining at 420°C), ferrierite zeolite in both ammonium and potassium forms (TOSOH), and 13X (Fluka), while precursors of W, Pt, Cu, and Ni were ammonium metatungstate, tetraamineplatinunchloride, cupric acetate, and nickel nitrate hexahydrate, respectively. Catalysts were identified as $Pt/WO_3/ZrO_2$, $Pt/WO_3/Zr(OH)_4$, Cu/K-Fer, Cu/H-Fer, and Ni/13X, respectively. The metal loadings were 2% platinum, 16% tungsten, 2.2% Cu, and 10% Ni. Reactant used was 99.5% pro-analysis anhydrous glycerin. Experiences in liquid phase were performed in a stainless steel reactor of 200 ml capacity, operated in a discontinuous form, loading 60 ml of glycerol and the finely divided catalyst, purged with nitrogen, reaching to operating temperature, and finally bubbling hydrogen in the liquid at 18 bar. Reagent and reaction products were analyzed by gas chromatography using a megabore DB-20 column and a FID. The gas phase experiments were performed in a tubular quartz down-flow fixed-bed reactor; the system has a vaporizer and a condenser previous and after the reactor, respectively. Reaction was monitored by chromatography, the flow of gases and non-condensates were analyzed on-line using a sampling valve to inject sample into a megabore GS-Alumina (J&W) column; the condensed fraction during reaction was analyzed at the end of it, as by the liquid phase experiences. Catalysts were characterized by temperature-programmed reduction (TPR), ammonia temperature-programmed desorption (NH_3-TPD), X-ray diffraction (XRD), and FTIR.

XRD patterns of catalysts prepared from $Zr(OH)_4$ and ZrO_2 show that $Zr(OH)_4$ was initially amorphous; only the tetragonal (T) crystalline structure of zirconia is formed after calcining at 500°C, while both T and M (monoclinic) structures appeared after calcining at 700°C. The addition of Pt on the W-impregnated material produced an interaction with the W species affecting an adequate definition of the 3 peaks of tetragonal phase of WO_3. According to IR characterization, the 3400 cm^{-1} band decreased due to the loss of hydroxyl groups by effect of calcining of $Zr(OH)_4$, which produces dehydration of material, formation of crystalline phases, and their subsequent stabilization. TPR profiles of materials prepared from $Zr(OH)_4$ and ZrO_2 were qualitatively similar: the addition of Pt after impregnating W favored the reduction of these species, appearing species that reduce at low temperature (300 and 400°C) and shifting to lower temperature the reduction of ones which reduced at 700°C; the sample with only one calcination displayed a well-defined reduction peak at 130°C assigned to Pt, while the preparation with two calcinations the Pt species were more difficult to reduce and did not have a well-defined peak. Finally, the NH_3-TPD profiles showed that the addition of Pt changes the amount of acid sites and produces a shift of the desorption peak maximum.

Catalytic behavior of $Pt/WO_3/ZrO_2$ and $Pt/WO_3/Zr(OH)_4$ was evaluated in the liquid phase hydrogenolysis. Operating conditions were selected to reach a low conversion of glycerol to analyze the selectivity to 1,2-PD and 1,3-PD. Table 2 shows selectivity to propanediols and the ratio between both them for catalysts having either only one final calcination or two calcinations, a first one after the W impregnation and the second one after Pt impregnation. It is remarkable that all samples allow forming 1,3-PD, at ratios higher or similar to 1,2-PD. Catalysts with two calcinations improved the selectivity to propanediols and reached a larger proportion of 1,3-PD. Different ratios allow to consider the effect of preparation of catalysts on their catalytic behavior which can be associated to the interaction of platinum and tungsten, the modification of active sites or the formation of new ones. It has been previously reported for this catalytic system [Vaudagna et al., 1997]. The production of 1,3-PD by hydrogenolysis of glycerol on Pt/WO_3-ZrO_2 was previously reported at higher pressures, 5.5 and 8.0 MPa, higher weight ratio of catalyst/glycerol, and in the presence of organic solvents [Kurosaka et al., 2008; Gong et al., 2009].

Parameter	$Pt/WO_3/Zr(OH)_4$[1]	$Pt/WO_3/Zr(OH)_4$[2]	$Pt/WO_3/ZrO_2$[1]	$Pt/WO_3/ZrO_2$[2]
S_{PD} (%)	26.2	29.7	24.4	30.3
$R_{1,3\text{-}PD/1,2\text{-}PD}$	1.05	1.56	0.79	1.86

Reaction conditions: 200°C, 1.8 MPa, 40 g of glycerol, 1 g of catalyst, 8 h-on-stream.
(1): Samples submitted to only final calcination; (2): Samples with two calcinations.

Table 2. Selectivity to propanediols (S_{PD}) and ratio between 1,3-PD and 1,2-PD($R_{1,3\text{-}PD/1,2\text{-}PD}$) for the hydrogenolysis of glycerol in liquid phase on $Pt/WO_3/Zr(OH)_4$ and $Pt/WO_3/ZrO_2$.

XRD patterns of ferrierite showed peaks of the orthorhombic crystalline structure of zeolite, while Cu/K-Fer and Cu/H-Fer also presented two bands at 2θ= 43.4° and 51.3° assigned to Cu^0 species [Guo et al., 2009]; the absence of diffraction peaks corresponding to CuO would allow to consider that copper species were completely reduced in the analyzed samples [Wang and Liu, 2007]. Impregnating nickel species on 13X did not modify the crystalline structure of 13X, appearing a peak at 44° assigned to metallic nickel species [Yu et al., 2010]. Considering TPR profiles, K-Fer and H-Fer did not consume hydrogen, while Cu/K-Fer has a small shoulder at 220°C, the main reduction peak at 260°C, and a small one which finished at 500°C; it indicates the presence of three species of copper. Cu/H-Fer only displayed the main reduction peak with a maximum at 260°C. Zeolite 13X showed a slight consumption of hydrogen between 650 and 750°C, while Ni/13X presented the main peak centered at 435°C and a second one mounted on the tail of the main peak; it allows to infer the existence of two nickel species. Considering NH_3-TPD profiles, H-Fer showed two bands with maxima at 330°C and 500°C; the addition of platinum increased acidity and shifted the maximum from 330 to 400°C. K-Fer showed a single band with maximum at 400°C, being acidity lower than H-Fer; Cu/K-Fer presented a similar profile that K-Fer. Cu/H-Fer has a similar acidity than H-Fer, being it higher than the Cu/K-Fer one. Zeolite 13X showed a broad band with maximum at 500°C; the addition of nickel decreased the acidity of the base material.
Catalytic behavior of materials was measured in the hydrogenolysis of glycerol in gas phase at 200°C and atmospheric pressure. Table 3 shows glycerol conversion and selectivities to acetol and propyleneglycol using Cu/H-Fer, Cu/K-Fer, Ni/13X, and copper chromite with and without barium. Cu/H-Fer was more active than Cu/K-Fer (conversion 22 and 13%, respectively), but Cu/K-Fer was more selective to acetol and 1,2-PD. Ni/13X was very active, 79.9% conversion, but few selective to acetol and 1,2-PD; methane was the only product detected in the gas stream while ethyleneglycol was the major one in the condensed fraction. Copper chromite stabilized with barium was the most active catalyst, reaching 83.0% conversion while both copper chromite samples were the most efficient in selectivity to acetol and 1,2-PD. Sample without barium is the most selective to acetol (67.8%), while the one with barium improved significantly the selectivity to 1,2-PD, up to 29.7%.

Catalyst	Cu/H-Fer	Cu/K-Fer	Ni/13X	Cu/Cr_2O_3	$Ba\text{-}Cu/Cr_2O_3$
X(%)	22.2	13.0	79.9	24.5	83.0
S_a (%)	4.0	37.1	2.0	67.8	43.3
$S_{1,2}$ (%)	0.7	3.7	0.9	12.2	29.7

Reaction conditions: 200°C, atmospheric pressure, 1 ml h^{-1} of 20% w/v glycerol solution, 70 ml min^{-1} H_2 stream, 200 mg of catalyst, and 3 h-on-stream.

Table 3. Conversion of glycerol (X) and selectivity to acetol (S_a) and 1,2-PD ($S_{1,2}$) during the hydrogenolysis of glycerol in gas phase on prepared and commercial samples.

2.2 Selective oxidation: Glycerol to dihydroxyacetone (DHA)

2.2.1 Characteristics, processes, reaction conditions, and catalysts

The oxidation of glycerol is an alternative to the problem of its growing availability allowing obtain compounds with added value. The oxidation reaction at atmospheric conditions needs adding either stoichiometric acids such as chromic acid or compounds as potassium permanganate, but it generates large amounts of undesirable products; it makes difficult the sustainability of the process which is practically unfeasible [Carrettin et al., 2003]. Primary products of oxidation of glycerol are glyceraldehyde, glyceric acid, tartronic acid and dihydroxyacetone; tartronic acid can be oxidized to glycolic, glycoxylic, oxalic, and mesoxalic acids, while the latter one can be obtained by oxidation of hydroxypyruvic acid, which is obtained by oxidation of DHA [Demirel-Gülen et al., 2005]. The main products have not been yet developed due to low selectivities and yields reached with the existing processes, which operate with low concentration solutions of glycerol [Corma et al., 2007]. Glyceric acid, used in medicine as metabolites in the glycolysis cycle and a precursor in the synthesis of amino acids, is mostly produced by a fermentation process [Kenji et al., 1989; Teruyuki and Yoshinori, 1989; Takehiro et al., 1993]. Hydroxypyruvic acid is obtained by oxidation of glycerol or sugars using mineral acids and it is the precursor of serine aminoacid [Corma et al., 2007]. Tartronic acid is widely used as a precursor of major products such as oxalic acid [Fordham et al., 1995]. Considering the oxidation of glycerol in liquid phase by heterogeneous catalysis, the nature of metal and the pH of medium affect the reaction selectivity. Kimura et al. [1993] demonstrated that the catalytic oxidation process using platinum (Pt) supported on carbon (C) was more efficient than the conventional fermentation; the DHA selectivity decreased drastically in a basic medium (pH 8) and was less than 10% in an acid pH (pH 2-3), increasing the selectivity to DHA to 80% by incorporating bismuth (Bi) to Pt/C. Abbadi and van Bekkum [1996] studied the oxidation of glycerol and DHA on catalysts of Bi-Pt/C at 65°C, reaching 95% conversion of glycerol and 93% selectivity to glyceric acid at pH 5-6; under acidic conditions, DHA, hydroxypyruvic acid, and oxalic acid were the main products, while at pH 8 glyceric acid was obtained and the production of DHA dropped drastically. García et al. [1995] favored the oxidation of primary or secondary hydroxyl group by controlling reaction conditions, the nature of metal and the pH of medium; glycerol conversion was 90% with 70% glyceric acid and 8% DHA using Pd/C, while the production of DHA was 37% with 75% conversion on Bi-Pt/C at pH 2. Using gold catalysts supported on carbon, the selectivity to glyceric acid decreased markedly when gold nanoparticles with average diameter of 6 nm are well dispersed on the surface, while the selectivity increased to 92-95% with nanoparticles larger than 20 nm, indicating the importance of preparation method on performance in the oxidation reaction of glycerol [Porta and Prati, 2004]. Monometallic catalysts such as Pt/C, Pd/C, and Au/C are able to produce glyceric acid [Porta and Prati, 2004; Bianchi et al., 2005], while bimetallic ones as Au-Pd/C and Au-Pt/C produce tartronic acid and glyceraldehyde, respectively [Bianchi et al., 2005].

DHA is a monosaccharide included into the group of ketoses, particularly in the triose one. Its chemical formula is $C_3H_6O_3$ and has no chiral center, making it the only one that has no optical activity and is not toxic. DHA is naturally found in plants such as sugar cane and beet, produced by the fermentation of glycerol. DHA in its pure form presents in two ways, as a monomer or dimer, predominantly the latter. In aqueous solution, DHA is found as monomer, which can gradually tautomerize to glyceraldehyde. The equilibrium between both compounds depends largely on pH; DHA is favored in an acidic medium (greater stability at pH 3) while glyceraldehyde in an alkaline one [Yaylayan et al., 1999; Zhua et al., 2003]. DHA is a chemical used in the cosmetic industry to make artificial tans; people with sensitive skin should limit exposure to sun due to the consequences that may ensue. A natural tan carries the risk of getting skin cancer due to the activation and deactivation of multiple cellular signals, so there is a growing demand for artificial tans. DHA is the active ingredient in those bronzers, varying its concentration between 2 and 5% [Brown, 2001]. Due to its potential capacity of taining, DHA is also used for the treatment of vitiligo, an autoimmune disease that affects the melanocytes (cells related to the coloration of skin). The method of treatment is very well used and accepted [Fesq et al., 2001]. DHA also finds an important application in the chemical industry as reagent of great versatility for the production of compounds like lactic acid, hydroxypyruvic acid or 1,2-PD [Corma et al., 2007; Hekmat et al., 2003; Bicker et al., 2005]. DHA is obtained through a microbiological process producing it by incomplete oxidation of glycerol by means of the glycerol-dehydrogenase enzyme [Young et al., 1980; Gupta et al., 2001].

2.2.2 Results and discussion using impregnated catalysts

An optimal catalyst should present a good glycerol conversion with high selectivity to DHA; this section presents the oxidation of glycerol, using both commercial and prepared catalysts. Commercial materials were copper chromite (43% of CuO and 39% of Cr_2O_3, 49 m^2g specific surface and 0.21 cm^3g^{-1} pore volume) and Raney nickel (PRICAT 9910, Johnson Matthey). Supported catalysts were prepared by impregnation by the incipient-wetness technique; commercial samples of coal (CARBONAC GA-160) and potassium ferrierite (17.8 SiO_2/Al_2O_3 molar ratio) were used as starting materials, while tetraammineplatinum-chloride hydrate and nickel nitrate hexahydrate were Pt and Ni precursors, respectively. Solutions with the desirable concentration were prepared. The impregnated materials were maintained at room temperature for 4 h, and then dried overnight in an oven at 110°C. Samples were calcined in a continuous air flow at 500°C; then, they were reduced in a hydrogen flow at 500°C. Catalysts were identified as Pt/C, Pt/K-Fer, and Ni/K-Fer. Catalysts were characterized by TPR, NH_3-TPD, XRD, and FTIR. The catalytic behavior during the oxidation reaction was measured in a system with a glass reactor that allows the semicontinuous operation; the reactor has an entrance by feeding the airflow which bubbles in the liquid and an exit that derives the unreacted airflow and any volatile product produced by reaction. A continuous stirring is made with a magnetic plate that also allows reach the selected temperature. Reaction conditions were: atmospheric pressure, 60°C, 30 ml of 20% glycerol solution, 300 mg of catalyst, pH 3, and 900 cm^3 min^{-1} air flow. The reactant and reaction products were analyzed by gas chromatography, using a 25 m long, 0.33 mm i.d. DB-20 column and a FID. The reaction of glycerol oxidation produces some acids which are not detected by FID; then, the method of the internal standard (1-butanol) to quantify glycerol and DHA and determine conversion and selectivity to DHA, was used.
Figure 1 displays TPR profiles corresponding to both Pt/K-Fer and the support. K-Fer does not shown any consumption of hydrogen, while a well-defined peak centered at 120°C appears in the Pt/K-Fer profile, being this peak assigned to the reduction of Pt species.
Figure 2 shows the catalytic behavior during the oxidation of glycerol on different catalysts. Copper chromite, Raney nickel, Ni/K-Fer, and Bi/K-Fer did not present activity for that oxidation reaction under the studied conditions. Pt/C showed activity and oxidized glycerol (34.9% conversion) but without producing DHA. It agrees with previous results indicating this catalyst as selective to glyceric acid [Kimura and Tsuto, 1993]. The best performance in

the selective oxidation to DHA was obtained with Pt/K-Fer, increasing glycerol conversion up to 48.3% and reaching 40.5% selectivity to DHA. In the presence of the same metal sites (Pt), changing the support (C or K-Fer) favored the oxidation of either primary or secondary hydroxyl group. It was reported the addition of a second metal as Bi on Pt impregnated on C promotes the selective oxidation of glycerol to DHA, being it associated to the changing on the environment of the active site [Kimura and Tsuto, 1993]. The goal of Pt/K-Fer is that allows obtain DHA with only platinum sites; it could be related to a selectivity of form presented by the ferrierite zeolite, being capable to favor the formation of DHA.

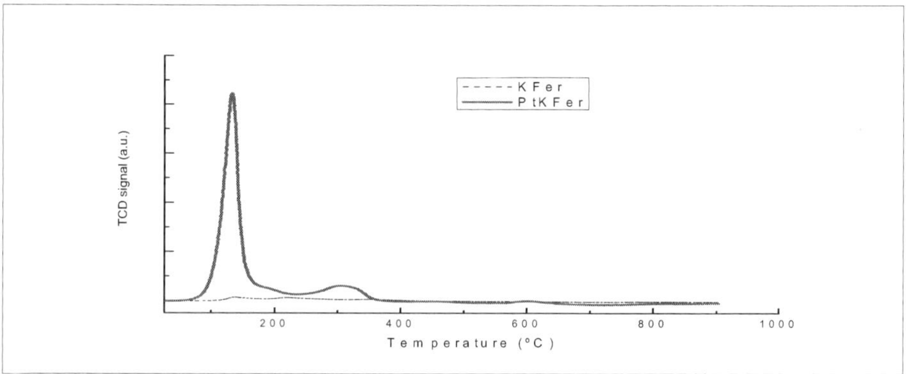

Fig. 1. TPR profiles corresponding to K-Fer zeolite and Pt/K-Fer catalyst.

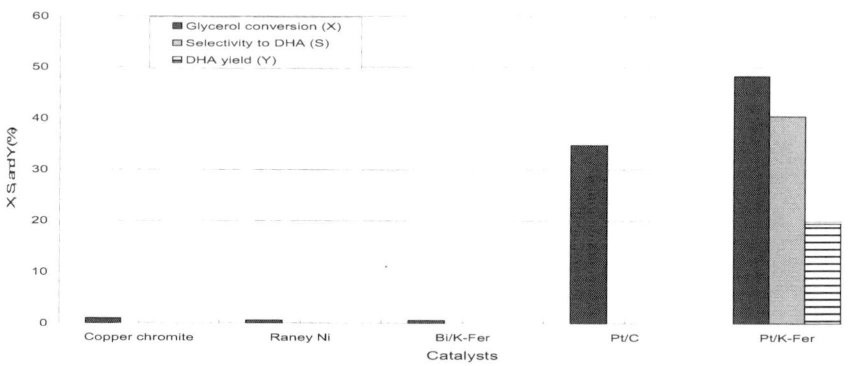

Fig. 2. Conversion of glycerol (X), selectivity to DHA (S), and DHA yield (Y) from the oxidation reaction of glycerol using commercial and prepared catalysts.
Reaction conditions: 300 mg of catalyst, 30 ml of 20%glycerol solution, 60 °C, atmospheric pressure, 900 $cm^3 min^{-1}$ air stream, pH 3 and 6 hour-on-stream.

Figure 3 shows the effect of pH of reaction medium on the amount of DHA during the glycerol oxidation on platinum containing catalysts. Pt/C did not practically show activity in the DHA production independently of variable or constant pH. Pt/K-Fer was active to produce DHA, reaching 9.8 and 31.8% of DHA when reaction takes place at variable and constant pH, respectively. At variable pH, the catalytic performance improved at pH close to 3. Figure 4 displays the effect of pre-reduction step of platinum containing catalysts on their perfomance during the glycerol oxidation. Pt/C without reduction was inactive to produce DHA, while reduced showed an incipient activity. Pt/K-Fer was active reaching 6.1% of DHA when reaction takes place on catalysts without pre-reduction and increasing to 32.8% for pre-reduced material. The presence of sites in the metallic form improved DHA production, and the pretreatment of catalysts significantly influenced their behavior.

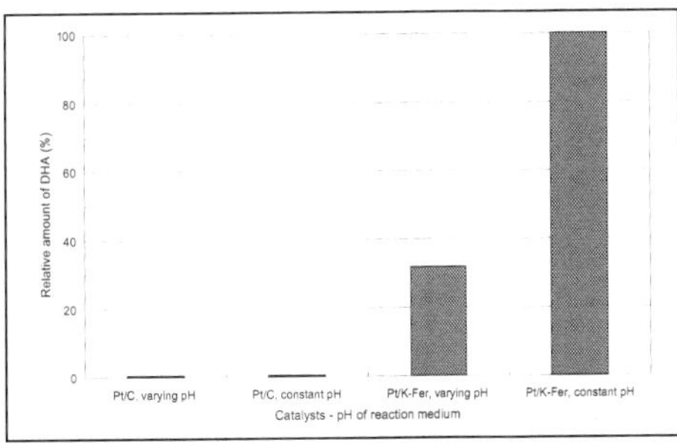

(*): 100% assigned to the maximum value reached.

Fig. 3. Relative concentration of DHA* in the reaction medium with different catalysts, at 60°C and pH varying or constant.

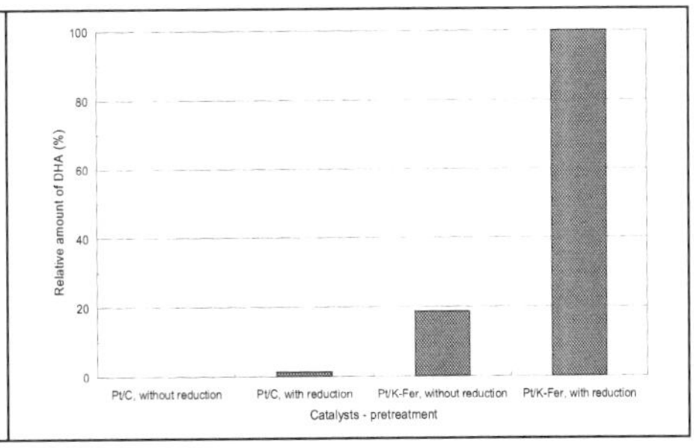

(*): 100% assigned to the maximum value reached.

Fig. 4. Relative concentration of DHA* in the reaction medium for catalysts with and without reduction, at 60°C and pH 3.

2.3 Steam reforming: Bio-hydrogen from glycerol

2.3.1 Processes and thermodynamic data

In the last few years, the potential of renewable hydrogen as a clean energy vector has been increasing attention from the business, scientific, and political worlds. Moreover, technical advances in fuel cell industry are increasing the hydrogen demand, the most simplest and abundant element. Currently, hydrogen is mainly obtained from feeding based on fossil fuels, making it necessary to find renewable sources of raw materials to produce it. Consequently, there has been a growing interest in environmentally clean renewable sources for hydrogen production. In this context, new technologies have been developed for glycerol reforming. A complete review related to the processes capable to convert glycerol into hydrogen, including steam reforming, partial oxidation, autothermal reforming, aqueous-phase reforming, and supercritical water reforming, was carried out by Adhikari et al. [2009]. By considering the steam reforming of glycerol to produce hydrogen, it is a highly endothermic process which involves complex reactions, and several by-products are formed during transformation. The yield of hydrogen varies significantly according to the operating conditions, and the process is thermodynamically favored by high temperatures, low pressures, and an excess of steam. Adhikari et al. [2007a] performed a thermodynamic equilibrium study of that steam-reforming in the 600-1000 K, 1-5 atm, and 1:1-9:1 water:glycerol feed ratio ranges, resulting temperature higher than 900 K, atmospheric pressure, and 9:1 water:glycerol molar ratio as the best conditions for producing hydrogen with minimum methane production. By thermodynamic studies in the 550-1200 K, 1-50 atm, and 1:1-12:1 water:glycerol molar ratio ranges, Wang et al. [2008] reported temperatures between 925 and 975 K and 9-12 water:glycerol ratios at atmospheric pressure as optimal conditions for hydrogen production, whereas higher temperatures and lower reactant ratios at 20-50 atm were suitable for the production of synthesis gas. Rossi et al. [2009] reported the increase of hydrogen production by increasing water:glycerol feed ratio or temperature. Chen et al. [2009] analyzed the adsorption-enhanced steam reforming of glycerol, stating that the use of a CO_2 adsorbent enhanced from 6 to 7 moles of hydrogen produced per mole of glycerol, while the most favorable temperature for steam reforming in the presence of a CO_2 adsorbent was 800-850 K, being about 100 K lower than that for reforming without CO_2 adsorption. Dou et al. [2010] evaluated the steam reforming of both pure glycerol and crude by-product of a biodiesel production plant at atmospheric pressure, with and without in situ CO_2 sorption, between 400 and 700°C; both crude glycerol and steam conversions and the hydrogen purity reached 100, 11, and 68%, respectively at 600°C.

2.3.2 Catalysts, reaction conditions, and deactivation processes

By considering catalytic conversion of glycerol to hydrogen, the gas phase reforming has now increased interest due to its operational characteristics and efficiency of reaction. Hirai et al. [2005] reported the production of hydrogen by steam-reforming of glycerol using ruthenium catalyst, being the main reaction product synthesis gas; this syngas can be converted into methanol which is used for methyl esterification of vegetable oils, then, 100% biomass-based bio-diesel fuel could be obtained. Among the catalysts studied, Ru/Y_2O_3 gave the best performance; at 500°C, the hydrogen yield increased as the ruthenium loading increased up to 3 wt.%, while a further increment in loading to 5 wt.% did not affect the behavior. Zhang et al. [2007] reported the hydrogen production by reforming of ethanol and glycerol over ceria-supported Ir, Co, and Ni catalysts, studying both nature of the active metals and reaction pathways; Ir/CeO_2 showed a quite promising catalytic performance with 100% glycerol conversion and hydrogen selectivity higher than 85% at 400°C. Adhikari et al. [2007b] evaluated the performance of a Ni/MgO catalyst during the steam reforming process and compared it to the thermodynamic data. Adhikari et al. [2008a] also studied the hydrogen production using Ni/CeO_2, Ni/MgO, and Ni/TiO_2 catalysts. Ni/CeO_2 displayed the best catalytic performance at 600°C, 12:1 water:glycerol molar ratio, and 0.5 mL/min feed flow rate, reaching the maximum hydrogen selectivity (74.7%) compared to Ni/MgO and Ni/TiO_2 which showed 38.6 and 25.3%, respectively. Nevertheless, the maximum hydrogen yield was obtained at 650°C with Ni/MgO, corresponding to 4 mol of H_2 out of 7 mol of stoichiometric maximum [Adhikari et al., 2008b]. Cui et al. [2009] evaluated the steam reforming of glycerol on non-substituted and partially Ce substituted $La_{1-x}Ce_xNiO_3$ mixed oxides; Ni species were easily reduced in the $La_{0.3}Ce_{0.7}NiO_3$, being this catalyst highly active with conversions approaching to the equilibrium at temperatures between 500 and 700°C, and forming the smallest amount of carbonaceous deposits. It confirmed the efficient operation and high stability of the non-noble, inexpensive catalyst of $La_{0.3}Ce_{0.7}NiO_3$. Slinn et al. [2008] studied the steam reforming of combined glycerol and water by-product streams of a biodiesel plant on a platinum alumina catalyst, reaching a high gas yield (almost 100%) with 70% selectivity (dry basis) at high temperatures; the optimum conditions were 860°C, 0.12 mol/min glycerol flow per kg of catalyst, and 2.5 steam/carbon ratio. Kunkes et al. [2009] reported an integrated catalytic approach for the production of hydrogen by glycerol reforming coupled with the water-gas shift reaction. This process uses two catalyst beds that can be tuned to yield hydrogen (and CO_2) or synthesis gas at 573 K and atmospheric pressure; the first bed is a carbon-supported bimetallic platinum-based catalyst to achieve conversion of glycerol to H_2/CO mixture, followed by a second bed of 1%Pt/CeO_2/ZrO_2 which is effective for the water-gas shift reaction. This integrated system displayed 100% carbon conversion of concentrated glycerol solutions into CO_2 and CO, with a 80% hydrogen yield of the ideally amount from the stoichiometric conversion of glycerol to H_2 and CO followed by equilibrated water-gas shift with the water present in the feed.

The steam reforming of glycerol has been also studied using other Ni-impregnated catalysts. Adhikari et al. [2007c] prepared fourteen catalysts by incipient wetness impregnating on ceramic foam monoliths (92% Al_2O_3 and 8% SiO_2), measuring glycerol conversion and selectivity to hydrogen in the 600-900°C range; Ni/Al_2O_3 and Rh/CeO_2/Al_2O_3 reached the best conversion and selectivity, respectively. By increasing water:glycerol molar ratio, both glycerol conversion and hydrogen selectivity increased; at 900°C, 9:1 water:glycerol molar ratio, and 0.15 ml/min feed flow rate, the hydrogen selectivity was 80 and 71% using Ni/Al_2O_3 and Rh/CeO_2/Al_2O_3, respectively. By increasing metal loading, conversion increased but hydrogen selectivity did not significantly change. Valliyappan et al. [2008] evaluated a commercial Ni/Al_2O_3 catalyst in the range of steam to glycerol weight ratio of 0:100–50:50 to produce either hydrogen or syngas; pure glycerol was completely converted to gas containing 92 mol% syngas at 50:50 steam:glycerol ratio. At 800°C and 25:75 ratio, a maximum 68.4 mol% hydrogen, and syn gas production of 89.5 mol%, were obtained. Iriondo et al. [2008] compared both aqueous phase and steam reforming of glycerol over alumina-supported nickel catalysts modified with Ce, Mg, Zr, and La; different catalyst functionalities were necessary to carry out aqueous-phase and steam reforming of glycerol. For the aqueous phase process, the addition of Ce, La, and Zr to Ni/Al_2O_3 improved conversion, although all samples showed an important deactivation associated to the oxidation of active metallic Ni during reaction. In the steam reforming process, Ce, La, Mg, and Zr on Ni based catalysts promoted the hydrogen selectivity; differences in activity were explained in terms of enhancement in: surface nickel concentration (Mg), capacity to activate steam (Zr), and stability of nickel phases under reaction conditions (Ce and La). Iriondo et al. [2009] also studied the performance of monometallic (Ni and Pt) and bimetallic (Pt–Ni) catalysts and the effect of lanthana modified alumina support during the glycerol steam reforming; the lanthana addition improved catalytic activity of Ni catalysts, reaching the best selectivity to hydrogen with an intermediate content of lanthana. Buffoni et al. [2009] prepared and characterized nickel catalysts supported on commercial Al_2O_3 with and without addition of ZrO_2 and CeO_2, and measured their catalytic behavior during the steam reforming at atmospheric pressure and 450–600°C. Profeti et al. [2009] studied the steam reforming of ethanol and glycerol on Ni/CeO_2–Al_2O_3 catalysts modified with noble metals (Pt, Ir, Pd, and Ru); the presence of CeO_2 dispersed on alumina prevented the formation of inactive nickel aluminate. The highest catalytic performance for the glycerol reforming was obtained with Ni-Pt catalyst, producing the highest H_2 yield and low amounts of CO. Sánchez et al. [2010] using a Ni-alumina catalyst for the steam reforming of glycerol at 600-700°C, atmospheric pressure, and 16:1 water:glycerol molar ratio, obtained 96.8% conversion after 4 hours-on-stream at 600°C, and increasing to 99.4% at 700°C. Dou et al. [2009] produced hydrogen by sorption-enhanced steam reforming of glycerol at atmospheric pressure and 400–700°C on a commercial Ni-based catalyst and a dolomite sorbent for the reforming reaction and in situ CO_2 removal, respectively; hydrogen productivity was increased by increasing temperature and methane became negligible above 500°C. The optimal temperature was 500°C, at which the CO_2 breakthrough time is the longest and the H_2 purity is the highest. Chiodo et al. [2010] investigated features of Rh and Ni supported catalysts in the steam reforming to produce syn-gas to feed a high-temperature fuel cell system; Rh/Al_2O_3 resulted more active and stable than Ni supported catalysts.

Operating conditions during the steam reforming of glycerol affect stability of Ni-supported catalysts, causing deactivation. Carbon deposition on the catalyst surface will result several undesirable reactions and products affecting the purity of the reforming products. Carbon occurrence may arise due to the decomposition of CO or CH_4 or the reaction of CO_2 or CO with H_2 [Adhikari et al., 2007c]; carbon formation is thermodynamically inhibited at temperature higher than 900 K, atmospheric pressure, and 9:1 water:glycerol molar ratio [Adhikari et al., 2007b]. At water:glycerol molar ratios lower than 3:1, the insufficient steam supply favored the methane decomposition forming solid carbon, decreasing the hydrogen production, and also causing catalyst deactivation [Rossi et al., 2009]. By considering thermodynamic analyses of adsorption-enhanced steam reforming of glycerol, Chen et al. [2009] analyzed the effects of reaction parameters on the carbon formation, concluding the use of a CO_2 adsorbent can suppress the carbon-formation reaction and substantially reduce the lower limit of the water:glycerol feed ratio. Slinn et al. [2008] reported a minimal degradation of a platinum-alumina catalyst after several days of continuous operating under the optimum conditions for glycerol reforming, only 0.4% of feed was deposited. Luo et al. [2008] stated the hydrogen production on a supported-Pt catalyst is accompanied by side reactions which form carbonaceous entities on the surface causing catalyst activity drop. Buffoni et al. [2009] prepared and characterized nickel catalysts supported on commercial Al_2O_3 and Al_2O_3 modified by addition of ZrO_2 and CeO_2, being Ni/CeO_2/Al_2O_3 the most stable system; it was associated to the Ce effect in inhibition of secondary dehydration reactions that form unsaturated hydrocarbons which are coke precursors and generate a fast catalyst deactivation. Profeti et al. [2009] using Ni/CeO_2–Al_2O_3 catalysts modified with noble metals (Pt, Ir, Pd, and Ru) explained the presence of them decreased the reduction temperatures of NiO species interacting with the support and stabilized the Ni sites in the reduced state along the reforming reaction, increasing conversion and decreasing coke formation. Finally, Chiodo et al. [2010] stated that independently of metal (Rh and Ni) impregnated and temperature in the steam reforming, reaction is affected by coke formation mainly promoted by the presence of olefins formed by glycerol thermal decomposition. By thermodynamic reasons the hydrogen production should be favored at high temperature, but operating at higher than 923 K promotes the formation of encapsulated carbon which negatively reflects on catalyst stability. Few information and non-systematic analysis of deactivation processes during the hydrogen production from glycerol are available.

2.3.3 Results and discussion with Ni-impregnated on alumina

An optimal catalyst should present a good glycerol conversion with high selectivity to hydrogen, maintaining it with time. This section presents the steam reforming of glycerol, using catalysts of Ni impregnated on alumina prepared by the incipient-wetness technique. The starting material was a commercial γ-Al_2O_3, previously calcined at 550°C, using nickel nitrate hexahydrate as the nickel precursor. Solutions with adequate concentrations were prepared to obtain 2.6, 5.8, and 9.9 wt.% nickel loading on the solid. Samples were identified as Ni*(x)*/Al_2O_3, being *"x"* the corresponding nickel loading. Catalysts were characterized by TPR, NH_3-TPD, XRD, and FTIR. The catalytic behavior during the steam reforming was measured in a continuous down-flow fixed-bed quartz tubular reactor; catalysts were calcined in-situ, cooled, reduced in a H_2 stream at 300 or 500ºC, and finally heated up to the reaction temperature in a N_2 stream. Feed was a glycerol aqueous solution (0.17 ml min^{-1}), co-feeding a N_2 stream. The catalytic performance was measured at atmospheric pressure, 600-700ºC, 3.4-10.0 WHSV, 20-60 ml min^{-1} N_2 flow rate, and 16:1-6:1 water:glycerol molar ratio. Reactant and reaction products were analyzed by gas chromatography; gases and non-condensable products were on-line analyzed detecting H_2, carbon oxides, and methane in a column filled with Molecular Sieve 13X 80/100 mesh (CRS), using either N_2 or helium as carrier and a TCD, while methane and other light products in a GS-Alumina Megabore column. Samples condensed during reaction were off-line analyzed in a DB-20 Megabore column. Hydrogen production and selectivity to carbon oxides and to methane were calculated from TCD data and using a standard sample containing CO, CH_4, CO_2, N_2, and H_2. More details about preparation, characterization, catalytic evaluation, and analysis were previously reported [Sánchez et al., 2010].

TPR results of catalysts with 5 and 15% Ni impregnated on alumina showed three different Ni species: free or superficial NiO (reduction temperature below 400ºC), NiO bonded to Al_2O_3 (reduction temperature between 400 and 690ºC), and NiO incorporated into the Al_2O_3 framework and forming $NiAl_2O_4$ (reduction temperature above 700ºC) [Rynkowski et al., 1993; Zhu et al., 2008]. The reduction temperature of Ni species varied between 575 and 660ºC by calcining between 300 and 550ºC [Rynkowski et al., 1993]. Al_2O_3 impregnated with 4 wt.% Ni showed low and high proportion of $NiAl_2O_4$ species after calcining at 400 or 650ºC, respectively; this behavior was related to the diffusion of Ni ions into the Al_2O_3 network [Scheffer et al., 1989]. Furthermore, by decreasing the metal loading, the reduction of Ni species needed higher temperature because of a larger metal-support interaction [Brito and Laine, 1993]. This increment in reduction temperature was understand considering a low Ni loading on the support generates a greater proportion of unreduced Ni species which are better stabilized in the γ-Al_2O_3 vacant sites [Uemura et al., 1986]. In the 2.6-9.9% Ni loading range, the main species was NiO, being $NiAl_2O_4$ only in a low proportion, possibly due to the calcining temperature used, while Ni*(2.6)*/Al_2O_3 did not display the free or superficial NiO species; the high reduction temperature of Ni species on Al_2O_3 indicates a difficult reduction of them [Sánchez et al., 2010]. NH_3-TPD results with Al_2O_3 showed a broad band with maximum centered around 340-350°C. [Soled et al., 1988]. The strength of acid sites has been classified as weak, medium, and high according to the desorption temperature below 250ºC, between 250 and 400ºC, and higher than 400ºC, respectively [Auroux et al., 2001; Iengo et al., 1998]. Using Ni/Al_2O_3, the maximum of desorption band was observed at 300ºC, and acid sites were mostly considered Lewis acid sites with medium and weak strength [Hardiman, 2007]. By increasing the Ni loading in the 2.6-9.9% range, total acidity increases without significant changes in the acid strength profiles. Similar XRD patterns to Al_2O_3 support and Ni/Al_2O_3 catalysts indicated the presence of the γ-Al_2O_3 phase [Zhu et al., 2008]. Ni/Al_2O_3 has a slight increase in intensity respect to the support, verifying the formation of nickel-aluminate spinel type structure ($NiAl_2O_4$) at 37º [Zhu et al., 2008]. However, γ-Al_2O_3 has a pseudo-spinel structure and their structural network parameters are very similar to those of $NiAl_2O_4$ species [Lo Jacono et al., 1971]. By calcining Ni/Al_2O_3 at 600ºC, the pattern displayed large peaks at 37 and 44.8° due to the presence of $NiAl_2O_4$, while NiO species (peak at 43.2°) were observed to a lesser extent [Auroux, et al., 2001]. After calcining at 300ºC, peaks at 43 and 65°, corresponding to crystalline NiO, are only visible on materials with loadings higher than 7.3% w/w Ni; smaller Ni loads only produced the corresponding profile to support, with defined peaks at 37, 45 and 67° [Mattos et al., 2004]. For catalyst prepared with 2.6-9.9% Ni loadings, calcination temperature may hinder the detection of $NiAl_2O_4$ and NiO species by XRD. FTIR spectra corresponding to Ni/Al_2O_3 assigned the 3500 cm^{-1} broad band to the interaction between OH groups and/or chemisorbed water on the support with the alumina free hydroxyls through hydrogen bonding [Stoilova et al., 2002]. The 3780 cm^{-1} band has been also related to the presence of OH groups [Knözinger and Ratnasamy, 1978]. By comparing γ-Al_2O_3 and Ni/Al_2O_3 in the high-frequency region, the surface metal oxide species reduced the intensity of the main band; it was related to a decrease in the hydration degree of support surface [Kapteijn et al., 1994]. In the low-frequency region, studies using NO adsorption attributed bands near 1800 cm^{-1} to nitrite species bounded to reduced Ni ions, bands between 1340 and 1460 cm^{-1} to linearly bounded nitrite groups, and bands between 1620 and 1640 cm^{-1} at bounded nitrate species [Centi et al., 1995]. However, bands at 1620 and 1470-1480 cm^{-1} were also previously identified and assigned to the support [Turek et al., 1992]. On catalysts containing 2.6-9.9% Ni loadings, the increase of Ni loading decreases the hydration degree of support surface, while the intensity of the low-frequency band only decreases at the highest Ni loading, possibly related to the large amount of metal Ni species.

Figures 5 and 6 show glycerol conversion and hydrogen yield, respectively, for catalysts with different Ni loadings and reaction conditions. Using Ni*(5.8)*/Al_2O_3, conversion was larger by increasing reaction temperature from 600 to 700°C; this behavior was more pronounced at 8 h-on-stream. Independent of operating time, in the ranges of WHSV, water:glycerol molar ratios, nitrogen flow rate, Ni loadings, and reduction temperature, conversion was practically complete. The hydrogen yield was referred to the performance at 700°C, 5.1 WHSV, 40 ml min^{-1} N_2 flow rate, and 16:1 water:glycerol molar ratio. The lower the WHSV or N_2 flow rate, the higher the hydrogen yield. Hydrogen production improved by decreasing the water:glycerol molar ratio to 6:1 or reducing catalyst at 500°C. Changing the Ni loading, hydrogen yield did not significantly modify. Ni/Al_2O_3 having 2.6, 5.8, and 9.9% Ni loadings showed high both conversion and hydrogen yield, being temperature the reaction parameter with the most important effect over the catalytic performance. Evaluating Rh, Pt, Pd, Ir, Ru, Ni, Ce, Rh/CeO_2, Pt/CeO_2, Pd/CeO_2, Ir/CeO_2, Ru/CeO_2, Ni/CeO_2 impregnated on a mixed Al_2O_3-SiO_2 support, Ni reached the best catalytic performance with 94% glycerol conversion and 80% selectivity to hydrogen at 900°C, feeding 0.15 ml min^{-1} and 9:1 water:glycerol molar ratio [Adhikari et al., 2007c]. By thermodynamic data, optimum conditions for hydrogen production were temperatures between 650 and 700°C, water:glycerin ratios of 9-12, and atmospheric pressure [Wang et al., 2008]. Using a commercial Ni/Al_2O_3 catalyst, maximum hydrogen production was obtained at 800°C [Valliyappan et al., 2008]. Then, the good stability observed can be explained considering the catalytic conditions employed, mainly the high water:glycerol molar ratio which improves stability. Neither glycerol conversion nor hydrogen production can be directly related to the Ni loading; there is a minimum loading above which addition of Ni does not improve the catalytic behavior. It agrees with previous results of steam reforming of glycerol on Ru/Y_2O_3, which reported an optimal loading of 3% Ru [Hirai et al., 2005]; the increase in metal loading on Ni/Al_2O_3 and $Rh/CeO_2/Al_2O_3$ increased conversion while hydrogen selectivity remained relatively unaffected, reaching the best performance with 3.5 wt% loading [Adhikari et al., 2007c]. Seeking to determine the active site, Ni*(9.9)*/Al_2O_3 has a significantly larger amount of NiO-bonded to Al_2O_3 sites than on Ni*(5.8)*/Al_2O_3, displaying both materials similar catalytic behavior; then, that higher proportion does not increase conversion and/or hydrogen yield. Furthermore, Ni*(2.8)*/Al_2O_3 was an active material and it did not present the free or superficial NiO sites; then, these sites are not required for the glycerol reforming reaction.

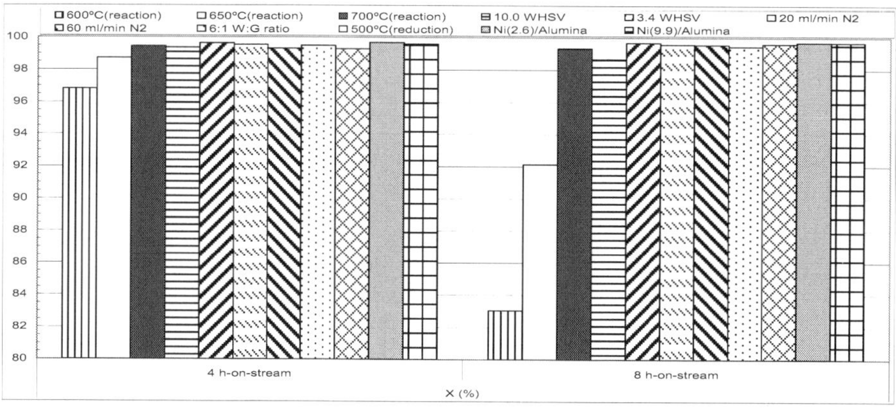

Fig. 5. Glycerol conversion (X) during the steam reforming of glycerol on Ni/Al_2O_3 catalysts under different pretreatment and operating conditions, at two times. *Unmodified reaction conditions: atmospheric pressure, and 0.17 ml min^{-1} glycerol aqueous solution flow rate.*

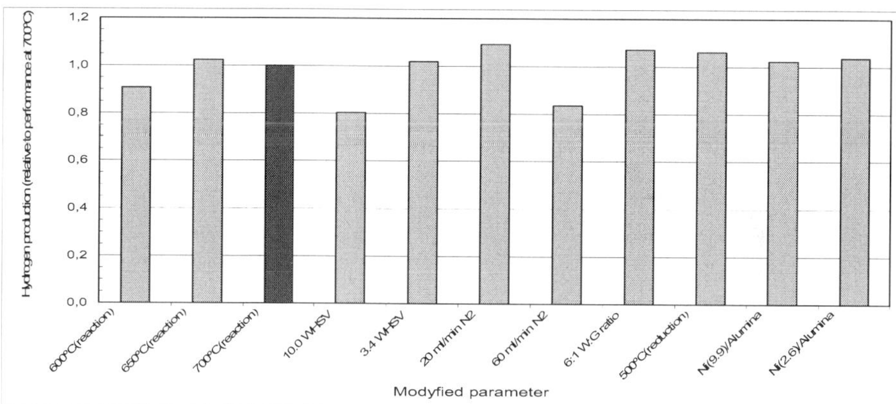

Fig. 6. Hydrogen yield (Y) during the steam reforming of glycerol on Ni/Al_2O_3 catalysts under the pretreatment and operating conditions detailed in Fig.5. *Unmodified reaction conditions: atmospheric pressure, 0.17 ml min^{-1} glycerol aqueous solution flow rate, and 4 hours-on-stream.*

Figure 7 shows by-product distribution in the non-condensed and gaseous stream detected by FID after 4 hours-on-stream for different catalysts and conditions. The main by-product was methane, followed by ethene, ethane, propene, isobutane, propane, and other ones in small proportions. Methane decreased by increasing WHSV, decreasing the water:glycerol molar ratio to 6:1 or increasing the Ni loading to 9.9%; the effect is more pronounced at 8 h. Distribution pattern only changed at 10.0 h^{-1}, being ethene the main by-product (47.1% at 4 h and 55.8% at 8 h) followed by methane (37.1% at 4 h and 27.2% at 8 h). This behavior could be associated with catalyst deactivation by carbonaceous deposit formation. Previous results reported Ni impregnated catalysts undergo a rapid deactivation [Wen et al., 2008]. Figure 8 displays the composition of non-condensed and gaseous stream analyzed on-line by TCD, as a function of time, during the steam reforming on Ni*(5.8)*/Al_2O_3 at 700°C, atmospheric pressure, 5.1 WHSV, 40 ml min^{-1} nitrogen flow, 16:1 water:glycerol molar ratio, and 0.17 ml min^{-1} glycerol aqueous solution flow rate. Products were hydrogen, carbon monoxide, and methane, varying their proportion along 1-6 operating cycles; a similar qualitative behavior was observed during each cycle: hydrogen decreasing whereas carbon monoxide and methane increased. The starting level of each cycle depends on the regenerating conditions. Hydrogen was higher than 60 mol% in all time range, reaching 93.8% as the highest value; carbon monoxide varied between 3.6 and 26.7%, and methane between 2.6 and 12.4%. Both carbon monoxide formation and methanation reactions can be controlled by selecting adequate pretreatment and operating conditions. Undesirable side reactions such as dimerization-oligomerization, thermal decomposition, cracking, disproportion, and hydrogen transfer promote carbon deposition which leading to blockage of catalyst pores with the corresponding loss of activity [Slinn et al., 2008]. At temperature higher than 720 K, glycerol is subjected to pyrolysis phenomena and it drastically decomposes before to reach the catalyst surface; olefins formed by that glycerol thermal decomposition promotes coke formation affecting the steam reforming process [Chiodo et al., 2010]. The highest stability of Ni/CeO_2/γ-Al_2O_3 has been related to a Ce effect which suppresses secondary dehydration reactions forming unsaturated hydrocarbons that are coke precursors generating fast catalyst deactivation [Buffoni et al., 2009]. By the same catalytic system, the addition of noble metal (Pt, Ir, Pd, and Ru) stabilized the Ni sites in the reduced state along the reforming reaction, increasing glycerol conversion and decreasing the coke formation [Profeti et al., 2009]. Small amounts of ethene and propene were detected between the reaction products (see Fig. 7); then, it could be associated to deactivation process.

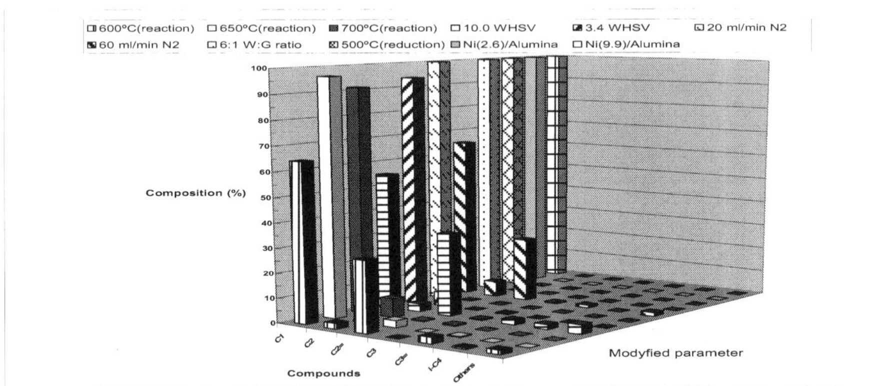

Fig. 7. By-product distribution corresponding to the gaseous fraction during the steam reforming of glycerol on Ni/Al_2O_3 catalysts under the pretreatment and operating conditions detailed in Fig. 5. *Unmodified reaction conditions: atmospheric pressure, 0.17 ml min*$^{-1}$ *glycerol aqueous solution flow rate, and 4 hours-on-stream.*

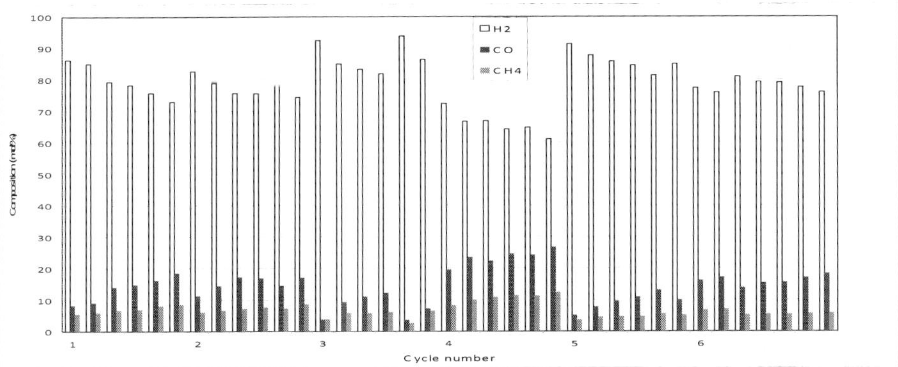

Fig. 8. Composition of the non-condensed and gaseous stream during the steam reforming of glycerol on Ni*(5.8)*/Al_2O_3 catalyst as a function of the time-on-stream for several cycles of reaction-regeneration. *Reaction conditions: 700°C, atmospheric pressure, 5.1 WHSV, 40 ml min^{-1} nitrogen flow, 16:1 water:glycerol molar ratio, and 0.17 ml min^{-1} glycerol aqueous solution flow rate.*

3. Conclusion

The production of biodiesel generates glycerol as by-product. The conversion of glycerol in added-value compounds has benefits because it comes from renewable raw materials, enabling a sustainable environmental development, and also allows improve the economics of biodiesel process. Glycerol is intermediate in the synthesis of some compounds used in industry, such as propyleneglycol, 1,3-propanediol, and ethyleneglycol by hydrogenolysis; acetol and acrolein by dehydration; dihydroxyacetone, glyceric acid, and hydropiruvic acid by oxidation; glycidol by epoxidation; glycerol carbonate by transesterification; mono- and diglycerides by selective etherification; polyglycerol by polymerization; and bio-hydrogen by steam reforming, partial oxidation, autothermal reforming, aqueous-phase reforming, and supercritical water reforming. All these uses allow considering glycerol as an important key-compound in the environment of future biorefinery. By considering the hydrogenolysis of glycerol, the production of propyleneglycol was reported on several supported catalysts, resulting attractive the improvement in selectivity by using zeolites as base materials, while the production of 1,3-propanediol took place on Pt/WO_3/ZrO_2, being important a complete study of the effect of preparation technique of this material over its catalytic performance; moreover, a comparison between both liquid and gaseous phase hydrogenolysis is needed. The selective oxidation of glycerol to dihydroxyacetone has been reached on a monometallic catalyst, Pt impregnated on potassium ferrierite, reaching both conversion and selectivity similar to the ones obtained with active bimetallic materials; the pH of reaction medium played an important role over the selectivity, while the structure of zeolite favored the DHA formation. Finally, the steam reforming of glycerol to produce hydrogen was conduced on Ni impregnated on alumina catalysts, which presented Ni species on the surface difficult to reduce and mainly as NiO, reaching high conversions and selectivity for conditions studied and being carbon monoxide the main by-product followed by methane; catalyst deactivation took place during reaction but regenerating is possible restoring the catalytic performance.

4. Acknowledgment

The author acknowledge the financial support of U.N.L. and CONICET. Finally, the author is also acknowledged to InTech by the possibility to edit this chapter.

5. References

Abbadi, A., van Bekkum, H. (1996) "Selective chemo-catalytic routes for the preparation of β-hydroxypyruvic acid". Applied Catalysis A: General 148, (1996), 113-122.

Adhikari, S.; Fernando, S.; Gwaltney, S.R.; Filip To, S.D.; Bricka, R.M.; Steele, P.H.; Haryanto, A. (2007) International Journal of Hydrogen Energy 32, (2007), 2875 - 2880.

Adhikari, S.; Fernando, S.; Haryanto, A. (2007) Energy & Fuels 21, (2007), 2306-2310.

Adhikari, S; Fernando, S.; Haryanto, A. (2007) Catalysis Today 129, (2007) 355-364.

Adhikari, S.; Fernando, S.D.; Filip To, S.D.; Bricka, R.M.; Steele, P.H.; Haryanto, A. (2008) Energy & Fuels 22, (2008), 1220-1226.

Adhikari, S.; Fernando, S.D.; Haryanto, A. (2008) Renewable Energy 33, (2008), 1097-1100.

Adhikari, S.; Fernando, S.D.; Haryanto, A. (2009) Energy Conversion and Management 50, (2009) 2600–2604.

Akiyama M., Sato S., Takahashi R., Inui K., Yokota M., (2009) "Dehydration–hydrogenation of glycerol into 1,2-propanediol at ambient hydrogen pressure", Applied Catalysis A: General 371, (2009), 60-66.

Alhanash A., Kozhevnikova E. F., Kozhevnikov I. V., (2008) "Hydrogenolysis of Glycerol to Propanediol Over Ru: Polyoxometalate Bifunctional Catalyst", Catalysis Letters 120, (2008), 307-311.

Ali, Y.; Hanna, M.; Cuppett, S. (1995) Journal of the American Oil Chemists' Society 72, (1995), 1557-1564.

Archambault, É. (2004) Science-Metrix Canadian R&D Biostrategy, National Research Council of Canada, Montreal, Canada, Apr. 2004.

A. Auroux, A.; Monaci, R.; Rombi, E.; Solinas, V.; Sorrento, A.; Santacesaria, E. (2001) Thermochimica Acta 379, (2001), 227-231.

Balaraju M., Rekha V., Sai Prasad P. S., Prasad R. B. N., Lingaiah N., (2008) "Selective Hydrogenolysis of Glycerol to 1, 2 Propanediol Over Cu–ZnO Catalysts", Catalysis Letters 126, (2008), 119-124.

Balaraju M., Rekha V., Sai Prasad P. S., Prabhavathi Devi B. L. A., Prasad R.B.N., Lingaiah N., (2009) "Influence of solid acids as co-catalysts on glycerol hydrogenolysis to propylene glycol over Ru/C catalysts", Applied Catalysis A: General 354, (2009), 82-87.

Balaraju M., Rekha V., Prabhavathi Devi B. L .A., Prasad R.B.N., Sai Prasad P. S., Lingaiah N., (2010) "Surface and structural properties of titania-supported Ru catalysts for hydrogenolysis of glycerol", Applied Catalysis A: General 384, (2010), 107-114.

Bianchi, L., Canton, P., Dimitratos, N., Porta, F., Prati, L. (2005) "Selective oxidation of glycerol with oxygen using mono and bimetallic catalysts based on Au, Pd and Pt metals". Catalysis Today 102–103, (2005), 203–212

Bicker, M., Endres, S., Ott, L., Vogel, H. (2005)"Catalytic conversion of carbohydrates in subcritical water: a new chemical process for lactic acid production". Journal of Molecular Catalysis A Chemical. 239, (2005), 151–7

Biebl H., Menzel K., Zeng A. P., Deckwer W. D., (1999) "Microbial production of 1,3-propanediol", Applied Microbiology and Biotechnology 52, (1999), 289-297.

Bolado S., Treviño R. E., García-Cubero M. T., González-Benito G., (2010) "Glycerol hydrogenolysis to 1,2-propanediol over Ru/C catalyst", Catalysis Communications 12,(2010), 122-126.

Brito, J.L.; Laine, J. (1993) Journal of Catalysis 139, (1993) 540-550.

Brown, D. (2001) "Skin pigmentation enhancers". Journal of Photochemistry and Photobiology B 63, (2001),148–61

Buffoni, I.N.; Pompeo, F.; Santori, G.F.; Nichio, N.N. (2009) Catalysis Communications 10, (2009), 1656–1660.

Carrettin, S., McMorn, P., Johnston, P., Griffin, K., Kiely, C. J., Hutchings, G. J. (2003) "Oxidation of glycerol using supported Pt, Pd and Au catalysts". Physical Chemistry Chemical Physics, 5, (2003), 1329–1336.

Casale B., Gomez A. M., (1993) US Patent 5.214.219, 1993.

Casale B., Gomez A. M, (1994) US Patent 5.276.181, 1994.

Centi, G.; Perathoner, S.; Biglino, D.; Giamello, E. (1995) Journal of Catalysis 152, (1995), 75-92.

Che T. M., Westiefeld N.J., (1987) US Patent 4.462.394, 1987.

Chen, H.; Zhang, T.; Dou, B.; Dupont, V.; Williams, P.; Ghadiri, M.; Ding, Y. (2009) International Journal of Hydrogen Energy 34, (2009), 7208 - 7222.

Chiodo, V.; Freni, S.; Galvagno, A.; Mondillo, N.; Frusteri, F. (2010) Applied Catalysis A: General 381, (2010), 1-7.

Chiu C. W., Dasari M. A, Sutterlin W. R., Suppes G. J., (2006) "Removal of Residual Catalyst from Simulated Biodiesel's Crude Glycerol for Glycerol Hydrogenolysis to Propylene Glycol", Industrial & Engineering Chemistry Research 45, (2006), 791-795.

Chiu C. W., Tekeei A., Ronco J. M., Banks M. L., Suppes G. J., (2008) "Reducing Byproduct Formation during Conversion of Glycerol to Propylene Glycol", Industrial & Engineering Chemistry Research 47, (2008), 6878-6884.

Corma A., Iborra S., Velty A., (2007) "Chemical Routes for the Transformation of Biomass into Chemicals", Chemical Reviews 107, (2007), 2411-2502.

Cui, Y.; Galvita, V.; Rihko-Struckmann, L.; Lorenz, H.; Sundmacher, K. (2009) Applied Catalysis B: Environmental 90, (2009), 29–37.

Dasari M. A., Kiatsimkul P. P., Sutterlin W. R., Suppes G. J., (2005)"Low-pressure hydrogenolysis of glycerol to propylene glycol", Applied Catalysis A: General 281, (2005), 225-231.

Demirel-Gülen, S., Lucas, M., Claus, P. (2005) "Liquid phase oxidation of glycerol over carbon supported gold catalysts". Catalysis Today 102–103, (2005), 166–172

Dou, B.; Dupont, V.; Rickett, G.; Blakeman, N.; Williams, P.T.; Chen, H.; Ding, Y.; Ghadiri, M. (2009) Bioresource Technology 100, (2009), 3540–3547.

Dou, B.; Rickett, G.L.; Dupont, V.; Williams, P.T.; Chen, H.; Ding, Y.; Ghadiri, M. (2010) Bioresource Technology 101, (2010), 2436–2442.

Drent E., Jager W. W., (2000) US Patent 6.080.898, 2000.

EU Directive 2001/0265 (COD) (2003) "The Promotion of the Use of Biofuels or Other Renewable Fuels for Transport", European Parliament, 05/2003.

Feng J., Fu H., Wang J., Li R., Chen H., Li X., (2008) "Hydrogenolysis of glycerol to glycols over ruthenium catalysts: Effect of support and catalyst reduction temperature", Catalysis Communications 9, (2008), 1458-1464.

Fesq, H., Brockow, K., Strom, K., Mempel, M., Ring, J., Abeck, D. (2001) "Dihydroxyacetone in a new formulation—a powerful therapeutic option in vitiligo". Dermatology 203, (2001), 241–243

Fordham, P., Besson, M., Gallezot, P. (1995) "Selective catalytic oxidation of glyceric acid to tartronic and hydroxypyruvic acids". Applied Catalysis A: General 133, (1995), I79-I84

Franke O., Stankowiak A., (2010) US Patent 7.812.200, 2010.

Fulton, L. Office of Energy Efficiency, Technology and R&D, (2004) International Energy Agency, Paris, France, Apr 2004.

Gandarias I., Arias P. L., Requies J., Güemez M. B., Fierro J. L. G., (2010) "Hydrogenolysis of glycerol to propanediols over a Pt/ASA catalyst: The role of acid and metal sites on product selectivity and the reaction mechanism", Applied Catalysis B: Environmental 97, (2010), 248-256.

Garcia, R., Besson, M., Gallezot, P. (1995) "Chemoselective catalytic oxidation of glycerol with air on platinum metals". Applied Catalysis A: General 127, (1995), 165-176

Gong L., Lu Y., Ding Y., Lin R., Li J., Dong W., Wang T., Chen W., (2009) "Solvent Effect on Selective Dehydroxylation of Glycerol to 1,3-Propanediol over a $Pt/WO_3/ZrO_2$ Catalyst", Chinese Journal of Catalysis 30, (2009), 1189-1191

Gong L., Lu Y., Ding Y., Lin R., Li J., Dong W., Wang T., Chen W., (2010) "Selective hydrogenolysis of glycerol to 1,3-propanediol over a $Pt/WO_3/TiO_2/SiO_2$ catalyst in aqueous media", Applied Catalysis A: General 390, (2010), 119-126.

Guo L., Zhou J., Mao J., Guo X., Zhang S., (2009) "Supported Cu catalysts for the selective hydrogenolysis of glycerol to propanediols", Applied Catalysis A: General 367, (2009), 93-98.

Guo X., Li Y., Shi R., Liu Q., Zhan E., Shen W., (2009) "Co/MgO catalysts for hydrogenolysis of glycerol to 1, 2-propanediol", Applied Catalysis A: General 371, (2009), 108-113.

Gupta, A., Singh, V.K., Qazi, G.N., Kumar, A. (2001) "Gluconobacter oxydans: its biotechnological applications". Journal of Molecular Microbiology and Biotechnology, 3, (2001), 445–456

Hardiman, K.M. (2007) Thesis of Philosophy, School of Chemical Sciences and Engineering, University of New South Wales, Sydney, Australia, Feb 2007.

Hekmat, D., Bauer, R., Fricke, J. (2003) "Optimization of the microbial synthesis of dihydroxyacetone from glycerol with Gluconobacter oxydans". Bioprocess and Biosystems Engineering. 26, (2003), 109–16

Hirai, T.; Ikenaga, N-O.; Miyake, T.; Suzuki, T. (2005) Energy & Fuels 9, (2005), 1761-1762.

Iengo, P.; Di Serio, M.; Sorrentino, A.; Solinas, V; Santacesaria, E. (1998) Applied Catalysis A 167, (1998), 85-101.

Huang L., Zhu Y., Zheng H., Ding G., Li Y., (2009) "Direct Conversion of Glycerol into 1,3-Propanediol over $Cu\text{-}H_4SiW_{12}O_{40}/SiO_2$ in Vapor Phase", Catalysis Letters 131, (2009), 312-320.

Huang Z., Cui F., Kang H., Chen J., Zhang X., Xia C., (2008) "Highly Dispersed Silica-Supported Copper Nanoparticles Prepared by Precipitation-Gel Method: A Simple but Efficient and Stable Catalyst for Glycerol Hydrogenolysis", Chemistry of Materials 20, (2008), 5090-5099.

Huang Z., Cui F., Kang H., Chen J., Xia C.; (2009) "Characterization and catalytic properties of the CuO/SiO_2 catalysts prepared by precipitation-gel method in the hydrogenolysis of glycerol to 1,2-propanediol:Effect of residual sodium", Applied Catalysis A: General 366, (2009), 288-298.

Iengo, P.; Di Serio, M.; Solinas, V.; Gazzoli, D.; Sálvio, G.; Santacesaria, E. (1998) Applied Catalysis A 170, (1998), 225-244.

Iriondo, A.; Barrio, V.L.; Cambra, J.F.; Arias, P.L.; Güemez, M.B.; Navarro, R.M.; Sánchez-Sánchez, M.C.; García Fierro, J.L. (2009) Catalysis Communications 10, (2009) 1275-1278.

Kapteijn, F.; van Langeveld, A.D.; Moulijn, J.A.; Andreini, A.; Vuurman, M.A.; Turek, A.M.; Jehng, J.M.; Wachs, I.E. (1994) Journal of Catalysis 150, (1994), 94-104.

Kenji, S., Hidehiko, T., Setsuo, F. (1989) JP Patent 01225486, 1989

Kim N. D., Oh S., Joo J. B., Jung K. S., Yi J., (2010) "Effect of preparation method on structure and catalytic activity of Cr-promoted Cu catalyst in glycerol hydrogenolysis", Korean Journal of Chemical Engineering 27, (2010), 431-434.

Kim N. D., Oh S., Joo J. B., Jung K. S., Yi J., (2010) "The Promotion Effect of Cr on Copper Catalyst in Hydrogenolysis of Glycerol to Propylene Glycol", Topics in Catalysis 53, (2010), 517-522.

Kimura, H., Tsuto, K. (1993) "Selective oxidation of glycerol on a platinum-bismuth catalyst". Applied Catalysis A: General, 96, (1993), 217-228

Knözinger, H.; Ratnasamy, P. (1978) Catalysis Reviews,Science and Engineering 17, (1978), 31-70.

Korolev Y. A., Greish A. A., Kozlova L. M., Kopyshev M. V., Litvin E. F., Kustov L. M., (2010) "Glycerol Dehydroxylation in Hydrogen on a Raney Cobalt Catalyst", Catalysis in Industry 2, (2010), 287-289.

Kunkes, E.L.; Soares, R.R.; Dumesic, J.A. (2009) Applied Catalysis B: Environmental 90, (2009), 693–698.

Kurian J. V., (2005) "A New Polymer Platform for the Future — Sorona® from Corn Derived 1,3-Propanediol", Journal of Polymers and the Environment 13, (2005), 159-167.

Kurosaka T., Maruyama H., Naribayashi I., Sasaki Y., (2008) "Production of 1,3-propanediol by hydrogenolysis of glycerol catalyzed by $Pt/WO_3/ZrO_2$", Catalysis Communications 9, (2008), 1360-1363.

Kusunoki Y., Miyazawa T., Kunimori K., Tomishige K., (2005) "Highly active metal–acid bifunctional catalyst system for hydrogenolysis of glycerol under mild reaction conditions", Catalysis Communications 6, (2005), 645-649.

Lahr D. G., Shanks B. H., (2003) "Kinetic Analysis of the Hydrogenolysis of Lower Polyhydric Alcohols: Glycerol to Glycols", Industrial & Engineering Chemistry Research 42, (2003), 5467-5472.

Lahr D. G., Shanks B. H., (2005) "Effect of sulfur and temperature on ruthenium-catalyzed glycerol hydrogenolysis to glycols", Journal of Catalysis 232, (2005), 386-394.

Liang L., Ma Z., Ding L., Qiu J., (2009) "Template Preparation of Highly Active and Selective Cu-Cr Catalysts with High Surface Area for Glycerol Hydrogenolysis", Catalysis Letters 130, (2009), 169-176.

Lo Jacono, M.; Schiavello, M.; Cimino, A. (1971) Journal of Physical Chemistry 75, (1971) 1044-1050.

Luo, N.; Fu, X.; Cao, F.; Xiao, T.; Edwards, P.P. (2008) Fuel 87, (2008), 3483-3489.

Ma L., He D., Li Z., (2008) "Promoting effect of rhenium on catalytic performance of Ru catalysts in hydrogenolysis of glycerol to propanediol", Catalysis Communications 9, (2008), 2489-2495

Ma L., He D., (2009) "Hydrogenolysis of Glycerol to Propanediols Over Highly Active Ru–Re Bimetallic Catalysts", Topics in Catalysis 52, (2009), 834–844.

Ma L., He D., (2010) "Influence of catalyst pretreatment on catalytic properties and performances of Ru-Re/SiO_2 in glycerol hydrogenolysis to propanediols", Catalysis Today 149, (2010), 148-156.

Maris E. P., Davis R.J., (2007) "Hydrogenolysis of glycerol over carbon-supported Ru and Pt catalysts", Journal of Catalysis 249, (2007), 328-337.

Maris E. P., Ketchie W. C., Murayama M., Davis R. J., (2007) "Glycerol hydrogenolysis on carbon-supported PtRu and AuRu bimetallic catalysts", Journal of Catalysis 251, (2007), 281-294.

Maervoet V. E. T., De Mey M., Beauprez J., De Maeseneire S., Soetaert W. K., (2011) "Enhancing microbial conversion of glycerol to 1,3-Propanediol using Metabolic Engineering", Organic Process Research & Development 15, (2011), 189-202.

Mane R. B, Hengne A. M., Ghalwadkar A. A., Vijayanand S., Mohite P. H., Potdar H. S., Rode C. V., (2010) "Cu:Al Nano Catalyst for Selective Hydrogenolysis of Glycerol to 1,2-Propanediol", Catalysis Letters 135, (2010), 141-147.

Marinoiu A., Ionita G., Gáspár C. L., Cobzaru C., Marinescu D., Teodorescu C., Oprea S., (2010) "Selective hydrogenolysis of glycerol to propylene glycol using heterogeneous catalysts", Reaction Kinetics, Mechanisms and Catalysis 99, (2010), 111-118.

Mattos, A.R.J.M.; Probst, S.H.; Afonso, J.C.; Schmal, M. (2004) Journal of Brazilian Chemical Society 15, (2004), 760-766.

Meher L. C., Gopinath R., Naik S. N., Dalai A. K., (2009) "Catalytic Hydrogenolysis of Glycerol to Propylene Glycol over Mixed Oxides Derived from a Hydrotalcite-Type Precursor", Industrial & Engineering Chemistry Research 48, (2009), 1840-1846.

Miyazawa T., Kusunoki Y., Kunimori K., Tomishige K., (2006) "Glycerol conversion in the aqueous solution under hydrogen over Ru/C + an ion-exchange resin and its reaction mechanism", Journal of Catalysis 240, (2006), 213-221.

Miyazawa T., Koso S., Kunimori K., Tomishige K., (2007) "Development of a Ru/C catalyst for glycerol hydrogenolysis in combination with an ion-exchange resin", Applied Catalysis A: General 318, (2007), 244-251.

Miyazawa T., Koso S., Kunimori S., Tomishige K., (2007) "Glycerol hydrogenolysis to 1,2-propanediol catalyzed by a heat-resistant ion-exchange resin combined with Ru/C", Applied Catalysis A: General 329, (2007), 30-35.

Nakagawa Y., Shinmi Y., Koso S., Tomishige K., (2010) "Direct hydrogenolysis of glycerol into 1,3-propanediol over rhenium-modified iridium catalyst", Journal of Catalysis 272, (2010), 191-194.

Oertel, D. (2007) TAB report 114, Office of Technology Assessment at the German Parlament (TAB), Berlin, Germany, Mar. 2007.

Perosa A., Tundo P., (2005) "Selective Hydrogenolysis of Glycerol with Raney Nickel", Industrial & Engineering Chemistry Research 44, (2005), 8535-8537.

Peterson, C.; Reece, D.; Thompson, J.; Beck, S.; Chase, C. (1996) Biomass and Bioenergy 10, (1996), 331-336.

Porta, F., Prati, L. (2004) "Selective oxidation of glycerol to sodium glycerate with gold-on-carbon catalyst: an insight into reaction selectivity". Journal of Catalysis 224, (2004), 397–403.

Profeti, L.P.R.; Ticianelli, E.A.; Assaf, E.M. (2009) International Journal of Hydrogen Energy 34, (2009) 5049-5060.

Rossi, C.C.R.S. ; Alonso, C.G.; Antunes, O.A.C.; Guirardello, R.; Cardozo-Filho, L. (2009) International Journal of Hydrogen Energy 34, (2009), 323-332.

Roy D., Subramaniam B., Chaudhari R. V., (2010) "Aqueous phase hydrogenolysis of glycerol to 1,2-propanediol without external hydrogen addition", Catalysis Today 156, (2010), 31-37.

Rynkowski, J.M.; Paryjczak, T.; Lenik, M. (1993) Applied Catalysis A 106, (1993), 73-82.

Sánchez, E.; D'Angelo, M.; Comelli, R. (2010) International Journal of Hydrogen Energy 35, (2010), 5902-5907.

Sauer M., Marx H., Mattanovich D., (2008) "Microbial Production of 1,3-Propanediol", Recent Patents on Biotechnology 2, (2008), 191-197.

Saxena R.K., Anand P., Saran S., Isar J., (2009) "Microbial production of 1,3-propanediol: Recent developments and emerging opportunities", Biotechnology Advances 27, (2009), 895-913.

Scheffer, B.; Molhoek, P.; Moulijn, (1989) J.A. Applied Catalysis 46, (1989) 11-30.

Schmidt S. R., Tanielyan S. K., Marin N., Alvez G., Augustine R. L., (2010) "Selective Conversion of Glycerol to Propylene Glycol Over Fixed Bed Raney® Cu Catalysts", Topics in Catalysis 53, (2010), 1214-1216.

Schuster L., Eggersdorfer M., (1997) US Patent 5.616.817, 1997.

Shinmi Y., Koso S., Kubota T., Nakagawa Y., Tomishige K., (2010) "Modification of Rh/SiO_2 catalyst for the hydrogenolysis of glycerol in water", Applied Catalysis B: Environmental 94, (2010), 318-326.

Slinn, M; Kendall, K; Mallon, C; Andrews, J. (2008). Bioresource Technology 99, (2008) 5851-5858.

Soled, S.L.; McVicker, G.B.; Murrell, L.L.; Sherman, L.G.; Dispenziere, N.C.; Hsu, S.L.; Waldman, D. (1988) Journal of Catalysis,111, (1988). 286-295.

Stankowiak A, Franke O., (2011) US Patent 7.868.212, 2011.

Stoilova, D.G.; Koleva, V.G.; Cheshkova, K.T. Z. (2002) Physical Chemistry 216, (2002), 737-747.

Susuki N., Yoshikawa Y., Takahashi M., Tamura M., (2010) US. Patent 7.799.957, 2010.

Susuki N., Tamura M., Mimura T., (2010) European Patent 2.239.247, 2010.

Takehiro, I., Hiroyuki, H., Akira, F., Yukinaga, Y. (1993) JP Patent 05331100, 1993

Teruyuki, N., Yoshinori, K. (1989) JP Patent 1168292, 1989

Tuck M. W. M., Tilley S. N., (2008) US Patent 7.355.083, 2008.

Turek, A.M.; Wachs, I.E.; DeCanio, E. (1992) Journal of Physical Chemistry 96, (1992), 5000-5007.

Uemura, Y.; Hatate Y.; Ikari, A. (1986) Journal of the Japan Petroleum Institute 29, (1986), 143-150.

Valliyappan, T.; Ferdous, Bakhshi, N.N.; Dalai, A.K. (2008) Topics in Catalysis 49, (2008), 59–67.

Vasiliadou E. S., Heracleous E., Vasalos I.A., Lemonidou A. A., (2009) "Ru-based catalysts for glycerol hydrogenolysis - Effect of support and metal precursor", Applied Catalysis B: Environmental 92, (2009), 90-99

Vaudagna S., Comelli R.A., Fígoli N., (1997) "Influence of the tungsten oxide precursor on WO_x-ZrO_2 and Pt/WO_x-ZrO_2 properties", Applied Catalysis A: 164, (1997), 265-280

Vicente, G.; Martínez, M. J. Aracil, (2004) Bioresource Technology 92, (2004), 297-305.

Wang K., Hawley, M.C., DeAthos S.J., (2003) "Conversion of Glycerol to 1,3-Propanediol via Selective Dehydroxylation", Industrial & Engineering Chemistry Research 42, (2003) 2913-2923.

Wang S., Liu H., (2007) "Selective hydrogenolysis of glycerol to propylene glycol on Cu-ZnO catalysts", Catalysis Letters 117, (2007), 62-67.

Wang, X.; Li, S.; Wang, H.; Liu, B.; Ma, X. (2008) Energy & Fuels 22, (2008), 4285–4291.

Wen, G.; Xu, Y.; Ma, H.; Xu, Z.; Tian, Z. (2008) International Journal of Hydrogen Energy 33, (2008), 6657-6666.

Werpy, T.; Petersen, G. (2004) Energy Efficiency and Renewable Energy, U.S. Department of Energy, Washington D.C., U.S.A., Aug. 2004.

Yaylayan, V.A., Harty-Majors, S., Ismail, A. (1999) "Investigation of DL-glyceraldehy-dihydroxyacetone interconversion by FTIR spectroscopy". Carbohydrate Research 318, (1999), 20–25.

Young, T., Eom, C.Y., Song, T., Cho, J.W., Kim, Y.M. (1980) "Dihydroxyacetone synthase from a methanol-utilizing carboxydobacterium, Acinetobacter sp. Strain JC1 DSM 3803". Journal of Bacteriology 179, (1980), 6041–6047

Yu W., Xu J., Ma H., Chen C., Zhao J., Miao H., Song Q., (2010) "A remarkable enhancement of catalytic activity for KBH_4 treating the carbothermal reduced Ni/AC catalyst in glycerol hydrogenolysis", Catalysis Communications 11, (2010), 493-497.

Yu W., Zhao J., Ma H., Miao H., Song Q., Xu J., (2010) "Aqueous hydrogenolysis of glycerol over Ni-Ce/AC catalyst: Promoting effect of Ce on catalytic performance", Applied Catalysis A: General 383, (2010), 73-78.

Yuan Z., Wu P., Gao J., Lu X., Hou Z., Zheng X., (2009) "Pt/Solid-Base: A Predominant Catalyst for Glycerol Hydrogenolysis in a Base-Free Aqueous Solution", Catalysis Letters 130, (2009), 261-265.

Yuan Z., Wang J., Wang L., Xie W., Chen P., Hou Z., Zheng X., (2010) "Biodiesel derived glycerol hydrogenolysis to 1,2-propanediol on Cu/MgO catalysts", Bioresource Technology 101, (2010), 7088-7092.

Yuan Z., Wang L., Wang J., Xia S., Chen P., Hou Z., Zheng X., (2011) "Hydrogenolysis of glycerol over homogenously dispersed copper on solid base catalysts", Applied Catalysis B: Environmental 101, (2011), 431-440.

Zhang, B.; Tang, X.; Li, Y.; Xu, Y.; Shen, W. (2007) International Journal of Hydrogen Energy 32, (2007), 2367 - 2373.

Zhao J., Yu W., Chen C., Miao H., Ma H., Xu J.,(2010)"Ni/NaX: A Bifunctional Efficient Catalyst for Selective Hydrogenolysis of Glycerol", Catalysis Letters 134, (2010), 184-189.

Zheng J., Zhu W., Ma C., Hou Y., Zhang W., Wang Z., (2010) "Hydrogenolysis of glycerol to 1,2-propanediol on the high dispersed SBA-15 supported copper catalyst prepared by the ion-exchange method", Reaction Kinetics, Mechanisms and Catalysis 99, (2010), 455-462.

Zheng Y., Chen X., Shen Y., (2008) "Commodity Chemicals Derived from Glycerol, an Important Biorefinery Feedstock", Chemical Reviews 108, (2008), 5253-5277.

Zhou Z., Li X., Zeng T., Hong W., Cheng Z., Yuan W., (2010) "Kinetics of Hydrogenolysis of Glycerol to Propylene Glycol over Cu-ZnO-Al_2O_3 Catalysts", Chinese Journal of Chemical Engineering, 18, (2010), 384-390.

Zhu, X.; Huo, P.; Zhang, Y.; Cheng, D.; Liu, C. (2008) Applied Catalysis B 81, (2008), 132-140.

Zhua, Y., Youssefb, D., Porte, C., Rannouc, A., Delplancke-Ogletree, M.P., Mi Lung-Somarriba, B. Lo. (2003) "Study of the solubility and the metastable zone of 1,3-dihydroxyacetone for the drowning out process". Journal of Crystal Growth 257, (2003), 370–377

17

Antioxidative and Anticorrosive Properties of Bioglycerol

Maria Jerzykiewicz and Irmina Ćwieląg-Piasecka
Faculty of Chemistry, Wroclaw University, Wroclaw
Poland

1. Introduction

Raw glycerol fraction (also called bioglycerol) is the second product of fatty acids triglycerides alcoholysis. However, by many biodiesel producers it is treated as a waste, unwanted material. Due to overproduction of the glycerol fraction there have been many attempts to find its proper utilization, especially that it can comprise up to 30% of the whole biodiesel production. Finding proper applications for bioglycerol became a research target of scientists from all over the world (project E! 3590 Use-Glycerol: New concept for utilization of glycerol fraction from biodiesel production, http://www.eurekanetwork.org/project/-/id/3590, Corma et al., 2007, Pagliaro et al., 2007). What is important, bioglycerol has an enormous chemical potential and should be considered as a valuable transesterification product and thus a versatile feedstock for the creation of new chemicals (Pagliaro & Rossi, 2010). Until recently bioglycerol purification was based on the separation of fatty acids from the mixture and recycling them to biodiesel production process. There were also a few tests made to use raw glycerol fraction omitting distillation procedures. One of them was to utilize bioglycerol as an additive during composting process, especially when potassium hydroxide as a catalyst in transesterification was used. This utilization was abandoned due to the technological problems and relatively too expensive comparing with the cost of compost. On the other hand, production of pharmaceutically pure glycerol is very time and energy consuming due to its high boiling point. There is also a propano-1, 2, 3-triol overproduction, thus its price is low. Market is already flooded with glycerol-containing products especially in food, pharmaceutical, cosmetic and leather industry. That is why there arises an issue of finding new ways of biodiesel utilization. It is essential to make this fraction as valuable as biodiesel. The principal directions of solving "bioglycerol overproduction problem" are oriented towards the development of production methods for the new platform chemicals (Behr et al., 2008, Corma et al. 2007, Gu et al., 2008, Zheng et al., 2008, Zhou et al. 2008). Therefore the final products should be obtained directly from raw bioglycerol. In such case the purification process would be shifted to the final product that would have much lower boiling point or different solubility and it would be cheaper. A few trials were made to elaborate production of glycerol formals, very valuable fuel additives (Trybula et al., 2010). There was also hydrogen produced on the basis of bioglycerol fermentation (Ito et al. 2005) or with the use of catalyst (Hirai et al., 2005, Huber et al. 2003). Until now the main attempts of the studies were focused on the most abundant constituent of the fraction - glycerol. The second direction of bioglycerol research concerns the fraction as an integrate material and takes advantage of its minor constituents. Fatty acids and alcohol are utilized by turning them back to the transesterification process. The residues of fatty acids and unreacted triglycerides can be turned into soaps or lubricants, thus the final product would contain glycerol as "an additive" (patents: Jerzykiewicz et al., 2006, PL 378802, PL 207449, Jerzykiewicz et al., 2008, PL 381066, PL 386312, Lukosek et al. 2007, PL 384164, PL 383841). Recently there were also investigations on trace components such as antioxidants performed.

The concept of the studies presented in this chapter was to find the antioxidants routes during the biodiesel production and to establish their structure. Are the antioxidants originated in oils transferred during transesterification process to the biodiesel or to bioglycerol? The antioxidants occurring in the transesterification products should have phenolic-like structure due to their vegetable oil origin. Virgin oils are rich in substances that exert antioxidant properties such as: tocopherols, phospholipids, carotenes and sterols (Velasco et al, 2005, Blekas et al., 1995, Koski et al. 2002). Chemical properties of the tocopherols lead to their better solubility in bioglycerol than biodiesel. If the glycerol fraction was the receiver of the antioxidants, these valuable compounds would be condensed there (up to 30% of the initial volume). Additionally, structural investigations would be necessary to confirm whether the conditions of performed transesterification influence the tocopherol activity. Thus apart from standard methods used for establishing the main bioglycerol constituents (gas chromatography, assays based on titration) more advanced methods are indispensable. One of the most reliable techniques for detection as well as quantitative characterization of the antioxidants is electron paramagnetic resonance (EPR) spectroscopy. Apart from direct EPR measurements of free radicals and other paramagnetic species the method allows studying diamagnetic compounds like phenolic antioxidants by the use of spin trapping or radical scavenging techniques.

2. Bioglycerol – What it is?

Bioglycerol is a mixture of: propano-1, 2, 3-triol, residues of alcohol, soaps, fatty acids esters, mono-, di-, triglycerides, phospholipids and some minor constituents. The most popular method helpful in determination of the fraction composition is gas chromatography. Presented in the chapter results of glycerol fraction composition were analyzed using gas chromatography (GC) (Hewlett-Packard, model 5890 GLC II). The chromatograph was equipped with a flame ionization detector (FID) and 11 m x 0.22 mm x 0.1 μm high temperature capillary Utra-2. Separation of esters C-18:1, 2, 3 was achieved by the usage of HP-FFAP capillary column. Argon was used as a carrier gas at a flow rate of 2 ml/min. The column temperature was increased from 100°C to 380°C, at the rate of 15°C/min. This last temperature was maintained for 2 minutes.

Data obtained from chromatography concern only glycerol, traces of different esters and between fractions. It has to be kept in mind that to establish more details about chemical properties of the fraction there are also titration methods necessary in order to determine saponification number and a soap content. As a standard procedure bioglycerol is preliminarily distilled to remove water and alcohol. Then the percentage data obtained from gas chromatography are recalculated according to the soap, water and alcohol contents.

An exemplary data of several bioglycerols (table 1), chosen as representative from about forty investigated samples, were derived from producers using different catalyst (NaOH for

W and K; KOH for the rest of the samples), different alcohol (ethanol for W, methanol for the other samples) and different oil (W- waste, used frying oil; Z, K - rapeseed oil, T - mixture of pure, waste oils and animal fats). The results show that the proportion of constituents can be very versatile. It depends mainly on the technology used by producers and is strongly influenced by the type and diversity of lipids used. Some producers apply pure rapeseed oil as an input material, some pure palm oil or a mixture of used, waste oils and animal fats. What is more, different alcohol (methanol or ethanol) and catalyst can be used. Finally the resulted bioglycerol mixtures might vary in fatty acid (more or less unsaturated), antioxidants and other natural compounds originated in oils, content.

Component [%]		Z			T	
	W	Z1	Z2	K	T1	T2
Glycerol	79	60	96	57	60	70
BF	0.35	0.21	0.69			
FAME C-16:0		0.21	0.2	0.42	0.24	0.11
Acid C-16	0.6			0.28	0.72	
FAME C-18:(1-3)	0.36	4.17	0.63	7.6	4.2	
BF		0.24	0.15			
Acid C-18	0.13			0.2	6.0	1.3
Acid C-18: (1:3)	2.45	1.41				
BF	0.54	0.71				
FAME C-20:0				0.2		
Monoglycerides		3.09		1.8	2.2	1.9
Diglycerides		0.04				
Triglycerides	0	0	0	0	0	0
Soaps	20.9	19.7	2.15	32.4	26.6	26.1
Water & alcohol	28	17.7	22.3	23	16.7	6.8
SN [mgKOH/g]	0.67	25.7	0	31.5	34.5	9

Table 1. Composition of exemplary glycerol fractions (BF- between fraction, SN - saponification number, FAME - fatty acid methyl/ethyl ester).

As it is shown in table 1 even composition of bioglycerols from the same producer (Z1 and Z2; T1 and T2) can be very different. In this case content of the fraction depends more on the specific conditions during the time of the production, than on the materials used. It suggests that the main significance is due to an accuracy of the technology processing.

Alcohol removed from bioglycerol is returned to the transesterification process. Some producers additionally perform purification processes such as acidification and separation of lipid residues such as fatty acids, which can be then recycled to the process. Application of this step leads to the formation of additional two fractions - fatty acids with low glycerol content (3-6%) and acidified, more condensed glycerol fraction.

3. Antioxidants content studies

The analytical methods based on the standard procedures determine only the main bioglycerol constituents content. In order to establish antioxidant presence and oxidation stability additional methods need to be employed. Popular techniques dealing with the problem are

UV-Vis or Electron Paramagnetic Resonance (EPR) spectroscopies (Papadimitrou et al., 2006, Jerzykiewicz et al., 2009, Jerzykiewicz et al., 2010, Jerzykiewicz et al. 2011).

3.1 Folin-Ciocalteu assay

The most common method for the quantitative studies of antioxidants in natural substances is the Folin-Ciocalteu (FC) assay (Roura et. al. 2006, George et al., 2005) concerning UV-Vis measurements of investigated substance with the Folin-Ciocalteu reagent (FCR). It is also called the Gallic Acid Equivalence method (GAE) due to the gallic acid application as the reference compound (Singleton and Rossi, 1965). However, caffeic acid (Koski et al. 2002) is also commonly used in such investigations. The FC analysis is based on the reaction of the reagent (a mixture of phosphomolybdate and phosphotungstate) with reductive (antioxidant) compounds. Measured at 765 nm signal intensity gives information about amount of the substance required to inhibit the oxidation of the FCR. Although it is used as the standard method in the investigation of phenolic antioxidants in natural mixtures of plant origin like oils and fruits, it is not a reliable one in the bioglycerol studies. The main disadvantage (known from 40ties of bygone century) of the method is that not only the phenolic antioxidants react with the FCR (Abul-Fadl, 1949, Ikawa et al., 2003, Everette et al., 2010). The reaction can be influenced (inhibited or accelerated) by several factors (George et al., 2005). In case of bioglycerols the reaction is mostly induced by metal ions (like potassium ions from catalyst), but there also exists a range of different factors (organic and inorganic compounds) which may falsify the results and hence make them incomparable among the bioglycerol samples. Additional problem concerns partial insolubility of the bioglycerol in water solution which extorts change in the Folin-Ciocalteu procedures. Generally for such samples an extraction (Koski et al. 2002) is performed, then extract is tested. However, there might be new problem generated, with the efficiency of the extraction. Another issue is the fact that not every type of antioxidant may be transferred to the solvent used in extraction and some losses of the investigated substances are possible. Too many factors influence the FC method and thus make it inadequate to be recommended for glycerol fractions analysis.

3.2 EPR spectroscopy

The method which allows investigations of the natural mixtures without their previous purification and separation is the Electron Paramagnetic Resonance (EPR) spectroscopy, called also Electron Spin Resonance (ESR). EPR is based on the interaction of unpaired electron of the compound with the magnetic field, which induces splitting of the energy of unpaired electron spin (Zeeman splitting effect). Created two (or more) energy levels enable absorption of microwave frequency radiation. The EPR signal is usually recorded as a first derivative (dA/dB), of the absorption (A). Shape of the absorption signal can be approximated by a Gaussian or a Lorentzian curve, or as the mixture of both. When measurements are performed there are only compounds with unpaired electron detectable (unlike the rest of the constituents). This property is essential in the studies of minute quantities of paramagnetic species and due to this very complex mixtures can be investigated without previous purification. The intensities and width of lines detected using EPR spectroscopy enable quantitative calculations of the paramagnetic species (N_x on figure 1). Application of quantitative standards gives the absolute values of spins per gram. The fundamental EPR parameter, g, exhibits the structural properties of the electron. For free electron the parameter is equal 2.0023 and variations from this value provide important

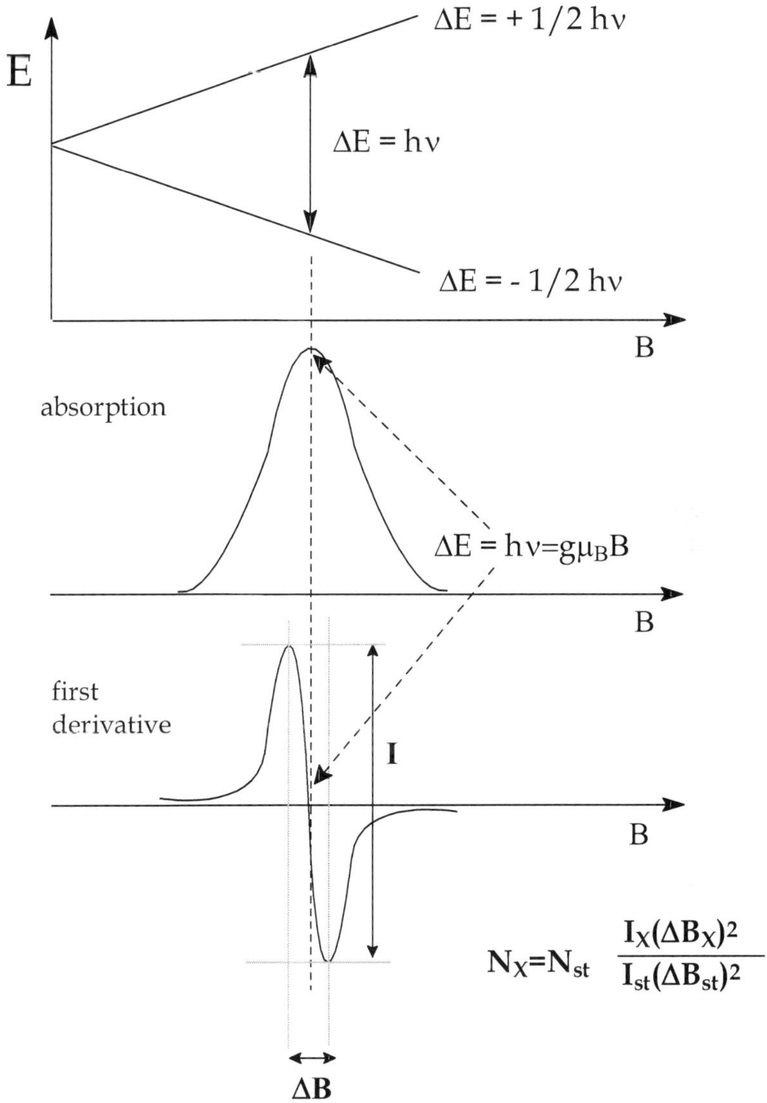

Fig. 1. Zeeman splitting of energy levels of an electron placed in magnetic field (A - absorption, B- the magnetic field, μ_B-Bohr magneton, ν – frequency, h- Planck constant). For quantitative calculations: I_X, ΔB_x– intensity and width of investigated line, I_{st}, ΔB_x–, N_{st}– intensity, width and spin concentration of standard line).

information about the structure of an electron neighborhood. For π-type of organic radicals g-parameter is generally higher than for σ-radicals (Gerson and Huber, 2003, Lund et al., 2011).

The g parameter can be calculated directly from the spectrum. Radical structure can be also described on EPR spectra by the hyperfine splitting occurring when spin of unpaired electron interacts with non-zero spin of nuclei. The number and intensity of the lines split depend on the value of the nucleus spin (1/2, 1, 2/3…) and number of the interacting nuclei.
The EPR spectroscopy has already been used as a helpful tool in the studies of oxidative properties of edible oils and other food products (Papadimitrou et al. 2006, Jung & Min, 1992, Vicente et al. 1995, Andersen & Skibsted, 2001). Investigations concerning oxidative properties of bioglycerols are not very common and widely applied. Authors performed such research for all the transesterification products and an input material. Presented herein results were executed on Bruker spectrometers from Elexsys series.
At the beginning of the oxidative stability research of investigated mixtures (different oils and its products: biodiesel and bioglycerol) were subjected to the direct EPR measurements. Especially interesting results were obtained for bioglycerols. EPR spectra revealed the existence of the free radicals of semiquinone-like structure (figure 2) in these fractions.

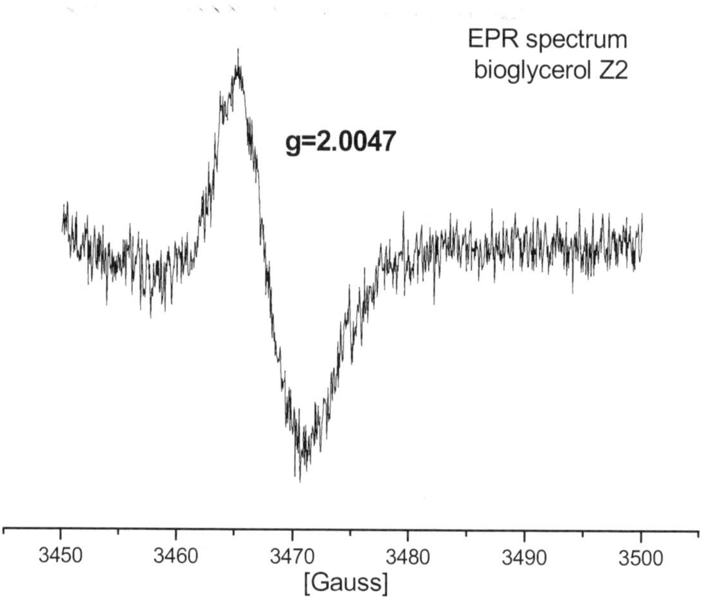

Fig. 2. EPR spectrum of glycerol fraction Z2. Five scans accumulated.

The calculated from the spectra g parameter was in a range from 2.0043 to 2.0047, which is characteristic for an unpaired electron situated on oxygen substituted to the aromatic ring (Gerson & Huber, 2003)– typical structure of poliphenolic antioxidant. This sort of the radical (semiquinone type) exists in the mixtures in quinone - semiquinone - hydroquinone equilibria. They can be easily shifted (i.e. by pH change) towards one of the forms, which in consequence changes radical concentration. They are also sensitive to metal ion binding and depending on metal type the radicals concentration can be increased (s and p shell metals) or decreased (d shell, i.e. Cu(II), Mn(II)) (Jerzykiewicz et al. 2002, Jerzykiewicz, 2004). Formation of radicals in bioglycerols may be the result of high pH during the transesterification process caused by the application of basic catalyst. Stability of detected in bioglycerols radicals is also an important feature apart from the g parameter and concentration. They were found to be very permanent and their EPR spectra remained unchanged even for a few years!

The radicals concentration was too small to be measured quantitatively (spectra had to be accumulated 5-10 times), but they were the direct proof of phenolic systems existence in glycerol fractions, which from now on became the subject of more advanced studies. The radical of phenolic origin was not observed for every bioglycerol. Samples of technical grade purity (or higher) did not exhibit any radical signal at all.

3.2.1 EPR studies of free radical scavenging

EPR spectroscopy allows to detect only paramagnetic species on the spectrum, but the method can be also employed in the investigation of non-paramagnetic compounds. For this purpose free radical scavenging EPR method is used, where standard, stable radical (TEMPO, galvinoxyl, TEMPOL, DPPH) is employed. The radical during reaction with the investigated substance becomes a non-radical species hence as a result the decrease (scavenging) of the radical signal is observed. Progress of the reaction is calculated by the double integration of the EPR spectrum. Properly performed experiment allows to obtain the content of the investigated compound with high accuracy. In antioxidants studies the most popular radical scavenged by these bioactive compounds is galvinoxyl called also Coppinger's radical (figure 3) (Gerson & Huber, 2003, Ramadan et al. 2003).

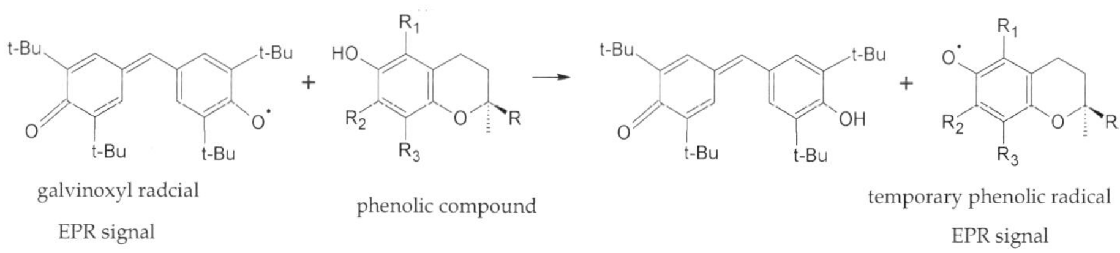

Fig. 3. Reaction of galvinoxyl radical (on the left) with phenolic (tocopherol like) compound.

The EPR signal of galvinoxyl radical (figure 4) is a well-known doublet of quintets: hyperfine splitting originates from four equivalent hydrogens situated on aromatic ring and from the hydrogen of C-H group joining the aromatic rings. The proper distinguishing of hyperfine parameters is not always possible, because the signals in the centre overlap and in some cases lines may not be separated. An attention should be paid when performing experiments in the solvent reacting with galvinoxyl (eg. toluene). Galvinoxyl radical in the described on the figure 3 reaction becomes diamagnetic and from the investigated phenolic system temporary radical is formed. This newly formed radical can also be recorded (figure 5), but mainly quenching of galvinoxyl radical is observed. As the result, in most of the situations the only change observed on the spectra is a decrease of the galvinoxyl radical signal intensity. A decrease of its intensity is the measure of reaction progress thus the antioxidant content in the investigated mixtures.

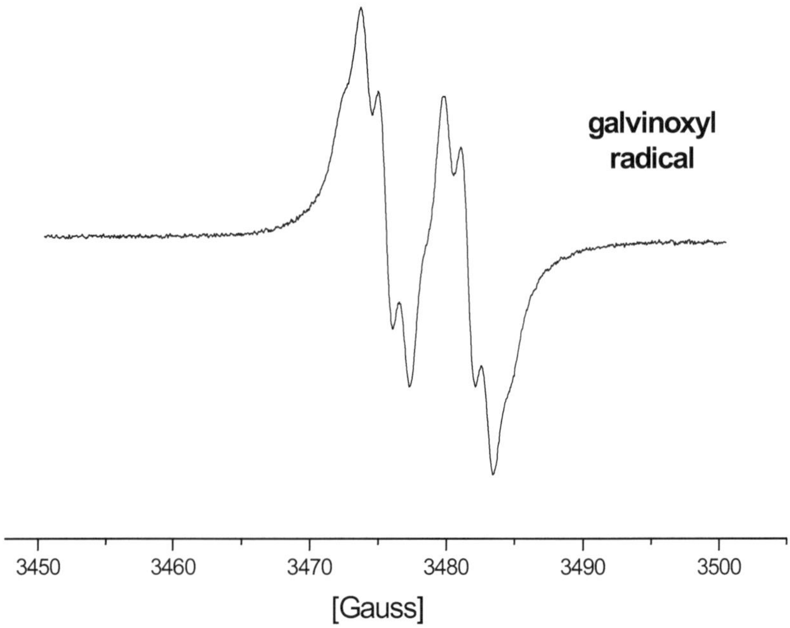

Fig. 4. EPR spectrum of galvinoxyl radical

Procedure of galvinoxyl scavenging assay using EPR spectroscopy

19 mg of sample diluted in the 200 µl of solvent
200 µl of galvinoxyl radical solution (1.1 mmol /l)
Ethyl alcohol as the solvent for bioglycerol, hexane for oil and biodiesel
The reference sample: 200 µl of solvent and 200 µl of galvinoxyl radical solution.

Prepared samples are measured in glass capillaries (0.8 mm i.d.) kept in standard quartz EPR test tubes. The EPR signals are recorded immediately and every few minutes after the beginning of the reaction, until disappearance of the signal. For establishing quantitative data EPR signal is measured after constant period of time from mixing for all the components. Measurements are performed using EPR spectrometers at the standard parameters for free radicals, modulation amplitude 1 G, at 100 kHz magnetic field modulation, X-band frequency counter at room temperature.
The intensity of galvinoxyl radical signal is calculated by the double integration of the spectra performed using common EPR programs (i.e. Bruker WinEPR Processing).

For the quantitative interpretation of the results a calibration curve is prepared. As a reference compound there can be phenolic acid applied such as caffeic or gallic acid. Results are expressed per the chosen acid equivalents (mmol/l or g/l).

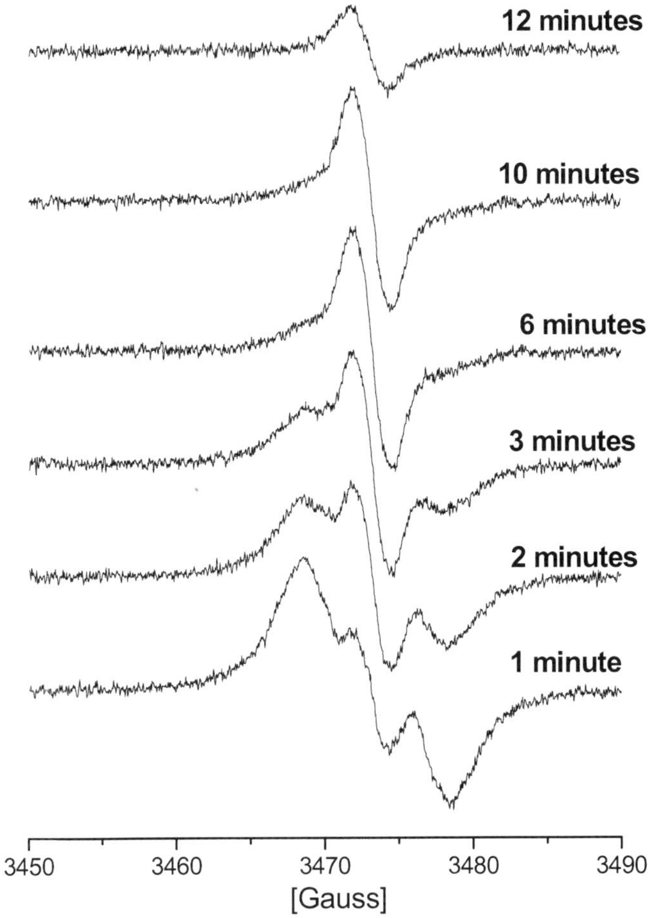

Fig. 5. EPR spectra of galvinoxyl radical reacting with glycerol fraction Z2 (ethanolic solution). Spectra were recorded at first and up to the 12th minute of the reaction.

For the presented bioglycerols the temporary radical was observed only for fractions with the highest phenolic compounds content. The g parameter of formed, unstable radical was 2.0045 – similar to the one obtained from direct studies of bioglycerol Z2, but the radicals recorded are not the same. The concentration of the transient radical is much higher, regarding dilution and recording of only one scan. What is more, scavenging measurements are performed in glass capillaries of much smaller volume than for the direct acquisition of pure bioglycerol sample in standard EPR probe (4 -5 mm diameter). The signal disappears in about 15 -20 minutes.

Intensity of the reaction depends on the concentration of antioxidant in the measured mixture. Figure 6 presents results of free radicals scavenging for four sets of samples: oils and their transesterification products (biodiesel and glycerol fractions). As the figure clearly indicates galvinoxyl radical was scavenged by glycerol fractions much more efficiently than by oils and biodiesel samples (figure 6). That proves unequivocally that during transesterification all antioxidants occurring naturally in oils are transferred to glycerol fraction. The antioxidants in bioglycerol are condensed due to the smaller volume in comparison with the initial oil volume.

The efficiency of free radical quenching was also distinct for samples originated in various biodiesel producers. As for presented sets, each consisting of: oil, bioglycerol and biodiesel it is easy to distinguish the difference (between bioglycerol from set 1 and 2 or 3). Despite the fact that the samples are from the same producer, they do not exhibit the same quenching abilities towards galvinoxyl. More detailed studies were performed for investigated mixtures from many different producers (figure 7). Concerning the ability for the radical scavenging bioglycerols can be divided into 3 groups. In the first group there are the most efficient galvinoxyl quenchers. For this group of samples phenolic type of radical was detected during the direct measurement and a temporary radical (figure 5) found during the experiment with galvinoxyl. In this group concentration of phenolic antioxidants was also the highest. Medium content of phenols, referring to the lower antioxidant concentration was observed for bioglycerols from most of the producers and finally, there was no, or very low scavenging activity observed in the third group for analytical grade or purified in laboratory glycerol fractions.

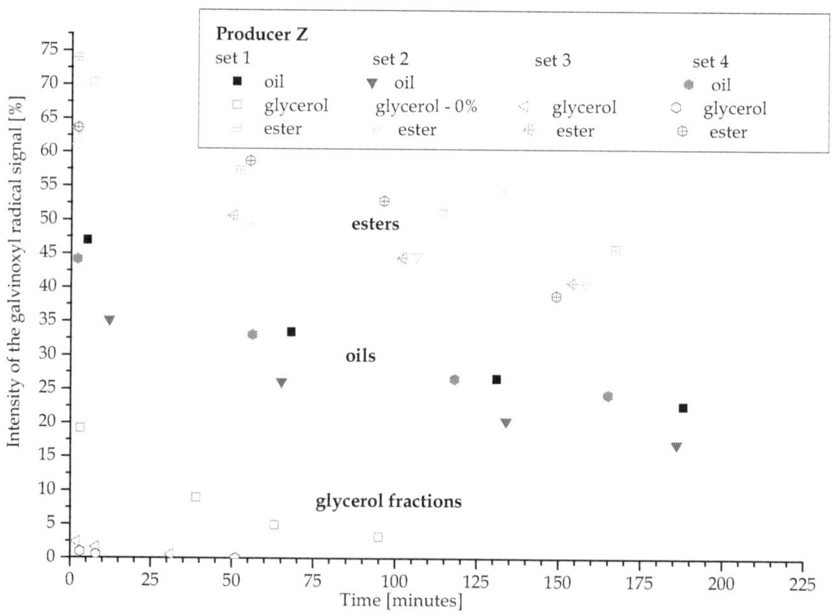

Fig. 6. Dependence of galvinoxyl radical signal intensity for oils and its products *verus* time.

Definitely it is quite easy to distinct differences which decide about scavenging properties between the first two groups and the third one. Fractions which are purified are devoided from antioxidants. Differentiation between group of high and medium scavenging properties is more complicated. Bioglycerols originated in waste oils or animal fats exhibited medium scavenging properties (second group of galvinoxyl scavengers). Explanation of this fact may be connected with the already small content of antioxidant in the input material. However, there was no correlation found between radical scavenging and soap content or saponification number estimated for bioglycerols. Glycerol fractions obtained from rapeseed oils were more diversified accordingly to the type of the process used by producer and technology rather than chemical constituency based on chromatography and other standard analyses.
Experiments with galvinoxyl not only allow to compare the radical scavenging activity between samples, but also to calculate the concentration of the antioxidants. For this purpose the calibration curve has to be prepared from the standard, model antioxidant. The most common reference compounds used are gallic and caffeic acids. There is no need to record the EPR spectra every few minutes to follow the exponential decay of galvinoxyl signal, but all the samples should be measured once at exactly the same time (in our case 5 minutes) from mixing them with galvinoxyl. Results of the quantitative EPR measurements are discussed in chapter 3.4.

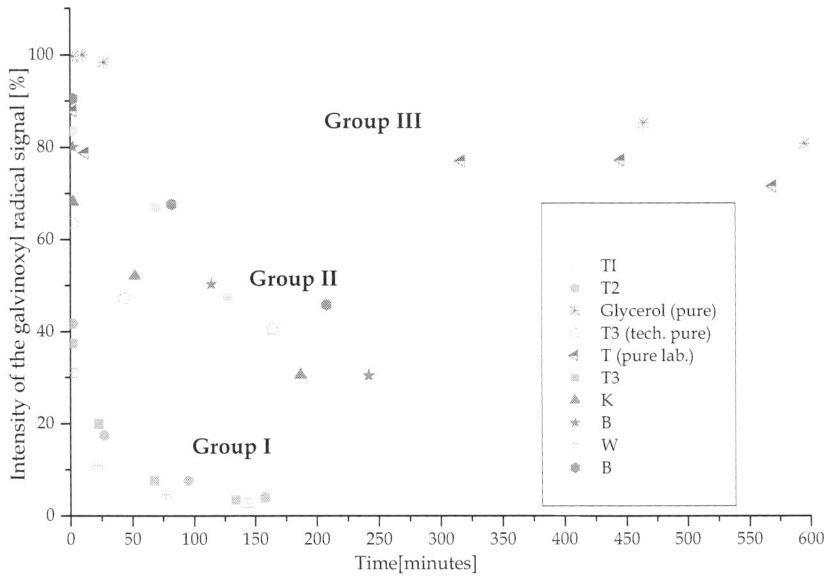

Fig. 7. Dependence of galvinoxyl radical signal intensity for the glycerol fractions from different producers *versus* time.

3.3 UV-Vis studies of free radical scavenging

Reaction of galvinoxyl with phenolic compounds can be also investigated by the use of UV-Vis spectroscopy. Preparation of the graph describing the reaction is the same as fo EPR method, but single instead of double integration of the signal is applied.

Procedure of galvinoxyl scavenging assay using UV-Vis spectroscopy
Solution A: 95 mg of bioglycerol sample dissolved in ethanol to the volume of 5 ml
Solution B: 10.6 µM solution of galvinoxyl in ethanol

Calibration curve: set of caffeic acid ethanolic solutions in concentration range from 7 to 350 µM. The standard solutions are used instead of A solution during measurements.

2 ml of solution B is placed in a cuvette and then 100 µl of solution A is added, stirred and measured after 1 minute.
An absorbance is measured at 429 nm at room temperature.

Figure 8 presents absorption of galvinoxyl signal at 429 nm (black line). Decrease of maximum absorption of galvinoxyl in time caused by the interaction of the radical with phenolic system is clearly presented after one and five minutes from the beginning of the reaction. Absorption value obtained from the spectrum is then recalculated to appropriate phenolic acid equivalent. As it is seen on the spectra (figure 8) new chemical species with absorption at 578.5 nm is created during the reaction. However, increase of the signal intensity is not linear and is not a good indicator of the reaction progress or antioxidant concentration. The main disadvantage of this method is higher volume of the solutions necessary for the measurements (1-2 ml).

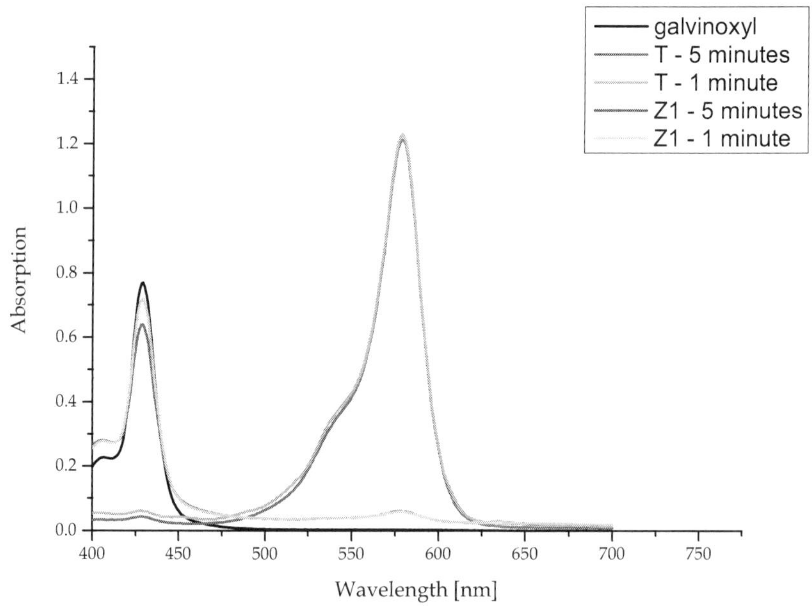

Fig. 8. UV-Vis spectra illustrating pure galvinoxyl ethanolic solution absorption (black line) and results of the reaction between radical and bioglycerol T and Z1 after 1 and 5 minutes (colored lines).

Similarly to the EPR technique also quantitative assay can be performed with caffeic or gallic acid solutions of different concentrations (calibration curves method). For this measurements there is no requirement of recording the total slope of the scavenging reaction *versus* time, but one point (as in EPR), which is obtained by measuring the absorption of galvinoxyl solution in ethanol at 429 nm. The most important is to measure the spectra at exactly the same time, for solutions of model antioxidant and for the investigated mixture (for example 1 or 5 minutes from mixing).

3.4 Comparison of different methods assaying antioxidant content

As it was mentioned above, reaction of galvinoxyl radical with phenolics was monitored using two different spectroscopies: EPR and UV-Vis. Quantitative analysis of phenolic compounds was conducted on the basis of calibration curves of phenolic acids. Results of both assays expressed in caffeic acid equivalents are given in table 2.

Bioglycerol	Phenolic antioxidant content [g/l]	
	UV-Vis	EPR
O	0.025	0.023
W	0.0060	0.0045
B	0.0091	0.012
A	0.010	0.0088

Table 2. Phenolic antioxidant content calculated for caffeic acid as the standard for bioglycerol from different producers (W- waste oil and NaOH as catalyst, rape seed oils: O – as catalyst NaOH, B – laboratory scale, A – commercial scale).

Results obtained from EPR and UV-Vis spectroscopies are comparable. However, it has to be kept in mind that both methods assign species based on their different properties. EPR detects only paramagnetic species, that is why EPR spectroscopy is more selective. Another advantage of EPR is that it uses smaller quantities of the investigated samples. However, EPR instrumentation is incomparably more expensive than in case of standard UV-Vis spectrometers. On the other hand EPR spectrometers have became more popular in industrial laboratories due to small, easy to use apparatus evaluable on the market. Till now the computer-size spectrometers are dedicated for analysis of food and beer quality and there is no obstacle to use them in oil products standard analysis.

Apart from spectroscopic methods also classical, analytical procedures were helpful in the studies of antioxidant properties of bioglycerols. One of these methods is a standard Herbert test, which also proved anticorrosive properties of bioglycerols. Before measurements all fractions were saponified (according to the Patent by Jerzykiewicz et al., PL378802) and solutions of different initial and final soap content were used for analysis. The standard Herbert tests were performed using special steel plates with the surface of 25 and 43 cm^2. The plates were initially weighted and then were kept at room temperature for the 120 h in the saponified solutions. As it was expected the corrosivity of all saponified bioglycerols decreased with the increase of the soap content, which was crucial for the results. The sample of low initial soap content, even having the same final soap content to another bioglycerol samples of different constitution, exhibited much better anticorrosive properties in Herbert test. The same bioglycerols were found to be the best free radicals scavengers in galvinoxyl experiments.

4. The antioxidants structure

Depending on the origin of the bioglycerol antioxidants present in the mixtures can have different structure (figure 9). Composition of tocopherols, tocotrienols and carotenoids, the substances responsible for antioxidant activity in oils, vary depending on the source of an input material. Naturally occurring antioxidants in oils belong mainly to the group of tocopherols. α-Tocopherol is considered to be predominant antioxidant in olive and sunflower oils (Velasco et al, 2005, Blekas et al., 1995, Koski et al. 2002) while γ-tocopherol prevails in rapeseed oil (Velasco et al., 2005, Koski et al. 2002). The content of antioxidants depends also on the type of oil purification procedures (cold-pressed, refined). Refining removes about 40% of tocopherols and 98% of carotenoids (Koski et al., 2002).

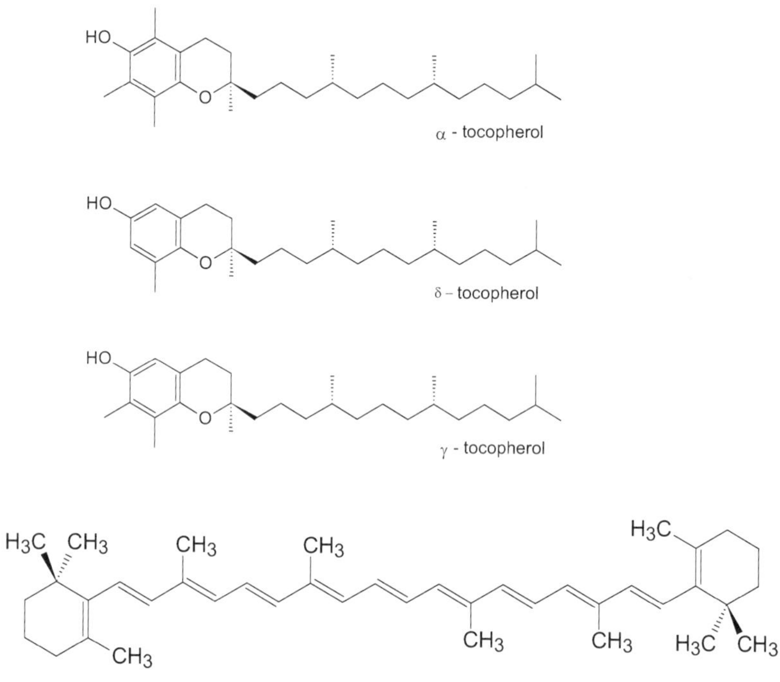

Fig. 9. Exemplary structures of antioxidants present in virgin oils.

Establishing the structure and type of antioxidants found in bioglycerols is an additional problem in the studies. What is more, the antioxidant structure might change during the transesterification process. To solve this puzzle EPR spectroscopy spin trapping technique was used accompanied by DFT calculations (Jerzykiewicz et al., 2010). The structure of active antioxidants was investigated by the studies of temporary radicals formed upon

oxidation reactions. During oxidative stress in such a complex mixture as glycerol fraction different, transient, unstable radicals can be formed. These radicals have their origin in the oils or fats constituents such as tocopherols, lipids and glycerol. All these species have different structure and can be easily distinguished by their EPR parameters. However, they are too reactive and recombine too quickly to be measured directly by EPR spectroscopy. Investigation of such unstable radicals is possible when spin trapping technique is applied.
Spin trapping technique is based on the reaction of the stable diamagnetic compound (a spin trap, non-radical species) with the short- living radical (figure 10). On the figure 11 there are the structures of different spin traps given, such as: PBN (N-t-butyl-α-phenylnitrone) and another commonly used nitrones: DMPO (5,5-Dimethyl-1-pyrroline N-oxide) (Dikalov & Mason, 2001) or POBN (α-(4-Pyridyl N-oxide)-N-tert-butylnitrone). The choice of a proper spin trap for the experiment depends mainly on its solubility and reactivity (McCormick et al. 1995). In case of research of substances originated in oils such as glycerol fractions the best spin trap applied is PBN due to its high solubility in lipids.

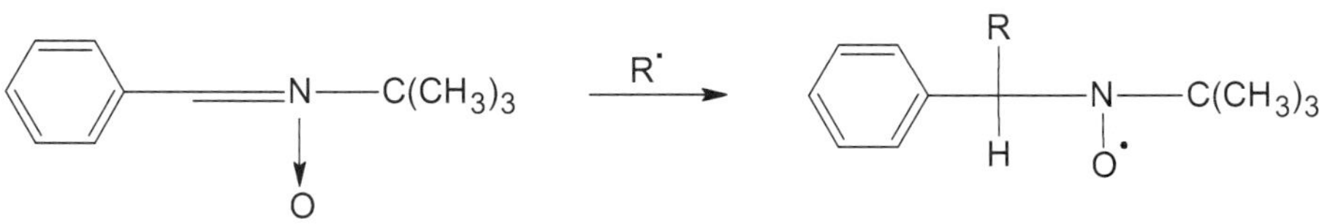

Fig. 10. The PBN (N-t-butyl-α-phenylnitrone) spin trap reaction with unstable radical (•R), as the result stable PBN/•R spin adduct is formed.

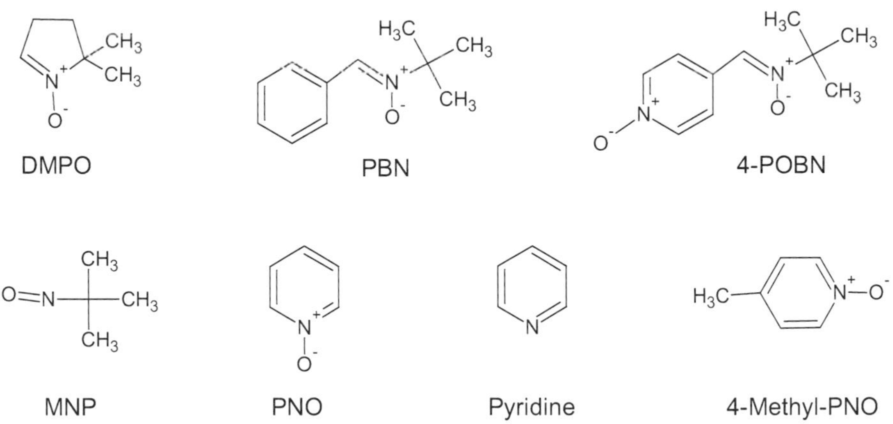

Fig. 11. Comparison of different spin traps formulas.

A spectrum of PBN adduct consists of triplets of doublets - three lines reflecting hyperfine interaction of unpaired electron with nitrogen nucleus and each of these lines split into two - interaction with hydrogen nucleus. The parameters of the splitting constants, A_N and A_H (for the nitrogen and hydrogen nuclei, respectively) depend on the type of radical trapped and are used as a source of structural information about it. Computer simulation of EPR

spectra allows to identify the nature of the radicals as oxygen-, nitrogen- or carbon-centered. The parameters are also helpful in differentiation between the size of the group attached to the spin trap (big or small adducts) (Janzen and Blackburn, 1968, Jerzykiewicz et al., 2011, Jerzykiewicz et al., 2010). What is more, spectra of adducts allow to perform quantitative studies of formed radicals.
In the described experiment apart from bioglycerol samples a few reference compounds such as: α- and δ- tocopherols, oils, different triglycerides, fatty acids and pure glycerol were used in analogous experiments.
The EPR spectra of bioglycerols exhibited typical splitting pattern for the PBN adducts (figure 12), but the hyperfine parameters changed in time of the experiment and were dependent on the glycerol fraction composition and antioxidant content. In the beginning of the oxidation all bioglycerols exhibited spectra with the same hyperfine parameters as recorded for the adduct of PBN with radical from α-tocopherol (table 3, figure 12). They were distinctively different from parameters obtained for PBN/•δ - tocopherol adduct. This fact indicates that antioxidants in bioglycerols have a structure of α - tocopherol. During the experiment, values of hyperfine coupling constants underwent change in comparison with the initial ones. Week or two from the beginning of the experiment, the spectra recorded for bioglycerols, where previously described free radicals scavenging investigation proved the lowest antioxidant content, exhibited spectra of different shape (broadening of the signal) and parameters values. On the contrary spectra of bioglycerols with the highest antioxidant content remained unchanged until the total disappearance of the signal (even 3 months). The change of hypefine parameters, found for the samples of smaller antioxidant content, especially reduction of A_H parameter (~2.0 G) indicate trapping of a radical group of a bigger size. Comparison of the spectra and parameters calculated from them with the values obtained for reference

Procedure of spin trapping assay

Solution A: PBN 0.067 M solution (in ethanol, acetone, DMSO or ethyl acetate)
Soultion B: 3 % H_2O_2 (in respective to PBN solution solvent)

250 mg of bioglycerol dissolved in 0.5 ml of solution A. To the homogenous solution 0.125 ml of solution B is added and stirred.

Prepared samples are measured in glass capillaries (0.8 mm i.d.) kept in standard quartz EPR test tubes. The EPR signals are recorded 10 minutes and 1 hour after mixing. Later mixture is sampled and measured every day and afterwards every few days till the complete disappearance of the signal. In order to establish the hyperfine parameters of the spectra simulation should be performed using appropriate programs (i.e. Bruker WinEPR Simfonia).
Solvents different than ethanol may be used, however it has to be kept in mind that DMSO and ethyl acetate form methyl and methoxyl radicals in described conditions. As a result adducts of these radicals could be formed with PBN instead of radicals from α-tocopherol or lipids.
Measurements are performed using EPR spectrometer at the standard parameters for free radicals at 100 kHz magnetic field modulation, X-band frequency counter at room temperature at five replications due to the low intensity of the signals.

compounds solutions showed that similar hyperfine parameters were observed for oxidized oil samples and triglycerides or fatty acids. Despite the high content of glycerol in every glycerol fraction, PBN adduct with glycerol was not observed on any of the spectra. Parameters for PBN adduct with pure glycerol are given in table 3.
Oxidation of bioglycerols was then dependent on the constituency of the fraction. Samples, where α - tocopherol content was the highest were protected by its antioxidant properties, whereas for the bioglycerols with medium and low α - tocopherol concentration oxidation was inhibited until the exhaustion of the antioxidant. When α - tocopherol was consumed oil residues of the fractions were oxidized and typical PBN/•lipid adducts were found on the EPR spectra. Similar experiments performed in DMSO as a solvent exhibited different type of adducts. Instead of adducts originated in antioxidants or lipids, adducts derived from the solvent were observed. Methyl radicals generated in DMSO and oxidized to the methoxyl ones were then trapped by PBN. It is essential to underline, that samples with high α - tocopherol content inhibited oxidation of methyl radicals, thus PBN adducts with methoxyl radicals were not observed on the EPR spectra.

	Antioxidant concentration	First day		1 week		3 weeks	
		$a_{iso}(^{14}N)$	$a_{iso}(^{1}H)$	$a_{iso}(^{14}N)$	$a_{iso}(^{1}H)$	$a_{iso}(^{14}N)$	$a_{iso}(^{1}H)$
T1	5.75	15.3	3.8	15.3	3.8	15.3	3.8
Z1	5.88	15.3	3.8	15.3	3.8	15.4	3.7
Z2	2.17	15.3	3.8	15.3	3.5	15.3	2.4
T2	0.20	15.3	3.8	15.3	3.2	15.5	1.8
L2	0.044	15.5	3.2	15.4	2.1	15.3	1.9
α-tocopherol	n/a	-	-	15.3	3.9	15.3	3.8
δ-tocopherol	n/a	-	-	15.5	3.3	15.4	2.2
O2	n/a	-	-	14.5	2.0	15.1	1.9
1,2,3-propanotriol	n/a	-	-	14.6	2.6	14.6	2.7

Table 3. Antioxidant concentration (mmol/l) calculated per caffeic acid equivalents and isotropic hyperfine coupling constants (in Gauss) of PBN adducts (ethanolic solutions) in different period of time after reaction started. (from Jerzykiewcz et al. 2010)

Additionally, for better understanding of the reactions occurring upon oxidative stress several standard mixtures, mimicking the bioglycerol systems were prepared in different solvents. The mixtures contained of pure glycerol, triglyceride (glyceryl trioleate or glyceryl trilinolenate), α - tocopherol; pure glycerol, fatty acid (linolenic acid or oleic acid), α-tocopherol, pure glycerol and α - tocopherol. Similar mixtures were prepared also without addition of α - tocopherol. The best bioglycerol mimicking properties were found for the samples consisted of: glycerol, triglyceride and α - tocopherol. When triglycerides were exchanged for fatty acids the oxidation processes observed for the mixture via PBN adducts exhibited much different pattern than for bioglycerols. Similarly samples of glycerol and α - tocopherol only, exhibited different EPR spectra than bioglycerol.

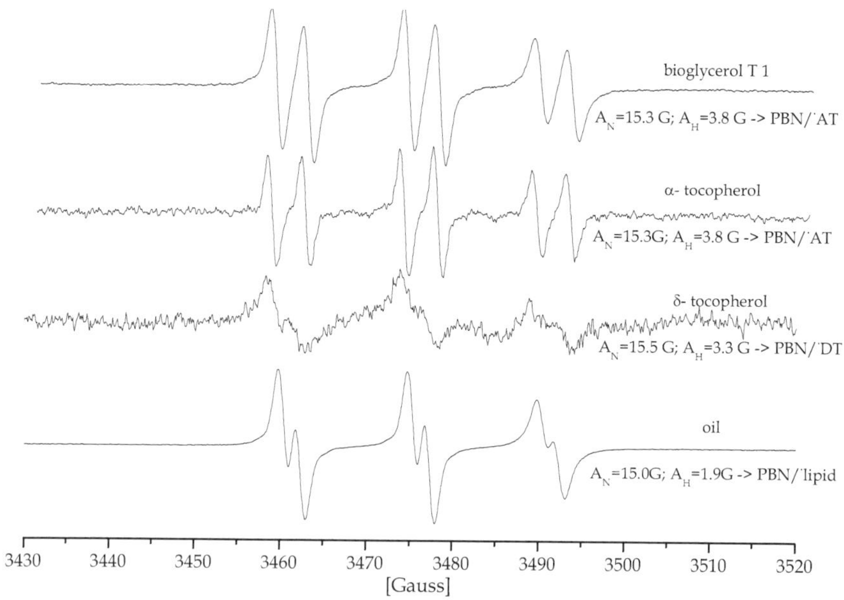

Fig. 12. EPR spectra of the PBN spin adducts formed upon oxidation of bioglycerol, α-, δ-tocopherol and oil.

Surprisingly, the DFT calculations of several possible PBN adducts with radicals of α - tocopherol origin indicated carbon-, not oxygen- centered type of the radicals trapped. This fact suggests a more complex processes associated with antioxidant activity perhaps through non-phenolic groups. This effect was stimulated in our case by the use of the radicals from H_2O_2 against α - tocopherol. Therefore, α-tocopherol exposed to reactive oxygen species, even after exhaustion of its phenolic antioxidant capability, underwent formation and trapping of different types of carbon - centered radicals. Of course it has to be kept in mind, that oxidation initiated by different reactions (i.e. photolysis) or in different solvents may result in creation of another kind of radicals (Rosenau et al., 2007).

When discussing the EPR parameters of the trapped radicals it is very important to take special consideration on the solvent used for measurements. Especially, when like in ethanol solution hydrogen bonds could interfere with the investigated radical. Thus, in the DFT calculations "solvent effect" was incorporated using Tomasi's polarized continuum method (PCM) and additionally the explicit solvent molecules were included.

The solvent effect was easily found on the spectra, according to the previous Works of Janzen group (Janzen et al., 1982) on different types of radicals trapped by PBN. Hyperfine parameters such as A_N and A_H depend on the polarization of the solvent ($E_T(30)$). The difference between parameters obtained for different solvents is small but noticeable. The bigger the polarization, the higher the EPR hyperfine parameters are. Respective values to A_N = 15.4 G and A_H = 3.8 G for ethyl alcohol ($E_T(30)$=51.9) are smaller for acetone A_N = 15.1 G and A_H = 3.5 G ($E_T(30)$=42.2). This is very important fact when considering experiments of the trapped radicals in different solvents.

5. Conclusions

Raw glycerol fraction (bioglycerol) is a product of transesterification process. This mixture (although treated by producers as unwanted waste) is a valuable source of chemicals. Apart from well-known main constituents like: glycerol, fatty acids, triglycerides or residues of alcohol it also consists of some minor compounds. With the use of EPR spectroscopy we proved that some of these minor components originated in oils are transferred during the biodiesel production process to bioglycerol. Reactions of these phenolic compounds consisted in glycerol fraction with galvinoxyl radical showed free radical scavenging properties of bioglycerol. This method developed with the usage of EPR or UV-Vis spectroscopy was much more helpful in the investigations of antioxidants concentration than popular Folin-Ciocalteu assay. Concentration of the antioxidants was different for bioglycerol samples obtained from various biodiesel producers and is strongly affected by the technology and an input material applied.

Structural investigations based on EPR spin trapping technique proved the existence of α - tocopherol type of antioxidants in bioglycerols. Upon oxidative stress unstable radicals react with PBN spin trap and create stable radicals adducts. The parameters of the adducts compared with several standard substances proved that the carbon - centered α- tocopherol radical was trapped. The structure of the radical was also confirmed by computational analysis (DFT method). Although the concentrations of α-tocopherol type phenolic compounds in bioglycerol is higher comparing with oils (where they originate from) it still exhibits antioxidant properties. It is important, according to a well-known fact (Blekas et al., 1995, Jung & Min, 1992) that too high concentration of antioxidants causes opposite effect, thus α-tocopherol can therefore behave as prooxidant accelerating oxidation reactions. For all the investigated samples this effect was not observed.

Results presented in the chapter show usefulness of EPR spectroscopy in the studies of antioxidant properties of bioglycerols both qualitatively and quantitatively and were confirmed by the other spectroscopic and non-spectroscopic methods.

The high content of antioxidants in bioglycerol (comparing with its source - oil) and the positive anticorrosive Herbert tests indicate new possibilities of usage for the fraction without very time and energy consuming purification procedures. Although bioglycerol is still a complex mixture of different components it is the potential substrate for the production of anticorrosive surfactants for technical demands. Products like shampoos obtained on the basis of the bioglycerol for car or track washing purposes will consist of glycerol, soaps and antioxidants. Additional usage offered is connected with lubricant industry. The anticorrosive lubricant may be a very valuable product, especially in machine tool and automotive industry.

6. References

Abul-Fadl, M. A.M. (1949). Colorimetric Determination of Potassium by Folin-Ciocalteu Phenol Reagent, *Biochemical Journal* , 44, 282-285, ISSN: 0264-6021

Andersen, M. L.& Skibsted L. H. (2001). Modification of the Levels of Polyphenols in Wort and Beer by Addition of Hexamethylenetetramine or Sulfite during Mashing *Journal of Agricultural and Food Chemistry, 49,* 5232-5237, ISSN: 0021-8561

Behr, A.; Eilting, J.; Irawadi, K.; Leschinski, J.& Lindner, F. (2008). Improved utilisation of renewable resources: New important derivatives of glycerol. *Green Chemistry,*Vol. 10, pp.13–30, ISSN 1463-9262

Blekas, G., Tsimidou, M. & Boskou, D. (1995). Contribution of α-tocopherol to olive oil stability. *Food Chemistry*, 52, 3, 289-294, ISSN: 0308-8146

Corma, A.; Iborra, S. & Velty, A. (2007). Chemical Routes for the Transformation of Biomass into Chemicals. *Chemical Reviews,* Vol. 107, pp. 2411-2502, ISSN: 0009-2665

McCormick M. L, Buettner G. R. & Britigan B. E. (1995). The Spin Trap a-(4-Pyridyl-1-oxide)-*N-tert*-butylnitrone Stimulates Peroxidase-mediated Oxidation of Deferoxamine IMPLICATIONS FOR PHARMACOLOGICAL USE OF SPIN-TRAPPING AGENTS, *The Journal of Biological Chemistry,* Vol. 270, No. 49, pp. 29265–29269, ISSN 0021-9258

Dikalov, S. I, Mason R.P. (2001). Spin trapping of polyunsaturated fatty acid-derived peroxyl radicals: reassignment to alkoxyl radical adducts. *Free Radical Biology and Medicine,* 30, 2, 187–197, ISSN: 0891-5849

Everette, J.D.; Bryant, Q.M; Green, A. M.; Abbey,Y. A.; Wangila, G. W. and Walker R. B. (2010). Thorough Study of Reactivity of Various Compound Classes toward the Folin-Ciocalteu Reagent, *Journal of Agricultural and Food Chemistry,* 58, 8139–8144, ISSN: 0021-8561

George, S.; Brat, P.; Alter, P. and Amiot, M. J. (2005). Rapid Determination of Polyphenols and Vitamin C in Plant-Derived Products, *Journal of Agricultural and Food Chemistry.*, 53, 5, pp. 1371-1373, ISSN: 0021-8561

Gerson, F. and Huber, W. (2003). *Electron Spin Resonance Spectroscopy of organic Radicals.,* WILEY-WCH, ISBN 3-527-30275-1

Gu, Y.; Azzouzi, A.; Pouilloux, Y.; Jerome, F. & Barrault, J. (2008). Heterogeneously catalyzed etherification of glycerol: new pathways for transformation of glycerol to more valuable chemicals. *Green Chemistry,* Vol.10, pp. 164-167, ISSN 1463-9262

Hirai, T.; Ikenaga, N.O.; Miyake, T. and Suzuki T. (2005). Production of Hydrogen by Steam Reforming of Glycerin on Ruthenium Catalyst, *Energy Fuels,* 19 (4), pp 1761–1762, ISSN: 0887-0624

Huber, G. W.; Shabaker, J. W. and Dumesic, J. A. (2003). Raney Ni-Sn Catalyst for H_2 Production from Biomass-Derived Hydrocarbons *Science* 300, 2075-2077, ISSN 0036-8075

Ikawa, M.; Schaper, T. D.; Dollard, C.A. and Sasner J. J. (2003). Utilization of Folin-Ciocalteu Phenol Reagent for the Detection of Certain Nitrogen Compounds, *Journal of Agricultural and Food Chemistry, 51,* 1811-1815, ISSN: 0021-8561

Ito, T; Nakashimada, Y.; Senba, K.; Matsui, T. and Nishio, N. (2005). Hydrogen and ethanol production from glycerol-containing wastes discharged after biodiesel manufacturing process, *Journal of Bioscience and Bioengineering,* 100, 3, pp. 260-265, ISSN: 1389-1723

Janzen E.G. and Blackburn B. J. (1968). Detection and identification of short-lived free radicals by an Electron Spin Resonance trapping technique., *Journal of the American Chemical Society.* 90, 5909, ISSN: 0002-7863

Janzen E. G.; Coulter, Oehler, U. and Bergsma, J. (1982). *Solvent Effects on the Nitrogen and β-Hydrogen Hyperfine Splitting Constants of Aminoxyl Radicals Obtained in Spin Trapping Experiments,* Canadian Journal of Chemistry 60, 2725 -2733, ISSN: 1480-3291

Jerzykiewicz, M. (2004). *Formation of New Radicals In Humic Acids upon Interaction Pb(II) Ions,* Geoderma 122, 305–309, ISSN: 0016-7061

Jerzykiewicz, M.; Jezierski, A.; Czechowski, F. and Drozd, J. (2002). Influence of metal ions binding on free radical concentration in humic acids. A quantitative electron paramagnetic resonance study., *Organic Geochemistry,* 33, 265 –268; ISSN: 0146-6380

Jerzykiewicz, M.; Cwielag, I. & Jerzykiewicz, W. (2009). The antioxidant and anticorrosive properties of crude glycerol fraction from biodiesel production. *Journal of Chemical Technology and Biotechnology,* Vol. 84, pp. 1196–1201, ISSN 1097-4660

Jerzykiewicz, M.; Ćwieląg-Piasecka, I.; Witwicki, M. & Jezierski, A. (2010). EPR spin trapping and DFT studies on structure of active antioxidants in biogycerol. *Chemical Physics Letters,* Vol. 497, pp. 135–141, ISSN: 0009-2614

Jerzykiewicz, M.; Ćwielag-Piasecka, I.; Witwicki, M. & Jezierski, A. (2011). α-Tocopherol impact on oxy-radical induced free radical decomposition of DMSO: spin trapping EPR and theoretical studies, *Chemical Physics,* 383, 27–34, ISSN: 0301-0104

Jerzykiewicz, W.; Naraniecki, B.; Lukosek, M.; Rolnik, K.; Kosno, J.; Fiszer, R.; Majchrzak, M. (2006). Method of glycerine fraction management. PL378802 - patent

Jerzykiewicz, W.; Naraniecki, B.; Lukosek, M.; Kosno, J.; Jerzykiewicz, M. (2006). Biodegradable anti-freeze agent. PL207449 - patent

Jerzykiewicz, W.; Naraniecki, B.; Lukosek, M.; Kosno, J.; Jerzykiewicz, M.; Fiszer, R. (2006). Biodegradable antifreeze agent method of production. PL 381066 -patent

Jerzykiewicz, W.; Naraniecki, B.; Lukosek, M.; Charciarek, A.; Zdunek, A. (2008). Washing agent. PL386312 - patent

Jung, Y. M. and Min, D. B. (1992). Effects of oxidizied α-, γ- and δ- tocopherols on the oxidative stability of purified soybean oil. *Food Chemistry.,* 45, 3, 183-187. ISSN: 0308-8146

Koski, A.; Psomiadou, E.; Tsimidou, M.; Hopia, A.; Kefalas, P.; Wahala, K. and Heinonen, M. (2002). Oxidative stability and minor constituents of virgin olive oil and cold-pressed rapeseed oil. *European Food Research and Technology,* 214(4), 294-298, ISSN 1438-2377

Lund, A.; Shiotani, M. and Shimada, S. (2011). Principles and Applications of ESR Spectroscopy, Springer, ISBN 978-1-40205343-6

Lukosek, M.; Jerzykiewicz, W.; Naraniecki B.; Charciarek A. (2007). Degreasing agent. PL384164 – patent

Lukosek, M.; Jerzykiewicz, W.; Naraniecki, B.; Tomik, Z.; Waćkowski, J.; Zdunek, A.; Fiszer, R. (2007). Washing and preservation agent. PL383841 – patent

Pagliaro, M.; Ciriminna, R.; Kimura, H.; Rossi, M. & Pina, C. D. (2007). *From Glycerol to Value-Added Products.* Angewandte Chemie International Edition, Vol. 46, pp. 4434–4440

Pagliaro, M. and Rossi, M. (2010). *The Future of glycerol: 2nd edition* RSC Green Chemistry No.8, Cambridge, UK, ISBN 978-1-84973-046-4

Papadimitriou, V.; Sotiroudis, T.G.; Xenakis, A.; Sofikiti, N.; Stavyiannoudaki, V. and Chaniotakis, N.A. (2006). Oxidative stability and radical scavenging activity of extra virgin olive oils: an electron paramagnetic resonance spectroscopy study. *Analitica Chimica Acta* 573-574:453-8, ISSN: 0003-2670

Ramadan, M. F.; Lothar, W. Kroh; Moürsel, J. T. (2003). Radical Scavenging Activity of Black Cumin (*Nigella sativa* L.), Coriander (*Coriandrum sativum* L.), and Niger

(*Guizotia abyssinica* Cass.) Crude Seed Oils and Oil Fractions. *Journal of Agricultural and Food Chemistry, 51,* 6961-6969 ISSN: 0021-8561

Rosenau, T.; Kloser, E.; Gille, L.; Mazzini, F. and Netscher, T. (2007). Vitamin E Chemistry. Studies into Initial Oxidation Intermediates of α-Tocopherol: Disproving the Involvement of 5a-C-Centered "Chromanol Methide" Radicals, *The Journal of Organic Chemistry,* 72 (9), pp 3268–3281, ISSN: 0022-3263

Roura, E.; Andre´s-Lacueva, C.; Estruch, R. and Lamuela-Ravent, R. M. (2006). Total Polyphenol Intake Estimated by a Modified Folin–Ciocalteu Assay of Urine, *Clinical Chemistry* 52, No. 4, 749, ISSN: 1018-5593

Singleton V. L. and. Rossi Jr J. A. (1965). Colorimetry of Total Phenolics with Phosphomolybdic-Phosphotungstic Acid Reagents *American Journal of Enology and Viticulture.* 16:3:144-158, ISSN: 002-9254

Trybula, S.; Terelak K.; Olejarz A.; Nowakowski S.; Zawadzka M. (2010). *Method of manufacturing of the biofuel compositions* Pl386584 - patent

Velasco, J.; Andersen, M.L. and Skibsted, L.H. (2005). Electron Spin Resonance spin trapping for analysis of lipid oxidation in oils: inhibiting effect of the spin trap α-phenyl-n-tert-butylnitrone on lipid oxidation., *Journal of Agricultural and Food Chemistry* 53, 1328-1336, ISSN: 0021-8561

Vicente, L.; Deighton, N.; Glidewell, S. M.; Empis, J. A. and Goodman, B. A. (1995). In situ measurement of free radical formation during the thermal decomposition of grape seed oil using "spin trapping" and electron paramagnetic resonance spectroscopy, *Zeitschrift für Lebensmitteluntersuchung und -Forschung A,* 200:44-46, ISSN: 1431-4630

Zheng, Y.; Chen, X. & Shen, Y. (2008). Commodity Chemicals Derived from Glycerol, an Important Biorefinery Feedstock. *Chemical Reviews,* Vol.108, pp. 5253–5277, ISSN: 0009-2665

Zhou, C.; Beltramini, J. N.; Fana, Y. & Lu, G. Q. (2008). Chemoselective catalytic conversion of glycerol as a biorenewable source to valuable commodity chemicals. *Chemical Society Reviews,* Vol.37, pp. 527–549, ISSN 0306-0012

18

Use of Glycerol in Biotechnological Applications

Volker F. Wendisch, Steffen N. Lindner and Tobias M. Meiswinkel
Chair of Genetics of Prokaryotes, Faculty of Biology & CeBiTec, Bielefeld University, Germany

1. Introduction

Since decades the limited access to petroleum oil is a major concern and substitutions for fossil fuels are needed. One promising substitute is biodiesel, which is widely produced from vegetable oils, e.g. from rape seeds, soybeans, sunflower seeds or animal fats. In the synthesis of biodiesel, oils and fats are transesterified to fatty acid methyl ester in the presence of sodium hydroxide or potassium hydroxide. In this process, glycerol is generated as stoichiometric byproduct with a ratio of 10% (w/w) with respect to biodiesel produced. In 2009, the biodiesel production of the world reached 16 million tons (Licht, 2010), with the lion's share produced by the European Union with 9 million tons (EBB, European Biodiesel Board 2010), followed by the United States with a production of 2.7 million tons (Licht, 2010). Hence, 1.6 million tons of glycerol were produced as obligatory by-product. Glycerol finds applications as an ingredient of various products, such as creams, food, feed, and pharmaceuticals, but the demand for glycerol in these processes is limited. Integrated conversion of raw glycerol from the biodiesel process to value-added products is a driver towards higher cost efficiency of biodiesel production.

Glycerol is a good source of carbon and energy for growth of several microorganisms and may be suitable for the biotechnological production of a number of chemicals in fermentative processes. To date several microbiological productions have been adjusted to glycerol as carbon and energy source or, if glycerol is close to the desired product, are based on glycerol as substrate anyway. For instance, the biotechnological production of 1,3-propanediol and dihydroxyacetone has predominantly been carried out from glycerol, since these processes are catalyzed in a two and one step reaction, respectively. Bacterial 1,3-propanediol production from glycerol is involving two enzymes. First glycerol is dehydrated by glycerol dehydratase to 3-hydroxypropionaldehyde, which is subsequently converted to 1,3-propanediol by 1,3-propanediol dehydrogenase. Predominantly, 1,3-propanediol production has been approved with *Clostridium* strains, *Klebsiella pneumoniae*, or *Escherichia coli* (Zhu et al. 2002, Biebl et al. 1998, Forsberg, 1987). 1,3-propanediol finds application in the production of the polyester polytrimethylene terephthalate. Strains of *Gluconobacter oxydans* are used for producing dihydroxyacetone (glycerone) from glycerol (Flickinger & Perlman, 1977). Dihydroxyacetone is used as a building block in organic chemistry and as a skin tanning agent in cosmetics.

Besides products that can be derived from glycerol in one or two reactions further products requiring more complex conversions have been investigated, e.g. succinic acid production with *Anaerobiospirillum succiniciproducens* (Lee et al. 2001) or *Escherichia coli* (Blankschien et al., 2010, Zhang et al., 2010), citrate production with *Yarrowia lipolytica* (Rywinska & Rymowicz, 2010), ethanol production with *Escherichia coli, Saccharomyces cerevisiae,* or *Hansenula polymorpha* (Hong et al., 2010, Shams Yazdani & Gonzalez, 2008, Yu et al., 2010), production of amino acids with *Corynebacterium glutamicum* (Rittmann et al., 2008), or propionate production with *Propionibacteria* (Himmi et al., 2000).

Crude glycerol preparations from biodiesel factories differ considerably from the pure chemical glycerol, e.g. by their salt content. The applicability of crude glycerol from biodiesel production plants has already been demonstrated for production strains of *Clostridium* and *Klebsiella* in 1,3-propanediol production (Gonzalez-Pajuelo et al., 2004, Mu et al., 2006), *Yarrowia lipolytica* in citrate production (Papanikolaou et al., 2002), *Kluyvera cryocrescens* and *Klebsiella pneumoniae* in ethanol production (Choi et al., 2011, Oh et al., 2011), as well as *Basfia succiniciproducens* in succinic acid production (Scholten et al., 2009).

This chapter will summarize state-of-the-art of glycerol-based biotechnological processes and will discuss future developments.

2. Use of glycerol as a carbon source in biotechnological applications

Glycerol has many applications, it is used for the production of food, cosmetics, paints, pharmaceutics, paper, textiles, leather and for the production of various chemicals (Wang et al., 2001). It can be used as a stabilizing agent for storage of cells and proteins. Physiologically, glycerol is essential for the biosynthesis of membranes, since it is the backbone of glycerolipids. And for its function as a component of lipids and fats it is an abundant source of carbon and energy in nature.

Formerly, glycerol was a valuable product derived from glucose via dihydroxyacetone-phosphate and glycerol-3-phosphate by glycerol-3-phosphate dehydrogenase (EC 1.1.1.94) and glycerol-3-phosphatase (EC 3.1.3.21). The yeast *Saccharomyces cerevisiae* which uses glycerol as an osmolyte under osmotic stress conditions was engineered for efficient glycerol production from glucose. *S. cerevisiae* possesses two isozymes of each, glycerol-3-phosphate forming glycerol-3-phosphate dehydrogenases and glycerol-3-phosphatases (Larsson et al., 1993, Pahlman et al., 2001). Sulphite treatment of yeasts enabled glycerol production (Petrovska et al., 1999), as did metabolic engineering on the glycerol production pathway, e.g. deletion of the triosephosphate isomerase gene (Overkamp et al., 2002), deletion of the alcohol dehydrogenase gene, overexpression of a glycerol-3-phosphate dehydrogenase gene (Navarro-Avino et al., 1999), or overexpression of a glycerol exporter gene (Cordier et al., 2007). Nowadays, glycerol arises from the biodiesel production. In 2010 glycerol formed as byproduct in the biodiesel process amounted to 1.6 million tons (Licht, 2010), which is extending the glycerol demand by far, thus, making microbial glycerol production unprofitable.

Glycerol as carbon source for growth

Glycerol can be used as a source of carbon and energy by many organisms. The initial step of glycerol utilization is its uptake into the cell. Albeit the small and uncharged molecule can diffuse through membranes without a transport system, many organisms possess glycerol transporters. In *Escherichia coli,* glycerol transport is mediated by the glycerol facilitator (Heller et al., 1980). In the wine bacterium *Pediococcus pentosaceus* active glycerol uptake has been reported (Pasteris & Strasser de Saad, 2008). Active glycerol transport has been

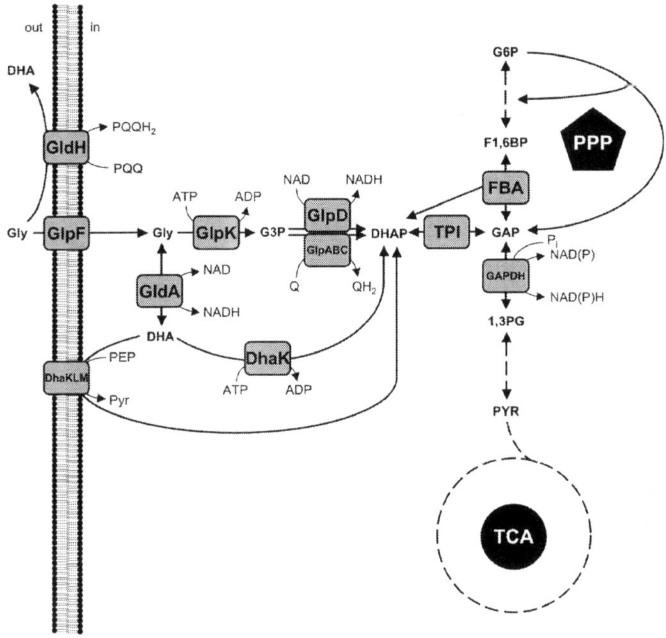

Fig. 1. Pathways of glycerol utilization. Abbreviations: 1,3-PG 1,3-phosphoglycerate, DHA dihydroxyacetone, DhaK ATP dependent dihydroxyacetone kinase, DhaKLM PEP dependent dihydroxyacetone kinase, DHAP dihydroxyacetone-phosphate, F1,6BP fructose-1,6-bisphosphate, G3P glycerol-3-phosphate, G6P glucose-6-phosphate, GAP glyceraldehyde-3-phosphate, GAPDH glyceraldehyde-3-phosphate dehydrogenase, GldA soluble glycerol dehydrogenase, GldH membrane bound glycerol dehydrogenase, GlpABC quinone dependent glycerol-3-phosphate dehydrogenase, GlpF glycerol facilitator, GlpK glycerol kinase, GlpD glycerol-3-phosphate dehydrogenase, Gly glycerol, PEP phosphoenolpyruvate, P_i inorganic phosphate, PPP pentose phosphate pathway, Pyr pyruvate, TCA tricarboxylic acid cycle, TPI triosephosphate isomerase.

described in yeasts, such as sodium dependent symport for *Debaryomyces hansenii* (Lucas et al., 1990) and proton dependent symport for *Pichia sorbitophila* (Lages & Lucas, 1995) and *Saccharomyces cerevisiae* (Lages & Lucas, 1997, Ferreira et al., 2005). In yeasts, active glycerol transport is mostly regarded of importance with respect to the use of glycerol as an osmolyte under osmotic stress conditions.

The imported glycerol can enter the metabolism at the level of the glycolytic intermediate dihydroxyacetone-phosphate. Two routes for the formation of dihydroxyacetone-phosphate from glycerol have been reported. In the first, ATP-dependent glycerol kinase (EC 2.7.1.30) phosphorylates glycerol to glycerol-3-phosphate, which is subsequently oxidized to dihydroxyacetone-phosphate by glycerol-3-phosphate dehydrogenase, which are either quinone or FAD-dependent (EC 1.1.5.3) or NAD-dependent (EC 1.1.1.8). In the second pathway, glycerol is oxidized to dihydroxyacetone by glycerol dehydrogenase (EC 1.1.1.6) before being phosphorylated to dihydroxyacetone-phosphate by ATP- or phosphoenolpyruvate-dependent dihydroxyacetone kinase (EC 2.7.1.29).

The glycerol-3-phosphate pathway is active in e.g. *Gluconobacter oxydans* and *Clostridium acetobutylicum* (Claret et al., 1994, Gonzalez-Pajuelo et al., 2006), whereas *Clostridium butyricum* uses the dihydroxyacetone pathway (Gonzalez-Pajuelo et al., 2006). In *E. coli, Klebsiella pneumoniae, Saccharomyces cerevisiae* and other yeasts, both pathways are present (Gonzalez et al., 2008, Wang et al., 2001, Ruch et al., 1974, Norbeck & Blomberg, 1997). In *S. cerevisiae,* glycerol is mainly utilized via glycerol-3-phosphate while the pathway via dihydroxyacetone is suggested to play a role during hyperosmotic stress for regulation of the intracellular glycerol concentration (Blomberg, 2000). In *E. coli,* the pathway via dihydroxyacetone operates under certain anaerobic conditions only (Gonzalez et al., 2008), but the major pathway is via glycerol-3-phosphate. In *K. pneumoniae,* the glycerol-3-phosphate pathway is active under aerobic and the dihydroxyaceton pathway is active under anaerobic conditions (Ruch et al., 1974).

2.1 Dihydroxyacetone

Dihydroxyacetone (DHA; 1,3-Dihydroxypropan-2-one, Glycerone) is produced biotechnologically with a global market of 2,000 tons per year (Pagliaro et al., 2007). DHA is produced biotechnologically, since its chemical synthesis is expensive and requires safety measures to cope with hazardous reactants (Hekmat et al., 2003). The most popular use of DHA is as coloring agent in sunless tanning products, such as creams and lotions (Levy, 1992). The tanning effect depends on Maillard-like reactions of DHA with the amino acids of the outer skin layer. Historically, also applications for medical treatments of glycogenesis, a glycogen storage disease, and diabetes mellitus have been described. Currently, the use of DHA as building block for chemical synthesis appears to have the highest potential for a production process based on biodiesel-derived glycerol (Enders et al., 2005, Zheng et al., 2008, Hekmat et al., 2003).

Gluconobacter oxydans is used for the production of DHA from glycerol. This bacterium belongs to the family of *Acetobacteraceae* (acetic acid bacteria), which are able to oxidize many carbohydrates and alcohols incompletely. To this end, *G. oxydans* possesses a variety of membrane-bound dehydrogenases. The membrane-bound pyrroloquinoline quinone (PQQ)-dependent glycerol dehydrogenase (EC 1.1.99.22) catalyzes oxidation of the secondary hydroxy group of glycerol to DHA (Matsushita et al., 1994). The enzyme is the protein product of *sldA* and *sldB* (Prust et al., 2005) and its localization allows extracellular oxidization of glycerol to DHA in the periplasma with a concurrent reduction of the membrane-localized PQQ. Besides membrane-bound glycerol dehydrogenase (Fig. 2), *G. oxydans* also possesses an intracellular catabolic pathway for the use of glycerol as a carbon source for growth (Fig. 1), in which glycerol is phosphorylated by ATP-dependent glycerol kinase to yield glycerol-3-phosphate which in turn is oxidized to dihydroxyacetone-phosphate by NAD-dependent glycerol-3-phosphate dehydrogenase (Claret et al., 1994). Since a functional glycolysis pathway is missing in *G. oxydans* and since its tricarboxylic acid cycle is incomplete, dihydroxyacetone-phosphate is metabolized via the pentose phosphate pathway (Greenfield & Claus, 1972).

Problems have occurred in the process of DHA production by *G. oxydans,* most importantly inhibition of the biotransformation process by the substrate glycerol and the product DHA as both inhibit growth and DHA production (Claret et al., 1992, Claret et al., 1993, Bauer et al., 2005). These problems have been met by optimizing production conditions including immobilization of *G. oxydans* cells to a polyvinyl alcohol matrix, which resulted in cells active for 14 days while maintaining glycerol dehydrogenase activity above 90 % (Wei et al., 2007).

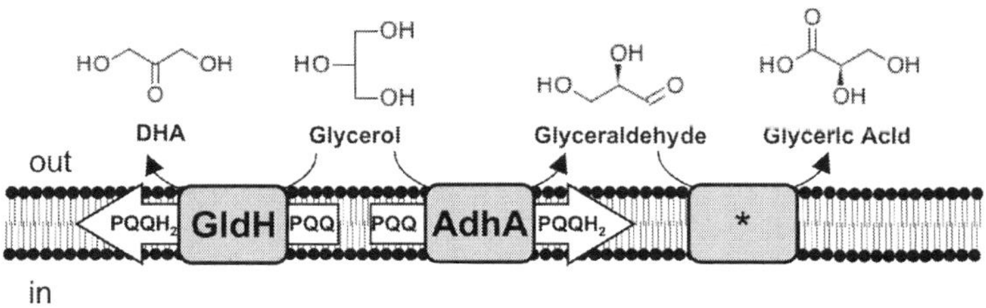

Fig. 2. Extracellular glycerol oxidation in *Gluconobacter oxydans*. Abbreviations: DHA dihydroxyacetone, PQQ oxidized pyrroloquinoline quinone, $PQQH_2$ reduced pyrroloquinoline quinone, AdhA alcohol dehydrogenase, * unidentified glyceric acid forming dehydrogenase.

Also fed-batch cultivations were shown to be supportive for DHA production compared to batch fermentations, as higher total glycerol amounts could be converted into DHA by avoiding inhibitory concentrations of glycerol, however, in this case production yields are limited with respect to DHA accumulation (Bories et al., 1991). Further enhancements of DHA production were achieved when using repeated fed-batch cultivations, in which DHA concentrations were kept below the inhibitory concentration. By this method immobilized cells were reused for up to 100 fed-batch cycles reducing costs and time for cleaning, sterilization, and inoculation (Hekmat et al., 2003). Other studies focused on optimizing culturing conditions for DHA production focused on cultivation media, aeration, or pH (Wethmar & Deckwer, 1999, Svitel & Sturdik, 1994, Tkac et al., 2001, Holst et al., 1985).
After the genome sequence of *G. oxydans* became public (Prust et al., 2005), metabolic engineering of *G. oxydans* has been reported for optimizing DHA production. Overexpression of the genes encoding the glycerol dehydrogenase, *sldAB*, led to increased biomass formation and higher DHA yields from glycerol (Herrmann et al., 2007). As the formation of glyceric acid as byproduct interferes with DHA production, disruption of *adhA*, the gene for the PQQ-dependent alcohol dehydrogenase, which is involved in the oxidation of glycerol to glyceric acid was shown to improve the product yield by abolishing glyceraldehyde and glyceric acid formation (Habe et al., 2010a). In addition, the mutant lacking *adhA* was less inhibited by high initial glycerol concentrations than the parent strain. In line with the notion that glyceric acid inhibits growth and DHA production by *G. oxydans*, it was found that addition of glyceric acid to the medium reduced growth and DHA production (Habe et al., 2010a) and that AdhA activity strongly increased when very high glycerol concentrations were used (Habe et al., 2009d). In the next step, the combination of *adhA* disruption and *sldAB* overexpression resulted in a strain with very high productivity and strongly increased tolerance towards glycerol (Li et al., 2010b). Since the production of DHA by the obligate aerobic *G. oxydans* is characterized by a very high demand for oxygen to oxidize reduced PQQ (Hekmat et al., 2003, Svitel & Sturdik, 1994), the gene encoding *Vitreoscilla* hemoglobin, an oxygen transporting protein, was shown to be beneficial (Li et al., 2010a).
Besides *G. oxydans* strains other acetic acid bacteria have also been reported for DHA production from glycerol, e.g. *Gluconobacter melanogenus* (Flickinger & Perlman, 1977) and *Acetobacter xylinum* (Nabe et al., 1979). Until today, production of DHA from biodiesel-derived raw glycerol has not yet been reported.

2.2 1,2-Propanediol

1,2-propanediol (propylenglycol, 1,2-PDO) is a commodity chemical with a wide range of applications, including polyester resins, plastics, antifreeze agents, de-icing products, detergents, or paints. The global demand for 1,2-PDO is estimated to be up to 1.6 million tons per year (Shelley, 2007). Chemically, 1,2-PDO is produced from propylene.

A variety of microorganisms have been reported as natural producers of 1,2-PDO, such as the bacteria *E. coli* (Hacking & Lin, 1976, Gonzalez et al., 2008), *Thermoanaerobacterium thermosaccharolyticum* (Cameron & Cooney, 1986), *Bacteroides ruminicola* (Turner & Roberton, 1979), *Salmonella typhimurium*, and *Klebsiella pneumoniae* (Badia et al., 1985), but also yeasts have been shown to produce 1,2-PDO (Suzuki & Onishi, 1968).

Two main routes for the microbial production of 1,2-PDO exist. First, 1,2-PDO may be formed from lactaldehyde, an intermediate of dissimilatory desoxy sugar (e.g. fucose, rhamnose) utilization generated either by fuculose-1-phosphate aldolase (EC 4.1.2.17) or by rhamnulose-1-phosphate aldolase (EC 4.1.2.19). Lactaldehyde is reduced to 1,2-PDO by NADH-dependent lactaldehyde reductase (FucO, EC 1.1.1.77), as shown e.g. for *Salmonella typhimurium* (Suzuki & Onishi, 1968): Due to high prices of fucose and rhamnose this pathway is not applicable.

The second route to 1,2-PDO diverts from glycolysis and, thus, 1,2-PDO production from glucose is feasible. The glycolytic intermediate dihydroxyacetone-phosphate is funneled into the methylglyoxal pathway by methylglyoxal synthase (EC 4.2.3.3) forming methylglyoxal. Methylglyoxal synthases from e.g. *Escherichia coli, Clostridium acetobutylicum,* and *Saccharomyces cerevisiae* have been purified and characterized (Cooper & Anderson, 1970, Freedberg et al., 1971, Hopper & Cooper, 1972, Huang et al., 1999, Murata et al., 1985), all of which were inhibited by phosphate. Methylglyoxal in turn is reduced to 1,2-PDO in two subsequent NADH/NADPH-dependent reactions. Two variants of methylglyoxyal reduction to 1,2-PDO are known with either acetol or lactaldehyde as intermediate. Acetol a may arise from methylglyoxal e.g. by aldehyde dehydrogenase (EC 1.1.1.2) or alcohol dehydrogenases (EC 1.1.1.1) from *E. coli* (Misra et al., 1996), by methylglyoxal reductase from *S. cerevisiae* (EC 1.1.1.283) (Nakamura et al., 1997) or acetol oxidoreductase from *E. coli* (Boronat & Aguilar, 1981). In the second step acetol is converted to 1,2-PDO by e.g. *E. coli* glycerol dehydrogenase (EC 1.1.1.6). The variant with lactaldehyde as intermediate involves e.g. glycol dehydrogenase from *Enterobacter aerogenes* (EC 1.1.1.185) (Carballo et al., 1993) or glycerol dehydrogenase from *E. coli* (EC 1.1.1.6) (Altaras & Cameron, 1999) for the first reduction step and e.g. *E. coli* 1,2-PDO reductase for the second reduction. Both ways necessitate two reduction equivalents per 1,2-PDO formed.

1,2-PDO production by *E. coli*

First approaches towards metabolic engineering of *E. coli* strains for production of 1,2-PDO from glucose by Cameron et al. involved overexpression of genes for either aldose reductase or glycerol dehydrogenase (Cameron et al., 1998). Later on, the same group coexpressed glycerol dehydrogenase genes from *E. coli* or *K. pneumoniae* together with the *E. coli* methylglyoxal synthase gene in *E. coli* and reached up to 0.7 g l^{-1}, an almost three-fold increase when compared to overexpression of *E. coli* or *K. pneumoniae* glycerol dehydrogenase gene alone (0.25 g l^{-1}) (Altaras & Cameron, 1999). Additional overexpression of yeast alcohol dehydrogenase or *E. coli* 1,2-PDO reductase further improved production performance and 1,2-PDO concentrations of 4.5 g l^{-1} and a yield of 0.19 g (g glucose)$^{-1}$ were achieved in a fed-batch fermentation process (Altaras & Cameron, 2000). Elimination of lactate dehydrogenase by gene deletion improved 1,2-PDO production by an *E. coli* strain overexpressing genes coding for methylglyoxal synthase from *Clostridium acetobutylicum* and glycerol dehydrogenase from *E. coli* (Berrios-Rivera et al., 2003).

Glycerol has a higher degree of reduction than glucose, thus, higher 1,2-PDO yields are theoretically possible using glycerol (0.72 g g^{-1}) as compared to glucose (0.63 g g^{-1}) (Clomburg & Gonzalez, 2011). Moreover, 1,2-PDO has been reported to be a natural product of anaerobic fermentation of glycerol in *E. coli* by Gonzalez et al. (Gonzalez et al., 2008).

Glycerol is converted to 1,2-PDO in a pathway consisting of glycerol dehydrogenase for oxidation of glycerol to dihydroxyacetone and phosphorylation of the latter to dihydroxyacetone-phosphate by phosphoenolypyruvate-dependent dihydroxyacetone kinase. Subsequently, dihydroxyacetone-phosphate is reduced to 1,2-PDO which requires two NADH. Thus, to generate NADH and ATP required in these reactions a portion of dihydroxyacetone-phosphate needs to be catabolized in glycolysis and onwards to ethanol (Gonzalez et al., 2008).

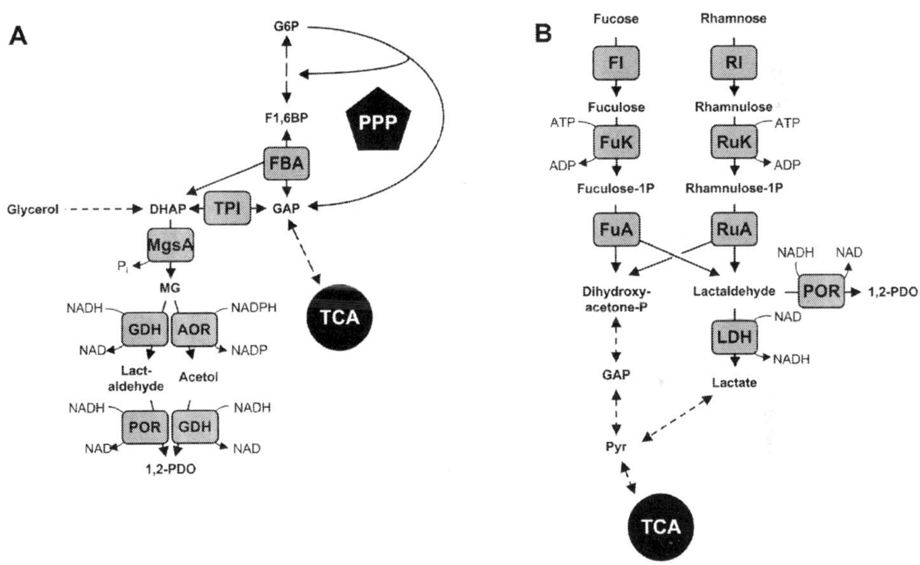

Fig. 3. 1,2-propanediol production pathways. A, 1,2-propanediol production from glycolytic intermediate dihydroxyacetone-phosphate; Abbreviations: 1,2-PDO 1,2-propanediol, AOR, acetol oxidoreductase, DHAP, dihydroxyacetone-phosphate, F1,6BP, fructose-1,6-bisphosphate, FBA, fructose-1,6-bisphosphate aldolase, G6P, glucose-6-phosphate, GAP, glyceraldehyde-3-phosphate, GDH, glycerol dehydrogenase, MG, methylglyoxal, MgsA, methylglyoxal synthase, P_i inorganic phosphate, POR 1,2-propanediol-oxidoreductase, TCA tricarboxylic acid cycle, TPI triosephosphate isomerase; B, 1,2-propanediol production from fucose and rhamnose; Abbreviations: FI fucose isomerase, FuA fuculose-1-phosphate aldolase, FuK fuculose kinase, LDH lactate dehydrogenase, RI rhamnose isomerase, RuK rhamnulose kinase, RuA rhamnulose-1-phosphate aldolase.

Recently, Clomburg et al. rationally engineered *E. coli* for effective 1,2-PDO production from glycerol. They introduced a 1,2-PDO production pathway via methylglyoxal synthase, glycerol dehydrogenase, and aldehyde oxidoreductase. Replacement of the phosphoenolpyruvate-dependent dihydroxyacetone kinase by ATP-dependent dihydroxyacetone kinase from *Citrobacter freudii* elevated dihydroxyacetone-phosphate availability and enhanced 1,2-PDO production from 0.02 to 0.15 g g^{-1}. Moreover, the formation of byproducts succinate, acetate, ethanol, and formate increased but lactate was reduced. To eliminate byproduct formation several gene deletions, e.g. of genes coding for lactate dehydrogenase, acetate kinase, phosphate acetyltransferase, formate hydrogen lyase, fumarate reductase, alcohol dehydrogenase, and pyruvate dehydrogenase, were tested alone or in combination. Ethanol formation could not be abolished as deletion of the alcohol dehydrogenase gene strongly decreased 1,2-PDO production (0.02 compared to 0.12 g g^{-1}) and glycerol consumption, but increased formation of acetate and succinate as byproducts. However, deletion of the genes for acetate kinase, phosphate acetyltransferase, and lactate dehydrogenase resulted in an increased product yield of 0.21 g g^{-1}, but entailed increased ethanol, formate, and pyruvate formation. The use of raw glycerol by this strain reduced formate formation and increased the 1,2-PDO yield (0.24 g g^{-1}) (Clomburg & Gonzalez, 2011). Taken together, engineered *E. coli* strains allow for 1,2-PDO yields from glycerol that are comparable to yields from glucose, making glycerol a feasible substrate for microbial 1,2-PDO production.

1,2-PDO production by other microorganisms

Recombinant strains of *S. cerevisiae* and of *C. glutamicum* have been developed for production of 1,2-PDO, as well. *S. cerevisiae* carrying multiple genome-integrated copies of *E. coli* methylglyoxal synthase and glycerol dehydrogenase genes was able to produce 1,2-PDO (Lee & da Silva, 2006). Plasmid-borne expression of the latter genes combined with the deletion of the endogenous triosephosphate isomerase gene resulted in the production of 1.1 g l^{-1} 1,2-PDO (Jung et al., 2008). Expression of *E. coli* methylglyoxal synthase gene and a *Corynebacterium glutamicum* aldo-keto reductase gene enabled *C. glutamicum* for the production of 1,2-PDO from glucose (1.8 g l^{-1}) (Niimi et al., 2011).

2.3 1,3-Propanediol

1,3-propanediol (1,3-PDO) can be used in several chemical applications. It is a substrate in the formulation reactions for polyesters, polyethers, polyurethanes, adhesives, composites, laminates, coatings, and moldings. In addition, 1,3-PDO itself is used e.g. as solvent or antifreeze agent. Most importantly, the production of polytrimethylene terephthalate (PTT) from 1,3-PDO and terephthalic acid is the driver of the growing market for 1,3-PDO. The polymer PTT is a promising polyester with numerous applications, e.g. as compound in textile, carpet, upholstery, or specialty resins. Due to its properties, PTT might be favored over polymers such as nylon, polyethylene terephthalate and polybutylene terephthalate. PTT has environmental benefits since it is biodegradable (for recent reviews see (da Silva et al., 2009, Liu et al., 2010, Carole et al., 2004). The chemical producers Shell and DuPont produce PTT from 1,3-PDO which is commercialized under the names Sorona, Hytrel (DuPont) and Corterra (Shell).

1,3-PDO is a success story of a glycerol-based biotechnological process. When 1,3-PDO was first discovered to be a fermentative product of glycerol in 1881 (Werkman & Gillen, 1932), it received little interest until the development of PTT in 1941 (Whinfield & Dickinson, 1946).

However, while terephthalic acid was readily available the expense of 1,3-PDO production limited efficient PTT production. Nowadays, chemical production of 1,3-PDO from ethylene oxide or acrolein as well as its biotechnological production enable supply of 1,3-PDO at low cost. It has to be noted that chemical synthesis of 1,3-PDO suffers from high energy consumption, toxic intermediates, and expensive catalysts as major drawbacks.
Currently, the market for 1,3-PDO is estimated to be about 50,000 tons per year (Liu et al., 2010), but due to a growing production of PTT a market volume of 230,000 tons is foreseen for 2020 (Carole et al., 2004). An indication of the growing interests and the potential of biotechnological 1,3-PDO production are the current decisions of the joint venture of DuPont and Tate & Lyle to extend their 1,3-PDO biotech plant in Louden, TN, USA, by 35 %. The actual capacity of the plant is 45,000 tons per year (Greenwood, 2010). Also the French company Metabolic Explorer (Clermont-Ferrand, France) decided to build a plant for biotechnological 1,3-PDO production in Malaysia in partnership with Malaysian Bio-XCell. The plant with a capacity to produce 50,000 tons 1,3-PDO annually is expected to start with a production of 8,000 tons per year (Degalard, 2011). While the process of DuPont and Tate & Lyle is based on sugars from corn hydrolysates, the Metabolic Explorer process will be based on crude glycerol.

Biotechnological production of 1,3-PDO

Several bacteria have been shown to naturally possess the ability of 1,3-PDO production, all of which belong either to enterobacteria or to firmicutes. Production of 1,3-PDO has been shown e.g. for strains of *Klebsiella pneumoniae, Clostridium butyricum, Clostridium acetobutylicum, Clostridium pasteurianum, Clostridium butylicum, Clostridium beijerinckii, Clostridium kainantoi, Lactobacillus brevis* and *Lactobacillus buchneri,* and *Enterobacter agglomerans* (Forage & Foster, 1982, Barbirato et al., 1996, Nakas et al., 1983, Schutz & Radler, 1984, Homann et al., 1990, Biebl, 1991, Biebl et al., 1992, Daniel et al., 1995).
Biosynthesis of 1,3-PDO from glycerol is catalyzed in a reducing pathway involving a cytosolic two-step process. First, glycerol dehydratase (EC 4.2.1.30) or diol dehydratase (EC 4.2.1.28) convert glycerol into 3-hydroxypropionaldehyde (3-HPA) (Schneider & Pawelkiewicz, 1966, Toraya et al., 1978). Glycerol dehydratase from *K. pneumoniae* has been reported to be inactivated by glycerol and necessitates a reactivating factor encoded by *gdrAB* (Tobimatsu et al., 1999). In the second step, 3-HPA is reduced to 1,3-PDO by NADPH- or NADH-dependent 1,3-PDO dehydrogenases (EC 1.1.1.202) e.g. by DhaT from *K. pneumoniae* or YqhD from *E. coli* (Johnson & Lin, 1987). Besides NADH-dependent DhaT, *K. pneumoniae* possesses a second NADPH dependent enzyme active as a 1,3-propanediol dehydrogenase. This enzyme was found, since a *dhaT* deletion mutant was still able to produce 1,3-PDO and 3-hydroxypropionic acid from glycerol albeit with reduced efficiency (Ashok et al., 2011). The sought enzyme was later identified as a homolog to *E. coli* YqhD and overexpression of the respective gene was shown to restore 1,3-PDO production by the *dhaT* deletion mutant (Seo et al., 2010).
The regeneration of reduction equivalents for 1,3-PDO production from glycerol is ensured in *K. pneumoniae* by simultaneous operation of the oxidative pathway of glycerol utilization. Here, NADH is generated by glycerol dehydrogenase during oxidation of glycerol to dihydroxyacetone and during catabolism of dihydroxyacetone phosphate, which is formed from dihydroxyacetone by dihydroxyacetone kinase (Seo et al., 2009). Thus, the demand of NADH limits the theoretical yield of 1,3-PDO production from glycerol. The requirement of some glycerol being oxidized and catabolized in glycolysis and the tricarboxylic acid cycle

entails the formation of unwanted byproducts such as acetic acid, lactic acid, formic acid, succinic acid, butyric acid, 2,3-butanediol, and ethanol. The main byproducts of *K. pneumoniae* strains are 2,3-butanediol, acetic acid, ethanol, and lactic acid (Menzel et al., 1997, Zhang et al., 2006, Kretschmann et al., 1993), whereas *C. butyricum* and metabolically engineered *C. acetobutylicum* strains mainly accumulate byproducts acetic acid and butyric acid during 1,3-PDO production (Papanikolaou et al., 2000, Papanikolaou et al., 2004, Saintamans et al., 1994, Gonzalez-Pajuelo et al., 2005, Soucaille, 2008, Sarcabal et al., 2007). With *C. pasteurianum,* butanol but not 1,3-PDO is the main fermentation product from glycerol, and ethanol, acetic acid, butyric acid and lactic acid are formed as further byproducts (Biebl, 2001). Metabolically engineered *E. coli* were reported to accumulate formic acid, acetic acid, lactic acid, and pyruvic acid as the major byproducts (Tong et al., 1991, Skraly et al., 1998) and accumulate growth inhibiting metabolites glycerol-3-phosphate and methylglyoxylate (Tkac et al., 2001, Zhu et al., 2001).
A major concern in 1,3-PDO production is the fact that the substrate glycerol, the intermediate 3-HPA, the product 1,3-PDO, and several byproducts inhibit growth and production. In *K. pneumoniae,* 1,3-PDO yields decrease with increasing glycerol concentrations and metabolic flux analyses revealed a higher carbon flux via the oxidative glycerol utilization pathway to the loss of 1,3-PDO (Xiu et al., 2011). In *C. butyricum,* growth is completely inhibited at 1,3-PDO concentrations higher than 60 g l^{-1}. Also the byproducts acetic acid (27 g l^{-1}) and butyric acid (19 g l^{-1}) abolished growth of this bacterium as did glycerol concentrations of 80 g l^{-1} or more (Biebl, 1991, Colin et al., 2000). Growth of *K. pneumoniae* is inhibited at glycerol concentrations above 110 g l^{-1} under aerobic and above 133 g l^{-1} under anaerobic conditions. Also the byproducts acetic acid (15 g l^{-1}), lactic acid (19 g l^{-1}), and ethanol (26 g l^{-1}) (15, 19, 26 g l^{-1} under anaerobic and 24, 26, and 17 g l^{-1} under aerobic conditions) inhibit growth of *K. pneumoniae* (Cheng et al., 2005). The accumulation of 3-HPA, the intermediate product of 1,3-PDO production has a toxic effect on growth and 1,3-PDO fermentation in *K. pneumoniae.* Both, glycerol dehydratase and 1,3-propanediol dehydrogenase are sensitive to 3-HPA. 1,3-propanediol dehydrogenase activity decreased as 3-HPA accumulated, leading to a further increase in 3-HPA concentrations (Hao et al., 2008a). Purified 1,3-propanediol dehydrogenase from *E. agglomerans* CNCM 1210 was shown to be inhibited by NAD^+ (K_i 0.29 mM) and 1,3-PDO (K_i 13.7 mM) and therefore might be limiting production yields of 1,3-PDO (Barbirato et al., 1997). Also the glycerol dehydratases from *C. freundii* and metagenome samples were shown to be inhibited by 1,3-PDO (Knietsch et al., 2003), moreover glycerol dehydratases from *K. pneumoniae* and *C. freundii* are inhibited by deactivation by glycerol (Tobimatsu et al., 1999, Tobimatsu et al., 2000, Kajiura et al., 2001, Seifert et al., 2001).
To overcome production limitations by e.g. substrate and product inhibition or byproduct inhibition, several approaches to optimize cultivation conditions were followed. Because the oxidative glycerol utilization pathway is necessitated for NADH regeneration but also leads to the formation of unwanted byproducts the cultivations of *K. pneumoniae* is preferably carried out under micro-aerobic conditions. Chen et al. compared cultivation conditions during 1,3-PDO production with *K. pneumoniae.* They found that final 1,3-PDO concentrations and yields were increased in batch fermentations under micro-aerobic conditions. Productivity increased from 0.8 to 1.57 under anaerobic and micro-aerobic conditions, respectively, and ethanol was reduced as well (Chen et al., 2003). Hao et al. postulated that the use of *K. pneumoniae* in a fed batch fermentation process using initial

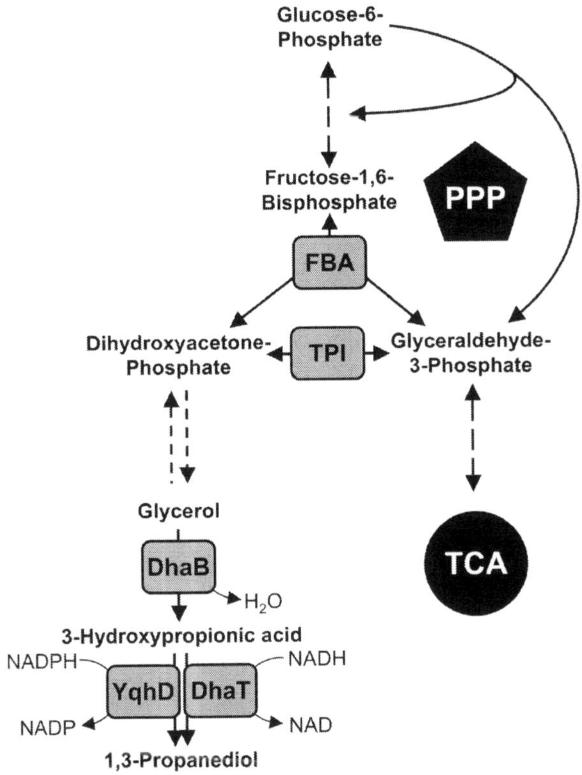

Fig. 4. 1,3-propanediol production pathway. Abbreviations: DhaB glycerol dehydratase, DhaT NADH dependent 1,3-propanediol dehydrogenase, FBA fructose-1,6-bisphosphate aldolase, PPP pentose phosphate pathway, TCA tricarboxylic acid cycle, TPI triosephosphate isomerase, YqhD NADPH dependent 1,3-propanediol dehydrogenase.

glycerol concentration of 30 g l^{-1} and subsequently keeping it to 7-8 g l^{-1} during exponential growth phase avoids toxic concentrations of 3-HPA (Hao et al., 2008a). For *Citrobacter freundii,* Pflugmacher et al. could show that cell immobilization to polyurethane carrier particles supports productivity to 8.2 g l^{-1} h^{-1} (Pflugmacher & Gottschalk, 1994). Gungormusler et al. reported 1,3-PDO production from raw glycerol with use *Clostridium beijerinckii* cells immobilized to ceramic rings and pumice stones in combination with a hydraulic retention time system. They could show an increase in productivity and yield for 1,3-PDO production and predicted a maximal 1,3-PDO production rate of 30 g l^{-1} h^{-1} (Gungormusler et al., 2011).

Metabolic engineering of *K. pneumoniae, C. acetobutylicum* and *E. coli* for 1,3-PDO production

Several metabolic engineering approaches were made for the optimization of 1,3-PDO production and reduction of byproduct formation with *K. pneumoniae* and *C. acetobutylicum*. *E. coli,* not a natural producer of 1,3-PDO, was engineered for 1,3-PDO production.

K. pneumoniae

K. pneumoniae was genetically manipulated to gain higher 1,3-PDO titers and production rates and to reduce process competing byproducts. When overexpressing the gene encoding the first enzyme of 1,3-PDO production glycerol dehydratase (*dhaB*) Zhao et al. (2009) found no effect on 1,3-PDO yields but reported decreased formation of the byproducts ethanol and 2,3-butanediol and an increase of acetic acid production (Ma et al., 2009). The overexpression of the 1,3-PDO dehydrogenase gene *dhaT* was shown to positively affect 1,3-PDO production in many studies and to reduce the formation of the toxic intermediate and substrate of 1,3-PDO dehydrogenase 3-HPA (Rao et al., 2010, Ma et al., 2009, Zhuge et al., 2010, Hao et al., 2008b). While aiming to reduce concentrations of the inhibitory intermediate 3-HPA Hao et al. (2008) overexpressed the 1,3-PDO dehydrogenase gene (*dhaT*) in *K. pneumoniae* TUAC01. During fermentation with 30 or 50 g l^{-1} glycerol 3-HPA accumulation was significantly reduced to 1.49 and 2.02 mM, respectively, compared to the parental strain which produced 7.55 and 12.57 mM, respectively (Hao et al., 2008b). Accordingly, Ma et al. (2010) overexpressed the 1,3-PDO dehydrogenase gene in *K. pneumoniae* and found a 3-fold decrease of 3-HPA production and an increase in 1,3-PDO production by 16.5% (Ma et al., 2010b). Similarly, Zhao et al. (2009) reported a strong decrease of the toxic intermediate 3-HPA by overexpression of the 1,3-PDO dehydrogenase gene leading to an increased molar yield of 0.64 mol mol^{-1} compared to 0.51 mol mol^{-1} of the parental strain and decreased lactic acid, ethanol and succinic acid concentrations in fed-batch fermentations (Ma et al., 2009). Zhuge et al. (2010) combined overexpression of *dhaT* and *yqhD* (encoding NADH- and NADPH-dependent 1,3-PDO dehydrogenases from *K. pneumoniae* and *E. coli*, respectively) in *K. pneumoniae*. The recombinant strain had a slightly elevated product titer (18.3 g l^{-1} compared to 17.1 g l^{-1}) and increased molar 1,3-PDO yield (0.51 to 0.57 mol mol^{-1}) in batch fermentations. Furthermore, the byproducts 3-HPA, succinic acid, lactic acid, acetic acid and ethanol were significantly reduced (Zhuge et al., 2010). When analyzing a *K. pneumoniae* mutant strain defective in the genes for NADH-dependent 1,3-PDO dehydrogenase and the oxydative glycerol utilization pathway Seo et al. (2009) reported that the mutant surprisingly retained the ability of 1,3-PDO production. 1,3-PDO yields were low but a strongly reduced byproduct formation was reported (Seo et al., 2009). Later, Seo et al. (2010) published the identification of a second 1,3-PDO dehydrogenase, which possesses high homology to *E. coli* YqhD and is, in contrast to the *dhaT* encoded enzyme, NADPH-dependent. Overproduction of the NADPH-dependent 1,3-PDO dehydrogenase from *K. pneumoniae* resulted in restoration of 1,3-PDO production in the mutant defective in the genes for NADH-dependent 1,3-PDO dehydrogenase and the oxydative glycerol utilization pathway and moreover byproduct formation remained low in the recombinant strain, as the oxidative pathway of glycerol utilization was absent. Although the 1,3-PDO concentrations were lower as compared to the parental strain (4.7 compared to 7.9 g l^{-1}), the molar yield of 0.54 mol mol^{-1} of the recombinant strain was higher compared to 0.48 mol mol^{-1} (Seo et al., 2010). To reduce byproduct formation Horng et al. (2010) constructed a *K. pneumoniae* mutant lacking the genes for glycerol dehydrogenase and dihydroxyacetone kinase. This mutant was reported to have ceased lactate, 2,3-butanediol, and ethanol byproduct formation and furthermore showed higher 1,3-PDO productivity when glycerol dehydratase and 1,3-PDO dehydrogenase were overexpressed (Horng et al., 2010). A lactate dehydrogenase mutant was constructed by

Xu et al. (2009) in *K. pneumoniae* HR526. The accumulation of lactate was reduced from 40 g l^{-1} in the parental strain to less than 3 g l^{-1} in the lactate dehydrogenase mutant, which showed very low lactate dehydrogenase activity. The mutant furthermore produced higher 1,3-PDO concentrations (95 g l^{-1} as compared to 102 g l^{-1}) with a higher yield (0.48 mol mol^{-1} to 0.52 mol mol^{-1}) and higher production rate (1.98 g l^{-1} h^{-1} to 2.13 g l^{-1} h^{-1}). Reduced lactate production increased NADH availability for 1,2-PDO production in *K. pneumoniae* (Xu et al., 2009) and *K. oxytoca* (Yang et al., 2007). Similarly, inactivation of the NADH-dependent aldehyde dehydrogenase gene in *K. pneumoniae* abolished ethanol formation, improved NADH availability and resulted in a 1,3-PDO production rate of 1.07 g l^{-1} h^{-1} and a yield of 0.70 mol mol^{-1} (Zhang et al., 2006).

E. coli

Wild-type *E. coli* is unable to convert glycerol to 1,3-PDO (Tong et al., 1991), but already in 1991 Tong et al. constructed a recombinant *E. coli* strain for the production of 1,3-PDO in an early application of metabolic engineering. A *K. pneumoniae* ATCC 25955 genomic library in *E. coli* was screened for anaerobic growth on glycerol and dihydroxyacetone and 1,3-PDO production. The selected recombinant possessed glycerol dehydratase, 1,3-PDO dehydrogenase, glycerol dehydrogenase, and dihydroxyacetone kinase from *K. pneumoniae* and produced 1 g l^{-1} 1,3-PDO with a yield of 0.46 mol mol^{-1} (Tong et al., 1991). Cofermentation of glycerol with glucose or xylose increased yields from 0.46 mol mol^{-1} to 0.63 mol mol^{-1} in the presence of glucose and to 0.55 mol mol^{-1} when cofermented with xylose (Tong & Cameron, 1992). Skraly et al. (1998) further optimized 1,3-PDO production with *E. coli* as they constructed an artificial operon containing the *K. pneumoniae* genes of 1,3-PDO production. They could show that the recombinant *E. coli* strain yielded 0.82 mol mol^{-1} (glycerol only) in a fed-batch process cofermenting glycerol and glucose, but only 9.3 g l^{-1} of glycerol where converted (Skraly et al., 1998). The additional expression of the glycerol dehydratase reactivating factor genes *gdrAB* together with the expression of *K. pneumoniae* genes encoding glycerol dehydratase and 1,3-PDO dehydrogenase yielded 8.6 g l^{-1} 1,3-PDO in a fed-batch fermentation (Wang et al., 2007). A higher final 1,3-PDO concentration of 13.2 g l^{-1} was obtained when they substituted *K. pneumoniae* 1,3-PDO dehydrogenase with the *E. coli* YqhD, which possesses NADPH-dependent 1,3-PDO dehydrogenase activity (Tan et al., 2007). *YqhD* from *E. coli* was also used in addition to the vitamin B12-independent glycerol dehydratase from *C. butyricum* in a two-stage fermentation process (aerobic biomass production from glucose followed by anaerobic 1,3-PDO production from glycerol). By this process 1,3-PDO concentrations of 104.4 g l^{-1} were reached, with a production rate of 2.62 g l^{-1} h^{-1} and a molar yield of 1.09 mol mol^{-1} (90.2% g g^{-1}) (referred to glycerol only) (Tang et al., 2009). Some efforts have also been made in the abolishment of toxic intermediate metabolites in *E. coli* during 1,3-PDO production. In *E. coli* high glycerol concentrations inhibit growth and 1,3-PDO production due to intracellular accumulation of glycerol-3-phosphate (Cozzarelli et al., 1965), the product of glycerol kinase, first enzyme in the oxidative pathway of glycerol utilization in *E. coli*. Zhu et al. (2002) could show that the glycerol-3-phosphate concentration increased when glycerol concentrations were elevated. The glycerol-3-phosphate accumulation was due to inefficient expression of the glycerol-3-phosphate dehydrogenase gene, and was overcome by usage of a glycerol kinase mutant, which showed 2.5-fold increased 1,3-PDO production (Zhu et al., 2002). The reduction of methylglyoxal formation by expression of the *Pseudomonas putida* glyoxalase I resulted in increased 1,3-PDO production by 50% (Zhu et al., 2001).

C. acetobutylicum

C. butyricum is a promising candidate for efficient 1,3-PDO production because it produces high concentrations of 1,3-PDO and possesses a vitamin B12-independent glycerol dehydratase circumventing the addition of expensive vitamin B12. Because tools for genetic manipulations of *C. butyricum* are unavailable, Gonzalez-Pajuelo et al. (2005) introduced the 1,3-PDO pathway from *C. butyricum* into *C. acetobutylicum*. The recombinant *C. acetobutylicum* strain (DG1(pSPD5)) accumulated 84 g l^{-1} 1,3-PDO in a fed-batch culture and reached a high production rate of 3 g l^{-1} h^{-1} (Gonzalez-Pajuelo et al., 2005).

Raw glycerol

Investigations of raw glycerol use for the production of 1,3-PDO have been made with *Clostridium* and *Klebsiella* strains. *K. pneumoniae* produced 1,3-PDO concentrations from raw glycerol from soybean oil biodiesel production close to that from pure glycerol (51.3 g l^{-1}) with a productivity of 1.7 g l^{-1} h^{-1} (compared to 2 g l^{-1} h^{-1}) (Mu et al., 2006). Similar concentrations (56 g l^{-1}) were reported by Hiremath et al. who used glycerol obtained from *Jatropha* seed biodiesel production (Hiremath et al., 2011). *C. beijerinckii* was also reported to produce 1,3-PDO from raw glycerol (Gungormusler et al., 2011). Notably, productivity was increased to 1.51 g l^{-1} h^{-1} when using raw glycerol as compared to 0.84 g l^{-1} h^{-1} when using pure glycerol (Um et al., 2010). Moon et al. compared 1,3-PDO production with *Klebsiella* and *Clostridium* strains from glycerol derived from biodiesel production from waste vegetable oil and soybean oil (Moon et al., 2010). Possibly due to inhibitory methanol concentrations, the use of soybean derived glycerol was better than use of raw glycerol derived from waste vegetable oil from different suppliers. In this study, a higher tolerance of *Klebsiella* strains to the different raw glycerols used compared to the *Clostridium* strains was observed (Moon et al., 2010). *C. butyricum* was also shown to utilize raw glycerol for 1,3-PDO production (Papanikolaou et al., 2000, Gonzalez-Pajuelo et al., 2004). Inhibitory effects of raw glycerol components on 1,3-PDO production with *C. butyricum* were found not to be due to NaCl or methanol, but rather due to oleic acid (Chatzifragkou et al., 2010).

2.4 Ethanol

Ethanol as a bio-fuel is mainly gained from sugarcane in Brazil, from corn in the USA and from sugar beets in the EU (da Silva et al., 2009). Since ethanol production is already done in a tens of billions scale, production deriving from crude glycerol may contribute only a small fraction. Nevertheless there has been considerable research on this topic in order to use crude glycerol efficiently to produce ethanol (Licht, 2010).

Unfortunately, the well known ethanol producer *Saccharomyces cerevisiae* grows very slow on glycerol and, thus, the growth had to be considerably improved, e.g. by selecting *S. cerevisiae* strain CBS8066-FL20 which grows much faster (0.2 h^{-1} rather than at 0.01 h^{-1}) (Ochoa-Estopier et al., 2011). Ethanol accumulation of the yeast *Hansenula polymorpha* was improved from 0.83 g l^{-1} to 2.74 g l^{-1} by expression of genes encoding pyruvate decarboxylase (*pdc*) and aldehyde dehydrogenase II (*adhB*) from *Zymomonas mobilis*. Combined with the expression of glycerol dehydrogenase (*dhaD*) and dehydroxyacetone kinase (*dhaKLM*) genes from *Klebsiella pneumonia* even more ethanol (3.1 g l^{-1}) was produced (Hong et al., 2010). Elementary mode analysis and metabolic evolution of *E. coli* mutants led to conversion of 40 g l^{-1} glycerol to ethanol reaching 90% of the theoretical yield (Trinh & Srienc, 2009).

E. coli strains have been engineered to produce ethanol and H_2 or ethanol and formate from crude gycerol. Due to overexpression of genes for glycerol dehydrogenase (*gldA*) and dihydroxyacetonekinase (*dhaKLM*) 95% of the theoretical maximum yield and specific production rates of 15-30 mmol (g cell)$^{-1}$ h^{-1} could be obtained (Shams Yazdani & Gonzalez, 2008). A newly isolated bacterium, *Kluyvera cryocrescens* S26, was able to produce 27 g l^{-1} ethanol with yield of 0.8 mol mol^{-1} and a productivity of 0.61 g l^{-1} h^{-1} (Choi et al., 2011).

2.5 Succinate

Succinic acid (succinate) is a so called platform chemical based on which a variety of other chemicals are produced, e.g. tetrahydrofuran, γ-butyrolactone, adipic acid, 1,4-butanediol and n-methyl-pyrrolidone. Based on this spectrum of products there are several markets succinic acid is involved in, such as pharmaceuticals, chemistry of biodegradable polymers, surfactants and detergents (Zeikus et al., 1999). Various microorganisms have been engineered for succinate production (Wendisch et al., 2006).

Succinic acid is an intermediate of the TCA cycle with four carbon atoms and two carboxylic groups. Today most of the produced succinic acid derives from the petrochemical industry and only a small part already comes from biotechnological processes. For the chemical synthesis the nonrenewable fossil fuel butane is the starting point leading through maleic anhydride to succinic acid (Zeikus et al., 1999).

Production of succinate from glycerol is interesting because both share the same level of reduction, thus, when produced from glycerol and CO_2 no further electron source is necessary. Various attempts have been made to use bacteria to efficiently produce succinate using natural succinic acid producers as well as metabolically engineered strains (Zeikus et al., 2007, Ingram et al., 2008, Lin et al., 2005, Samuelov et al., 1991, Okino et al., 2005, Singh et al., 2009, Van der Werf et al., 1997, Zhang et al., 2009). Among the natural producers, *Anaerobiospirillum succiniciproducens* was shown to use glycerol as sole or combined carbon source to efficiently produce succinic acid (Lee et al., 2004). They found a high succinic acid yield of 133% (or 160% when yeast extract was additionally fed) for glycerol concentrations of 6.5 g l^{-1}. Glycerol entailed less formation of acetic acid as byproduct and, thus, an easier downstream processing (Lee et al., 2001). Succinate production by the related bacterium *Basfia succiniciproducens* from crude glycerol in a continuous cultivation process allowed for product yields of about 1 g g^{-1} (Scholten et al., 2009).

Succinate production from glycerol by *E. coli* appeared to be favored under aerobic conditions as indicated by elementary mode analysis (Chen et al., 2010). Using microaerobic conditions, a recombinant producing pyruvate carboxylase from *Lactococcus lactis* and lacking pathways to byproducts showed succinic acid yields on glycerol comparable to those on glucose (Blankschien et al., 2010). About 80% of the maximum theoretical yield could be achieved by inserting three key mutations affecting phosphoenolpyruvate carboxykinase (*pck*), part of the phosphotransferase system (*ptsI*) and the pyruvate formate lyase (*pflB*) (Zhang et al., 2010).

2.6 Citrate

Citric acid is produced by fermentation at a scale of about 1,600,000 t/a (Papanikolaou et al., 2002, Berovic & Legisa, 2007), and it is sold for about 0.8 €/kg (Weusthuis et al., 2011). The main markets are the food and the pharmaceutical industries as well as applications in cosmetics, detergents and cleaning products. Citric acid is produced almost exclusively by *Aspergillus niger* in a submerged fermentation process using starch- or sucrose-based media like molasses (Soccol et al., 2006).

Different strains of the yeast *Yarrowia lipolytica* have been investigated for citric acid production using glycerol as a carbon source. For *Yarrowia lipolytica* Wratislavia AWG7 a maximal yield of 0.67 g g^{-1} was reported in continuous culture using glycerol as carbon source with only low contamination by the common byproduct isocitric acid (Rywinska et al., 2011, Rywinska & Rymowicz, 2010).
In 2002, Papanikolaou et al. reported about the *Y. lipolytica* strain LGAM S(7)1 being capable of growing on crude glycerol as carbon source and producing up to 35 g l^{-1} citric acid (Papanikolaou et al., 2002) . In 2010 and 2011, much higher citric acid concentrations of 112 g l^{-1} and a yield of 0.6 g g^{-1} using crude glycerol and the acetate-negative mutant strain *Y. lipolytica* A-101-1.22 (Rymowicz et al., 2010) or the acetate-negative strain *Y. lipolytica* N15 could be achieved (Kamzolova et al., 2011).

2.7 Amino acids

Amino acids are a multi-billion dollar business (Wendisch, 2007). They are used as flavor enhancers (L-glutamate), feed additives (L-lysine, L-methionine, L-threonine, L-tryptophane), to produce sweeteners such as aspartam (L-aspartate, L-phenylalanine), and in various pharmaceutical applications. The biggest products are L-glutamate (2,160,000 tons per year) and L-lysine (1,330,000 tons per year) (Ajinomoto, 2010a, Ajinomoto, 2010b). Pathways for amino acid synthesis start from intermediates of glycolysis (e.g. L-serine from 3-phosphoglycerate, L-valine from pyruvate), glycolysis and pentose phosphate pathway (aromatic amino acids L-tyrosine, L-phenylalanine, and L-tryptophane, from phosphoenolpyruvate (PEP) and erythrose-4-phosphate), and tricarboxylic acid cycle intermediates (L-glutamate, L-glutamine from 2-oxoglutarate, L-lysine, L-aspartate from oxaloacetate) (Schneider & Wendisch, 2001, Wendisch, 2007, Gopinath et al., 2011). In principle, production of amino acids from glycerol should be possible. In the following section the biosynthesis of L-glutamate, L-lysine, and L-phenylalanine are described, since strains for production of these from glycerol have been described.
Corynebacterium glutamicum is a natural L-glutamate producer (Eggeling & Bott, 2005), but the excretion of L-glutamate needs to be "triggered", e.g. by limitation of biotin (Shiio et al., 1962). Biotin is essential for the activity of acetyl-CoA carboxylase, necessary for fatty acid synthesis, and hence for membrane precursors, thus effect of biotin limitation on L-glutamate production is thought to be due to a higher permeability of the cell membrane (Shimizu & Hirasawa, 2007). Also addition of detergents like Tween 40 (Takinami et al., 1965), antibiotics like penicillin (Nara et al., 1964), and ethambutol (Radmacher et al., 2005, Stansen et al., 2005) trigger L-glutamate production. In *C. glutamicum* L-glutamate is mainly synthesized by NADPH dependent glutamate dehydrogenase from the tricarboxylic acid cycle intermediate 2-oxoglutarate (Bormann et al., 1992). This holds true for high ammonia concentrations, when ammonia is low L-glutamate is synthesized via L-glutamine by glutamine synthetase and glutamine-2-oxoglutarate aminotransferase. Crucial for L-glutamate production is the anaplerosis of the tricarboxylic acid cycle by either pyruvate carboxylase or PEP carboxylase. Pyruvate carboxylase has been shown to be indispensable under detergent triggered production conditions (Peters-Wendisch et al., 2001) and vice versa under biotin limiting conditions PEPcarboxylase is responsible for anaplerosis (Sato et al., 2008, Delaunay et al., 2004, Lapujade et al., 1999). *C. glutamicum* was engineered for glycerol utilization by expression of the genes for glycerol facilitator, glycerol kinase, and glycerol-3-phosphate dehydrogenase from *E. coli* (Rittmann et al., 2008). Under ethambutol

triggered L-glutamate production conditions recombinant *C. glutamicum* showed reduced L-glutamate yields from glycerol compared to glucose, 0.11 g g^{-1} compared to 0.20 g g^{-1}, respectively (Rittmann et al., 2008).

Production of L-lysine, which is used as a feed additive, is also carried out with *C. glutamicum* (Wendisch, 2007, Eggeling & Bott, 2005). The precursors of L-lysine production are the tricarboxylic acid cycle intermediate oxaloacetate and the glycolytic intermediate pyruvate. Deregulation of the L-lysine production pathway by introduction of feedback resistant variants of the key enzyme aspartate kinase, which usually is inhibited by L-lysine and L-threonine (Kalinowski et al., 1991) enables *C. glutamicum* for L-lysine production. Further increases were made by overexpression of the gene for pyruvate carboxylase (Peters-Wendisch et al., 2001), which provides L-lysine precursor oxaloacetate by the anaplerotic reaction from pyruvate (Peters-Wendisch et al., 1998). The anaplerotic reaction from PEP was shown to be dispendable for L-lysine production from glucose, however, it might play an important role if glucose is phosphorylated by use of ATP or polyphosphate and not PEP, which was shown to enhance L-lysine production and might elevate PEP availability (Lindner et al., 2011). Vice versa also an inactivation of the gene for PEP carboxykinase, catalyzing decarboxylation of oxaloacetate to PEP, entailed increased L-lysine production (Riedel et al., 2001). NADPH supply is very important for L-lysine production, as four molecules of NADPH are needed for one molecule L-lysine. The main path of NADPH generation is the oxidative branch of the pentose phosphate pathway (Marx et al., 1996), where NADP is reduced to NADPH by glucose-6-phosphate dehydrogenase and 6-phosphogluconate dehydrogenase, thus to enhance L-lysine production numerous attempts have been made towards increasing the pentose phosphate pathway flux, hence NADPH availability, hence L-lysine production. A deletion of the phosphoglucose isomerase gene drives the complete flux from glucose-6-phosphate into the pentose phosphate pathway and was shown to increase L-lysine production but to the cost of reduced growth (Marx et al., 2003). Redirection of the glycolytic flux towards the entry of the pentose phosphate pathway was furthermore achieved by overexpression of the fructose-bisphosphatase gene (Becker et al., 2005, Georgi et al., 2005) as well as use of feedback resistant variants of glucose-6-phosphate dehydrogenase and 6-phosphogluconate dehydrogenase (Becker et al., 2007, Ohnishi et al., 2005). Also the increase of NADP availability by overexpression of a NAD kinase gene resulted in increased L-lysine production (Lindner et al., 2010). To establish L-lysine production from glycerol Rittmann et al. introduced the *Escherichia coli* glycerol utilization genes in a metabolic engineered *C. glutamicum* L-lysine producing strain (deregulated L-lysine pathway and higher anaplerotic from pyruvate to oxaloacetate). L-lysine yields were slightly lower from glycerol as glucose, 0.19 g g^{-1} compared to 0.26 g g^{-1}, respectively (Rittmann et al., 2008). Glycerol has also been used as a source of carbon for the production of the polymer of L-lysine ε-Poly-L-lysine with *Streptomyces sp.*(Chen et al., 2011b, Chen et al., 2011a). ε-Poly-L-lysine is an antimicrobial agent against bacteria, yeasts, and viruses (Shima et al., 1984) and therefore interesting for the pharmaceutical industry (Shih et al., 2004) and it is used as food preservative.

The main use of the aromatic amino acid phenylalanine is in production of the sweetener aspartam. Biosynthesis of aromatic amino acids from PEP and erythrose-4-phosphate involves the shikimic acid pathway and dedicated terminal biosynthesis pathways for tryptophan, tyrosine and phenylalanine (Sprenger, 2007). Biosynthesis of aromatic amino

acids e.g. in *Escherichia coli* and *C. glutamicum* was engineered e.g. by gene deregulation (Berry, 1996, Herry & Dunican, 1993), by gene copy number increase (Chan et al., 1993), and by the use of feedback-resistant enzyme variants, e.g. variants of 3-deoxy-D-arabino-heptulosonate 7-phosphate synthase, the first enzyme of the shikimic acid pathway, variants of anthranilate synthase of the tryptophan pathway in *E. coli* (Tribe & Pittard, 1979) or anthranilate phosphoribosyltransferase of the tryptophan pathway in *C. glutamicum* (O'Gara & Dunican, 1995). In addition, strains were engineered for increased supply of the precursors PEP and erythrose-4-phosphate. In *C. glutamicum,* PEP availability was increased in PEP carboxylase mutants and erythrose-4-phosphate concentrations were elevated by overexpression of the transketolase gene (Ikeda & Katsumata, 1999, Ikeda et al., 1999, Katsumata & Kino, 1989). Similar approaches were made in *E. coli,* where PEP carboxylase or pyruvate kinase gene knock outs and overexpression of PEP carboxykinase increased PEP supply (Miller et al., 1987, Gosset et al., 1996, Chao & Liao, 1993, Backman, 1992). Furthermore, overexpression of genes encoding PEP synthase, PEP carboxykinase, and the use of an ATP-dependent glucose phosphorylation system instead of the PEP-dependent phosphotransferase system had positive effects on the availability of shikimic acid pathway precursor PEP (Patnaik et al., 1995, Liao, 1996, Gulevich et al., 2004). In *E. coli,* the availability of erythrose-4-phosphate could be increased by overproduction of transketolase and transaldolase or by phosphoglucose isomerase gene disruption (Draths & Frost, 1990, Draths et al., 1992, Lu & Liao, 1997, Mascarenhas et al., 1991, Frost, 1992). Up to now, only L-phenylalanine production from glycerol has been shown, but results might be transferable to the other aromatic amino acids. Similar final concentrations of L-phenylalanine were reported for an engineered *E. coli* strain regardless of the use of glycerol, glucose or sucrose as carbon source. Notably, a higher yield was reported when glycerol was used (0.58 g g^{-1}) as compared to the use of sucrose (0.25 g g^{-1}) (Khamduang et al., 2009).
Polyamines may be derived from amino acids (Schneider & Wendisch, 2010). While strain development for sugar-based production of polyamines such as the diamine 1,4-diaminobutane, which is used e.g. in the polyamide market, has been successful (Schneider & Wendisch, 2010), glycerol-based production of poylamines has not yet been reported.

2.8 2,3-Butanediol

2,3-Butanediol (2,3-BDO) is used as a solvent, fuel, and for the production of polymers and chemicals (Perego et al., 2003, Saha & Bothast, 1999). Bacterial 2,3-BDO production has been shown e.g. with strains of *Klebsiella pneumoniae, Klebsiella oxytoca, Enterobacter aerogenes, Bacillus polymyxa,* and *Bacillus licheniformis* (Grover et al., 1990, Perego et al., 2000, De Mas et al., 1988, Nilegaonkar et al., 1996, Jansen et al., 1984). Biosynthesis of 2,3-BDO is funneled from pyruvate in three steps. First, acetolactate synthase (EC 2.2.1.6) catalyses the condensation of two pyruvate molecules to acetolactate with concomitant CO_2 liberation. Second, acetolactate is decarboxylated by acetolactate decarboxylase (EC 4.1.1.5) to acetoin. Third, acetoin is reduced to 2,3-butanediol by 2,3-BDO dehydrogenase (acetoin reductase; EC 1.1.1.4) (Juni, 1952). Thus for 2,3-BDO production all substrates first need to be converted to pyruvate, the intermediate of glycolysis.
2,3-BDO is a product of mixed acid fermentation and, thus, associated with byproduct formation. Byproduct reduction approaches were made with *K. oxytoca* mutants defective in genes encoding lactate dehydrogenase and phosphotransacetylase, reducing lactate and acetate byproduct formation by 88% and 92%, respectively, but increasing 2,3-BDO production only by 7.8% (Ji et al., 2008). Also formation of ethanol, a major byproduct of 2,3-

BDO production with *K. oxytoca,* could be eliminated by insertion mutagenesis of the aldehyde dehydrogenase gene and 2,3-BDO production from glucose in a fed-batch process was improved to yield 130 g l^{-1} 2,3-BDO with a productivity of 1.63 g l^{-1} h^{-1} and a yield of 0.48 g g^{-1} (Ji et al., 2010).

Many substrates have been used for the production of 2,3-BDO. Use of starch as a substrate for 2,3-BDO production has been shown with *K. pneumoniae* by overexpression of a secretory α-amylase (Wei et al., 2008). With *B. licheniformis* corn starch hydrolysates were applied to 2,3-BDO production (Perego et al., 2003). With *E. aerogenes,* food industry wastes such as starch hydrolysates, raw and decoloured molasses, and whey permeate were used for the fermentation of 2,3-BDO (Perego et al., 2000). The use of lignocellulosic compounds for 2,3-BDO has also been reported, e.g. corncob hydrolysates were used in processes with *K. oxytoca* (Cheng et al., 2010) and *K. pneumoniae* (Ma et al., 2010a).

Glycerol was used for the production of 2,3-BDO as well. Because 1,3-propanediol production is preferably carried out from glycerol e.g. by *K. pneumoniae* and because 2,3-BDO is a known byproduct of this process (Biebl et al., 1998), glycerol might be a good substrate for 2,3-BDO production. Production of 2,3-BDO from glycerol by *K. pneumoniae* G31 resulted in final concentrations of 49.2 g l^{-1}. The medium pH had a large influence on 2,3-BDO fermentation from glycerol with 2,3-BDO production being favored at alkaline pH (Petrov & Petrova, 2009). In addition, intense aeration increased 2,3-BDO synthesis and reduced byproducts (Petrov & Petrova, 2010).

2.9 Hydrogen

Hydrogen production is highly desirable as a source of clean energy to be used, e.g. in fuel cells. Processes for the use of glycerol or crude glycerol respectively are under investigation. Besides microbial strategies to generate H_2 from crude glycerol there are also promising chemicals techniques such as steam reforming, partial oxidation, auto thermal reforming, aqueous-phase reforming, and supercritical water reforming (Xiaohu Fan, 2010). Currently, only low concentrations of glycerol can be used in microbial H_2 production process to avoid that other products like 1,3-propanediol or ethanol are produced along with H_2. *Enterobacter aerogenes* HU-101 showed hydrogen yields of 1.12 mol mol^{-1} using crude glycerol, but at relatively low glycerol concentrations of 1.7 g l^{-1} (Ito et al., 2005). Mixed cultures isolated from soil or wastewater converted crude glycerol to H_2 with a yield 0.31 mol mol^{-1} and to 1,3-Propanediol with a yield of 0.59 mol mol^{-1}. These values are lower than the ones on glucose but similar to the ones with pure glycerol, suggesting that inhibiting substances in crude glycerol may not be a problem in this process (Selembo et al., 2009). Production rates of 0.68 ± 0.16 mmol H_2 l^{-1} h^{-1} could be achieved by an evolved *E. coli* BW25113 *frdC* negative strain along with some ethanol production (Hu & Wood, 2010).

2.10 Glyceric acid

Glyceric acid is a known byproduct of dihydroxyacetone production from glycerol with *Gluconobacter oxydans.* The path from glycerol to glyceric aid, which might be suitable for chemical applications (Habe et al., 2009a), occurs via two dehydrogenases. First, alcohol dehydrogenase oxidizes glycerol to glyceraldehyde which is subsequently oxidized further to glyceric acid by a so far unidentified enzyme (Habe et al., 2009d). In a screen of various acetic acid bacteria *Acetobacter tropicalis* was the best glyceric acid producing strain (Habe et al., 2009b). *A. tropicalis* produced 101.8 g l^{-1} glyceric acid while *Gluconobacter frateurii* accumulated 136.5 g l^{-1} (Habe et al., 2009d). The involvement of a membrane-bound alcohol dehydrogenase in glyceric acid production was investigated with *G. oxydans* IFO12528. Gene disruption of the alcohol dehydrogenase entailed severely reduced glyceric acid concentrations, indicating a role of the alcohol dehydrogenase in glyceric acid production (Habe et al., 2009d). *G. frateurii* was engineered for glyceric acid production by disruption of the glycerol dehydrogenase gene *sldA*, thus, eliminating dihydroxyacetone production. The growth retardation of this strain on glycerol alone was overcome by addition of sorbitol to the medium. A higher glyceric acid concentration of 89.1 g l^{-1} was reached with the *sldA* mutant compared to the parental strain (54.7 g l^{-1}) as production dihydroxyacetone as byproduct was avoided (Habe et al., 2010b). Glyceric acid production from raw glycerol pretreated with activated charcoal by *Gluconobacter sp.* NBRC3259 reached comparable concentration of glyceric acid (45.9 g l^{-1} and 54.7 g l^{-1}) and of dihydroxyacetone (28.2 g l^{-1} and 33.7 g l^{-1}) as production from pure glycerol (Habe et al., 2009c).

2.11 Biosurfactants

Surfactants are used in numerous applications such as cleaners, emulsifiers, in coatings, laundry detergents, or in paints. The global surfactant market is predicted to reach 17.9 billion dollars by 2015 (Global Industry Analysts, 2010). Biosurfactants consist of a hydrophilic part and a hydrophobic/lipophilic part, making them amphiphiles/tensides. The hydrophilic part can consist of a sugar, peptide or protein, while the hydrophobic part contains fatty acids or fatty alcohols. The great advantage of biosurfactants over chemically produced tensides is that they are biodegradable, hence environmentally friendly, less toxic, and they can be produced from renewable resources. Natural biosurfactant producers are bacteria, yeast, and fungi (Mulligan, 2005). Biosurfactants are already used in many applications, e.g. in the food, agriculture, chemical, pharmaceutical, and cosmetic industries (for recent review see (Pacwa-Plociniczak et al., 2011)). Glycerol is the backbone of the lipid component of biosurfactants and use of pure and raw glycerol in biosurfactant production has been investigated with a variety of organisms. The yeast *Pseudozyma antarctica* produced 16.3 g l^{-1} glycolipid biosurfactants from glycerol (Morita et al., 2007). Glycolipid surfactants production was also shown with *Candida bombicola* and product concentrations of 12.7 g l^{-1} of sophorolipids could be obtained from glycerol and oleic acid (Ashby & Solaiman, 2010). When a biodiesel co-product stream consisting of 40% glycerol, 34% hexane-solubles, and 26% water was used production of sophorolipids by *C. bombicola* was strongly increased (from 9 g l^{-1} to 60 g l^{-1}) (Ashby et al., 2005). Production of glucosylmannosyl-glycerolipid with *Microbacterium spec.* was reported to be 1.5-fold higher when glycerol instead of glucose is used in media containing the complex medium compounds peptone and yeast extract (Lang et al., 2004, Wicke et al., 2000). When several carbon sources were analyzed for rhamnolipid production by *Pseudomonas aeruginosa* EM1, glycerol was the best carbon source besides glucose, yielding 4.9 and 7.5 g l^{-1} respectively (Wu et al., 2008). Glycerol-based biosurfactant production with *Pseudomonas aeruginosa* UCP0992 yielded 8.0 g l^{-1} biosurfactants (Silva et al., 2010), processes with *Rhodococcus erythropolis* yielded 1.7 g l^{-1} biosurfactants (Ciapina et al., 2006) and with *Ustilago maydis* 32.1 g l^{-1} glycolipid biosurfactants from 50 g l^{-1} glycerol could be produced (Liu et al., 2011).

3. Conclusions

Glycerol availability has increased tremendously as it arises as byproduct of the biodiesel process. Besides using glycerol as a chemical in creams and other small-scale applications, glycerol may be used as starting material for large-scale biotechnological processes. Several microbiological process are based on glycerol as substrate anyway, e.g. the biotechnological production of 1,3-propanediol and dihydroxyacetone. In addition, as glycerol is a good source of carbon and energy for growth of several microorganisms it may be suitable for the biotechnological production of a number of chemicals in fermentative processes. Microbial catalysts have been developed for the production of succinic acid, citric acid, glyceric acid, propionic acid, ethanol, 1,2-propanediol, 2,3-butanediol, biosurfactants and amino acids. Successful implementation of these processes critically depends on strain optimization, recalcitrance to inhibitors present in crude glycerol preparations from biodiesel factories and measures to reduce glycerol price volatility. If successful, coupling biodiesel production to the production of value-added products from the side-stream glycerol "on the spot" would represent an excellent example of applying the biorefinery concept to a large-scale process.

4. Acknowledgements

Work in the laboratory of the authors is supported in part by grants from the BMBF (0315589G, 0315598E, 316017A), ERA-IB (22009508B) and ESF (PAK529).

5. References

Ajinomoto, (2010a) Feed-Use Amino Acids Business. Available from World Wide Web: http://www.ajinomoto.com/ir/pdf/Feed-useAA-Oct2010.pdf. Cited 18 March 2011.

Ajinomoto, (2010b) Food Products Business. Available from World Wide Web: http://www.ajinomoto.com/ir/pdf/Food-Oct2010.pdf. Cited 18 March 2011.

Altaras, N. E. & D. C. Cameron, (1999) Metabolic engineering of a 1,2-propanediol pathway in *Escherichia coli*. *Appl Environ Microbiol* 65: 1180-1185.

Altaras, N. E. & D. C. Cameron, (2000) Enhanced production of (R)-1,2-propanediol by metabolically engineered *Escherichia coli*. *Biotechnol Prog* 16: 940-946.

Ashby, R. D., A. Nunez, D. K. Y. Solaiman & T. A. Foglia, (2005) Sophorolipid biosynthesis from a biodiesel co-product stream. *J Am Oil Chem Soc* 82: 625-630.

Ashby, R. D. & D. K. Solaiman, (2010) The influence of increasing media methanol concentration on sophorolipid biosynthesis from glycerol-based feedstocks. *Biotechnol Lett* 32: 1429-1437.

Ashok, S., S. M. Raj, C. Rathnasingh & S. Park, (2011) Development of recombinant *Klebsiella pneumoniae* Δ*dhaT* strain for the co-production of 3-hydroxypropionic acid and 1,3-propanediol from glycerol. *Appl Microbiol Biotechnol* 90: 1253-1265.

Backman, K., (1992) Method of biosynthesis of phenylalanine. U.S. Patent No. US 5169768.

Badia, J., J. Ros & J. Aguilar, (1985) Fermentation mechanism of fucose and rhamnose in *Salmonella typhimurium* and *Klebsiella pneumoniae*. *J Bacteriol* 161: 435-437.

Barbirato, F., J. P. Grivet, P. Soucaille & A. Bories, (1996) 3-Hydroxypropionaldehyde, an inhibitory metabolite of glycerol fermentation to 1,3-propanediol by enterobacterial species. *Appl Environ Microbiol* 62: 1448-1451.

Barbirato, F., A. Larguier, T. Conte, S. Astruc & A. Bories, (1997) Sensitivity to pH, product inhibition, and inhibition by NAD^+ of 1,3-propanediol dehydrogenase purified from *Enterobacter agglomerans* CNCM 1210. *Arch Microbiol* 168: 160-163.

Bauer, R., N. Katsikis, S. Varga & D. Hekmat, (2005) Study of the inhibitory effect of the product dihydroxyacetone on *Gluconobacter oxydans* in a semi-continuous two-stage repeated-fed-batch process. *Bioprocess Biosyst Eng* 28: 37-43.

Becker, J., C. Klopprogge, A. Herold, O. Zelder, C. J. Bolten & C. Wittmann, (2007) Metabolic flux engineering of L-lysine production in *Corynebacterium glutamicum*--over expression and modification of G6P dehydrogenase. *J Biotechnol* 132: 99-109.

Becker, J., C. Klopprogge, O. Zelder, E. Heinzle & C. Wittmann, (2005) Amplified expression of fructose 1,6-bisphosphatase in *Corynebacterium glutamicum* increases in vivo flux through the pentose phosphate pathway and lysine production on different carbon sources. *Appl Environ Microbiol* 71: 8587-8596.

Berovic, M. & M. Legisa, (2007) Citric acid production. *Biotechnol Annu Rev* 13: 303-343.

Berrios-Rivera, S. J., K. Y. San & G. N. Bennett, (2003) The effect of carbon sources and lactate dehydrogenase deletion on 1,2-propanediol production in *Escherichia coli*. *J Ind Microbiol Biotechnol* 30: 34-40.

Berry, A., (1996) Improving production of aromatic compounds in *Escherichia coli* by metabolic engineering. *Trends Biotechnol* 14: 250-256.

Biebl, H., (1991) Glycerol Fermentation of 1,3-Propanediol by *Clostridium butyricum* - Measurement of Product Inhibition by Use of a Ph-Auxostat. *Appl Microbiol Biot* 35: 701-705.

Biebl, H., (2001) Fermentation of glycerol by *Clostridium pasteurianum* - batch and continuous culture studies. *J Ind Microbiol Biot* 27: 18-26.

Biebl, H., S. Marten, H. Hippe & W. D. Deckwer, (1992) Glycerol Conversion to 1,3-Propanediol by Newly Isolated *Clostridia*. *Appl Microbiol Biot* 36: 592-597.

Biebl, H., A. P. Zeng, K. Menzel & W. D. Deckwer, (1998) Fermentation of glycerol to 1,3-propanediol and 2,3-butanediol by Klebsiella pneumoniae. *Appl Microbiol Biotechnol* 50: 24-29.

Blankschien, M. D., J. M. Clomburg & R. Gonzalez, (2010) Metabolic engineering of *Escherichia coli* for the production of succinate from glycerol. *Metab Eng* 12: 409-419.

Blomberg, A., (2000) Metabolic surprises in *Saccharomyces cerevisiae* during adaptation to saline conditions: questions, some answers and a model. *FEMS Microbiol Lett* 182: 1-8.

Bories, A., C. Claret & P. Soucaille, (1991) Kinetic-Study and Optimization of the Production of Dihydroxyacetone from Glycerol Using *Gluconobacter oxydans*. *Process Biochem* 26: 243-248.

Bormann, E. R., B. J. Eikmanns & H. Sahm, (1992) Molecular analysis of the *Corynebacterium glutamicum gdh* gene encoding glutamate dehydrogenase. *Mol Microbiol* 6: 317-326.

Boronat, A. & J. Aguilar, (1981) Experimental evolution of propanediol oxidoreductase in *Escherichia coli*. Comparative analysis of the wild-type and mutant enzymes. *Biochim Biophys Acta* 672: 98-107.

Cameron, D. C., N. E. Altaras, M. L. Hoffman & A. J. Shaw, (1998) Metabolic engineering of propanediol pathways. *Biotechnol Prog* 14: 116-125.

Cameron, D. C. & C. L. Cooney, (1986) A Novel Fermentation - the Production of R(-)-1,2-Propanediol and Acetol by *Clostridium thermosaccharolyticum*. *Bio-Technol* 4: 651-654.

Carballo, J., A. Bernardo, M. J. Prieto & R. M. Sarmiento, (1993) Kinetics of alpha-dicarbonyls reduction by L-glycol dehydrogenase (NAD+) from *Enterobacter aerogenes*. *Ital J Biochem* 42: 79-89.

Carole, T. M., J. Pellegrino & M. D. Paster, (2004) Opportunities in the industrial biobased products industry. *Appl Biochem Biotechnol* 113-116: 871-885.

Chan, E. C., H. L. Tsai, S. L. Chen & D. G. Mou, (1993) Amplification of the Tryptophan Operon Gene in *Escherichia coli* Chromosome to Increase L-Tryptophan Biosynthesis. *Appl Microbiol Biot* 40: 301-305.

Chao, Y. P. & J. C. Liao, (1993) Alteration of growth yield by overexpression of phosphoenolpyruvate carboxylase and phosphoenolpyruvate carboxykinase in *Escherichia coli*. *Appl Environ Microbiol* 59: 4261-4265.

Chatzifragkou, A., D. Dietz, M. Komaitis, A. P. Zeng & S. Papanikolaou, (2010) Effect of biodiesel-derived waste glycerol impurities on biomass and 1,3-propanediol production of *Clostridium butyricum* VPI 1718. *Biotechnol Bioeng* 107: 76-84.

Chen, X., L. Tang, S. Li, L. Liao, J. Zhang & Z. Mao, (2011a) Optimization of medium for enhancement of epsilon-poly-L-lysine production by *Streptomyces sp.* M-Z18 with glycerol as carbon source. *Bioresour Technol* 102: 1727-1732.

Chen, X., D. J. Zhang, W. T. Qi, S. J. Gao, Z. L. Xiu & P. Xu, (2003) Microbial fed-batch production of 1,3-propanediol by *Klebsiella pneumoniae* under micro-aerobic conditions. *Appl Microbiol Biot* 63: 143-146.

Chen, X. S., S. Li, L. J. Liao, X. D. Ren, F. Li, L. Tang, J. H. Zhang & Z. G. Mao, (2011b) Production of epsilon-poly-L: -lysine using a novel two-stage pH control strategy by *Streptomyces sp.* M-Z18 from glycerol. *Bioprocess Biosyst Eng* 34: 561-567.

Chen, Z., H. Liu, J. Zhang & D. Liu, (2010) Elementary mode analysis for the rational design of efficient succinate conversion from glycerol by *Escherichia coli*. *J Biomed Biotechnol* 2010: 518743.

Cheng, K. K., H. J. Liu & D. H. Liu, (2005) Multiple growth inhibition of *Klebsiella pneumoniae* in 1,3-propanediol fermentation. *Biotechnol Lett* 27: 19-22.

Cheng, K. K., Q. Liu, J. A. Zhang, J. P. Li, J. M. Xu & G. H. Wang, (2010) Improved 2,3-butanediol production from corncob acid hydrolysate by fed-batch fermentation using *Klebsiella oxytoca*. *Process Biochem* 45: 613-616.

Choi, W. J., M. R. Hartono, W. H. Chan & S. S. Yeo, (2011) Ethanol production from biodiesel-derived crude glycerol by newly isolated *Kluyvera cryocrescens*. *Appl Microbiol Biotechnol* 89: 1255-1264.

Ciapina, E. M., W. C. Melo, L. M. Santa Anna, A. S. Santos, D. M. Freire & N. Pereira, Jr., (2006) Biosurfactant production by *Rhodococcus erythropolis* grown on glycerol as sole carbon source. *Appl Biochem Biotechnol* 131: 880-886.

Claret, C., A. Bories & P. Soucaille, (1992) Glycerol Inhibition of Growth and Dihydroxyacetone Production by *Gluconobacter oxydans*. *Current Microbiology* 25: 149-155.

Claret, C., A. Bories & P. Soucaille, (1993) Inhibitory Effect of Dihydroxyacetone on *Gluconobacter oxydans* - Kinetic Aspects and Expression by Mathematical Equations. *J Ind Microbiol* 11: 105-112.

Claret, C., J. M. Salmon, C. Romieu & A. Bories, (1994) Physiology of *Gluconabacter oxydans* during Dihydroxyacetone Production from Glycerol. *Appl Microbiol Biot* 41: 359-365.

Clomburg, J. M. & R. Gonzalez, (2011) Metabolic engineering of *Escherichia coli* for the production of 1,2-propanediol from glycerol. *Biotechnol Bioeng* 108: 867-879.

Colin, T., A. Bories & G. Moulin, (2000) Inhibition of *Clostridium butyricum* by 1,3-propanediol and diols during glycerol fermentation. *Appl Microbiol Biotechnol* 54: 201-205.

Cooper, R. A. & A. Anderson, (1970) Formation and Catabolism of Methylglyoxal during Glycolysis in *Escherichia coli. Febs Letters* 11: 273-&.

Cordier, H., F. Mendes, I. Vasconcelos & J. M. Francois, (2007) A metabolic and genomic study of engineered *Saccharomyces cerevisiae* strains for high glycerol production. *Metab Eng* 9: 364-378.

Cozzarelli, N. R., J. P. Koch, S. Hayashi & E. C. Lin, (1965) Growth stasis by accumulated L-alpha-glycerophosphate in *Escherichia coli. J Bacteriol* 90: 1325-1329.

da Silva, G. P., M. Mack & J. Contiero, (2009) Glycerol: a promising and abundant carbon source for industrial microbiology. *Biotechnol Adv* 27: 30-39.

Daniel, R., K. Stuertz & G. Gottschalk, (1995) Biochemical and Molecular Characterization of the Oxidative Branch of Glycerol Utilization by *Citrobacter freundii. Journal of Bacteriology* 177: 4392-4401.

De Mas, C., N. B. Jansen & G. T. Tsao, (1988) Production of optically active 2,3-butanediol by *Bacillus polymyxa. Biotechnol Bioeng* 31: 366-377.

Degalard, P., (2011) METabolic EXplorer develops its first plant in Malaysia. Available from World Wide Web: http://www.ubifrance.com/my/Posts-1483-METabolic-EXplorer-develops-its-first-plant-in-Malaysia. Cited 22 June 2011.

Delaunay, S., P. Daran-Lapujade, J. M. Engasser & J. L. Goergen, (2004) Glutamate as an inhibitor of phosphoenolpyruvate carboxylase activity in *Corynebacterium glutamicum. J Ind Microbiol Biotechnol* 31: 183-188.

Draths, K. M. & J. W. Frost, (1990) Synthesis Using Plasmid-Based Biocatalysis - Plasmid Assembly and 3-Deoxy-D-Arabino-Heptulosonate Production. *Journal of the American Chemical Society* 112: 1657-1659.

Draths, K. M., D. L. Pompliano, D. L. Conley, J. W. Frost, A. Berry, G. L. Disbrow, R. J. Staversky & J. C. Lievense, (1992) Biocatalytic Synthesis of Aromatics from D-Glucose - the Role of Transketolase. *Journal of the American Chemical Society* 114: 3956-3962.

EBB, (European Biodiesel Board 2010) 2009-2010: EU biodiesel industry restained growth in challenging times. Annual biodiesel production statistics. Available in: http://www.ebb-eu.org/EBBpressreleases/EBB press release 2009 prod 2010_capacity FINAL.pdf.

Eggeling, L. & M. Bott, (2005) Handbook of *Corynebacterium glutamicum. CRC Press, Boca Raton, USA.*

Enders, D., M. Voith & A. Lenzen, (2005) The dihydroxyacetone unit--a versatile C(3) building block in organic synthesis. *Angew Chem Int Ed Engl* 44: 1304-1325.

Ferreira, C., F. van Voorst, A. Martins, L. Neves, R. Oliveira, M. C. Kielland-Brandt, C. Lucas & A. Brandt, (2005) A member of the sugar transporter family, Stl1p is the glycerol/H^+ symporter in *Saccharomyces cerevisiae. Mol Biol Cell* 16: 2068-2076.

Flickinger, M. C. & D. Perlman, (1977) Application of Oxygen-Enriched Aeration in the Conversion of Glycerol to Dihydroxyacetone by *Gluconobacter melanogenus* IFO 3293. *Appl Environ Microbiol* 33: 706-712.

Forage, R. G. & M. A. Foster, (1982) Glycerol fermentation in *Klebsiella pneumoniae*: functions of the coenzyme B12-dependent glycerol and diol dehydratases. *J Bacteriol* 149: 413-419.

Forsberg, C. W., (1987) Production of 1,3-Propanediol from Glycerol by *Clostridium acetobutylicum* and Other *Clostridium* Species. *Appl Environ Microbiol* 53: 639-643.

Freedberg, W. B., W. S. Kistler & E. C. Lin, (1971) Lethal synthesis of methylglyoxal by *Escherichia coli* during unregulated glycerol metabolism. *J Bacteriol* 108: 137-144.

Frost, J., (1992) Enhanced production of common aromatic pathway compounds. U.S. Patent No. US 5168056.

Georgi, T., D. Rittmann & V. F. Wendisch, (2005) Lysine and glutamate production by *Corynebacterium glutamicum* on glucose, fructose and sucrose: Roles of malic enzyme and fructose-1,6-bisphosphatase. *Metab Eng* 7: 291-301.

Global Industry Analysts, I., (2010) MCP-2146: Surface active agents - a global strategic business report. Available from World Wide Web: http://www.strategyr.com/pressMCP-2146.asp. Cited 22 June 2011.

Gonzalez-Pajuelo, M., J. C. Andrade & I. Vasconcelos, (2004) Production of 1,3-propanediol by *Clostridium butyricum* VPI 3266 using a synthetic medium and raw glycerol. *J Ind Microbiol Biotechnol* 31: 442-446.

Gonzalez-Pajuelo, M., I. Meynial-Salles, F. Mendes, J. C. Andrade, I. Vasconcelos & P. Soucaille, (2005) Metabolic engineering of *Clostridium acetobutylicum* for the industrial production of 1,3-propanediol from glycerol. *Metab Eng* 7: 329-336.

Gonzalez-Pajuelo, M., I. Meynial-Salles, F. Mendes, P. Soucaille & I. Vasconcelos, (2006) Microbial conversion of glycerol to 1,3-propanediol: physiological comparison of a natural producer, *Clostridium butyricum* VPI 3266, and an engineered strain, *Clostridium acetobutylicum* DG1(pSPD5). *Appl Environ Microbiol* 72: 96-101.

Gonzalez, R., A. Murarka, Y. Dharmadi & S. S. Yazdani, (2008) A new model for the anaerobic fermentation of glycerol in enteric bacteria: trunk and auxiliary pathways in *Escherichia coli*. *Metab Eng* 10: 234-245.

Gopinath V, Meiswinkel TM, Wendisch VF, Nampoothiri KM (2011) Amino acid production from rice straw and wheat bran hydrolysates by recombinant pentose-utilizing Corynebacterium glutamicum. Appl Microbiol Biotechnol. doi:10.1007/s00253-011-3478-x

Gosset, G., J. Yong-Xiao & A. Berry, (1996) A direct comparison of approaches for increasing carbon flow to aromatic biosynthesis in *Escherichia coli*. *J Ind Microbiol* 17: 47-52.

Greenfield, S. & G. W. Claus, (1972) Nonfunctional tricarboxylic acid cycle and the mechanism of glutamate biosynthesis in *Acetobacter suboxydans*. *J Bacteriol* 112: 1295-1301.

Greenwood, A., (2010) DuPont, Tate & Lyle JV expand US propandiol plant. Available from World Wide Web: http://www.icis.com/Articles/2010/05/04/9356156/dupont-tate-lyle-jv-expand-us-propandiol-plant.html. Cited 22 June 2011.

Grover, B. P., S. K. Garg & J. Verma, (1990) Production of 2,3-Butanediol from Wood Hydrolysate by *Klebsiella-Pneumoniae*. *World J Microb Biot* 6: 328-332.

Gulevich, A., I. Biryukova, D. Zimenkov, A. Skorokhodova, A. Kivero, A. Belareva & S. Mashko, (2004) Method for producing l-amino acid using bacterium having enhanced expression of *pckA* gene. World Patent No.WO 090125.

Gungormusler, M., C. Gonen & N. Azbar, (2011) Continuous production of 1,3-propanediol using raw glycerol with immobilized *Clostridium beijerinckii* NRRL B-593 in comparison to suspended culture. *Bioprocess Biosyst Eng*.

Habe, H., T. Fukuoka, D. Kitamoto & K. Sakaki, (2009a) Biotechnological production of D-glyceric acid and its application. *Appl Microbiol Biotechnol* 84: 445-452.

Habe, H., T. Fukuoka, D. Kitamoto & K. Sakaki, (2009b) Biotransformation of glycerol to D-glyceric acid by Acetobacter tropicalis. *Appl Microbiol Biotechnol* 81: 1033-1039.

Habe, H., T. Fukuoka, T. Morita, D. Kitamoto, T. Yakushi, K. Matsushita & K. Sakaki, (2010a) Disruption of the membrane-bound alcohol dehydrogenase-encoding gene improved glycerol use and dihydroxyacetone productivity in *Gluconobacter oxydans*. *Biosci Biotechnol Biochem* 74: 1391-1395.

Habe, H., Y. Shimada, T. Fukuoka, D. Kitamoto, M. Itagaki, K. Watanabe, H. Yanagishita & K. Sakaki, (2009c) Production of glyceric acid by *Gluconobacter sp.* NBRC3259 using raw glycerol. *Biosci Biotechnol Biochem* 73: 1799-1805.

Habe, H., Y. Shimada, T. Fukuoka, D. Kitamoto, M. Itagaki, K. Watanabe, H. Yanagishita, T. Yakushi, K. Matsushita & K. Sakaki, (2010b) Use of a *Gluconobacter frateurii* mutant to prevent dihydroxyacetone accumulation during glyceric acid production from glycerol. *Biosci Biotechnol Biochem* 74: 2330-2332.

Habe, H., Y. Shimada, T. Yakushi, H. Hattori, Y. Ano, T. Fukuoka, D. Kitamoto, M. Itagaki, K. Watanabe, H. Yanagishita, and others, (2009d) Microbial production of glyceric acid, an organic acid that can be mass produced from glycerol. *Appl Environ Microbiol* 75: 7760-7766.

Hacking, A. J. & E. C. Lin, (1976) Disruption of the fucose pathway as a consequence of genetic adaptation to propanediol as a carbon source in *Escherichia coli*. *J Bacteriol* 126: 1166-1172.

Hao, J., R. Lin, Z. Zheng, Y. Sun & D. Liu, (2008a) 3-Hydroxypropionaldehyde guided glycerol feeding strategy in aerobic 1,3-propanediol production by *Klebsiella pneumoniae*. *J Ind Microbiol Biotechnol* 35: 1615-1624.

Hao, J., W. Wang, J. Tian, J. Li & D. Liu, (2008b) Decrease of 3-hydroxypropionaldehyde accumulation in 1,3-propanediol production by over-expressing *dhaT* gene in *Klebsiella pneumoniae* TUAC01. *J Ind Microbiol Biotechnol* 35: 735-741.

Hekmat, D., R. Bauer & J. Fricke, (2003) Optimization of the microbial synthesis of dihydroxyacetone from glycerol with *Gluconobacter oxydans*. *Bioprocess Biosyst Eng* 26: 109-116.

Heller, K. B., E. C. Lin & T. H. Wilson, (1980) Substrate specificity and transport properties of the glycerol facilitator of *Escherichia coli*. *J Bacteriol* 144: 274-278.

Herrmann, U., C. Gatgens, U. Degner & S. Bringer-Meyer, (2007) Biotransformation of glycerol to dihydroxyacetone by recombinant *Gluconobacter oxydans* DSM 2343. *Appl Microbiol Biot* 76: 553-559.

Herry, D. M. & L. K. Dunican, (1993) Cloning of the trp gene cluster from a tryptophan-hyperproducing strain of *Corynebacterium glutamicum*: identification of a mutation in the trp leader sequence. *Appl Environ Microbiol* 59: 791-799.

Himmi, E. H., A. Bories, A. Boussaid & L. Hassani, (2000) Propionic acid fermentation of glycerol and glucose by *Propionibacterium acidipropionici* and *Propionibacterium freudenreichii ssp. shermanii*. *Appl Microbiol Biotechnol* 53: 435-440.

Hiremath, A., M. Kannabiran & V. Rangaswamy, (2011) 1,3-Propanediol production from crude glycerol from *Jatropha* biodiesel process. *N Biotechnol* 28: 19-23.

Holst, O., H. Lundback & B. Mattiasson, (1985) Hydrogen-Peroxide as an Oxygen Source for Immobilized *Gluconobacter oxydans* Converting Glycerol to Dihydroxyacetone. *Appl Microbiol Biot* 22: 383-388.

Homann, T., C. Tag, H. Biebl, W. D. Deckwer & B. Schink, (1990) Fermentation of Glycerol to 1,3-Propanediol by *Klebsiella* and *Citrobacter* Strains. *Appl Microbiol Biot* 33: 121-126.

Hong, W. K., C. H. Kim, S. Y. Heo, L. H. Luo, B. R. Oh & J. W. Seo, (2010) Enhanced production of ethanol from glycerol by engineered *Hansenula polymorpha* expressing pyruvate decarboxylase and aldehyde dehydrogenase genes from *Zymomonas mobilis*. *Biotechnol Lett* 32: 1077-1082.

Hopper, D. J. & R. A. Cooper, (1972) The purification and properties of *Escherichia coli* methylglyoxal synthase. *Biochem J* 128: 321-329.

Horng, Y. T., K. C. Chang, T. C. Chou, C. J. Yu, C. C. Chien, Y. H. Wei & P. C. Soo, (2010) Inactivation of *dhaD* and *dhaK* abolishes by-product accumulation during 1,3-propanediol production in *Klebsiella pneumoniae*. *J Ind Microbiol Biotechnol* 37: 707-716.

Hu, H. & T. K. Wood, (2010) An evolved *Escherichia coli* strain for producing hydrogen and ethanol from glycerol. *Biochem Biophys Res Commun* 391: 1033-1038.

Huang, K., F. B. Rudolph & G. N. Bennett, (1999) Characterization of methylglyoxal synthase from *Clostridium acetobutylicum* ATCC 824 and its use in the formation of 1, 2-propanediol. *Appl Environ Microbiol* 65: 3244-3247.

Ikeda, M. & R. Katsumata, (1999) Hyperproduction of tryptophan by *Corynebacterium glutamicum* with the modified pentose phosphate pathway. *Appl Environ Microbiol* 65: 2497-2502.

Ikeda, M., K. Okamoto & R. Katsumata, (1999) Cloning of the transketolase gene and the effect of its dosage on aromatic amino acid production in *Corynebacterium glutamicum*. *Appl Microbiol Biotechnol* 51: 201-206.

Ingram, L. O., K. Jantama, M. J. Haupt, S. A. Svoronos, X. L. Zhang, J. C. Moore & K. T. Shanmugam, (2008) Combining metabolic engineering and metabolic evolution to develop nonrecombinant strains of *Escherichia coli* C that produce succinate and malate. *Biotechnology and Bioengineering* 99: 1140-1153.

Ito, T., Y. Nakashimada, K. Senba, T. Matsui & N. Nishio, (2005) Hydrogen and ethanol production from glycerol-containing wastes discharged after biodiesel manufacturing process. *Journal of Bioscience and Bioengineering* 100: 260-265.

Jansen, N. B., M. C. Flickinger & G. T. Tsao, (1984) Production of 2,3-butanediol from D-xylose by *Klebsiella oxytoca* ATCC 8724. *Biotechnol Bioeng* 26: 362-369.

Ji, X. J., H. Huang, S. Li, J. Du & M. Lian, (2008) Enhanced 2,3-butanediol production by altering the mixed acid fermentation pathway in *Klebsiella oxytoca*. *Biotechnol Lett* 30: 731-734.

Ji, X. J., H. Huang, J. G. Zhu, L. J. Ren, Z. K. Nie, J. Du & S. Li, (2010) Engineering *Klebsiella oxytoca* for efficient 2, 3-butanediol production through insertional inactivation of acetaldehyde dehydrogenase gene. *Appl Microbiol Biotechnol* 85: 1751-1758.

Johnson, E. A. & E. C. Lin, (1987) *Klebsiella pneumoniae* 1,3-propanediol:NAD+ oxidoreductase. *J Bacteriol* 169: 2050-2054.

Jung, J. Y., E. S. Choi & M. K. Oh, (2008) Enhanced production of 1,2-propanediol by tpi1 deletion in *Saccharomyces cerevisiae*. *J Microbiol Biotechnol* 18: 1797-1802.

Juni, E., (1952) Mechanisms of Formation of Acetoin by Bacteria. *Journal of Biological Chemistry* 195: 715-726.

Kajiura, H., K. Mori, T. Tobimatsu & T. Toraya, (2001) Characterization and mechanism of action of a reactivating factor for adenosylcobalamin-dependent glycerol dehydratase. *Journal of Biological Chemistry* 276: 36514-36519.

Kalinowski, J., J. Cremer, B. Bachmann, L. Eggeling, H. Sahm & A. Puhler, (1991) Genetic and biochemical analysis of the aspartokinase from *Corynebacterium glutamicum. Mol Microbiol* 5: 1197-1204.

Kamzolova, S. V., A. R. Fatykhova, E. G. Dedyukhina, S. G. Anastassiadis, N. P. Golovchenko & I. G. Morgunov, (2011) Citric Acid Production by Yeast Grown on Glycerol-Containing Waste from Biodiesel Industry. *Food Technol Biotech* 49: 65-74.

Katsumata, R. & K. Kino, (1989) Process for producing amino acids by fermentation. Japan Patent 01317395 A (P2578488).

Khamduang, M., K. Packdibamrung, J. Chutmanop, Y. Chisti & P. Srinophakun, (2009) Production of L-phenylalanine from glycerol by a recombinant *Escherichia coli. J Ind Microbiol Biotechnol* 36: 1267-1274.

Knietsch, A., S. Bowien, G. Whited, G. Gottschalk & R. Daniel, (2003) Identification and characterization of coenzyme B12-dependent glycerol dehydratase- and diol dehydratase-encoding genes from metagenomic DNA libraries derived from enrichment cultures. *Appl Environ Microbiol* 69: 3048-3060.

Kretschmann, J., F.-J. Carduck, W.-D. Deckwer, C. Tag & B. H, (1993) Fermentive production of 1,3-propanediol. U.S. Patent No. US 5254467.

Lages, F. & C. Lucas, (1995) Characterization of a Glycerol H^+ Symport in the Halotolerant Yeast *Pichia sorbitophila. Yeast* 11: 111-119.

Lages, F. & C. Lucas, (1997) Contribution to the physiological characterization of glycerol active uptake in *Saccharomyces cerevisiae. Biochim Biophys Acta* 1322: 8-18.

Lang, S., W. Beil, H. Tokuda, C. Wicke & V. Lurtz, (2004) Improved production of bioactive glucosylmannosyl-glycerolipid by sponge-associated *Microbacterium* species. *Mar Biotechnol (NY)* 6: 152-156.

Lapujade, P., J. L. Goergen & J. M. Engasser, (1999) Glutamate excretion as a major kinetic bottleneck for the thermally triggered production of glutamic acid by *Corynebacterium glutamicum. Metab Eng* 1: 255-261.

Larsson, K., R. Ansell, P. Eriksson & L. Adler, (1993) A gene encoding sn-glycerol 3-phosphate dehydrogenase (NAD^+) complements an osmosensitive mutant of *Saccharomyces cerevisiae. Mol Microbiol* 10: 1101-1111.

Lee, P. C., W. G. Lee, S. Y. Lee & H. N. Chang, (2001) Succinic acid production with reduced by-product formation in the fermentation of *Anaerobiospirillum succiniciproducens* using glycerol as a carbon source. *Biotechnol Bioeng* 72: 41-48.

Lee, S. Y., S. H. Hong, S. H. Lee & S. J. Park, (2004) Fermentative production of chemicals that can be used for polymer synthesis. *Macromol Biosci* 4: 157-164.

Lee, W. & N. A. Dasilva, (2006) Application of sequential integration for metabolic engineering of 1,2-propanediol production in yeast. *Metab Eng* 8: 58-65.

Levy, S. B., (1992) Dihydroxyacetone-Containing Sunless or Self-Tanning Lotions. *J Am Acad Dermatol* 27: 989-993.

Li, M., J. Wu, J. Lin & D. Wei, (2010a) Expression of Vitreoscilla hemoglobin enhances cell growth and dihydroxyacetone production in *Gluconobacter oxydans. Curr Microbiol* 61: 370-375.

Li, M. H., J. Wu, X. Liu, J. P. Lin, D. Z. Wei & H. Chen, (2010b) Enhanced production of dihydroxyacetone from glycerol by overexpression of glycerol dehydrogenase in an alcohol dehydrogenase-deficient mutant of *Gluconobacter oxydans. Bioresour Technol* 101: 8294-8299.

Liao, J., (1996) Microorganisms and methods for overproduction of DAHP by cloned *pps* gene. World Patent No. WO 9608567.

Licht, F. O., (2010) World Ethanol and Biofuels Report. 8.

Lin, H., G. N. Bennett & K. Y. San, (2005) Fed-batch culture of a metabolically engineered *Escherichia coli* strain designed for high-level succinate production and yield under aerobic conditions. *Biotechnology and Bioengineering* 90: 775-779.

Lindner, S. N., H. Niederholtmeyer, K. Schmitz, S. M. Schoberth & V. F. Wendisch, (2010) Polyphosphate/ATP-dependent NAD kinase of *Corynebacterium glutamicum*: biochemical properties and impact of *ppnK* overexpression on lysine production. *Appl Microbiol Biotechnol* 87: 583-593.

Lindner, S. N., G. M. Seibold, A. Henrich, R. Kramer & V. F. Wendisch, (2011) Phosphotransferase System-Independent Glucose Utilization in *Corynebacterium glutamicum* by Inositol Permeases and Glucokinases. *Appl Environ Microbiol* 77: 3571-3581.

Lindner SN, Seibold GM, Kramer R, Wendisch VF (2011) Impact of a new glucose utilization pathway in amino acid-producing Corynebacterium glutamicum. Bio engineered Bugs 2 (5)

Liu, H., Y. Xu, Z. Zheng & D. Liu, (2010) 1,3-Propanediol and its copolymers: research, development and industrialization. *Biotechnol J* 5: 1137-1148.

Liu, Y., C. M. Koh & L. Ji, (2011) Bioconversion of crude glycerol to glycolipids in *Ustilago maydis*. *Bioresour Technol* 102: 3927-3933.

Lu, J. L. & J. C. Liao, (1997) Metabolic engineering and control analysis for production of aromatics: Role of transaldolase. *Biotechnol Bioeng* 53: 132-138.

Lucas, C., M. Dacosta & N. Vanuden, (1990) Osmoregulatory Active Sodium-Glycerol Cotransport in the Halotolerant Yeast *Debaryomyces-Hansenii*. *Yeast* 6: 187-191.

Ma, C. Q., A. L. Wang, Y. Wang, T. Y. Jiang, L. X. Li & P. Xu, (2010a) Production of 2,3-butanediol from corncob molasses, a waste by-product in xylitol production. *Appl Microbiol Biot* 87: 965-970.

Ma, X. Y., L. Zhao, Y. Zheng & D. Z. Wei, (2009) Effects of over-expression of glycerol dehydrogenase and 1,3-propanediol oxidoreductase on bioconversion of glycerol into 1,3-propandediol by *Klebsiella pneumoniae* under micro-aerobic conditions. *Bioproc Biosyst Eng* 32: 313-320.

Ma, Z., Z. Rao, B. Zhuge, H. Fang, X. Liao & J. Zhuge, (2010b) Construction of a novel expression system in *Klebsiella pneumoniae* and its application for 1,3-propanediol production. *Appl Biochem Biotechnol* 162: 399-407.

Marx, A., A. A. de Graaf, W. Wiechert, L. Eggeling & H. Sahm, (1996) Determination of the fluxes in the central metabolism of *Corynebacterium glutamicum* by nuclear magnetic resonance spetroscopy combined with metabolite balancing. *Biotechnol Bioeng* 49: 111-129.

Marx, A., S. Hans, B. Mockel, B. Bathe & A. A. de Graaf, (2003) Metabolic phenotype of phosphoglucose isomerase mutants of *Corynebacterium glutamicum*. *J Biotechnol* 104: 185-197.

Mascarenhas, D., D. J. Ashworth & C. S. Chen, (1991) Deletion of *pgi* alters tryptophan biosynthesis in a genetically engineered strain of *Escherichia coli*. *Appl Environ Microbiol* 57: 2995-2999.

Matsushita, K., H. Toyama & O. Adachi, (1994) Respiratory chains and bioenergetics of acetic acid bacteria. *Adv Microb Physiol* 36: 247-301.

Menzel, K., A. P. Zeng & W. D. Deckwer, (1997) High concentration and productivity of 1,3-propanediol from continuous fermentation of glycerol by *Klebsiella pneumoniae*. *Enzyme Microb Tech* 20: 82-86.

Miller, J. E., K. C. Backman, M. J. Oconnor & R. T. Hatch, (1987) Production of Phenylalanine and Organic-Acids by Phosphoenolpyruvate Carboxylase-Deficient Mutants of *Escherichia coli*. *J Ind Microbiol* 2: 143-149.

Misra, K., A. B. Banerjee, S. Ray & M. Ray, (1996) Reduction of methylglyoxal in *Escherichia coli* K12 by an aldehyde reductase and alcohol dehydrogenase. *Mol Cell Biochem* 156: 117-124.

Moon, C., J. H. Ahn, S. W. Kim, B. I. Sang & Y. Um, (2010) Effect of biodiesel-derived raw glycerol on 1,3-propanediol production by different microorganisms. *Appl Biochem Biotechnol* 161: 502-510.

Morita, T., M. Konishi, T. Fukuoka, T. Imura & D. Kitamoto, (2007) Microbial conversion of glycerol into glycolipid biosurfactants, mannosylerythritol lipids, by a basidiomycete yeast, *Pseudozyma antarctica* JCM 10317(T). *J Biosci Bioeng* 104: 78-81.

Mu, Y., H. Teng, D. J. Zhang, W. Wang & Z. L. Xiu, (2006) Microbial production of 1,3-propanediol by *Klebsiella pneumoniae* using crude glycerol from biodiesel preparations. *Biotechnol Lett* 28: 1755-1759.

Mulligan, C. N., (2005) Environmental applications for biosurfactants. *Environ Pollut* 133: 183-198.

Murata, K., Y. Fukuda, K. Watanabe, T. Saikusa, M. Shimosaka & A. Kimura, (1985) Characterization of methylglyoxal synthase in *Saccharomyces cerevisiae*. *Biochem Biophys Res Commun* 131: 190-198.

Nabe, K., N. Izuo, S. Yamada & I. Chibata, (1979) Conversion of Glycerol to Dihydroxyacetone by Immobilized Whole Cells of *Acetobacter xylinum*. *Appl Environ Microbiol* 38: 1056-1060.

Nakamura, K., S. Kondo, Y. Kawai, N. Nakajima & A. Ohno, (1997) Amino acid sequence and characterization of aldo-keto reductase from bakers' yeast. *Biosci Biotechnol Biochem* 61: 375-377.

Nakas, J. P., M. Schaedle, C. M. Parkinson, C. E. Coonley & S. W. Tanenbaum, (1983) System development for linked-fermentation production of solvents from algal biomass. *Appl Environ Microbiol* 46: 1017-1023.

Nara, T., S. Kinoshita & H. Samejima, (1964) Effect of Penicillin on Amino Acid Fermentation. *Agr Biol Chem Tokyo* 28: 120-124.

Navarro-Avino, J. P., R. Prasad, V. J. Miralles, R. M. Benito & R. Serrano, (1999) A proposal for nomenclature of aldehyde dehydrogenases in *Saccharomyces cerevisiae* and characterization of the stress-inducible *ALD2* and *ALD3* genes. *Yeast* 15: 829-842.

Niimi, S., N. Suzuki, M. Inui & H. Yukawa, (2011) Metabolic engineering of 1,2-propanediol pathways in *Corynebacterium glutamicum*. *Appl Microbiol Biotechnol* 90: 1721-1729.

Nilegaonkar, S. S., S. B. Bhosale, C. N. Dandage & A. H. Kapadi, (1996) Potential of *Bacillus licheniformis* for the production of 2,3-butanediol. *J Ferment Bioeng* 82: 408-410.

Norbeck, J. & A. Blomberg, (1997) Metabolic and regulatory changes associated with growth of *Saccharomyces cerevisiae* in 1.4 M NaCl. Evidence for osmotic induction of glycerol dissimilation via the dihydroxyacetone pathway. *J Biol Chem* 272: 5544-5554.

O'Gara, J. P. & L. K. Dunican, (1995) Mutations in the *trpD* gene of *Corynebacterium glutamicum* confer 5-methyltryptophan resistance by encoding a feedback-resistant anthranilate phosphoribosyltransferase. *Appl Environ Microbiol* 61: 4477-4479.

Ochoa-Estopier, A., J. Lesage, N. Gorret & S. E. Guillouet, (2011) Kinetic analysis of a *Saccharomyces cerevisiae* strain adapted for improved growth on glycerol: Implications for the development of yeast bioprocesses on glycerol. *Bioresour Technol* 102: 1521-1527.

Oh, B. R., J. W. Seo, S. Y. Heo, W. K. Hong, L. H. Luo, M. H. Joe, D. H. Park & C. H. Kim, (2011) Efficient production of ethanol from crude glycerol by a *Klebsiella pneumoniae* mutant strain. *Bioresour Technol* 102: 3918-3922.

Ohnishi, J., R. Katahira, S. Mitsuhashi, S. Kakita & M. Ikeda, (2005) A novel *gnd* mutation leading to increased L-lysine production in *Corynebacterium glutamicum*. *FEMS Microbiol Lett* 242: 265-274.

Okino, S., M. Inui & H. Yukawa, (2005) Production of organic acids by *Corynebacterium glutamicum* under oxygen deprivation. *Appl Microbiol Biotechnol* 68: 475-480.

Overkamp, K. M., B. M. Bakker, P. Kotter, M. A. Luttik, J. P. Van Dijken & J. T. Pronk, (2002) Metabolic engineering of glycerol production in *Saccharomyces cerevisiae*. *Appl Environ Microbiol* 68: 2814-2821.

Pacwa-Plociniczak, M., G. A. Plaza, Z. Piotrowska-Seget & S. S. Cameotra, (2011) Environmental applications of biosurfactants: recent advances. *Int J Mol Sci* 12: 633-654.

Pagliaro, M., R. Ciriminna, H. Kimura, M. Rossi & C. Della Pina, (2007) From glycerol to value-added products. *Angew Chem Int Ed Engl* 46: 4434-4440.

Pahlman, A. K., K. Granath, R. Ansell, S. Hohmann & L. Adler, (2001) The yeast glycerol 3-phosphatases Gpp1p and Gpp2p are required for glycerol biosynthesis and differentially involved in the cellular responses to osmotic, anaerobic, and oxidative stress. *J Biol Chem* 276: 3555-3563.

Papanikolaou, S., M. Fick & G. Aggelis, (2004) The effect of raw glycerol concentration on the production of 1,3-propanediol by *Clostridium butyricum*. *J Chem Technol Biot* 79: 1189-1196.

Papanikolaou, S., L. Muniglia, I. Chevalot, G. Aggelis & I. Marc, (2002) *Yarrowia lipolytica* as a potential producer of citric acid from raw glycerol. *J Appl Microbiol* 92: 737-744.

Papanikolaou, S., P. Ruiz-Sanchez, B. Pariset, F. Blanchard & M. Fick, (2000) High production of 1,3-propanediol from industrial glycerol by a newly isolated *Clostridium butyricum* strain. *J Biotechnol* 77: 191-208.

Pasteris, S. E. & A. M. Strasser de Saad, (2008) Transport of glycerol by *Pediococcus pentosaceus* isolated from wine. *Food Microbiol* 25: 545-549.

Patnaik, R., R. G. Spitzer & J. C. Liao, (1995) Pathway engineering for production of aromatics in *Escherichia coli*: Confirmation of stoichiometric analysis by independent modulation of AroG, TktA, and Pps activities. *Biotechnol Bioeng* 46: 361-370.

Perego, P., A. Converti, A. Del Borghi & P. Canepa, (2000) 2,3-butanediol production by *Enterobacter aerogenes*: selection of the optimal conditions and application to food industry residues. *Bioprocess Eng* 23: 613-620.

Perego, P., A. Converti & M. Del Borghi, (2003) Effects of temperature, inoculum size and starch hydrolyzate concentration on butanediol production by *Bacillus licheniformis*. *Bioresour Technol* 89: 125-131.

Peters-Wendisch, P. G., C. Kreutzer, J. Kalinowski, M. Patek, H. Sahm & B. J. Eikmanns, (1998) Pyruvate carboxylase from *Corynebacterium glutamicum*: characterization, expression and inactivation of the *pyc* gene. *Microbiology* 144: 915-927.

Peters-Wendisch, P. G., B. Schiel, V. F. Wendisch, E. Katsoulidis, B. Mockel, H. Sahm & B. J. Eikmanns, (2001) Pyruvate carboxylase is a major bottleneck for glutamate and lysine production by *Corynebacterium glutamicum*. *J Mol Microbiol Biotechnol* 3: 295-300.

Petrov, K. & P. Petrova, (2009) High production of 2,3-butanediol from glycerol by *Klebsiella pneumoniae* G31. *Appl Microbiol Biotechnol* 84: 659-665.

Petrov, K. & P. Petrova, (2010) Enhanced production of 2,3-butanediol from glycerol by forced pH fluctuations. *Appl Microbiol Biotechnol* 87: 943-949.

Petrovska, B., E. Winkelhausen & S. Kuzmanova, (1999) Glycerol production by yeasts under osmotic and sulfite stress. *Can J Microbiol* 45: 695-699.

Pflugmacher, U. & G. Gottschalk, (1994) Development of an Immobilized Cell Reactor for the Production of 1,3-Propanediol by *Citrobacter-Freundii*. *Appl Microbiol Biot* 41: 313-316.

Prust, C., M. Hoffmeister, H. Liesegang, A. Wiezer, W. F. Fricke, A. Ehrenreich, G. Gottschalk & U. Deppenmeier, (2005) Complete genome sequence of the acetic acid bacterium *Gluconobacter oxydans*. *Nat Biotechnol* 23: 195-200.

Radmacher, E., K. C. Stansen, G. S. Besra, L. J. Alderwick, W. N. Maughan, G. Hollweg, H. Sahm, V. F. Wendisch & L. Eggeling, (2005) Ethambutol, a cell wall inhibitor of *Mycobacterium tuberculosis*, elicits L-glutamate efflux of *Corynebacterium glutamicum*. *Microbiology* 151: 1359-1368.

Rao, Z. M., Z. Ma, B. Zhuge, H. Y. Fang, X. R. Liao & J. Zhuge, (2010) Construction of a Novel Expression System in *Klebsiella pneumoniae* and its Application for 1,3-Propanediol Production. *Appl Biochem Biotech* 162: 399-407.

Riedel, C., D. Rittmann, P. Dangel, B. Mockel, S. Petersen, H. Sahm & B. J. Eikmanns, (2001) Characterization of the phosphoenolpyruvate carboxykinase gene from *Corynebacterium glutamicum* and significance of the enzyme for growth and amino acid production. *J Mol Microbiol Biotechnol* 3: 573-583.

Rittmann, D., S. N. Lindner & V. F. Wendisch, (2008) Engineering of a glycerol utilization pathway for amino acid production by Corynebacterium glutamicum. *Appl Environ Microbiol* 74: 6216-6222.

Ruch, F. E., J. Lengeler & E. C. Lin, (1974) Regulation of glycerol catabolism in *Klebsiella aerogenes*. *J Bacteriol* 119: 50-56.

Rymowicz, W., A. R. Fatykhova, S. V. Kamzolova, A. Rywinska & I. G. Morgunov, (2010) Citric acid production from glycerol-containing waste of biodiesel industry by *Yarrowia lipolytica* in batch, repeated batch, and cell recycle regimes. *Appl Microbiol Biotechnol* 87: 971-979.

Rywinska, A., P. Juszczyk, M. Wojtatowicz & W. Rymowicz, (2011) Chemostat study of citric acid production from glycerol by *Yarrowia lipolytica*. *J Biotechnol* 152: 54-57.

Rywinska, A. & W. Rymowicz, (2010) High-yield production of citric acid by *Yarrowia lipolytica* on glycerol in repeated-batch bioreactors. *J Ind Microbiol Biotechnol* 37: 431-435.

Saha, B. C. & R. J. Bothast, (1999) Production of 2,3-butanediol by newly isolated *Enterobacter cloacae*. *Appl Microbiol Biotechnol* 52: 321-326.

Saintamans, S., P. Perlot, G. Goma & P. Soucaille, (1994) High Production of 1,3-Propanediol from Glycerol by *Clostridium butyricum* Vpi-3266 in a Simply Controlled Fed-Batch System. *Biotechnol Lett* 16: 831-836.

Samuelov, N. S., R. Lamed, S. Lowe & J. G. Zeikus, (1991) Influence of CO_2-HCO_3- Levels and Ph on Growth, Succinate Production, and Enzyme-Activities of *Anaerobiospirillum succiniciproducens. Appl Environ Microb* 57: 3013-3019.

Sarcabal, P., C. Croux & P. Soucaille, (2007) Method for preparing 1,3-propanediol by a recombinant micro-organism in the absence of coenzyme b12 or one of its precursors. U.S. Patent No. US 7267972

Sato, H., K. Orishimo, T. Shirai, T. Hirasawa, K. Nagahisa, H. Shimizu & M. Wachi, (2008) Distinct roles of two anaplerotic pathways in glutamate production induced by biotin limitation in *Corynebacterium glutamicum. J Biosci Bioeng* 106: 51-58.

Schneider, J. & V. F. Wendisch, (2010) Putrescine production by engineered *Corynebacterium glutamicum. Appl Microbiol Biotechnol.* 88(4):859–868

Schneider J, Niermann K, Wendisch VF (2011) Production of the amino acids l-glutamate, l-lysine, l-ornithine and l-arginine from arabinose by recombinant Corynebacterium glutamicum. J Biotechnol 154 (2-3):191-198.

Schneider, Z. & J. Pawelkiewicz, (1966) The properties of glycerol dehydratase isolated from *Aerobacter aerogenes*, and the properties of the apoenzyme subunits. *Acta Biochim Pol* 13: 311-328.

Scholten, E., T. Renz & J. Thomas, (2009) Continuous cultivation approach for fermentative succinic acid production from crude glycerol by *Basfia succiniciproducens* DD1. *Biotechnol Lett* 31: 1947-1951.

Schutz, H. & F. Radler, (1984) Anaerobic Reduction of Glycerol to Propanediol-1.3 by *Lactobacillus brevis* and *Lactobacillus buchneri. Syst Appl Microbiol* 5: 169-178.

Seifert, C., S. Bowien, G. Gottschalk & R. Daniel, (2001) Identification and expression of the genes and purification and characterization of the gene products involved in reactivation of coenzyme B-12-dependent glycerol dehydratase of *Citrobacter freundii. European Journal of Biochemistry* 268: 2369-2378.

Selembo, P. A., J. M. Perez, W. A. Lloyd & B. E. Logan, (2009) Enhanced hydrogen and 1,3-propanediol production from glycerol by fermentation using mixed cultures. *Biotechnol Bioeng* 104: 1098-1106.

Seo, J. W., M. Y. Seo, B. R. Oh, S. Y. Heo, J. O. Baek, D. Rairakhwada, L. H. Luo, W. K. Hong & C. H. Kim, (2010) Identification and utilization of a 1,3-propanediol oxidoreductase isoenzyme for production of 1,3-propanediol from glycerol in *Klebsiella pneumoniae. Appl Microbiol Biotechnol* 85: 659-666.

Seo, M. Y., J. W. Seo, S. Y. Heo, J. O. Baek, D. Rairakhwada, B. R. Oh, P. S. Seo, M. H. Choi & C. H. Kim, (2009) Elimination of by-product formation during production of 1,3-propanediol in *Klebsiella pneumoniae* by inactivation of glycerol oxidative pathway. *Appl Microbiol Biotechnol* 84: 527-534.

Shams Yazdani, S. & R. Gonzalez, (2008) Engineering *Escherichia coli* for the efficient conversion of glycerol to ethanol and co-products. *Metab Eng* 10: 340-351.

Shelley, S., (2007) A renewable route to propylene glycol. *Chem Eng Prog* 103: 6-9.

Shih, I. L., Y. T. Van & M. H. Shen, (2004) Biomedical applications of chemically and microbiologically synthesized poly(glutamic acid) and poly(lysine). *Mini Rev Med Chem* 4: 179-188.

Shiio, I., S. I. Otsuka & M. Takahashi, (1962) Effect of biotin on the bacterial formation of glutamic acid. I. Glutamate formation and cellular premeability of amino acids. *J Biochem* 51: 56-62.

Shima, S., H. Matsuoka, T. Iwamoto & H. Sakai, (1984) Antimicrobial Action of Epsilon-Poly-L-Lysine. *J Antibiot* 37: 1449-1455.

Shimizu, H. & T. Hirasawa, (2007) Production of Glutamate and Glutamate-Related Amino Acids: MolecularMechanism Analysis and Metabolic Engineering. *In Amino Acid Biosynthesis – Pathways, Regulation and Metabolic Engineering (Wendisch V.F., ed), Springer, Heidelberg, Germany*: DOI 10.1007/7171_2006_1064.

Silva, S. N., C. B. Farias, R. D. Rufino, J. M. Luna & L. A. Sarubbo, (2010) Glycerol as substrate for the production of biosurfactant by *Pseudomonas aeruginosa* UCP0992. *Colloids Surf B Biointerfaces* 79: 174-183.

Singh, A., M. D. Lynch & R. T. Gill, (2009) Genes restoring redox balance in fermentation-deficient *E. coli* NZN111. *Metab Eng* 11: 347-354.

Skraly, F. A., B. L. Lytle & D. C. Cameron, (1998) Construction and characterization of a 1,3-propanediol operon. *Appl Environ Microbiol* 64: 98-105.

Soccol, C. R., L. P. S. Vandenberghe, C. Rodrigues & A. Pandey, (2006) New perspectives for citric acid production and application. *Food Technol Biotech* 44: 141-149.

Soucaille, P., (2008) Process for the biological production of 1, 3-propanediol from glycerol with high yield. World Patent No. WO08052595

Sprenger, G., (2007) Aromatic Amino Acids *In* Amino Acid Biosynthesis - Pathways, Regulation and Metabolic Engineering (Wendisch, VF., ed), Springer, Berlin, Germany, pp. 93-128.

Stansen, C., D. Uy, S. Delaunay, L. Eggeling, J. L. Goergen & V. F. Wendisch, (2005) Characterization of a *Corynebacterium glutamicum* lactate utilization operon induced during temperature-triggered glutamate production. *Appl Environ Microbiol* 71: 5920-5928.

Suzuki, T. & H. Onishi, (1968) Aerobic Dissimilation of L-Rhamnose and Production of L-Rhamnonic Acid and 1,2-Propanediol by Yeasts. *Agr Biol Chem Tokyo* 32: 888-&.

Svitel, J. & E. Sturdik, (1994) Product Yield and by-Product Formation in Glycerol Conversion to Dihydroxyacetone by *Gluconobacter oxydans*. *J Ferment Bioeng* 78: 351-355.

Takinami, K., H. Yoshii, H. Tsuri & H. Okada, (1965) Biochemical Effects of Fatty Acid and Its Derivatives on L-Glutamic Acid Fermentation .3. Biotin-Tween 60 Relationship in Accumulation of L-Glutamic Acid and Growth of *Brevibacterium lactofermentum*. *Agr Biol Chem Tokyo* 29: 351-&.

Tan, T. W., F. H. Wang, H. J. Qu, D. W. Zhang & P. F. Tian, (2007) Production of 1,3-propanediol from glycerol by recombinant *E. coli* using incompatible plasmids system. *Mol Biotechnol* 37: 112-119.

Tang, X., Y. Tan, H. Zhu, K. Zhao & W. Shen, (2009) Microbial conversion of glycerol to 1,3-propanediol by an engineered strain of *Escherichia coli*. *Appl Environ Microbiol* 75: 1628-1634.

Tkac, J., M. Navratil, E. Sturdik & P. Gemeiner, (2001) Monitoring of dihydroxyacetone production during oxidation of glycerol by immobilized *Gluconobacter oxydans* cells with an enzyme biosensor. *Enzyme Microb Tech* 28: 383-388.

Tobimatsu, T., H. Kajiura & T. Toraya, (2000) Specificities of reactivating factors for adenosylcobalamin-dependent diol dehydratase and glycerol dehydratase. *Archives of Microbiology* 174: 81-88.

Tobimatsu, T., H. Kajiura, M. Yunoki, M. Azuma & T. Toraya, (1999) Identification and expression of the genes encoding a reactivating factor for adenosylcobalamin-dependent glycerol dehydratase. *J Bacteriol* 181: 4110-4113.

Tong, I. T. & D. C. Cameron, (1992) Enhancement of 1,3-propanediol production by cofermentation in *Escherichia coli* expressing *Klebsiella pneumoniae dha* regulon genes. *Appl Biochem Biotechnol* 34-35: 149-159.

Tong, I. T., H. H. Liao & D. C. Cameron, (1991) 1,3-Propanediol production by *Escherichia coli* expressing genes from the *Klebsiella pneumoniae dha* regulon. *Appl Environ Microbiol* 57: 3541-3546.

Toraya, T., S. Honda, S. Kuno & S. Fukui, (1978) Coenzyme B12-dependent diol dehydratase: regulation of apoenzyme synthesis in *Klebsiella pneumoniae* (*Aerobacter aerogenes*) ATCC 8724. *J Bacteriol* 135: 726-729.

Tribe, D. E. & J. Pittard, (1979) Hyperproduction of tryptophan by *Escherichia coli*: genetic manipulation of the pathways leading to tryptophan formation. *Appl Environ Microbiol* 38: 181-190.

Trinh, C. T. & F. Srienc, (2009) Metabolic engineering of *Escherichia coli* for efficient conversion of glycerol to ethanol. *Appl Environ Microbiol* 75: 6696-6705.

Turner, K. W. & A. M. Roberton, (1979) Xylose, arabinose, and rhamnose fermentation by *Bacteroides ruminicola*. *Appl Environ Microbiol* 38: 7-12.

Um, Y., S. A. Jun, C. Moon, C. H. Kang, S. W. Kong & B. I. Sang, (2010) Microbial Fed-batch Production of 1,3-Propanediol Using Raw Glycerol with Suspended and Immobilized *Klebsiella pneumoniae*. *Appl Biochem Biotech* 161: 491-501.

Van der Werf, M. J., M. V. Guettler, M. K. Jain & J. G. Zeikus, (1997) Environmental and physiological factors affecting the succinate product ratio during carbohydrate fermentation by *Actinobacillus sp.* 130Z. *Arch Microbiol* 167: 332-342.

Wang, Z. X., J. Zhuge, H. Fang & B. A. Prior, (2001) Glycerol production by microbial fermentation: a review. *Biotechnol Adv* 19: 201-223.

Wang F, Qu H, Zhang D, Tian P, Tan T (2007) Production of 1,3-propanediol from glycerol by recombinant E. coli using incompatible plasmids system. Mol Biotechnol 37 (2):112-119.

Wei, D. Z., Y. Zheng, H. Y. Zhang, L. Zhao, L. J. Wei & X. Y. Ma, (2008) One-step production of 2,3-butanediol from starch by secretory over-expression of amylase in *Klebsiella pneumoniae*. *J Chem Technol Biot* 83: 1409-1412.

Wei, S., Q. Song & D. Wei, (2007) Repeated use of immobilized *Gluconobacter oxydans* cells for conversion of glycerol to dihydroxyacetone. *Prep Biochem Biotechnol* 37: 67-76.

Wendisch, V. F., (2007) Amino Acid Biosynthesis - Pathways, Regulation and Metabolic Engineering. In: Microbiology Monographs. A. Steinbüchel (ed). Berlin: Springer, pp.

Wendisch, V. F., M. Bott & B. J. Eikmanns, (2006) Metabolic engineering of *Escherichia coli* and *Corynebacterium glutamicum* for biotechnological production of organic acids and amino acids. *Curr Opin Microbiol* 9: 268-274.

Werkman, C. H. & G. F. Gillen, (1932) Bacteria Producing Trimethylene Glycol. *J Bacteriol* 23: 167-182.

Wethmar, M. & W. D. Deckwer, (1999) Semisynthetic culture medium for growth and dihydroxyacetone production by *Gluconobacter oxydans*. *Biotechnol Tech* 13: 283-287.

Weusthuis, R. A., I. Lamot, J. van der Oost & J. P. Sanders, (2011) Microbial production of bulk chemicals: development of anaerobic processes. *Trends Biotechnol* 29: 153-158.

Whinfield, J. & J. Dickinson, (1946) Improvements relating to the manufacture of highly polymeric substances. UK Patent No. GB 578079

Wicke, C., M. Huners, V. Wray, M. Nimtz, U. Bilitewski & S. Lang, (2000) Production and structure elucidation of glycoglycerolipids from a marine sponge-associated *Microbacterium* species. *J Nat Prod* 63: 621-626.

Wu, J. Y., K. L. Yeh, W. B. Lu, C. L. Lin & J. S. Chang, (2008) Rhamnolipid production with indigenous *Pseudomonas aeruginosa* EM1 isolated from oil-contaminated site. *Bioresour Technol* 99: 1157-1164.

Xiaohu Fan, R. B. a. Y. Z., (2010) Glycerol (Byproduct of Biodiesel Production) as a Source for Fuels and Chemicals - Mini Review. *The Open Fuels & Energy Science Journal* 3: 17-22.

Xiu, Z. L., Y. H. Wang & H. Teng, (2011) Effect of aeration strategy on the metabolic flux of *Klebsiella pneumoniae* producing 1,3-propanediol in continuous cultures at different glycerol concentrations. *J Ind Microbiol Biot* 38: 705-715.

Xu, Y. Z., N. N. Guo, Z. M. Zheng, X. J. Ou, H. J. Liu & D. H. Liu, (2009) Metabolism in 1,3-propanediol fed-batch fermentation by a D-lactate deficient mutant of *Klebsiella pneumoniae*. *Biotechnol Bioeng* 104: 965-972.

Yang, G., J. Tian & J. Li, (2007) Fermentation of 1,3-propanediol by a lactate deficient mutant of *Klebsiella oxytoca* under microaerobic conditions. *Appl Microbiol Biotechnol* 73: 1017-1024.

Yu, K. O., S. W. Kim & S. O. Han, (2010) Engineering of glycerol utilization pathway for ethanol production by *Saccharomyces cerevisiae*. *Bioresour Technol* 101: 4157-4161.

Zeikus, J. G., M. K. Jain & P. Elankovan, (1999) Biotechnology of succinic acid production and markets for derived industrial products. *Appl Microbiol Biot* 51: 545-552.

Zeikus, J. G., J. B. McKinlay & C. Vieille, (2007) Prospects for a bio-based succinate industry. *Appl Microbiol Biot* 76: 727-740.

Zhang, X., K. Jantama, J. C. Moore, L. R. Jarboe, K. T. Shanmugam & L. O. Ingram, (2009) Metabolic evolution of energy-conserving pathways for succinate production in *Escherichia coli*. *Proc Natl Acad Sci U S A* 106: 20180-20185.

Zhang, X., K. T. Shanmugam & L. O. Ingram, (2010) Fermentation of glycerol to succinate by metabolically engineered strains of *Escherichia coli*. *Appl Environ Microbiol* 76: 2397-2401.

Zhang, Y., Y. Li, C. Du, M. Liu & Z. Cao, (2006) Inactivation of aldehyde dehydrogenase: a key factor for engineering 1,3-propanediol production by *Klebsiella pneumoniae*. *Metab Eng* 8: 578-586.

Zheng, Y., X. Chen & Y. Shen, (2008) Commodity chemicals derived from glycerol, an important biorefinery feedstock. *Chem Rev* 108: 5253-5277.

Zhu, M. M., P. D. Lawman & D. C. Cameron, (2002) Improving 1,3-propanediol production from glycerol in a metabolically engineered *Escherichia coli* by reducing accumulation of sn-glycerol-3-phosphate. *Biotechnol Prog* 18: 694-699.

Zhu, M. M., F. A. Skraly & D. C. Cameron, (2001) Accumulation of methylglyoxal in anaerobically grown *Escherichia coli* and its detoxification by expression of the *Pseudomonas putida* glyoxalase I gene. *Metab Eng* 3: 218-225.

Zhuge, B., C. Zhang, H. Y. Fang, J. A. Zhuge & K. Permaul, (2010) Expression of 1,3-propanediol oxidoreductase and its isoenzyme in *Klebsiella pneumoniae* for bioconversion of glycerol into 1,3-propanediol. *Appl Microbiol Biot* 87: 2177-2184.

19

Improved Utilization of Crude Glycerol By-Product from Biodiesel Production

Alicja Kośmider, Katarzyna Leja and Katarzyna Czaczyk
Poznań University of Life Sciences
Poland

1. Introduction

During the last ten years a significant increase in biodiesel production and its commercial applications was observed (Rahmat et al., 2010). Nowadays, biodiesel is only one alternative fuel which may replace crude oil because it can be use in vehicles with a diesel engine without modifications of major engine or fuel elements (Johnson & Taconi, 2007). Presently, the most often used biodiesel fuels are vegetable oil fatty acid methyl or ethyl esters produced by transesterification (Andre et al., 2010; Sendzikiene et al., 2007). For every three mol of ethyl esters one mol of crude glycerol is produced, which is equivalent to approximately 10 wt% of the total biodiesel production (Karinen & Krause, 2006; Pagliaro at al., 2009; Rahmat et. al., 2010). It is estimated that by 2016 the world biodiesel market will achieve the quantity of 37 billion gallons, which means that significantly more than 4 billion gallons of crude glycerol will be produced every year. The potential sale of this fraction might have an influence on the total price of biodiesel and make it cheaper (Fan et al., 2010; L. Wang et al., 2006). Pure glycerol may be used in many branches of industry, for example in food products (to sucrage liqueurs), cosmetics (as a moisturizing factor), textile industry, in pharmaceuticals, cellulosic industries; moreover, one can use it in nitrocellulose production as well as a supplement in fodder for pigs, swine, and hogs (Pagliaro et al., 2007; Z.X. Wang et. al., 2001). In contrast, use of raw glycerol is strictly limited because of its composition and a presence of pollutant substances. The main pollutants of this raw material include spent catalysts, residual methanol, mineral salts, heavy metals, mono- and diacylglycerols, free fatty acids and soaps (Dasari, 2007). The main biodiesel producers, thanks to the adequate installations inside their production plants, are able to purify raw glycerol. This is done via filtration, chemical steps, and filtration vacuum distillation. Then the technical grade glycerol (>97% pure, used for industrial type applications) or even refined USP grade glycerol (>99.7% pure, used in cosmetics, pharmaceuticals or food) is obtained. Unfortunately, such installations are too expensive for small or medium production plants. One solution to this problem is to sell raw glycerol to refineries in order to increase its value. Nevertheless, glycerol producers must pay for transportation of this glycerol fraction. Because transportation cost often equals or exceeds the price of raw glycerol, it does not make sense. Accordingly, this solution cannot be accepted (Johnson & Taconi, 2007). Therefore, new uses for crude glycerol must be sought and, luckily, many innovative methods of utilizations of this waste are under investigation. This chapter summarizes currently available studies and discusses possible ways of crude glycerol utilization, undertaken with the aim to improve economic viability of the biodiesel industry.

2. Crude glycerol as an animal feedstuff

An increase of price of maize (which results from an increase of biodiesel production) and accumulation of huge amount of crude glycerol, as a by-product during biodiesel production, resulted in new ideas: some scientists checked the possibility of application of this cheap crude glycerol as an animal feed ingredient instead of maize (Cerrate et al., 2006; Donkin, 2008; Dozier et al., 2008; Lammers et al., 2008a; Lammers et al. 2008b; Mourot et al., 1994). Biodiesel can be produced by a variety feedstock's such as mustard, rapeseed, canola, crambe, soybean oil, palm oil, sunflower oil, and waste cooking oils (Gerpen, 2005; F. Ma & Hanna, 1999; Moser, 2009; Thompson & He, 2006). The feedstock source and manufacturing process of biodiesel production are the key factors determining the composition of crude glycerol and therefore its nutritional value (Hansen et al., 2009; Thompson & He, 2006). Thereupon, it is indispensable to analyze physical, chemical and nutrient properties of crude glycerol with the prospect to include it into the animal diet. Hansen et al. (2009) conducted analytical tests of 11 crude glycerol samples collected from seven Australian biodiesel manufacturers. On the basis of conducted analyses they found that chemical composition of tested samples of crude glycerol varied considerably. The content of glycerol oscillated between 38 and 96%. In one sample the content of ash was more then 29%. This data confirms prior Tyson's et al. (2004) report which informed that content of mineral salts from transestrification can be 10 up to 30% of the crude glycerol by weight depending on the feedstock and process. It is known that high levels of potassium or sodium salts in the diet may result in electrolyte imbalance in animals (Dasari, 2007). Another potential hazardous compound in crude glycerol is methanol. Hansen et al. (2009) detected more than 4% of methanol in one of their research sample of crude glycerol and more than 11% of methanol in two other samples. The result of the methanol metabolic pathway is the accumulation of formate which excess cause toxic effect of methanol. Methanol poisoning may cause central nervous system injury, weakness, headache, vomits, blindness or Parkinsonian-like motor diseases (Dorman et al., 1993). FDA's Center for Veterinary Medicine decided that acceptable level of methanol in crude glycerol, which is used as a supplement in forage, cannot exceed 150 ppm, unless biodiesel producers prove that it has no negative influence on animal health (Dasari, 2007). Crude glycerol can be an attractive energy source for animal feed because of the similar energy value in comparison with corn and soybean meal, but the users of crude glycerol should bear in mind that it can pose a potential danger to the animals when the qualities of this biodiesel byproduct are not monitored properly (Dasari, 2007; Kerr & Dozier, 2008).

2.1 Glycerol in swine diets

Kijora et al. (1995) conducted two experiments with 48 fattening pigs to test the influence of glycerol as a component in diets. The pigs were fed up to 30% glycerol in barley-soya bean oil meal diets. In both these experiments, glycerol was used instead of barley. In the first experiment a fattening pig, which weights ca. 32 kg, eats fodder consisting of 5 and 10% of glycerol. In the second experiment, a fattening pig, which weight ca. 31.2 kg, is given fodder consisting of 5, 10, 20, and 30% of glycerol. In both of these experiments the control groups of pigs eat fodder without glycerol. The first experiment showed that pigs which eat fodder containg 5 and 10% of glycerol, had higher daily body weight gain. Probably it resulted from sweeter taste and better structure in fodder with glycerol. In the second experiment, no significant improvement in performance was detected in diets with 0, 5, 10, 20, 30% inclusion levels of glycerol. However, a diet containing 30% glycerol resulted in the significantly different feed conversion ratio in comparison with all other groups. These experiments did not demonstrate any pathological changes in kidney and liver in animals fed with fodder consisted glycerol. Moreover, diets have no influence on meat quality or carcass yield. Kijora et al. (1995) recommended supplementation of fodder with 10% of glycerol. In subsequent experiments, Kijora & Kupsch (1996) checked the possibility of application as a fodder supplement two technological distinct glycerols obtained in the process of biodiesel production. The dry weights of these glycerols were 77.6% and 99.7% and the ash content was 18.7 % and 4.8 %. Glycerol was added to fodder in quantity of 5 and 10%. The results from this experiment were collated with the results from the earlier experiments. The scientist found out that pigs in growing period ate 7.5% more fodder with glycerol (regardless of the type and its percentage content) then pigs which ate fodder without glycerol. Moreover, it was demonstrated that an increase of daily body weight strictly depended on the real consumption of glycerol. This effect was not observed in the finishing period. In 1997, Kijora et al. collated influence of application of glycerol, free fatty acids, and vegetable oil, as fodder supplements, on swine carcass backfat thickness and backfat composition in fattening pigs. In this experiment five different fodders were composed. To test each diet six pigs were used. The animals were fed with adequate amount of fodder during 14 weeks. After this time, the group of pigs which eat fodder with 10% of glycerol demonstrated the highest daily body weight gain. However, in collated to the feed conversion ratio there were any significant variations between these animal groups. These researches stated that supplemented animal's diet of glycerol caused decrease of content of polyenic acid in backfat in collated to other diets. The highest contents of palmitic and stearic acid were also observed in backfat pigs which eat fodder without additional fat. Doppenberg and Van Der Aar (2007) during experiments observed that sweet taste of fodder caused that pigs eat more and it resulted in higher daily body weight gain. They also stated that the maximum level of glycerol content in fodder was 5%. Lammers et al. (2007c) used 96 nursery pigs during 33 days of experiments in order to evaluate the influence of diet on weight gain. All pigs were 21 day-old and had the same weight. Animals had free access *ad libitum* to corn soybean isocaloric or isolysinic diets containing 0, 5, or 10% glycerol. No difference in pig performance according to diet was observed. The conclusion was that glycerol can be used as a fodder supplement for young pigs. Zijlstra et al. (2009) in their research used 72 weaned pigs in which three pelleted weat-based diets containing 0, 4 or 8% glycerol were applied. It occurred that glycerol used instead of wheat (up to 8%) can enhance growth performance of weaned pigs. Schieck et al. (2010) tested results of application of corn-soybean based diets containing 0, 3, 6 or 9% glycerol in lactating sows. No difference was observed in sows' performance according to diet. In conclusion they stated that glycerol (up to 9%) can be used as a supplement in diet of lactating sows as an alternative energy sources instead of maize.

2.2 Glycerol in poultry diets

Simon et al. (1996) in their research tested an influence of supplementation of glycerol on body weight gain, feed conversion ratio and N-balance in broiler chickens. The animals had *ad libitum* access to fodders with 0, 5, 10, 15, 20 and 25% of glycerol. It was stated that in result of a 31-day diet with forage supplemented of 5 and 10% of glycerol advantageous effects occurred with respect to the above parameters. The highest glycerol doses, the body weight gain, feed conversion ratio and N-balance in broiler chickens decreased. Pathological changes in kidney and liver were observed when the highest dose of glycerol was added. What is noteworthy, when 10% of glycerol was added to forage, chickens needed more water (Simon, 1996). In further experiments, Simon et al. (1997) observed an advantageous correlation between addition of glycerol to low-carbohydrate diet and an increase of nitrogen retention in the body. This effect was not observed in high-carbohydrate diets. Cerrate et. al. (2006) evaluated the usefulness of crude glycerol from biodiesel production as an energy source in broiler diets. Experiments were divided into two steps, each diet lasted 42 days. In the first step, 0, 5, and 10% of crude glycerol were added to forage. In the second step, 2.5 and 5% of glycerol were used. With an increase of crude glycerol in forage, the quantity of maize was decreased and quantity of soybean meal and poultry oil, to maintain these diets isocaloric and isonitrogenous, increased. The conclusion after the first step of this research was that the diet supplemented with 5% of glycerol had no influence in broilers performance in collate with died without glycerol. It was also observed that the forage with 10% of glycerol was not gladly consumed by animals what resulted in decrease in body weight gain. The second step of these experiments demonstrated that supplementation of forage with 2.5 and 5% of glycerol had no influence on body weight gain. However, significantly greater breast yield accounted as a percent of the dressed carcass for broilers fed with glycerol was observed. Abd-Elsamee et al. (2010) in their experiments use forage with 0, 2, 4, 6, and 8% of crude glycerol in broilers diets. The highest body weight and the body weight gain were observed when the forage with 6% of glycerol was used. What is important, the diet with 8% of glycerol did not cause any negative results on broiler chick performance, nutrient utilization or carcass characteristics. An influence of glycerol on egg performance and nutrient utilization in laying hens was tested by Świątkiewicz & Koreleski (2008). They tested forages in which corn starch was relieved by 2, 4, and 6% of glycerol. This forage was used in the diet of chicken between 28 and 53 weeks old. Effects showed that up to 6% of glycerol in the diet of laying hens had no negative influence on the performance or quality of eggs, nutrient retention, and metabolizability of energy. Other experiments demonstrated that supplementation of forage with 15% of glycerol had no negative influence on egg production, egg weight or egg mass of laying hens(Lammers et al., 2008b).

3. Catalytic conversion of glycerol

Recently, an increasing interest in the production of value-added chemicals from glycerol is observed. This tendency may lead to a decrease in biodiesel prices. Moreover, it can improve the glycerol market. Glycerol can be completely converted, among other compounds, into esters, ethers, propanediols, epichlorohydrin, acrolein and dihydroxyacetone.

3.1 Glycerol esters

Monoglycerides, polyglycerol esters and their derivatives can be obtained by the direct esterification of glycerol with carboxylic acids or by the transesterification of glycerol with carboxylic methyl esters or with triglicerides (Behr et al., 2007; Guerrero-Pérez et al., 2009). These compounds have wide applications as emulsifiers in food, cosmetic and pharmaceutical industries. Diaz et al. (2000, 2001) and Pérez-Pariente et al. (2003) reported the synthesis of monoglycerides by esterification of glycerol with lauric and oleic acids with functionalized ordered mesoporous materials containing R-SO_3H groups as catalysts. Diaz et al. (2005) investigated the influence of the alkyl chain length of HSO_3-R-MCM-41 on the esterification with the fatty acid mentioned above. On the basis of conducted research, the optimum balance between the nature of the organic groups supporting the sulfonic acid, the distance between sulfonic groups, and the porosity of material was determined. R. Nakamura et al. (2008) conducted the esterification of glycerol with lauric acid catalyzed by multi-valet metal salts to form mono- and dilaurins. They found that $ZrOCl_2 \cdot 8H_2O$ and $AlCl_3 \cdot 6H_2O$ are chloride's catalysts which are active in the formation of monolaurin, and Fe2 $(SO_4)_n \cdot H_2O$ and Zr $(SO_4)_n \cdot H_2O$ are sulfate's catalysts which have beneficial effects fo dilaurin production. These compounds have numerous applications in pharmaceutical industry, e.g. monolaurin is biologically active against HIV virus.

Another interesting compound is glycerol carbonate which is the cyclic ester of glycerol with carbonic acid (Behr et al. 2008). Glycerol carbonate has a tremendous potential in the chemical industry as a novel component of gas separation membranes, as a solvent, e.g. in colours, glues, cosmetics and pharmaceuticals, and as a source of new polymeric materials such as glycidol used in the production of polyurethanes and polycarbonates (Guerrero-Perez et al., 2009; Johnson & Taconi, 2007; Pagliaro et al., 2007; Rokicki et al., 2005). Vieville et al. (1998) described production of glycerol carbonate by direct carboxylation of glycerol with carbon dioxide in the presence of zeolites or ion exchange resins. Aresta et al. (2006) reported the first evidence of direct carboxylation of glycerol with carbon dioxide under Sn-complexes catalysis. However, those catalytic conversions need high pressures or supercritical conditions. Kim et al. (2007) reported the first enzymatic example of glycerol carbonate synthesis. They demonstrated the procedure of transesterification of renewable glycerol and dimethyl carbonate in the presence of immobilized lipase isolated from *Candida antarctica*. Enzymatic synthesis of glycerol carbonate has a huge potential because of the mild reaction conditions and high selectivity (Carrea & Riva, 2000).

3.2 Glycerol ethers

It is well known that glycerol cannot be directly added to fuel because of its polymerization in high temperatures and partial oxidation to toxic acrolein (Pagliaro et al., 2007). However, glycerol can be transformed into fuel additives, thanks to selective etherification. The glycerol tert-butyl ethers (GTBEs) can be synthesized by the reaction of glycerol with tert-butanol as well as with isobutene (Behr et. al., 2008). GTBEs have been studied as analogs of environmentally unfriendly methyl tert-butyl ethers (MTBEs) or ethyl tert-butyl ethers (ETBEs) which are currently added to fuels (Karinen & Krause, 2006; Monbaliu et al., 2010). The application of GTBEs leads to reduction of particulate matter, carbon monoxide, hydrocarbons and unregulated aldehydes in emissions (Pagliaro et al., 2007). The addition of these ethers into biodiesel also decreases the cloud point to a value similar to conventional diesel (Rahmat et al., 2010). Klepáčová et al. (2003) demonstrated the etherification of glycerol with tert-butanol at the presence of catex Amberlyst 15 as catalyst. On the basis of conducted experiments, almost 96% conversion of glycerol was detected at 90°C, reaction time 180 min and at the molar ratio tert-butanol/glycerol = 4:1. Klepáčová et al. (2005) also presented a comparative study of etherification of glycerol with tert-butanol and isobutene without solvent in a liquid phase. It was found that conversions of glycerol in the same temperature and with the same catalyst were always higher when isobutene was used. The 100% conversion of glycerol with selectivity of glycerol di- and tri- ethers higher than 92% in the etherification of glycerol with isobutene and over strong acid macroreticular ion-exchange resins as catalyst was achieved. Similar results were obtained by Melero et al. (2008) who reported the results of etherification of glycerol with isobutene and sulfonic mesostructured silicas used as catalysts. They also obtained 100% conversion of glycerol and selectivity of glycerol di- and tri- ethers up to ca. 90% without undesirable isobutylene formation. Another method of crude glycerol utilization is telomerization with 1,3-butadiene to form C_8 chain ethers, which have a wide range of applications such as useful building blocks for commercially valuable products such as detergents and surfactants (Behr et al. 2009). Most recently, direct telomerization of pure as well as crude glycerol with 1,3-butadiene carried out over palladium complexes as catalysts was reported (Palkovits et al., 2008a, 2008b).

3.3 Propanediols

Glycerol can be used in chemical production of a value-added compound called 1,2-propanediol known also as propylene glycol. It is widely used as a humectant food additive (E1520), to maintain moisture in medicines, cosmetics and tobacco products, and as a solvent for food colors and flavorings. 1,2-propanediol can be also used as a substitute for ethylene glycol in anti-freeze, especially since ethylene glycol unlike 1,2-propanediol is toxic and is banned in Europe (Johson & Taconi, 2007). Dasari et al. (2005) evaluated the production of 1,2-propanediol by glycerol hydrogenation in presence of nickel, palladium, platinum, copper, and copper-chromite catalysts at 200°C and less than 14 bar hydrogen pressure. They obtained 1,2-propanediol yields > 73% when copper-chromite was used as a catalyst. Perosa & Tundo (2005) converted glycerol to 1,2-propanediol in presence of raney-nickel as hydrogenation catalyst, at 150°C and low hydrogen pressure 10 bar. In these conditions, they detected 93% selectivity toward 1,2-propanediol and small amounts of ethanol and CO_2. When temperature of reactions increased to 190°C, it proceeded faster, but the selectivity of 1,2-propanediol dropped to 70-80% and ethanol and CO_2 as the sole by-products. They also observed that addition of a phosphonium salts improved the selectivity and rate toward 1,2-propanediol. S. Wang & Liu (2007) converted glycerol to 1,2-propanediol on Cu-ZnO catalysts in temperatures between 180-240°C and high hydrogen pressure 42 bar. They found that glycerol conversion and 1,2-propanediol selectivity depends on Cu and ZnO particle sizes. The 83.6% 1,2-propanediol selectivity at 22.5% glycerol conversion was achieved at 200°C in presence Cu-ZnO catalyst with relatively small Cu particles. Marinoiu et al. (2009) conducted the hydrogenolysis process with nickel catalyst which resulted in up to 98% 1,2-propanediol selectivity and 30% glycerol conversion in moderate temperature and pressures (200°C and 20-25 bar, respectively). Recently, Wu et al. (2011) reported hydrogenolysis of glycerol to 1,2-propanediol *via* hydrogen spillover by using Cu-Ru nanoparticle catalyst supported on nanotubes. Another propanediol with numerous applicatons is 1,3-propanediol. It is used as a monomer in the synthesis of a new type of polyesters such as polytrimethylene and terephthalate. It also found an application as a chemical intermediate used in the manufacture of polymers, cosmetics, medicines and heterocyclic compounds (Saxena et al., 2009). The catalytic conversion of glycerol to 1,3-propanediol was examined by Kurosaka et al. (2008) where glycerol hydrogenolysis was catalyzed by Pt/WO_3 supported on ZrO_2 with yields up to 24% toward 1,3-propanediol. Nakagawa et al. (2010) reported direct hydrogenolysis of glycerol over rhenium-modified iridium nanoparticle catalyst with 1,3-propanediol yield of 38% at 81% of glycerol conversion. These results are promising, nevertheless, special attention is paid to microbiological conversion of glycerol to 1,3-propanediol.

3.4 From glycerol to epichlorohydrin

The availability of large amount of crude glycerol has encouraged the development technologies that can use glycerol as a raw material in the production of epoxides. One such product is epichlorohydrin (Lewandowski et al., 2008). Epichlorohydrin is largely used in the production of epoxy resins (Herliati et al., 2011). It also found application in the production of pharmaceuticals, textile conditioners, dyes and paper sizing agents (Gerrero-Peréz et al., 2009). Dow Chemical Company has made epichlorohydrin from crude glycerol by using a GTE process which proceeds in two main steps (Bell et al., 2008). The process is comprising hydrochlorination of glycerol with hydrogen chloride gas at elevated temperature and pressure over carbocylic acid as catalyst resulting in a 30-50:1 mixture of 1,3-dichloropropan-2-ol and 2,3-dichloropropan-1-ol, followed by reaction with base to give epichlorohydrin (Bell et al., 2008). The Solvay Company has produced the epichlorohydrin at the reserved name Epicerol. First glycerol is chlorinated with anhydrous hydrogen chloride at moderate temperature to give 1,3-dichloropropan-2-ol and then by the addition of sodium hydroxide Epicerol is formed (Behr et al, 2008; Pagliaro et al., 2007).

3.5 Dehydration of glycerol to acrolein

Acrolein is a valuable versatile intermediate used in the production of acrylic acid, acrylic acid esters, detergents or super absorber polymers which can be used as retention agents in the production of paper (Corma et al., 2008; Fan et al., 2010; Ott et al., 2006). Ott et al. (2006) reported the usage of sub- or supercritical water as the reaction media. The maximum 75 mol% selectivity of acrolein with 50% of glycerol conversion was obtained at 360°C, 25 MPa and with the addition of zinc sulfate. Watanabe et al. (2007) dehydrated glycerol to acrolein in hot-compressed water. About 80 mol% selectivity of acrolein at 90% glycerol conversion at 400°C, 34.5 MPa over H_2SO_4 as a catalyst was achieved. Tsukuda et al. (2007) studied acrolein production over several solid catalysts. The acrolein selectivity >80 mol% with almost 100% glycerol conversion was achieved over silicotungstic acid supported on silica with mesopores of 10 nm, at 275°C and ambient pressure. Atia et al. (2008) investigated conversion of glycerol to acrolein using various heteropolyacid catalysts as active compounds. In this study, alumina was found to be superior to silica as support material. The 75% selectivity of acrolein at 100% glycerol conversion over silicotungstic acid supported over alumina and aluminosilicate was achieved. Corma et al. (2008) converted glycerol to acrolein by reacting gas-phase glycerol/water mixtures with a zeolite catalyst. The highest yield to acrolein was found at 350°C with a ZSM5 zeolite-based catalyst. They also detected that by increasing the temperature from 350°C to 500°C the conversion of glycerol from acrolein toward acetaldehyde was favored. Yan & Suppes (2009) investigated low-pressure packed-bed gas-phase dehydration of glycerol to acrolein. At the 0.85 MPa pressure, 260°C and over H_3PO_4/activated carbon catalyst, the 85% selectivity and almost 67% yield of acrolein was achieved. Ulgen & Hoelderich (2011) reported the dehydration of glycerol to acrolein with 85% selectivity in the presence of novel WO_3/TiO_2 catalysts in a continuous flow fixed bed reactor.

3.6 Oxidation of glycerol to dihydroxyacetone

Dihydroxyacetone is a value-added chemical currently used in cosmetics as the main active ingredient in all sunless tanning skincare preparations (Nguyen & Kochevar, 2003). It also serves as a building block in the synthesis of various fine chemicals (Enders et al., 2005;

Zheng et al., 2008). Dihydroxyacetone can be produced from glycerol *via* selective oxidation of its secondary hydroxyl groups (Pagliaro et al., 2007). Garcia et al. (1995) studied chemoselective oxidation of glycerol with air on platinum metals. On the basis of conducted experiments it was found that the main product of glycerol oxidation in presence of Pd/C or Pt/C catalyst is glyceric acid (70 and 55% selectivity, respectively). However, deposition of bismuth on platinum particles orientates selectivity toward secondary hydroxyl groups. In this case 50% selectivity of dihydroxyacetone at 70% conversion of glycerol was achieved. Recently, W. Hu et al. (2010) investigated the selective oxidation of glycerol to dihydroxyacetone in semibatch reactor over Pt-Bi catalyst. The optimization study revealed that the maximum dihydroxyacetone yield of 48% at 80% glycerol conversion at 80°C, 0.2 MPa and initial pH=2 was achieved. Still better results were obtained by Kimura et al. (1993) who found that incorporation of bismuth in platinum and usage of fixed-bed reactor causes 80% dihydroxyacetone selectivity at 40% glycerol utilization. Nowadays, microbial route of dihydroxyacetone production by using *Gluconobacter oxydans* is found to be more favorable as compared to chemical methods (Mishra et al., 2008).

4. Crude glycerol utilization in biotechnology

Bioconversion of glycerol, used as a carbon source for microorganisms' growth, makes it possible to eliminate problems of applying this raw material in chemical catalysis processes (such as high temperature and pressure, use of high specific cofactors). Because of higher level of glycerol reduction in comparison to conventional raw materials in microbial media, higher productivity of glycerol conversion is expected (Gonzalez et al., 2008; Himmi et al., 2000). The conversion of glycerol in glycolytic pathway into phosphoenolopyruvate or pyruvate induces production of double amount of reduction equivalents in comparison to glucose or xylose metabolism. Thus, glycerol delivers more energy indispensable to subsequent reactions of conversion (Barbirato et al., 1997; Yazdani & Gonzalez, 2007). From an economical point of view, it is more profitable to use glycerol in oxygen-free processes because of lower costs of equipment and lower consumption of energy (Yazdani & Gonzalez, 2007). Glycerol can be converted among other compounds, into 1,3-propanediol, propionic acid, succinic acid, citric acid, dihydroxyacetone, hydrogen, ethanol, pigments, polyhydroxyalcanoates and biosurfactants.

4.1 1,3-propanediol

1,3-propanediol is a typical and the oldest product of glycerol fermentation (Katrlík et al., 2007). A number of microorganisms which can grow anaerobically on glycerol are known among others *Clostridium diolis, Clostridium acetobutylicum, Clostridium butylicum, Clostridium perfingens, Clostridium butyricum, Clostridium pasteurianum, Enterobacter aerogenes, Enterobacter agglomerans, Klebsiella oxytoca, Klebsiella pneumoniae, Citrobacter freundii, Lactobacillus collinoides, Lactobacillus reuterii, Lactobacillus buchnerii, Pelobacter carbinolicus, Rautella planticola,* and *Bacillus welchii* (da Silva et al., 2009, Drożdżyńska et al., 2011). Papanikolau et al. (2004) investigated 1,3-propanediol biosynthesis by *Clostridium butyricum.* In batch fermentation 47.1 g/L of 1,3-propanediol was obtained. In a continuous process at the 0.04/h dilution rate, up to 44 g/L of 1,3-propanediol was formed. Yang et al. (2007) converted glycerol to 1,3-propanediol by using lactate-deficient mutant of *Klebsiella oxytoca.* On the basis of these experiments 83.56 g/L of 1,3-propanediol with yield of 0.62 mol/mol of glycerol and productivity of 1.61 g/L/h. In 2007, Cheng et al. reported the first pilot-scale

1,3-propanediol production using *Klebsiella pneumoniae*. 1,3-propanediol concentration in 5000 L fermentation reached almost 59 g/L with yield of 0.53 mol/mol of glycerol and productivity of 0.92 g/L/h. Guo et al. (2009) reported that *Klebsiella pneumoniae* CPS-deficient mutant (unable to produce capsular polysaccharides) in fed-batch fermentation was able to synthesize 78.13 g/L of 1,3-propanediol with yield and productivity of 0.53 mol/mol of glycerol and 1.95 g/L/h, respectively. Tang et al. (2009) studied bioconversion of glycerol to 1,3-propanediol using engineered strain of *Escherichia coli*. In this study, vector comprising *yqhD* gene of *Escherichia coli*, together with *dhaB1* and *dhaB2* genes of *Clostridium butyricum*, was created. The fermentation process was divided into two stages. First, *Escherichia coli* mutant was cultivated on culture medium with glucose, at 30°C and in the aerobic conditions to obtain high density of culture next followed by anaerobic glycerol fermentation at 42°C in order to enhance 1,3-propanediol production. This strategy allowed to obtain 104.4 g/L of 1,3-propanediol with yield of 1.09 mol/mol of glycerol and productivity of 2.61 g/L/h. Nowadays, Du Pont and Genencor International, Inc. can produce up to 135 g/L of 1,3-propanediol from glucose by genetically engineered strains of *Escherichia coli* (Drożdzyńska et al., 2011; C.E. Nakamura & Whited, 2003; Saxena et al., 2009). However, 1,3-propanediol biosynthesis from crude glycerol, because of its low price, seems to be economically favorable (Maervoet et al., 2011).

4.2 Propionic acid

Microorganisms which are not able to produce 1,3-propanediol, can offer other industry useful metabolite products (da Silva et al., 2009; Yazdani and Gonzalez, 2007). One possible use of crude glycerol is its utilization to propionic acid in fermentation processes. The microorganisms in this process are *Propionibacterium* (Himmi et al., 2000; A. Zhang & Yang, 2009; Zhu et al., 2010). Propionic acid is widely used in chemical, pharmaceutical and food industries. An important application of this acid is use as a fodder preservative and as a fixing agent for cheeses and baker's products (Kumar et al., 2006). Himmi et al. (2000) in their experiment checked an influence of two different bacterial media on the effectiveness of fermentation. The first medium was with glucose as a carbon source and the other with glycerol. The researchers used two bacterial strains, *Propionibacterium acidipropionici* and *Propionibacterium freudenreichii* ssp. *shermanii*. The highest production of propionic acid (in case of both strains) was observed in fed-batch cultivation. The cultivation was fed by glycerol. The productivity was about 49% higher (0.79 mol of propionic acid per 1 mol of glycerol) when *Propionibacterium acidipropionici* strain was used and about 45% (0.58 mol of propionic acid per 1 mol of glycerol) when *Propionibacterium freudenreichii* ssp. *shermanii* strain was used in comparison to cultivation feeding by glucose. Zhu et al. (2010) converted glycerol to propionic acid using propionic acid-tolerant strain of *Propionibacterium acidipropionici* in a large scale of 10m^3 bioreactor. The highest propionic acid production reached ca. 47g/L at 240h of process duration. A. Zhang & Yang (2009) studied the production of propionic acid with metabolically engineered *Propionibacterium acidipropionici* (ACK-Tet). It was found that adapted ACK-Tet mutant produced propionic acid with much higher yield than that from glucose (0.54-0.71g/g vs. 0.35 g/g, respectively). The maximum propionic acid concentration of about 106 g/L from glycerol fermentation was obtained.

4.3 Succinic acid

Succinic acid is used in production of synthetic gum and biodegradable polymers, such as polybutyrate succinate and polyamides (Zeikus et al., 1999; Song & Lee, 2006). Lee et al.

(2001) investigated bioconversion of glycerol to succinic acid by *Anaerobispirillum succiniproducens*. During this experiment the efficiency of succinic acid production was about 133% when glycerol was used as a raw material in comparison to use of glucose. The weight ratio of succinic acid to acetic acid (by-product) was 25.8:1 and it was 6.5 times higher than in case of using glucose as a carbon source. The consumption of glycerol by *Anaerobispirillum succiniproducens* cells depended on the quantity of yeast extract in bacterial medium. The highest succinic acid production was observed in cultivation feeding with glycerol and yeast extracts. In this experiment the efficiency of succinic acid production was about 160% and the weight ratio of succinic acid to acetic acid (by-product) was 37.1:1. Scholten & Dägele (2008) reported succinic acid production from crude glycerol by facultative anaerobic rod-shaped bacteria belonging to the family *Pasteurellaceae* with similarity of the genus *Mannheimia*. In experiments as a carbon source three different variants of crude glycerol were used: C1 with 90% of pure glycerol, C2 with 42% of pure glycerol, and 76% of pure glycerol. The results were compared with the results obtained from experiment in which medium contained pure glycerol (99%). After 9 hours of fermentation, the quantity of succinic acid was equal to 7.6 g/L (when glycerol C1 was used), 8.4 g/L (when glycerol C2 was used), and 7.4 g/L (when glycerol C3 was used). When pure glycerol was used the production was on a lower level - 6.2 g/L. Scholten et al. (2009) conducted the continuous fermentation of crude glycerol to succinic acid using *Basfia succiniproducens* DD1 isolated from the rumen of a cow. The highest concentration of 5.21 g/L of succinic acid at the 0.018/h dilution rate was achieved. Recently, several studies have been reported regarding the succinate production from glycerol by metabolically engineered strains of *Escherichia coli* (Blankschien et al.; 2010; X. Zhang et al., 2009, 2010).

4.4 Citric acid

Citric acid is used as a natural food preservative and also as appropriate flavoring feature of food and drinks. Other applications of this acid include a soft detergent, stabilizer and antioxidant (Soccol et al., 2006). In general, citric acid is formed *via* submerged microbial fermentation on molasses using *Aspergillus niger* (Ali et al., 2002; Papagianni, 2007). In recent years, several investigations concerned the use of *Yarrowia lipolytica* in crude glycerol conversion to citric acid (Lewinson et al.; 2007; Rymowicz et al., 2010; Rywińska et al., 2011; Rywińska & Rymowicz, 2011). Lewinson (2007) reported 21.6 g/L of citric acid production in batch fermentation of glycerol with 54% yield using *Yarrowia lipolytica* NRRL YB-423. Imandi et al. (2007) achieved 77.4 g/L of citric acid from crude glycerol using statistically optimized culture medium for *Yarrowia lipolytica* NCIM 3589. Rywińska et al. (2010) studied the efficiency of crude glycerol fermentation toward citric acid in fed-batch system using two acetate negative-mutants of *Yarrowia lipolytica*.. In these experiments 155.2 and 157.5 g/L of citric acid was formed from 300g/L of glycerol using *Yarrowia lipolytica* Wratislavia 1.31 and Wratislavia AWD7 mutants, respectively. Rymowicz et al. (2010) converted crude glycerol to citric acid by *Yarrowia lipolytica* A-101-1.22 in different types of cultures. The citric acid production was 112 g/L, 124.2 g/L and 96-107 g/L in batch, repeated batch and cell recycled systems, respectively.

4.5 Dihydroxyacetone

Bacteria *Gluconobacter oxydans* are able to microbiological bioconversion of glycerol to dihydroxacetone (Bauer et al., 2005; Claret et al., 2004; Z.C. Hu et al., 2010; Li et al., 2010; L.

Ma et al., 2010). Bories et al. (1991) stated that biosynthesis of dixydroxyacetone by *Gluconobacter oxydans* makes cells of this microorganisms grow slower. The concentration of dihydroxacetone on the level of 61 g/L completely inhibits growth of the cells. Z.C. Hu et al. (2010) worked on optimization of cultivating media with an aim to increase dihydroxyacetone production by *Gluconobacter oxydans* ZJB09112. The cultivations were carried out in shake flask and bubble column bioreactors. After feed bath cultivation (five-time glycerol feeding) on the optimized medium 161.9±5.9 g/L of dihydroxyacetone at 88.7±3.2% glycerol conversion rate was obtain. Gätgens et al. (2007) reported that over expression of glycerol dehydrogenase in *Gluconobacter oxydans* DSM2343 enhanced dihydroxyacetone production. They achieved up to about 30 g/L of dihydroxyacetone from 50 g/L of glycerol which was 20-40% more in comparison with control strains. Li et al. (2010) also investigated the over expression of glycerol dehydrogenase (GDH) in an alcohol dehydrogenase (ADH)-deficient mutant of *Gluconobacter oxydans* M5AM/GDH in order to improve dihydroxyacetone productivity. It was found that absence of ADH together with over expression of GDH gene substantially improved dihydroxyacetone formation. In four repeated biotransformations biosynthesis of dihydroxyacetone reached 385 g/L.

4.6 Hydrogen and ethanol co-production

Since hydrogen is expected in the future to be a clean energy source and ethanol probably be used as gasoline's additive, there are several important issues regarding co-production of these metabolites. Glycerol fermentation to hydrogen and ethanol using *Enterobacter aerogenes* HU-101 was studied by Ito et al. (2005). In continuous culture with packed-bed reactor with self-immobilized cells and using culture medium with pure glycerol, the maximum hydrogen production rate reached 80 mmol/L/h with ethanol yield of 0.8 mol/mol of glycerol. It was much higher a quantity than obtained from crude glycerol (30 mmol/L/h). Nevertheless, the application of porous ceramics as support material to fix cells in the reactor improved the results in case of crude glycerol fermentation process. In this case, hydrogen production rate was 63 mmol/L/h with ethanol yield of 0.85 mol/mol. In turn, Sakai & Yagishita (2007) reported crude glycerol bioconversion to hydrogen and ethanol by bioelectrochemical cells of wild type of *Enterobacter aerogenes* NBCR 12010 using thionine as an exogenous electron transfer mediator. On the basis of conducted experiments, one can say that the yield of hydrogen as well as ethanol reached more than 80%. Yazdani & Gonzalez (2008) studied co-production of ethanol and hydrogen from crude glycerol using engineered strain of *Escherichia coli* SY03. They inactivated fumarate reductase and phosphate acetyltransferase in order to minimize succinate and acetate co-production and overexpressed glycerol dehydrogenase and dihydroxyacetone kinase which are responsible for glycerol utilization to dihydroxyacetone phosphate (glycolytic pathway intermediate). On the basis of conducted experiments they reported that the ethanol-hydrogen co-production exceeded 95% yield of theoretical and specific rates.

4.7 Pigments

There are several studies regarding glycerol utilization to pigments, such as prodigiosin or carotenoids (da Silva et al., 2009). The red pigment prodigiosin is well known as an antifungal, immunosuppressive and anti-proliferative agent (Casullo de Araújo et al., 2010; Khanafari et al., 2006). Tao et al. (2005) described prodigiosin biosynthesis from glycerol by *Serratia marcescens* mutant obtained *via* ultra violet light mutation and rational screening methods.

They achieved best results in a two-step feeding strategy. In the first step, *Serratia marcescens* mutant was cultivated on culture medium with glucose in order to obtain high density of culture. In the second step, glycerol was used to induce prodigiosin production. This strategy allowed to obtain 583 mg/L of prodigiosin, which was 7.8 times higher a quantity than originally obtained from the parental strain. Among carotenoids, astaxanthin is widely used in aquaculture, pharmaceuticals, supplements and natural coloring (da Fonseca et al., 2011). Having antioxidant activity, astaxanthin helps to prevent, for instance, photo-oxidation from ultra violet light, inflammations, infectious ulcers from *Helicobacter pylori* or diseases related with aging (da Fonseca et al., 2011; O'Connor & O'Brien, 1998). In 1998, Kusdiyantini et al. reported utilization of glycerol to astaxanthin using *Phaffia rhodozyma*. After 168 h of batch fermentation astaxanthin concentration reached 33.7 mg/L. Razavi & Marc (2006) investigated optimal culture conditions for the total carotenoid production from technical glycerol using *Sporobolomyces ruberrimus* H110. In this study, the best results were obtained at 19°C and pH=6.0. In these conditions, the maximum total carotenoid concentration (torularhodin + β-carotene) of 3.84 mg/g of yeast was achieved. Recently, Yimyoo et al. (2011) demonstrated the potential of glycerol bioconversion by *Rhodosporidium paludigenum* as a new carotenoid producer. The maximum total carotenoid concentration (torularhodin + torulene + β-carotene) reached 3.43 mg/L (0.45 mg/ g of yeast) after 132 h of process duration, at 32°C and pH=6 in a medium containing 40 g/L of glycerol.

4.8 Polyhydroxyalcanoates

Polyhydroxyalcanoates (PHAs) belongs to microbial polyesters which are synthesized intracellulary by numerous bacteria under nutrient-limiting conditions, such as carbon and energy storage material (Solaiman et al., 2006; Tian et al., 2009). Since PHAs posseses biodegradable and biocompatible thermoplastic properties, increased attention is paid as to how replace petroleum-derived polymers by PHAs (Albuquerque et al., 2007; Zinn & Hany, 2005). Several bacteria have been explored in order to facilitate PHAs production from glycerol. In 2009, Ibrahim & Steinbüchel investigated several fed-batch cultivation strategies to improve production of polyhydroxybutyrate (PHB) by *Zobellella denitrificans* MV1. The PHB concentration reached up to 54.3±7.9 g/L at cell dry weight of 81.2±2.5 g/L after 50 h of fermentation in presence of 2% NaCl and with optimized feeding of glycerol and ammonia. In this study, PHB content of 66.9±7.6% of dry cell weight was observed. Cavalheiro et al. (2009) reported PHB production from crude glycerol by using *Cupriavidus necator* DSM 545. In this study, the maximum cell dry weight of 82.5 g/L with PHB accumulation of 38% was obtained. In turn, Kawata & Aiwa (2010) obtained 39% PHB yield from 3% crude glycerol culture after 47 h of *Halomonas* sp. KM-1 cultivation. Promising results were obtained by Shrivastav et al. (2010) who used *Halomonas hydrothermalis* in the production of PHB from crude glycerol. In this study, isolated SM-P-3M was found to be the most effective with PHA content of 75% of dry cell weight.

4.9 Biosurfactants

Surfactants have numerous household and industrial applications. Most of them, however, are based on petroleum and are chemically produced (Makkar et al., 2011). Since biosurfactants are less-toxic, biodegradable and have some unique surface-active properties, they might be an excellent substitutes for surfactants derived from petroleum (Banat et al., 2010). A lot of research in recent years was made in a field of biosurfactants production from glycerol (da Silva et al., 2009). Accordingly, in 2002, Rahman et al. obtained 1.77 g/L of rhamnolipid surfactant from glycerol, using *Pseudomonas aeruginosa* DS10-129. In turn, G.L. Zhang et al (2005) achieved rhamnolipids' concentration of 15.4 g/L, using *Pseudomonas aeruginosa* when cultured on basal mineral medium containing glycerol as a sole carbon source. Da Rosa et al. (2010) investigated the influence of culture medium components on rhamnolipid production by *Pseudomonas aeruginosa* LBM10, using experimental designs and response surface methodology. On the basis of conducted analyses, the maximum rhamnolipid concentration of 4.15 g/L at glycerol concentration of 13.2 g/L, C/N ratio of 12.8 and C/P ratio of 40 was obtained. Morita et al. (2007) investigated microbial conversion of crude glycerol to mannosylerythritol lipids (MELs) as glycolipid biosurfactants, using basidiomycete yeast, *Pseudozyma antarctica*. In this case the concentration of MEL reached 16.3 g/L by intermittent feeding of glycerol. Recently, Liu et al. (2011) optimized culture medium composition and environmental factors in order to improve glycolipids production from crude glycerol in *Ustilago maydis*. As a result of these actions, 32.1 g/L of total glycolipids after 8.2-day fed-batch bioprocess was obtained.

5. Conclusions

Energy from the renewable sources is one of the solutions connected with depletion of the world's nonrenewable energy sources such as carbon, petroleum gas or crude oil. This kind of energy is also beneficial for the natural environment (it can decrease an emission of CO_2). However, the cost of energy production from renewable sources is very high and it exceeds prices of traditional fuels (Fernando et al., 2007; Karinen & Krause, 2006; Wilke & Vorlop, 2004). Thus, alternative solutions are still sought in many places. Nowadays, biodiesel costs more than fuels produced from crude oil. For instance, prices of raw materials (which differ depending on the kind of raw material used), technology of production, taxes, and tax allowances - all have a major influence on the final cost of biodiesel (Hass et al., 2006). Obviously, the production of biofuels could become profitable but only on condition that producers would sell both the biofuels and by-products obtained in the process of manufacture, thus allowing for the development of processes which today are regarded as inexpensive. We may conclude that crude glycerol utilization into higher value products is critical when we want to make biodiesel production more sustainable and profitable.

6. References

Abd-Elsamee, M.O.; Abdo, Z.M.A.; EL-Manylawi, M.A.F. & Salim, I.H. (2010). Use of crude glycerine in broiler diets. *Egyptian Poultry Science Journal,* Vol.30, No.1, pp. 281-295, ISSN 1110-5623

Albuquerque, M.G.E; Eiroa, M.; Torres, C.; Nunes, B.R. & Reis, M.A.M. (2007). Strategies for the development of a side stream process for polyhydroxyalkanoate (PHA) production from sugar cane molasses. *Journal of Biotechnology,* Vol.130, pp. 411-42, ISSN 0168-1656

Ali, S.; Ikram. H.; Qadeer, M.A. & Iqbal, J. (2002). Production of citric acid by *Aspergillus niger* using cane molasses in a stirred fermentor. *Electronic Journal of Biotechnology,* Vol.5, No.3, pp. 258-271, ISSN 0717-3458

André, A.; Diamantopoulou, P.; Philippoussis, A.; Sarris, D.; Komaitis, M. & Papanikolaou, S. (2010). Biotechnological conversions of bio-diesel derived waste glycerol into added-value compounds by higher fungi: production of biomass, single cell oil and oxalic acid. *Industrial Crops and Products,* Vol.31, pp. 407-416, ISSN 0926-6690

Aresta, M.; Dibenedetto, A.; Nicito, F. & Pastore C. (2006). A study on the carboxylation of glycerol to glycerol carbonate with carbon dioxide: The role of the catalyst, solvent and reaction conditions. *Journal of Molecular Catalysis A: Chemical,* Vol.257, pp. 149-153, ISSN 1381-1169

Atia, H.; Armbruster, U. & Martin, A. (2008). Dehydration of glycerol in gas phase using heteropolyacid catalysts as active compounds. *Journal of Catalysis,* Vol.258, pp. 1349-1353, ISSN 0021-9517

Banat, I.M.; Franzetti, A.; Gandolfi, I.; Bestetti, G.; Martinotti, M.G.; Fracchia, L.; Smyth, T.J. & Marchant, R. (2010). Microbial biosurfactant production, applications and future potential. *Applied Microbiology and Biotechnology,* 87, No.2, pp.427-444, ISSN 1432-0614

Barbirato, F.; Himmi, E.H.; Conte, T. & Bories, A. (1997). Propionic acid fermentation from glycerol: comparison with conventional substrates, *Applied Microbiology and Biotechnology,* Vol.47, pp. 441-446, ISSN 1432-0614

Bauer, R.; Katsikis, N.; Varga, S. & Hekmat. D. (2005). Study of the inhibitory effect of the product dihydroxyacetone on *Gluconobacter oxydans* in a semi-continuous two-stage repeated-fed-batch process. *Bioprocess and Biosystems Engineering,* Vol.5, pp. 37-43, ISSN 1615-7605

Behr, A.; Eilting, J.; Irawadi, K.; Leschinski, J. & Lindner, F. (2007). Improved utilisation of renewable resources: New important derivatives of glycerol. *Green Chemistry,* Vol.10, No.1, pp. 13-30, ISSN 1463-9270

Behr, A.; Leschinski, J.; Awungacha, C.; Simic, S. & Knoth, T. (2009). Telomerization of butadiene with glycerol: Reaction control through process engineering, solvents, and additives. *ChemSusChem,* Vol.2, No.1., pp. 71-76, ISSN 1864-564X

Bell, B.M.; Briggs, J.R.; Campbell, R.M.; Chambers, S.M.; Gaarenstroom, P.D.; Hippler, J.G.; Hook, B.D.; Kearns, K.; Kenney, J.M.; Kruper, W.J.; Schreck, D.J.; Theriault, C.N. & Wolfe, C.P. (2008). Glycerin as a renewable feedstock for epichlorohydrin production. The GTE process. *Clean,* Vol.36, No.8, pp. 657-661, ISSN 1863-0669

Blankschien, M.D.; Clomburg, J.M. & Gonzalez, R. (2010). Metabolic engineering of *Escherichia coli* for the production of succinate from glycerol. *Metabolic Engineering,* Vol.12, No.5, pp. 409-419, ISSN 1096-7176

Bories, A.; Claret, C. & Soucaille, P. (1991). Kinetic study and optimization of the production of dihydroxyacetone from glycerol using *Gluconobacter oxydans. Process Biochemistry,* Vol.26, pp. 243-248, ISSN 1359-5113

Carrea, G & Riva, S. (2000). Properties and synthetic applications of enzymes in organic solvents. *Angewandte Chemie International Edition,* Vol.39., No.13, pp. 2226-2254, ISSN 1521-3773

Casullo de Araújo, H.W.; Fukishima, K. & Campos Takaki, G.M. (2010). Prodigiosin production by *Serratia marcescens* USC 1549 using renewable-resources as a low cost substrate. *Molecules,* Vol.15, No.10, pp. 6931-6940, ISSN 1420-3049

Cavalheiro, J.M.B.T.; de Almeida, M.C.M.D.; Grandfils, C. & da Fonseca, M.M.R. (2009). Poly(3-hydroxybutyrate) production by *Cupriavidus necator* using waste glycerol. *Process Biochemistry,* Vol.44, pp. 509-515, ISSN 1359-5113

Cerrate, S.; Yan, F.; Wang, Z.; Coto, C.; Sacakli, P. & Waldroup, P.W. (2006). Evaluation of glycerine from biodiesel production as a feed ingredient for broilers. *International Journal of Poultry Sciences,* Vol.5, pp. 1001-1007, ISSN 1349-0486

Cerrate, S.; Yan, F.; Wang, Z.; Coto, C.; Sacakli, P. & Waldroup, P.W. (2006). Evaluation of glycerine from biodiesel production as a feed ingredient for broilers. *International Journal of Poultry Science,* Vol.5, No.11, pp. 1001-1007, ISSN 1682-8356

Cheng, K.K..; Zhang, J.A.; Liu, D.H.; Sun. Y.; Liu, H.J.; Yang, M.D. & Xu, J.M. (2007). Pilot-scale production of 1,3-propanediol using *Klebsiella pneumoniae. Process Biochemistry,* Vol.42, No.4, pp. 740-744, ISSN 1359-5113

Claret, G.; Salmon, J.M.; Romieu, C. & Bories, A. (2004). Phisiology of *Gluconobacter oxydans* during dihydroksyacetone production from glycerol. *Applied Microbiology and Biotechnology,* Vol.41, pp. 359-365, ISSN 1432-0614

Corma, A.; Huber, G.W.; Sauvanaud, L & O'Connor, P. (2008). Biomass to chemicals: Catalytic conversion of glycerol/water mixtures into acrolein, reaction network. *Journal of Catalysis,* Vol.257, pp. 163-171, ISSN 0021-9517

da Fonseca, R.A.S.; da Silva Rafael, R.; Kalil, S.J.; Burkert, C.A.V. & Burkert, J.F.M. (2011). Different cells disruption methods for astaxanthin recovery by Phaffia rodozyma. *African Journal of Biotechnology,* Vol.10, No.7., pp. 1165-1171, ISSN 1684-5315

da Rosa, C.F.C.; Michelon, M.; de Medeiros Burkert, J.F.; Kalil, S.J. & Burkert, C.A.V. (2010). Production of rhamnolipid-type biosurfactant by *Pseudomonas aeruginosa* LBM10 grown on glycerol. *African Journal of Biotechnology,* Vol.9, No.53, pp. 9012-9017, ISSN 1684-5315

da Silva, G.P.; Mack, M. & Contiero, J. (2009). A promising and abundant carbon source for industrial microbiology. *Biotechnology Advances,* Vol.27, pp.30-39, ISSN 0734-9750

Dasari, M. (2007). Crude glycerol potential described. *Feedstuffs,* Vol.79, No.43, October 15

Dasari, M.; Kiatsimkul, P-P.; Sutterlin, W.R. & Suppes, G.J. (2005). Low-pressure hydrogenolysis of glycerol to propylene glycol. *Applied Catalysis A: General,* Vol.281, pp. 225-231, ISSN 0926-860X

Díaz, I.; Márquez-Alvarez, C.; Mohíno, F.; Pérez-Pariente, J. & Sastre, E. (2000). Combined alkyl and sulfonic acid functionalization of MCM-41-Type Silica: Part 2. Esterification of glycerol with fatty acids. *Journal of Catalysis,* Vol.193, pp. 295-302, ISSN 0021-9517

Díaz, I.; Márquez-Alvarez, C.; Mohíno, F.; Pérez-Pariente, J. & Sastre, E. (2001). A novel synthesis route of well ordered, sulfur bearing MCM-41-Type Silica: Part 2. Esterification of glycerol with fatty acids. *Microporous and Mesoporous Materials,* Vol.44-45, pp. 295-302, ISSN 1387-1811

Díaz, I.; Mohíno, F.; Blasco, T.; Sastre, E. & Pérez-Pariente, J. (2005). Influence of the alkyl chain lengh of HSO_3-R-MCM-41 on the esterification of glycerol with fatty acids. *Microporous and Mesoporous Materials,* Vol.80, pp. 33-42, ISSN 1387-1811

Donkin, S.S. (2008). Glycerol from biodiesel production: the new corn for dairy cattle. *Revista brasileira de zootecnia,* Vol.37, pp. 280-286, ISSN 1806-9290

Doppenberg, J. & Van Der Aar, P. (2007). The nutritional value of biodiesel by-products (Part 2: Glycerine). *Feed Business Asia.* March/April, pp. 42-43

Dorman, D.C.; Nassise, M.P.; Ekuta, J.; Bolon, B.; Medinsky, M.A. (1993). Acute methanol toxicity in minipigs. *Fundamental and Applied Toxicology,* Vol. 20, pp. 341-347, ISSN O272-0590

Dozier, W.A. III; Kerr, B.J.; Corzo, M.T.; Kidd, T.E.; Weber, K.; Bregendahl, K. (2008). Apparent metabolizable energy of glycerin for broiler chickens. *Poultry Science*, Vol.87, pp. 317-322, ISSN 1537-0437

Drożdżyńska, A.; Leja, K. & Czaczyk, K. (2011). Biotechnological production of 1,3-propanediol from crude glicerol. *BioTechnologia - Journal of Biotechnology, Computational Biology and Bionanotechnology*, Vol.92, No.1, pp. 92-100, ISSN 0860-7796

Enders, D.; Voith, M. & Lenzen, A. (2005). The dihydroxyacetone unit - a versatile C_3 building block in organic synthesis. *Angewandte Chemie International Edition*, Vol.44, pp. 1304-1325, ISSN 1521-3773

Fan, X.; Burton, R. & Zhou Y. (2010). Glycerol (byproduct of biodiesel production) as a source for fuels and chemicals – mini review. *The Open Fuels & Energy Science Journal*, Vol.3, pp.17-22, ISSN 1876-973X

Fernando, S.; Adhikari, S.; Kota, K. & Bandi, R. 92007). Glycerol based automotive fuels from future biorefineries. *Fuel*, Vol.86, pp. 2806-2809, ISSN 0016-2361

Garcia, R.; Besson, M. & Gallezot, P. (1995). Chemoselective catalytic oxidation of glycerol with air on platinum metals. *Applied Catalysis A: General*, Vol.127, No.1-2, pp. 165-176, ISSN 0926-860X

Gätgens, C.; Degner, U.; Bringer-Meyer, S. Hermann, U. (2007). Biotransformation of glycerol to dihydroxyacetone by recombinant *Gluconobacter oxydans* DSM 2343. *Applied Microbiology and Biotechnology*, Vol.76, pp. 553-559, ISSN 1432-0614

Gonzalez, R., Murarka, A.; Dharmadi, Y. & Yazdani, S.S. (2008). A new model for the anaerobie fermentation of glycerol in enteric bacteria: trunk and auxiliary pathways in *Escherichia coli*. *Metabolic Engineering*, Vol.10, No.5, pp. 234-245, ISSN 1096-7176

Grepen, J.V. (2005). Biodiesel processing and production. *Fuel Processing Technology*, Vol.86, pp. 1097-1107, ISSN 0378-3820

Guerrero-Pérez, M.O.; Rosas, J.M.; Bedia, J.; Rodríguez-Mirasol, J. & Cordero, T. (2009). Recent inventions in glycerol transformations and processing. *Recent Patents on Chemical Engineering*, Vol.2, pp. 11-21, ISSN 1874-4788

Guo, N.N.; Zheng, Z.M.; Mai, Y.L.; Liu, H.J. & Liu, D.H. (2009). Consequences of *cps* mutation of *Klebsiella pneumoniae* on 1,3-propanediol fermentation. *Applied Microbiology and Biotechnology*, Vol.86, No.2, pp. 701-707, ISSN 1432-0614

Hansen, C.F.; Hernandez, A.; Mullan, B.P.; Moore, K.; Trezona-Murray M.; King, R.H. & Pluske, J.R. (2009). A chemical analysis of samples of crude glycerol from the production of biodiesel in Australia, and the effects of feeding crude glycerol to growing-finishing pigs on performance, plasma metabolites and meat quality at slaugher. *Animal Production Science*, Vol.49, pp. 154-161, ISSN 1836-0939

Hass, M.J.; McAloon, J.; Yee, W.C. & Foglia T.A. (2006). A process model to estimate biodiesel production costs. *Bioresource Technology*, Vol.97, pp. 671-678, ISSN 0960-8524

Herliati, R.Y.; Intan, A.S.; Abidin, Z.Z. & Kuang, D. (2011). Preliminary design of semi-batch reactor for synthesis 1,3-dichloro-2-propanol using Aspen Plus. *International Journal of Chemistry*, Vol.3, No.1, pp. 196-201, ISSN 1916-9701

Himmi, E.H.; Bories, A.; Boussaid, A. & Hassani, L. (2000). Propionic acid fermentation of glycerol and glucose by *Propionibacterium acidipropionici* and *Propionibacterium freudenreichii ssp. shermanii. Applied Microbiology and Biotechnology,* Vol.53, pp. 435-440, ISSN 1432-0614

Hu, W.; Knight, D.; Lowry, D. & Varma, A. (2010). Selective oxidation of glycerol to dihydroxyacetone over Pt-Bi/C catalyst: Optimization of catalyst and reaction conditions. *Industrial & Engineering Chemistry Research,* Vol.49., No.21, pp. 10876-10882, ISSN 1520-5045

Hu, Z.C.; Liu, Z.Q.; Zheng, Y.G. & Shen, Y.C. (2010). Production of 1,3-dihydroxyacetone from glycerol by *Gluconobacter oxydans* ZJB09112. *Journal of Microbiology and Biotechnology,* Vol.20, No.2, pp. 340-345, ISSN 1738-8872

Ibrahim, M.H.A. & Steinbüchel, A. (2009). Poly(3-hydroxybutyrate) production from glycerol by *Zobellella denitrificans* MW1 via high-cell-density fed-batch fermentation and simplified solvent extraction. *Applied and Environmental Microbiology,* Vol.75, No.9, pp. 6222-6231, ISSN 0099-2240

Imandi, S.B.; Bandaru, V.V.R.; Somalanka, S.R. & Garapati, H.R. (2007). Optimization of medium consistuents for the production of citric acid from byproduct glycerol using Doehlert experimental design. *Enzyme and Microbial Technology,* Vol.40, pp. 1367-1372, ISSN 0141-0229

Ito, T.; Nakashimada, Y.; Senba, K.; Matsui, K. & Nishio, N. (2005). Hydrogen and etanol

Johnson, D.T & Taconi, K.A. (2007). The glycerin glut: options for the value-added conversion of crude glycerol resulting from biodiesel production. *Environmental Progress,* Vol.26, No.4, pp. 338-348, ISSN 1547-5921

Karinen, R.S. & Krause A.O.I. (2006). New biocomponents from glycerol. *Applied Catalysis A: General,* Vol.306, pp. 128-133, ISSN 0926-860X

Katrlík, J.; Vostiar, I.; Sefcovicová, J.; Tkác, J.; Mastihuba, V.; Valach, M.; Stefuca, V. & Gemeiner, P. (2007). A novel microbial biosensor based on cells of Gluconobacter oxydans for the selective determination of 1,3-propanediol in the presence of glycerol and its application to bioprocess monitoring. *Analytical and Bioanalytical Chemistry,* Vol.388, No.1, pp. 287-295, ISSN 1618-2642

Kawata, Y. & Aiba, S. (2010). Poly(3-hydroxybutyrate) production by isolated *Halomonas* sp. KM-1 using waste glycerol. *Bioscience Biotechnology & Biochemistry,* Vol.74, No.1., pp. 175-177, ISSN 0916-8451

Kerr, B.J. & Dozier, W.A. III (2008). Crude glycerin for monogastric feeds. *Render Magazine,* Vol.37, No.4, pp.10-11, ISSN 0090-8932

Khanafari, A.; Assadi, M.M. & Fakhr, F.A. (2006). Review of prodigiosin, pigmentation in *Serratia marcescens. Online Journal of Biological Sciences,* Vol.6, No.1., pp. 1-13, ISSN 1608-4217

Kijora, C. & Kupsch, R.D. (1996). Evaluation of technical glycerols from biodiesel production as a feed component in fattening of pigs. *Fett-Lipid,* Vol.98, pp. 240-245, ISSN 1521-4133

Kijora, C.; Bergner, H.; Kupsch, R.D. & Hagemann, L. (1995). Glycerol as a feed component in diets of fattening pigs. *Archives of Animal Nutrition-Archiv Fur Tierernahrung,* Vol.47, pp. 345-360, ISSN 1745-039X

Kijora, C.; Kupsch, R.D., Bergner, H., Wenk, C. & Prabucki, A.L. (1997). Comparative investigation on the utilization of glycerol, free fatty acids in combination with glycerol and vegetable oil in fattening of pigs. *Journal of Animal Physiology and*

Animal Nutrition-Zeitschrift Fur Tierphysiologie Tiernahrung Und Futtermittelkunde, Vol.77, pp. 127-138, ISSN 1439-0396

Kim, S.C.; Kim, Y.H.; Lee, H.; Yoon, D.Y. & Song, B.K. (2007) Lipase-catalyzed synthesis of glycerol carbonate from renewable glycerol and dimethyl carbonate through transesterification. *Journal of Molecular Catalysis B: Enzymatic,* Vol.49, pp. 75-78, ISSN 1381-1177

Kimura, H.; Tsuto, K.; Wakisaka, T.; Kazumi, Y. & Inaya, Y. (1993). Selective oxidation of glycerol on a platinum-bismuth catalyst. *Applied Catalysis A: General,* Vol.96. No.2., pp. 217-228, ISSN 0926-860X

Klepáčová, K.; Mravec, D. & Bajus, M. (2005). Tert butylation of glycerol catalyzed by ion-exchange resins. *Applied Catalysis A: General,* Vol.294, pp. 141-147, ISSN 0926-860X

Klepáčová, K.; Mravec, D.; Hájeková, E. & Bajus, M. (2003). Etherification of glycerol. *Petroleum and Coal,* Vol.45, No.1-2, pp. 54-57, ISSN 1337-7027

Kumar, S. & Babu, B.V. (2006). A brief review on propionic acid: a reneval energy source. *Proceedings of the National Conference on Environmental Conservation (NCEC),* 1-3 September, pp. 459-464

Kurosaka, T.; Maruyama, H.; Naribayashi, I. & Yoshiyuki, S. (2008). Production of 1,3-propanediol by hydrogenolysis of glycerol catalyzed by $Pt/WO_3/ZrO_2$. *Catalysis Communications,* Vol.9, No.6, pp. 1360-1363, ISSN 1566-7367

Kusdiyantini, E.; Gaudin, P.; Goma, G. & Blanc, P.J. (1998). Growth kinetics and astaxanthin production of *Phaffia rhodozyma* on glycerol as a carbon source during batch fermentation. *Biotechnology Letters,* Vol.20, pp. 929-934, ISSN 1573-6776

Lammers, P.J.; Honeyman, M.S.; Kerr, B.J. & Weber, T.E. (2007). Growth and performance of nursery pigs feed crude glycerol. Iowa State University Animal Report 2007. Available from http://www.ans.iastate.edu/report/air/2007pdf/R2224.pdf

Lammers, P.J.; Kerr, B.J., Weber, T.E.; Dozier, W.A. III; Kid, M.T., Bregendahl, K.& Honeyman, M.S. (2008a). Digestible and metabolizable energy of crude glycerol for growing pigs. *Journal of Animal Science,* Vol.86, pp. 602-608, ISSN 1525-3163

Lammers, P.J.; Kerr, B.J.; Honeyman, M.S.; Stadler, K.; Dozier, W.A. III; Weber, T.E.; Kidd, M.T.; Bregendahl, K. (2008b). Nitrogen-corrected apparent metabolizable energy value of crude glycerol for laying hens. *Poultry Science,* Vol.87, pp. 104-107, ISSN 1537-0437

Lee, PC.; Lee, W.G. & Chang, H.N. (2001). Succinic acid production with reduced by-product formation in the fermentation of *Anaerobispirillum succiniproducens* using glycerol as a carbon source. *Biotechnology and Bioengineering,* Vol.72, pp. 41-48, ISSN 1097-0290

Levinson, W.E.; Kurtzman, C.P. & Kuo, T.M. (2007). Characterization of *Yarrowia lipolytica* and related species for citric acid production from glycerol. *Enzyme and Microbial Technology,* Vol.41, pp. 292-295, ISSN 0141-0229

Lewandowski, G; Bartkowiak, M. & Milchert, E. (2008). Low-waste technology of glycerine epichlorohydrin production. *Oxidation Communications,* Vol.31, No.1, pp. 108-115, ISSN 0209-4541

Li, M.; Wu, J.; Liu, X.; Lin, J.; Wei, D. & Chen, H. (2010). Enhanced production of dihydroxyacetone from glycerol by overexpression of glycerol dehydrogenase in an alcohol dehydrogenase-deficient mutant of *Gluconobacter oxydans. Bioresource Technology,* Vol.101, No.21, pp. 8294-8299, ISSN 0960-8524

Liu, Y.; Koh, C.M.J. & Ji, L. (2011). Bioconversion of crude glycerol to glycolipids in *Ustilago maydis*. *Bioresource Technology*, Vol.102, pp. 3927-3933, ISSN 0960-8524

Ma, F. & Hanna, M.A. (1999). Biodiesel production: a review. *Bioresource Technology*, Vol.70, pp. 1-5, ISSN 0960-8524

Ma, L.; Wenyu, L.; Xia, Z. & Wen, J. (2010). Enhancement of dihydroxyacetone production by a mutant of *Gluconobacter oxydans*. *Biochemical Engineering Journal*, Vol.49, No.1., pp. 61-67, ISSN 1369-703X

Maervoet, V.E.T.; Mey, M.D.; Beauprez, J.; Maeseneire, S.D. & Soetaert, W.K. (2011). Enhancing the microbial conversion of glycerol to 1,3-propanediol using metabolic engineering. *Organic Process Research & Developments*, Vol.15, pp.189-202, ISSN 1520-586X

Makkar, R.S.; Cameotra, S.S. & Banat, I.M. (2011). Advances In utilization of renewable substrates for biosurfactant production. *AMB Express*, Vol.1, No.5, ISSN 2191-0855

Marinoiu, A.; Ionita, G.; Gáspár, C.-L.; Cobzaru, C. & Oprea, S. (2009). Glycerol hydrogenolysis to propylene glycol. *Reaction Kinetics and Catalysis Letters*, Vol.97,No.2, pp.315-320, ISSN 1878-5204

Melero, J.A., Vivente, G.; Morales, G.; Paniagua, S.; Moreno, J.M.; Roldán, R.; Ezquerro, C. & Pérez, C. (2008). Acid-catalyzed etherification of bio-glycerol and isobutylene over sulfonic mesostructured silicas. *Applied Catalysis A: General*, Vol.346, pp. 44-51, ISSN 0926-860X

Mishra, R.; Jain, S.R. & Kumar, A. (2008). Microbial production of dihydroxyacetone, *Biotechnology Advances*, Vol.26, No.4., pp. 293-303, ISSN 0734-9750

Monbaliu, J-C. M.R.; Winter, M.; Chevalier, B.; Schmidt, F.; Jiang. Y.; Hoogendoorn, R.; Kousemaker, M. & Stevens, C.V. (2010). Feasibility study for industrial production of fuel additives from glycerol. *Chimica Oggi/Chemistry Today*, Vol.28, No.4, pp. 8-11, ISSN 0392-839X

Morita, T.; Konishi, M.; Fukuoka, T.; Imura, T. & Kitamoto, D. (2007). Microbial conversion of glicerol into glycolipid biosurfactants, mannosylerythritol lipids, by a basidiomycete yeast, *Pseudozyma antarctica* JCM 10317. *Journal of Bioscience and Bioengineering*, Vol.104, No.1, pp. 78-81, ISSN 1389-1723

Moser, B.R. (2009). Biodiesel production, properties and feedstocks. *In vitro Cellular & Developmental Biology – Plant*, Vol.45, pp. 226-266, ISSN 1054-5476

Mourot, J.; Aumaitre, A.; Mounier, A.; Peiniau, P. & Francois, A.C. (1994). Nutritional and physiological effects of dietary glycerol in the growing pig. Consequences on fatty tissues and post mortem muscular parameters. *Livestock Production Science*, Vol.38, pp. 237-244, ISSN O301-6226

Nagakawa, Y.; Shinmi, Y.; Koso, S. & Keiichi, T. (2010). Direct hydrogenolysis of glycerol into 1,3-propanediol over rhenium-modified iridium catalyst. *Journal of Catalysis*, Vol.272, No.2, pp. 191-194, ISSN 0021-9571

Nakamura, C.E. & Whited, G. (2003). Metabolic engineering for the microbial production of 1,3-propanediol. *Current Opinion in Biotechnology*, Vol.14, pp. 454-459, ISSN 0958-1669

Nakamura, R.; Komura, K. & Sugi, Y. (2008). The esterification of glycerine with lauric acid catalyzed by multi-valent metal salts. Selective formation of mono- and dilaurins. *Catalysis Communications*, Vol.9, pp. 511-515, ISSN 1566-7367

Nguyen, B.C. & Kochevar, E. (2003). Factors influencing sunless tanning with dihydroxyacetone. *British Journal of Dermatology,* Vol.149, No.2, pp. 332-340, ISSN 1365-2133

O'Connor, I. & O'Brien, N. (1998). Modulation of UVA light-induced oxidative stress by β-carotene, lutein and astaxanthin in cultured fibroblasts. *Journal of Dermatological Science,* Vol.16, No.3, pp. 226-230, ISSN 1574-0757

Ott, L.; Bicker, M. & Vogel, H. (2006). Catalytic dehydratation of glycerol in sub- and supercritical water: a new chemical process for acrolein production. *Green Chemistry,* Vol.8, pp. 214-220, ISSN 1463-9270

Pagliaro, M.; Ciriminna, R.; Kimura, H.; Rossi, M. & Pina C.D. (2007). From glycerol to value-added Products. *Angewandte Chemie International Edition,* Vol. 46, pp. 4434-4440, ISSN 1521-3773

Pagliaro, M.; Ciriminna, R.; Kimura, H.; Rossi, M. & Pina C.D. (2009). Recent advances in the conversion of bioglycerol into value-added products. *European Journal of Lipid Science Technology,* Vol.111, pp. 788-799, ISSN 1438-9312

Palkovits, R.; Nieddu, I.; Klein, G.R.J.M. & Weckhuysen, B.M. (2008a). Highly active catalysts for the telomerization of crude glycerol with 1,3-butadiene. *ChemSusChem,* Vol.1, pp. 193-196, ISSN 1864-564X

Palkovits, R.; Nieddu, I.; Kruithof, C.A.; Klein, G.R.J.M. & Weckhuysen, B.M. (2008b). Palladium -based telomerization of 1,3-butadiene with glycerol using methoxy-functionalized triphenylphosphine ligands. *Chemistry- A European Journal,* Vol.14, pp. 8995-9005, ISSN 1521-3765

Papagianni, M. (2007). Advances in citric acid fermentation by *Aspergillus niger*: Biochemical aspects, membrane transport and modeling. *Biotechnology Advances,* Vol.25, pp. 244-263, ISSN 0734-9750

Papanikolau, S.; Fick, M. & Aggelis, G. (2004). The effect of raw glycerol concentration on the production of 1,3-propanediol by *Clostridium butyricum*. *Journal of Chemical Technology & Biotechnology,* Vol.79, No.11., pp. 1189-1196, ISSN 1097-4660

Pérez-Pariente, J.; Díaz, I.; Mohíno, F. & Sastre, E. (2003). Selective synthesis off fatty monoglycerides by using functionalized mesoporous catalysts. *Applied Catalysis A: General,* Vol. 254, pp. 173-188, ISSN 0926-860X

Perosa, A. & Tundo, P. (2005). Selective hydrogenolysis of glycerol with raney-nikel. *Industrial & Engineering Chemistry Research,* Vol.44, No.23, pp. 8534-8537, ISSN 1520-5045

production from glicerol-containing wastes discharged after biodiesel manufacturing process. *Journal of Bioscience and Bioengineering,* Vol.100, pp. 260-265, ISSN 1389-1723

Rahman, K.S.M.; Rahman, T.J.; McClean, S.; Marchant, R. & Bant., I.M. (2002). Rhamnolipid biosurfactant production by strains of *Pseudomonas aeruginosa* using low-cost raw materials. *Biotechnology Progress,* Vol.18, pp. 1277-1281, ISSN 1520-6033

Rahmat, N.; Abdullah, A. Z. & Mohammed A. R. (2010). Recent progress on innovative and potential technologies for glycerol transformation into fuel additives: a critical review. *Renewable and Suitable Energy Reviews,* Vol.14, pp. 987-1000, ISSN 1364-0321

Razavi, S.H. & March, I. (2006). Effect of temperature and pH on the growth kinetics and carotenoid production by *Sporobolomyces ruberrimus* H110 using technical glycerol as carbon source. *Iranian Journal of Chemistry & Chemical Engineering,* Vol.25, No.3, pp. 59-64, ISSN 1021-9986

Rokicki, G.; Rakoczy, P., Parzuchowski, P. & Sobiecki, M. (2005). Hyperbranched aliphatic polyethers obtained from environmentally benign monomer: glycerol carbonate. *Green Chemistry*, Vol.7, No.7, pp.529-539, ISSN 1463-9270

Rymowicz, W.; Fatykhova, A.R.; Kamzolova, S.V.; Rywińska, A. & Morgunov, I.G. (2010). Citric acid production from glycerol-containing waste of biodiesel industry by Yarrowia lipolytica in batch, repeated batch, and cell recycle regimes. *Applied Microbiology and Biotechnology*, Vol.87.,No.3, pp. 545-552, ISSN 1432-0614

Rywińska, A. & Rymowicz, W. (2011). Continuous production of citric acid from raw glycerol by *Yarrowia lipolytica* in cell recycle cultivation. *Chemical Papers*, Vol.65, No.2, pp. 119-123, ISSN 1336-9075

Rywińska, A.; Juszczyk, P.; Wojtatowicz, M. & Rymowicz, W. (2011). Chemostat study of citric acid production from glicerol by *Yarrowia lipolytica*. *Journal of Biotechnology*, Vol. 152, No.1-2, pp. 54-57, ISSN 0168-1656

Rywińska, A.; Rymowicz, W. & Marcinkiewicz, M. (2010). Valorization of raw glicerol for citric acid production by *Yarrowia lipolytica* yeast. *Electronic Journal of Biotechnology*, Vol.13, No.4, pp. 1-9, ISSN 0717-3458

Sakai, S. & Yagishita, T. (2007). Microbial production of hydrogen and ethanol from glycerol-containing wastes discharged from a biodiesel fuel production plant in a bioelectrochemical reactor with thionine. *Biotechnology and Bioengineering*, Vol.98, No.2., pp. 340-348, ISSN 1097-0290

Saxena, R.K.; Anand, D.; Saran, S. & Isar, J. (2009). Microbial production of 1,3-propanediol: Recent developments and opportunities. *Biotechnology Advances*, Vol.27, No.6, pp. 895-913, ISSN 0734-9750

Schieck, S.J.; Kerr, B.J.; Baidoo, S.K.; Shurson, G.C. & Johnston, L.J. (2010). Use of crude glycerol, a biodiesel coproduct, in diets for lactating sows. *Journal of Animal Science*, Vol.87, pp. 4042-4049, ISSN 1525-3163

Scholten, E. & Dägele, D. (2008). Succinic acid production by a newly isolated bacterium. *Biotechnology Letters*, Vol.30, No.12, pp. 2143-2146, ISSN 1573-6776

Scholten, E.; Renz, T. & Thomas, J. (2009). Continuous cultivation approach for fermentative succinic acid production from crude glycerol by *Basfia succiniproducens* DD1. *Biotechnology Letters*, Vol.31, pp. 1947-1951, ISSN 1573-6776

Sendzikiene, E.; Makareviciene, V. & Janulis, P. (2007). Oxidation stability of biodiesel fuel produced from fatty wastes. *Polish Journal of Environmental Studies*, Vol.16, No.1, pp. 17-22, ISSN 1230-1485

Shrivastav, A.; Mishra, S.K.; Shethia, B.; Pancha, I.; Jain, D. & Mishra, S. (2010). Isolation of promising bacterial strains from soil and marine environment for polyhydroxyalkanoates (PHAs) production utilizing *Jatropha* biodiesel byproduct. *International Journal of Biological Macromolecules*, Vol.47, pp. 283-287, ISSN 0141-8130

Simon, A. (1996). Administration of glycerol to broilers in drinking water. *Landbauforshung Volkenrode*, Vol.169, pp. 168-170, ISSN 0458-6859

Simon, A.; Bergner, H. & Schwabe, M. (1996). Glycerol - feed ingredient for broilers chickens. *Archives of Animal Nutrition*, Vol.49, pp. 103-112, ISSN 1477-2817

Simon, A.; Schwabe, M. & Bergner, H. (1997). Glycerol supplementation to broilers rations with low crude protein content. *Archives of Animal Nutrition*, Vol.50, pp. 271-282, ISSN 1477-2817

Soccol, C.R.; Vandenberghe, L.P.S. Rodrigues, C. & Pandey, A. (2006). New perspectives for citric acid production and application. *Food Technology and Biotechnology,* Vol.44, pp. 141-149, ISSN 1330-9862

Solaiman, D.K.Y.; Ashby, R.D.; Foglia, T.A. & Marmer, W.M. (2006). Conversion of agricultural feedstock and coproducts into poly(hydroxyalcanoates). *Applied Biotechnology and Microbiology,* Vol.71, pp. 783-789, ISSN 1432-0614

Song, H. & Lee, S.Y. (2006). Production of succinic acid by bacterial fermentation. *Enzyme and Microbial Technology,* Vol.39, pp. 352-361, ISSN 0141-0229

Świątkiewicz, S. & Koreleski, J. (2009). Effect of crude glycerin level in the diet of laying hens on egg performance and nutrient utilization. *Poultry Sciences,* Vol.88, pp. 615-618, ISSN 1537-0437

Tang, X.M.; Tan, Y.S.; Zhu, H.; Zhao, K. & Shen, W. (2009). Microbial conversion of glycerol to 1,3-propanediol by an engineered strain of *Escherichia coli*. *Applied Microbiology and Biotechnology,* Vol.75, No.6, pp. 1628-1634, ISSN 1432-0614

Tao, J.; Wang, X.; Shen, Y. & Wei, D. (2005). Strategy for the improvement of prodigiosin production by a *Serratia marcescens* through fed-batch fermentation. *World Journal of Biotechnology and Microbiology,* Vol.21, No.6-7, pp. 969-972, ISSN 1573-0972

Thompson, J.C. & He, B. (2006). Characterization of crude glycerol from biodiesel production from multiple feedstocks. *Applied Engineering of Agriculture,* Vol.22, No.2, pp. 261-265, ISSN 0883-8542

Tian, P.Y.; Shang, L.; Ren, H., Mi, Y.; Fan, D.D. & Jiang, M. (2009). Biosynthesis of polyhydroxyalcanoates: current research and development. *African Journal of Biotechnology,* Vol.8, No.5, pp. 709-714, ISSN 1684-5315

Tsukuda, E.; Sato, S.; Takahashi, R. & Sodesawa, T. (2007). Production of acrolein from glicerol over silica-supported heteropoly acids. *Catalysis Communications,* Vol.8, pp. 1349-1353, ISSN 1566-7367

Tyson, K.S.; Bozell, J.; Wallance, R.; Petersen, E. & Moens, L. (2004). Biomass oil analysis: research needs and recommendations. Technical Report. Available from http://www.nrel.gov/docs/fy04osti/34796.pdf

Ulgen, A. & Hoelderich, W.F. (2011). Conversion of glycerol to acrolein in the presence of WO_3/TiO_2 catalysts. *Applied Catalysis A: General,* ISSN 0926-860X, article in press, doi:10.1016/j.apcata.2011.04.005

Vieville, C.; Yoo, S.; Pelet, S. & Mouloungui, Z. (1998). Synthesis of glycerol carbonate by direct carbonation of glycerol in supercritical CO_2 in the presence of zeolites an ion exchange resins. *Catalysis Letters,* Vol.56, pp. 245-247, ISSN 1572-897X

Wang, L.; Du, W.; Liu, D.H.; Li, L.L. & Dai, N.M. (2006). Lipase-catalysed biodiesel production from soybean oil deodorizer distillate with absorbent present in tert-butanol system. *Journal of Molecular Catalysis B: Enzymatic,* Vol.43, pp. 29-31, ISSN 1381-1177

Wang, S. & Liu, H. (2007). Selective hydrogenolysis of glycerol to propylene glycol on Cu-ZnO catalysts. *Catalysis Letters,* Vol.117, No.1-2, pp. 62-67, ISSN 1572-879X

Wang, Z.X.; Zhue, J.; Fang, H. & Prior, B.A. (2001). Glycerol production by microbial fermentation: a review. *Biotechnology Advances,* Vol.19, pp. 201-223, ISSN 0734-9750

Watanabe, M.; Iida, T.; Aizawa, Y.; Aida, T.M. & Inomata, H. (2007). Acrolein synthesis from glycerol in hot-compressed water. *Bioresource Technology,* Vol.98, pp. 1285-1290, ISSN 0960-8524

Willke T. & Vorlop K.D. (2004). Industrial bioconversion of renewable resources as an alternative to conventional chemistry. *Applied Microbiology and Biotechnology*, Vol.66, pp. 131-142, ISSN 1432-0614

Wu, Z.; Mao, Y.; Wang, X. & Zhang, M. (2011). Preparation of Cu-Ru/carbon nanotube catalyst for hydrogenolysis of glycerol to 1,2-propanediol *via* hydrogen spillover. *Green Chemistry*, Vol.13, pp. 1311-1316, ISSN 1463-9270

Yan, W. & Suppes, G.J. (2009). Low-pressure packed-bed gas-phase dehydration of glycerol to acrolein. *Industrial & Engineering Chemistry Research*, Vol.48, No.7, pp. 3279-3283, ISSN 1520-5045

Yang, G.; Tian, J. & Li, J. (2007). Fermentation of 1,3-propanediol by lactate deficient mutant of *Klebsiella oxytoca* under microaerobic. conditions. *Applied Microbiology and Biotechnology*, Vol.73, pp. 1017-1024, ISSN 1432-0614

Yazdani, S.S. & Gonzalez, R. (2007). Anaerobic fermentation of glicerol: a path to economic viability for the biofuels industry. *Current Opinion in Biotechnology*, Vol.18, pp. 213-219, ISSN 0958-1669

Yazdani, S.S. & Gonzalez, R. (2008). Engineering *Escherichia coli* for the efficient conversion of glycerol to ethanol and co-products. *Metabolic Engineering*, Vol.10, No.6, pp. 340-351, ISSN 1096-7176

Yimyoo, T.; Yongmanitchai, W. & Limtong, S. (2011). Carotenoid production by *Rhodosporidium paludigenum* DMKU3-LPK4 using glycerol as the carbon source. *Kasetsart Journal Natural Sciences*, Vol.45, pp. 90-100, ISSN 0075-5192

Zeikus, J.G.; Jain, M.K. & Elankovan, P. (1999). Biotechnology of succinic acid production and markets for derived industrial products. *Applied Microbiology and Biotechnology*, Vol.51, pp. 545-552, ISSN 1432-0614

Zhang, A. & Yang, S-T. (2009). Propionic acid production from glycerol by metabolically engineered *Propionibacterium acidipropionici*. *Process Biochemistry*, Vol.44, No.12, pp. 1346-1351, ISSN 1359-5113

Zhang, G.L.; Wu, Y.T.; Qian, X.P. & Meng, Q. (2005). Biodegradation of crude oil by *Pseudomonas aeruginosa* in the presence of rhamnolipids. *Journal of Zhejiang University- Science B*, Vol.6, No.8, pp. 725-730, ISSN 1862-1783

Zhang, X.; Jantama, K.; Shanmugam, K.T. & Ingram, O. (2009). Reengineering *Escherichia coli* for succinate production in mineral salts medium. *Applied and Environmental Microbiology*, Vol.75, No.24, pp. 7807-7813, ISSN 0099-2240

Zhang, X.; Shanmugam, K.T. & Ingram, O. (2010). Fermentation of glycerol to succinate by metabolically engineered strains of *Escherichia coli*. *Applied and Environmental Microbiology*, Vol.76, No.8, pp. 2397-2401, ISSN 0099-2240

Zheng, Y.G.; Chen, X.L. & Shen, Y.C. (2008). Commodity chemicals derived from glycerol, an important biorafinery feedstock. *Chemical Reviews*, Vol.108, pp. 5253-5277, ISSN 1520-6890

Zhu, Y.; Li, J.; Tan, M.; Liu, L.; Jiang, L.; Sun, J.; Lee, P.; Du, G. & Chen. J. (2010). Optimization and scale up of propionic acid production by propionic acid-tolerant *Propionibacterium acidipropionici* with glycerol as the carbon source. *Bioresource Technology*, Vol.101, No.22, pp. 8902-8906, ISSN 0960-8524

Zijlstra, R.T.; Menjivar, K.; Lawrence, E. & Beltranena, E. (2009). The effect of feeding crude glycerol on growth performance and nutrient digestibility in weaned pigs. *Canadian Journal of Animal Science*, Vol.89, pp. 85-89, ISSN 0008-3984

Zinn, M. & Hany, R. (2005). Tailored material properties of polyhydroxyalcanoates through biosynthesis and chemical modification. *Advanced Engineering Materials,* Vol.7, pp. 408-411, ISSN 1527-2648

20

Utilization of Crude Glycerin in Nonruminants

Brian J. Kerr[1], Gerald C. Shurson[2],
Lee J. Johnston[2] and William A. Dozier, III[3]
[1]*USDA-Agricultural Research Service*
[2]*University of Minnesota*
[3]*Auburn University*
United States of America

1. Introduction

During digestion in non-ruminants, intestinal absorption of glycerol has been shown to range from 70 to 90% in rats (Lin, 1977) to more than 97% in pigs and laying hens (Bartlet and Schneider, 2002). Glycerol is water soluble and can be absorbed by the stomach, but at a rate that is slower than that of the intestine (Lin, 1977). Absorption rates are high, likely due to glycerin's small molecular weight and passive absorption rather than forming a micelle that is required for absorption of medium and long chain fatty acids (Guyton, 1991). Once absorbed, glycerol can be converted to glucose via gluconeogenesis or oxidized for energy production via glycolysis and citric acid cycle with the shuttling of protons and electrons between the cytosol and mitochondria depicted in Figure 1 (Robergs and Griffin, 1998). Glycerol metabolism largely occurs in the liver and kidney where the amount of glucose carbon arising from glycerol depends upon metabolic state and level of glycerol consumption (Lin, 1977; Hetenyi et al., 1983; Baba et al., 1995). With gluconeogenesis from glycerol being limited by the availability of glycerol (Cryer and Bartley, 1973; Tao et al., 1983), crude glycerin has the potential of being a valuable dietary energy source for monogastrics.

2. Crude glycerin: Caloric value for swine and poultry

Pure glycerin is a colorless, odorless, and a sweet-tasting viscous liquid, containing approximately 4.3 Mcal of gross energy (GE)/kg as-is basis (Kerr et al., 2009). However, crude glycerin can range from 3 to 6 Mcal GE/kg, depending upon its composition (Brambilla and Hill, 1966; Lammers et al., 2008b; Kerr et al., 2009). The difference in GE of crude glycerin compared with pure glycerin is not surprising, given that crude glycerin typically contains about 85% glycerin, 10% water, 3% ash (typically Na or K chloride), and a trace amount of free fatty acids. As expected, high amounts of water negatively influence GE levels while high levels of free fatty acids elevate the GE concentration. Various

---

NOTE: In the current text, use of the word "glycerin" refers to the chemical compound or feedstuff while 'glycerol' refers to glycerin on a biochemical basis relative to its function in living organisms. In addition, because glycerin is marketed on a liquid basis, all data are presented on an 'as-is' basis.

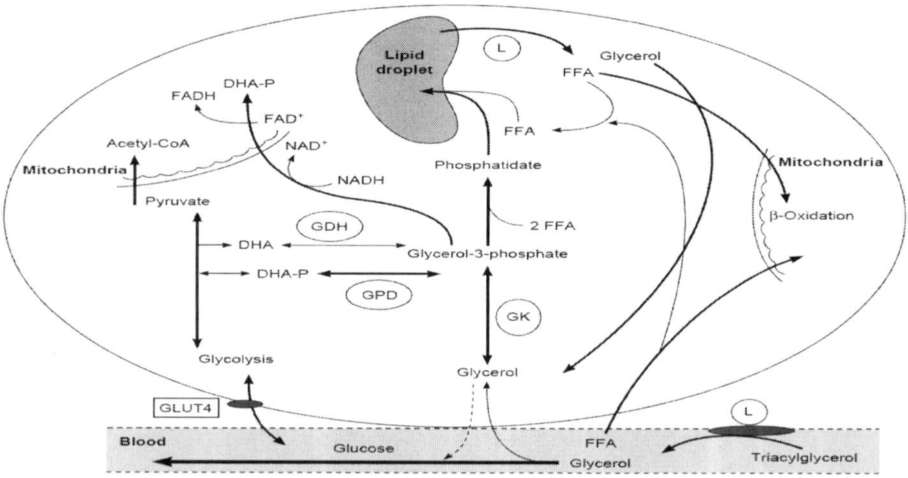

Fig. 1. Biochemical reactions involved in glycerol synthesis and metabolic conversation to glycerol-3-phosporate, phosphatidate and triacylglycerol.
DHA= dihydroxyacetone; DHA-P = dihydroxyacetone phosphate; FAD^+ = oxidised from flavin adenine dinucleotide; FADH = reduced from of flavin adenine dinucleotide; FFA = free fatty acid; GHD = glycerol dehydrogenase; GK = glycerol kinase; GLUT4 = glucose transport protein; GPD = glycerol phosphate dehydrogenase; L = lipase; NAD^+ = oxidised from of nicotinamide adenine dincleotide; NADH = reduced from of nicotinamide adenine dinucleotide.

experiments evaluating glycerin have assumed the metabolizable energy (swine nutrition terminology) or apparent metabolizable energy (poultry nutrition terminology), hereafter just called metabolizable energy (ME), of glycerin to be approximately 95% of its GE in dietary formulation (Brambilla and Hill, 1966; Lin et al., 1976; Rosebrough et al., 1980; Cerrate et al., 2006). Empirical determinations of ME content in crude glycerin have been lacking in non-ruminants until recently.

Bartlet and Schneider (2002) reported ME values of refined glycerin in broiler, laying hen, and swine diets, and showed that the ME value of glycerin decreased as the level of dietary glycerin increased (Table 1). On average, these values were 3,993, 3,929, and 3,292 kcal/kg for broilers, laying hens, and swine, respectively. Since pre-cecal digestiblity of glycerin is approximately 97% (Bartlet and Schneider, 2002), a possible explanation for the observed decrease in ME value may be a result of increased blood glycerol levels (Kijora et al., 1995; Kijora and Kupsch, 2006; Simon et al., 1996) after glycerin absorption, such that complete renal reabsorption is prevented and glycerol excretion in the urine is increased (Kijora et al., 1995; Robergs and Griffin, 1998).

Dietary glycerin, %	Broiler, kcal/kg	Laying hen, kcal/kg	Swine, kcal/kg
5	4,237	4,204	4,180
10	4,056	4,108	3,439
15	3,686	3,475	2,256

[1] Bartlet and Schneider, 2002

Table 1. Metabolizable energy of refined glycerin, as-is basis[1]

Lammers et al. (2008b) obtained a crude glycerin co-product (87% glycerin) and determined in nursery and finishing pigs that its ME was 3,207 kcal/kg, and did not differ between pigs weighing 10 or 100 kg (Table 2). Strictly based on glycerin content, this would equate to 3,688 kcal ME/kg on a 100% glycerin basis (3,207 kcal ME/kg/87% glycerin), which would be slightly lower than the 3,810 kcal ME/kg (average of the 5 and 10% inclusion levels) reported by Bartlet and Schneider (2002), but similar to the 3,656 kcal ME/kg as reported by Mendoza et al. (2010) using a 30% inclusion level of glycerin.

Trial	Pigs	Initial BW, kg	DE, kcal/kg	SEM	ME, kcal/kg	SEM
1[2]	18	11.0	4,401	282	3,463	480
2[3]	23	109.6	3,772	108	3,088	118
3[4]	19	8.4	3,634	218	3,177	251
4[4]	20	11.3	4,040	222	3,544	237
5[4]	22	99.9	3,553	172	3,352	192

[1] All experiments represent data from 5 d energy balance experiments following a 10 d adaptation period (Lammers et al., 2008b).
[2] Included pigs fed diets containing 0, 5, and 10% crude glycerin.
[3] Included pigs fed diets containing 0, 5, 10, and 20% crude glycerin.
[4] Included pigs fed diets containing 0 and 10% glycerin.

Table 2. Digestible and metabolizable energy of crude glycerin fed to pigs, as-is basis[1]

Similar to data reported by Bartlet and Schneider (2002) in 35 kg pigs, increasing crude glycerin from 5 to 10 or 20% in 10 kg pigs (Lammers et al., 2008b) quadratically reduced ME (3,601, 3,239, and 2,579 kcal ME/kg, respectively), which suggests that high dietary concentrations of crude glycerin may not be fully utilized by 10 kg pigs. In contrast, dietary concentrations of crude glycerin had no effect on ME determination in 100 kg pigs (Lammers et al., 2008b). The ratio of DE:GE is an indicator of how well a product is digested, and for the crude glycerin evaluated by Lammers et al. (2008b), it equaled 92% suggesting that crude glycerin is well digested. Similarly, Bartlet and Schneider (2002) reported that greater than 97% of the glycerin is digested before the cecum. In addition, the ratio of ME:DE indicates how well energy is utilized once digested, and for the crude glycerin evaluated by Lammers et al. (2008b) the ratio was 96%, which is identical to the ME:DE ratio for soybean oil, and is comparable to the ratio of ME:DE (97%) for corn grain (NRC, 1998), all of which support the assertion that crude glycerol is well utilized by the pig as a source of energy.

The energy value of crude glycerin in poultry has also been recently evaluated. Bartlet and Schneider (2002) reported that the ME content for refined glycerin is 3,929 and 3,993 kcal/kg for laying hens and broilers, respectively (Table 1). Studies by Lammers et al. (2008a) using laying hens, and Dozier et al. (2008) using broilers, reported a ME value of 3,805 and 3,434 kcal/kg, respectively, for the same lot of crude glycerin (87% glycerin). These estimates equate to 4,376 and 3,949 kcal/kg for laying hens and broilers, respectively, on a 100% purity basis, and compare favorably to the Bartlet and Schneider (2002) values for broilers, but higher than their value for laying hens. Contrary to the observations of Bartlet and Schneider (2002), Dozier et al. (2008) and Lammers et al. (2008a) reported no reduction in ME of crude glycerin as dietary inclusion level increased. However, Dozier et al. (2008) used ≤ 9% crude glycerin (equivalent to ≤ 7.8% pure glycerin) and Lammers et al. (2008a) used ≤ 15% crude glycerin (equivalent to ≤ 13.0% pure glycerin), which were slightly less than the inclusion levels (up to 15% refined glycerin) studied by Bartlet and Schneider (2002). Swiatkiewicz and Koreleski (2009) recently determined the ME of crude glycerin to be 3,970 kcal/kg in diets containing up to 6% crude glycerin fed to laying hens, but did not report the purity of the crude glycerin source.

Similar to other co-products used to feed livestock, the chemical composition of crude glycerin can vary widely (Thompson and He, 2006; Kijora and Kupsch, 2006; Hansen et al., 2009; Kerr et al., 2009). The consequences of this variation in energy value to animals have not been well described for crude glycerin. Recently, 10 sources of crude glycerin from various biodiesel production facilities in the U.S. were evaluated for energy utilization in non-ruminants (Table 3). The crude glycerin sources originating from soybean oil averaged 84% glycerin, with minimal variability noted among 6 of the sources obtained. Conversely, sources from commercial plants using tallow, yellow grease, and poultry oil as initial lipid feedstock ranged from 52 to 94% glycerin. The crude glycerin co-products derived from either non-acidulated yellow grease or poultry fat had the lowest glycerin content, but had the highest free fatty acid composition. The high fatty acid content of the non-acidulated yellow grease product was expected because the acidulation process results in greater separation of methyl esters which subsequently results in a purer form of crude glycerin containing lower amounts of free fatty acids (Ma and Hanna, 1999; Van Gerpen, 2005; Thompson and He, 2006). In contrast, the relatively high free fatty acid content in the crude glycerin obtained from the plant utilizing poultry fat as a feedstock source is difficult to explain because details of the production process were not available. Moreover, both of these two crude glycerin co-products (derived from non-acidulated yellow grease and poultry fat) had higher methanol concentrations than the other glycerin sources. Recovery of

Sample ID[3]	Glycerin	Moisture	Methanol	pH	NaCl	Ash	Fatty acids
USP	99.62	0.35	ND[2]	5.99	0.01	0.01	0.02
Soybean oil	83.88	10.16	0.0059	6.30	6.00	5.83	0.12
Soybean oil[4]	83.49	13.40	0.1137	5.53	2.84	2.93	0.07
Soybean oil	85.76	8.35	0.0260	6.34	6.07	5.87	ND
Soybean oil	83.96	9.36	0.0072	5.82	6.35	6.45	0.22
Soybean oil	84.59	9.20	0.0309	5.73	6.00	5.90	0.28
Soybean oil	81.34	11.41	0.1209	6.59	6.58	7.12	0.01
Tallow	73.65	24.37	0.0290	3.99	0.07	1.91	0.04
Yellow grease	93.81	4.07	0.0406	6.10	0.16	1.93	0.15
Yellow grease[5]	52.79	4.16	3.4938	8.56	1.98	4.72	34.84
Poultry fat	51.54	4.99	14.9875	9.28	0.01	4.20	24.28

[1] Samples analyzed as described in Lammers et al. (2008b) courtesy of Ag Processing Inc., Omaha, NE, 68154. Glycerin content determined by difference as: 100 - % methanol - % total fatty acid - % moisture - % ash.

[2]ND = not detected.

[3] USP=USP grade glycerin or initial feedstock lipid source.

[4] Soybean oil from extruded soybeans. All other soybean oil was obtained by hexane extraction of soybeans.

[5] Crude glycerin that was not acidulated.

Table 3. Chemical analysis of crude glycerin, % as-is basis[1]

methanol is also indicative of production efficiency because it is typically reused during the production process (Ma and Hanna, 1999; Van Gerpen, 2005; Thompson and He, 2006). The high amount of methanol content in crude glycerin from non-acidulated yellow grease was expected because this product has not been fully processed at the production facility. Why the crude glycerin obtained from the plant utilizing poultry fat had relatively high methanol content is unclear as no processing information was obtained from the plant, but it may be due to the lower overall efficiency of the production process at this plant (Ma and Hanna, 1999; Van Gerpen, 2005; Thompson and He, 2006).

The average ME of the 11 sources of glycerin described in Table 3 was 3,486 kcal/kg (Table 4; Kerr et al., 2009), with little differences among the sources with two exceptions. The two co-products with high levels of free fatty acids (co-products obtained from non-acidulated yellow grease and poultry fat) had higher ME values than the other crude glycerin co-products, which was not surprising given that these two co-products also had a higher GE concentration than the other crude glycerin co-products. The ME:GE ratio among all glycerin co-products was similar averaging 85%, which is similar to that reported by others (88%, Lammers et al., 2008b; 88%, Bartlet and Schneider, 2002; 85%, Mendoza et al., 2010). Because the GE of the crude glycerin can differ widely among co-products, comparison of ME as a percentage of GE provides valuable information on the caloric value of crude glycerin for non-ruminants, with a high ME:GE ratio indicating that a given crude glycerin source is well digested and utilized.

When the same glycerin co-products evaluated in swine by Kerr et al. (2009) were fed to broilers (Dozier et al., 2011) the ME averaged 3,646 kcal/kg (Table 4). When evaluating ME as a percent of GE in broilers, crude glycerin co-products originating from soybean oil resulted in similar values compared with co-products produced from tallow and acidulated yellow grease. In contrast, crude glycerin sources with high free fatty acid content had a

		Broiler, AME[1]		Swine, ME[2]	
Sample ID[3]	GE, kcal/kg	kcal/kg	% of GE	kcal/kg	% of GE
USP	4,325	3,662	84.7	3,682	85.2
Soybean oil	3,627	3,364	92.8	3,389	93.4
Soybean oil[4]	3,601	3,849	106.9	2,535	70.5
Soybean oil	3,676	3,479	94.6	3,299	89.9
Soybean oil	3,670	3,889	106.0	3,024	82.5
Soybean oil	3,751	3,644	97.2	3,274	87.3
Soybean oil	3,489	3,254	93.3	3,259	93.5
Tallow	3,173	3,256	102.6	2,794	88.0
Yellow grease	4,153	4,100	98.7	3,440	92.9
Yellow grease[5]	6,021	4,135	68.7	5,206	86.6
Poultry fat	5,581	3,476	62.3	4,446	79.7

[1] Dozier et al., 2011.
[2] Kerr et al., 2009.
[3] USP=USP grade glycerin or initial feedstock lipid source.
[4] Soybean oil from extruded soybeans. All other soybean oil was obtained by hexane extraction of soybeans.
[5] Crude glycerin that was not acidulated.

Table 4. Energy values of crude glycerin co-products in broilers and swine, as-is basis lower ME as a percentage of GE compared to the other glycerin co-products. If one excludes these two high free fatty acid products from the data set, ME as a percentage of GE averaged 97% (Dozier et al., 2011) which compares favorably to the 96% (5 and 10% inclusion levels only) reported by Bartlet and Schneider (2002), the 105% reported in laying hens by Lammers et al. (2008a), and the 95% reported in broilers by Dozier et al. (2008). Similar to data in swine, this indicates that crude glycerin is well digested and utilized by poultry.

The reduced ability of broilers to efficiently utilize glycerin co-products having relatively high free fatty acid content as indicated by their lower ME:GE ratio warrants additional discussion. Wiseman and Salvador (1991) reported a linear reduction of ME content in broiler diets containing increasing concentrations of free fatty acids, which was supported by others (Artman, 1964; Sklan, 1979) who reported that free fatty acids reduce the rate of absorption compared with lipid sources containing triglycerides and free fatty acids. This reduced absorption in products containing free fatty acids may be partially due to the absence of a monoglyceride backbone to aid absorption because the relatively low concentration of monoglycerides in the duodenum, which may depress the amount of fatty acids entering micellular solution. Furthermore, 2-monoglycerides promote water solubility which results in a mixed bile salt-monoglyceride fatty acid micelle (Hofmann and Borgstrom, 1962; Johnston, 1963; Senior, 1964) which can aid in lipid absorption.

Because more than one chemical component can influence energy content of feed ingredients, stepwise regression was used to predict GE and ME values, and ME as a percentage of GE among glycerin sources for both swine (Kerr et al., 2009) and broiler (Dozier et al., 2011) experiments utilizing the same crude glycerin co-products. If the GE of a crude glycerin is not known, the data indicate it can be predicted by: GE, kcal/kg = - 236 + (46.08 × % of glycerin) + (61.78 × % of methanol) + (103.62 × % of fatty acids), ($R^2 = 0.99$). In swine, ME content could subsequently be predicted by multiplying GE by 84.5% with no adjustment for composition (Kerr et al., 2009). For poultry, ME content could subsequently be predicted as: GE, kcal/kg × (91.63% – (0.61 × % free fatty acids) – (1.17 × % methanol) + (0.60 × % water)). Because free fatty acids, methanol and water may not be known, ME in poultry could also be predicted by multiplying GE by 97.4% if total fatty acid concentration is less than 0.5%, or by multiplying GE by 65.6% if total fatty acid concentrations range from 25 to 35% (Dozier et al., 2011). Additional research is needed to refine and validate these equations relative to glycerin, methanol, ash, and total fatty acid concentrations for both broilers and pigs.

3. Crude glycerin as a feed ingredient for swine

In swine, German researchers (Kijora and Kupsch, 2006; Kijora et al., 1995, 1997) have suggested that up to 10% crude glycerin can be fed to pigs with little effect on pig performance. Likewise, Mourot et al. (1994) indicated that growth performance of pigs from 35 to 102 kg was not affected by the addition of 5% glycerin (unknown purity) to the diet. The impact of dietary glycerin on carcass quality in pigs has been variable. Kijora et al. (1995) and Kijora and Kupsch (2006) showed no consistent effect of 5 or 10% crude glycerin addition to the diet on carcass composition or meat quality parameters, while in an additional study, pigs fed 10% crude glycerin exhibited a slight increase in backfat, 45 min pH, flesh color, marbling, and leaf fat (Kijora et al., 1997). Although they did not note any significant change in the saturated fatty acid profile of the backfat, there was a slight increase in oleic acid, accompanied by a slight decrease in linoleic and linolenic acid concentrations, resulting in a decline in the polyunsaturated to monounsaturated fatty acid ratio in backfat. Likewise, Mourot et al. (1994) reported no consistent change in carcass characteristics due to 5% crude glycerin supplementation of the diet, but did note an increase in oleic acid and a reduction in linoleic acid in backfat and *semimembranosus* muscle tissue. Kijora and Kupsch (2006) found no effect of glycerin supplementation on water loss of retail pork cuts. However, Mourot et al. (1994) reported a reduction in 24-h drip loss (1.75 versus 2.27%) and cooking loss was also reduced (25.6 vs 29.4%) from the the *Longissimus dorsi* and *semimembranosus* muscles due to dietary supplementation with 5% glycerin. Likewise, Airhart et al. (2002) reported that oral administration of glycerin (1 g/kg BW) 24 h and 3 h before slaughter tended to decrease drip and cooking loss of *Longissimus dorsi* muscle.

Recently, there has been increased interest in utilization of crude glycerin in swine diets due to the high cost of feedstuffs typically used in swine production. For newly weaned pigs, it appears that crude glycerin can be utilized as an energy source up to 6% of the diet, but crude glycerin does not appear to be a lactose replacement (Hinson et al., 2008). In 9 to 22 kg pigs, Zijlstra et al. (2009) reported that adding up to 8% crude glycerol to diets as a wheat replacement, improved growth rate and feed intake, but had no effect on gain:feed. In 28 to 119 kg pigs, supplementing up to 15% crude glycerol to the diet quadratically increased average daily gain and linearly increased average daily feed intake, but the net effect on feed efficiency was a linear reduction (Stevens et al., 2008). These authors also reported that crude glycerin supplementation appeared to increase backfat depth and Minolta L* of loin muscle, but decreased loin marbling and the percentage of fat free lean with increasing dietary glycerin levels. In 78 to 102 kg pigs, increasing crude glycerin from 0 or 2.5% to 5% reduced average daily feed intake when fat was not added to the diet, but had no effect when 6% fat was supplemented (Duttlinger et al., 2008a). This decrease in feed intake resulted in depressed average daily gain, but had no effect on feed efficiency. In contrast, Duttlinger et al. (2008b) reported supplementing up to 5% crude glycerin to diets had no effect on growth performance or carcass traits of pigs weighing 31 to 124 kg.

Supplementing 3 or 6% crude glycerin in pigs from 11 to 25 kg body weight increased average daily gain even though no effect was noted on feed intake, feed efficiency, dry matter, nitrogen, or energy digestibility (Groesbeck et al., 2008). Supplementing 5% pure glycerin did not affect pig performance from 43 to 160 kg, but pigs fed 10% glycerin had reduced growth rate and feed efficiency compared to pigs fed the control or 5% glycerin supplemented diets (Casa et al., 2008). In addition, diet did not affect meat or fat quality, or meat sensory attributes. In 51 to 105 kg pigs, including up to 16% crude glycerin did not affect pig growth performance or meat quality parameters (Hansen et al., 2009). Lammers et al. (2008c) fed pigs (8 to 133 kg body weight) diets containing 0, 5, or 10% crude glycerin and reported no effect of dietary treatment on growth performance, backfat depth, loin eye area, percentage fat free lean, meat quality, or sensory characteristics of the *Longissimus dorsi* muscle. In addition, dietary treatment did not affect blood metabolites or frequency of histological lesions in the eye, liver, or kidney, and only a few minor differences were noted in the fatty acid profile of loin adipose tissue. Likewise, Mendoza et al. (2010) fed heavy pigs (93 to 120 kg) up to 15% refined glycerin and reported no effect on growth performance, carcass characteristics, or meat quality. Schieck et al. (2010b) fed pigs either a control diet (16 weeks, 31 to 128 kg), 8% crude glycerin during the last 8 weeks (45 to 128 kg) or 8% crude glycerin for the entire 16 week period (31 to 128 kg) and reported that feeding crude glycerin during the last 8 weeks before slaughter supported similar growth performance, with little effect on carcass composition or pork quality, except for improvement in belly firmness,

Glycerin equivalency[2]	Daily gain	Daily feed intake	Gain:feed ratio
Ziljstra et al., 2009 / Wheat-soybean meal-fish meal-lactose / 9-22 kg			
4.0[3]	105	109	98
8.0[3]	108	105	104
Hinson et al., 2008 / Corn- soybean meal / 10-22 kg			
5.0	98	100	99
Goresbeck et al., 2008 / Corn- soybean meal / 11-25 kg			
2.7	107	103	103
5.4	108	104	103
Kijora et al., 1995 / Barley- soybean meal / 31-82 kg			
4.8	105	108	97
9.7	112	112	100
19.4	96	103	94
29.4	82	105	78
Kijora and Kupsch, 2006 / Barley- soybean meal / 24 to 95 kg			
2.9	103	108	97
4.9	102	106	97
7.6	102	101	101
8.3	102	107	97
10.0	103	104	100
Kijora et al., 1997 / Barley- soybean meal / 27-100 kg			
10.0	106	110	96
Kijora et al., 1995 / Barley- soybean meal / 32-96			
4.6	114	110	103
9.7	119	113	106
Mourot et al., 1994 / Wheat- soybean meal / 35-102 kg			
5.0	97	101	96
Lammers et al., 2008c / Corn- soybean meal (whey in Phase 1) / 8-133 kg			
4.2	101	102	97
8.5	100	103	97
Stevens et al., 2008 / Corn- soybean meal / 28-119 kg			
4.2	103	103	100
8.4	103	104	99
12.6	100	108	92
Duttlinger et al., 2008b / Corn- soybean meal / 31-124 kg			
2.5	99	99	99
5.0	99	101	98
Hansen et al., 2009 / Wheat-barley-lupin, soybean meal -blood meal-meat meal / 51-105 kg			
3.0	98	104	93
6.1	87	93	95
9.1	96	102	94
12.2	91	98	93

Schieck et al., 2010b / Corn-soybean meal / 31-127 kg			
6.6	104	105	98
Duttlinger et al., 2008a / Corn - soybean meal / 78-102 kg			
2.5	97	99	98
5.0	95	97	98
Casa et al., 2008 / Corn-barley-wheat bran- soybean meal / 43-159 kg			
5.0	101	100	101
10.0	96	100	95
Mendoza et al., 2010 / Corn- soybean meal / 93-120 kg			
5.0	106	105	101
10.0	100	101	98
15.0	95	100	95

[1] Percentage relative to pigs fed the diet containing no supplemental glycerin. Percentage difference does not necessarily mean there was a significant difference from pigs fed the diet containing no supplemental glycerin. Main dietary ingredients and weight range of pigs tested are also provided with each citation.

[2] Represents a 100% glycerin basis. In studies utilizing crude glycerin, values adjusted for purity of glycerin utilized.

[3] Unknown purity, but product contained 6.8% ash and 15.6% ether extract.

Table 5. Relative performance of pigs fed supplemental glycerin[1]

compared to pig fed the corn-soybean meal control diet. Longer term feeding (16 weeks) resulted in a slight improvement in growth rate, but a small depression in feed efficiency. Some minor differences in carcass composition were noted, but there was no impact on pork quality. When considering the results from all of these studies (Table 5), there appears to be no consistent (positive or negative) effect of feeding up to 15% crude glycerin on growth performance, carcass composition, or pork quality in growing-finishing pigs compared with typical cereal grain-soybean meal based diets.

Only one study has been reported relative to feeding crude glycerin to lactating sows. In that study, lactating sows fed diets containing up to 9% crude glycerin performed similar to sows fed a standard corn-soybean meal diet (Schieck et al., 2010a).

4. Crude glycerin as a feed ingredient for poultry

Several researchers have reported that glycerin is an acceptable feed ingredient for poultry (Campbell and Hill, 1962; Brambilla and Hill, 1966; Lin et al., 1976; Lessard et al., 1993; Simon et al., 1996, 1997; Cerrate et al., 2006; Swiatkiewicz and Koreleski, 2009; Min et al., 2010). Adding glycerin up to 5% of the diet had no adverse effects on growth performance or carcass yield in broilers (Lessard et al., 1993; Simon et al., 1996; Cerrate et al., 2006). Increasing dietary glycerin above 10%, however, can adversely affect growth performance and meat yield of broiler chickens (Simon et al., 1996; Cerrate et al., 2006), although this may be due to reduced flowability of feed observed when 10% glycerin was supplemented (Cerrate et al., 2006).

Although designed as an energy balance trial, Lammers et al. (2008a) reported no impact on egg production of layer chickens during the 8-day experiment. In an extensive study with laying hens, Swiatkiewica and Koreleski (2009) reported no effects of feeding up to 6% dietary crude glycerin on laying performance or egg quality parameters. In turkeys,

Rosebrough et al. (1980) found no adverse effects on egg production, egg weight, or feed utilization in hens fed a pure source of glycerin as an energy source over a 16-wk period. In conclusion, there appears to be no consistent (positive or negative) impact of feeding up to 10% crude glycerin on growth performance in growing broilers (Table 6), or in laying hens.

Glycerin equivalency[2]	Gain	Feed intake	Gain:feed ratio
Campbell and Hill, 1962 / Semipurified ingredients / 1-28 days of age			
20.0	99	103	97
Brambila and Hill, 1966 / Semipurified ingredients / 1-28 days of age			
3.0	111	-	-
Lin et al., 1976 / Semipurified ingredients/ 1-21 days of age			
20.3	98	105	93
42.1	56	60	79
Simon et al., 1996 / Corn-slybean meal-fish meal / 1-31 days of age			
5.0	103	103	99
10.0	104	104	100
15.0	97	103	95
20.0	89	100	89
25.0	75	75	79
Simon et al., 1997 / Corn-soybean meal / 1-23 days of age			
10.0[3]	109	108	101
Cerrate et al., 2006 / Corn-soybean meal-poultry meal / 1-42 days of age			
2.0[4]	104	103	99
4.0[4]	103	103	99
Cerrate et al., 2006 / Corn-soybean meal-poultry meal / 1-42 days of age			
4.0[4]	100	99	101
8.0[4]	94	97	97

[1] Percentage relative to broilers fed the diet containing no supplemental glycerin. Percentage difference does not necessarily mean there was a significant difference from broilers fed the diet containing no supplemental glycerin. Main dietary ingredients and age of broilers tested are also provided with each citation.
[2] Represents a 100% glycerin basis. In studies utilizing crude glycerin, values adjusted for purity of glycerin utilized.
[3] Average of chicks fed the 15% and 18% crude protein diets with amino acid supplementation only.
[4] An assumed purity of 80%.

Table 6. Relative performance of broilers fed supplemental glycerin[1]

5. Special considerations

Biodiesel can be produced from a variety of feedstocks, such as oils from soy, canola, and corn, waste cooking oils, and animal fats (Ma and Hanna, 1999; Van Gerpen, 2005; Thompson and He, 2006). Consequently, the composition of crude glycerin can vary, but typically ranges from: 78 to 85% glycerin, 8 to 15% water, 2 to 10% salt (NaCl or KCl), 0.5% free fatty acids (although non-acidulated co-products may be up to 35% FFA), and ≤ 0.5% methanol (Table 3). In addition to the variation in energy content, the amount of salt and methanol in crude glycerin may require modifications in diet formulation. Depending on the salt level in the crude glycerin, supplemental levels of dietary salt may need to be limited depending upon the species being fed. However, data suggests that in swine and poultry, up to 3% dietary NaCl will have no adverse effects on animal performance as long as adequate water is freely available (adapted from NRC, 1980), although the impact of increased water intake on increased manure volume and composition (Sutton et al., 1976) or wet litter (Hogge et al., 1999) needs to be considered.

Utilization of crude glycerin may also affect the ability of feed to flow in bulk bins and automatic feeding systems as suggested by Cerrate et al. (2006) and Hansen et al. (2009). We also noted that 10 and 20% glycerin levels seemed to affect feed flow (Lammers et al. 2008b; Kerr et al., 2009), especially in feeds containing dried whey. No scientific measures were taken in any of the above experiments; so, the potential interactions among the level of glycerin supplementation, diet type, and feed handling system flowability of feed are yet to be characterized. Such information will allow establishment of practical limits for crude glycerin supplementation.

Methanol levels in crude glycerin warrant special consideration. Methanol is a potentially toxic compound and has been reviewed in detail by others (Roe, 1982; Medinsky and Dorman, 1995; Skrzydlewska, 2003). Methanol can be introduced orally, by respiration, or through the skin, and is distributed by the blood to all organs and tissues in proportion to their water content (Liesivuori and Savolainen, 1991). Metabolic elimination of methanol is much slower than that of ethanol, and its metabolism is illustrated in Figure 2 (adapted from Skrzydlewska, 2003).

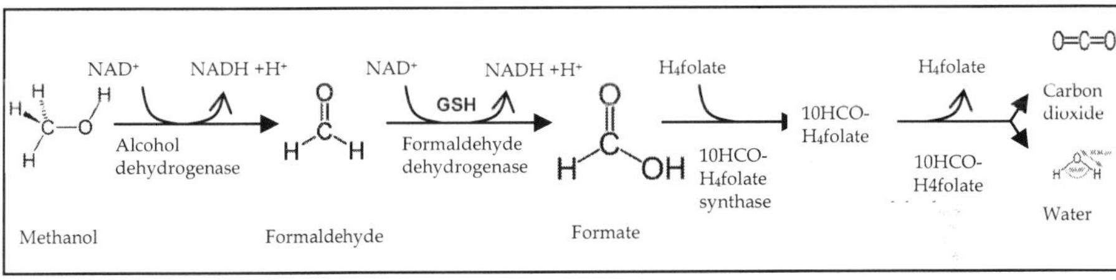

Fig. 2. Methanol Metabolism

Small amounts of methanol are excreted in the kidney and lung, but the majority is metabolized by the liver and released as CO_2. Acute methanol intoxication is manifested initially by signs of narcosis followed by a latent period in which formic acid accumulates causing metabolic acidosis (reduced blood pH, depletion of blood bicarbonate, visual degeneration, and abdominal, leg, and back pain). Chronic exposure to methanol causes headache, insomnia, gastrointestinal problems, and blindness. Animals differ widely in their ability to metabolize methanol depending upon enzyme activity and hepatic folate levels (Roe, 1982; Black et al., 1985; Medinsky and Dorman, 1995; Skrzydlewska, 2003). Little research on methanol metabolism or toxicity has been conducted in pigs. Makar et al. (1990) reported that the pig, compared to all other species studies, has extremely low levels of folates and very low levels of a key enzyme in the folate pathway, 10-formyl H_4folate dehydrogenase, suggesting the ability of the pig to dispose of formate is limited, and slower than that observed in rats or monkeys. However, Dorman et al. (1993) indicated that methanol- and formate-dosed minipigs did not develop optic nerve lesions, toxicologically significant formate accumulation, or metabolic acidosis, indicating that female minipigs do not appear to be overtly sensitive to methanol toxicity.

When considering the potential for methanol and formate toxicity, it is interesting to note that in some countries, formaldehyde, a methanol metabolite, can be used as a silage preservative, and formic acid can be used in finished feeds to reduce bacterial loads. Formic acid or formate salts have also been used safely in diets for swine (Overland et al., 2000; Canibe et al., 2005) and formaldehyde in diets for laying hens (Khan et al., 2006). It is also interesting to note that calcium formate has been used as a dietary calcium supplement for humans (Hanzlik et al., 2005).

As a general purpose feed ingredient, glycerin is regulated in the U.S. under 21CFR583.1320 requiring that levels of methanol in methyl esters of higher fatty acids should not exceed 0.015%. Recently, however, crude glycerin has been defined by the Association of American Feed Control Officials (AAFCO, 2010) and can be fed to non-ruminants up to 10% of the complete feed as long as it contains not less than 80% glycerin, not more than 15% water, not more than 0.15% methanol, up to 8% salt, up to 0.1% sulfur, and not more than 5 ppm heavy metals. German regulations (Normenkommission fur Einzelfuttermittel im Zentralausschuss der Deutschen Landwirtschaf, 2006) allow 0.5% (5,000 ppm) methanol in crude glycerin.

6. Conclusions

With a ME value of crude glycerin (adjusted to 85% glycerin) approximating 3,200 kcal/kg in swine and 3,600 kcal/kg in poultry (depending upon source), crude glycerin is an excellent source of calories in diets for non-ruminants. In general, feeding levels of up to 10% crude glycerin appear to have no consistent, positive or negative, effects on growth performance, carcass composition, lactation performance, or egg or meat quality. Levels of sodium- or potassium chloride, however, must be monitored to make formulation adjustments to supplemental salt additions, if necessary, to avoid increased manure volume for swine and wet litter for poultry. Concentrations of methanol in crude glycerin need to be monitored closely to ensure pig and poultry producers are in compliance with governmental regulations for feeding crude glycerin. Lastly, effects on feed handling and manufacturing characteristics need to be considered when determining inclusion rates of crude glycerin in practical diets for swine and poultry because of reduced feed flowability at high dietary inclusion rates.

7. References

AAFCO. 2010. Official Publication. Association of American Feed Control Officials, Inc., West Lafayette, IN.

Airhart, J. C., T. D. Bidner, and L. L. Southern. 2002. Effect of oral glycerol administration with and without dietary betaine on carcass composition and meat quality of late-finishing barrows. J. Anim. Sci. 80(Suppl. 2):71. (Abstr.).

Artman, N. R. 1964. Interactions of fats and fatty acids as energy sources for the chick. Poult. Sci. 43:994-1004.

Baba, H., X. J. Zhang, and R. R. Wolfe. 1995. Glycerol gluconeogenesis in fasting humans. Nutr. 11:149-153.

Bartlet, J., and D. Schneider. 2002. Investigation on the energy value of glycerol in the feeding of poultry and pig. Pages 15-36 in Union for the Promotion of Oilseeds-Schriften Heft 17.

Black, K. A., J. T. Eells, P. E. Neker, C. A. Hawtrey, and T. R. Tephly. 1985. Role of hepatic tetrahydrofolate in the species difference in methanol toxicity. Proc. Natl. Acad. Sci. 82:3854-3858.

Brambilla, S., and F. W. Hill. 1966. Comparison of neutral fat and free fatty acids in high lipid-low carbohydrates diets for the growing chicken. J. Nutr. 88:84-92.

Campbell, A. J., and F. W. Hill. 1962. The effects of protein source on the growth promoting action of soybean oil, and the effect of glycerine in a low fat diet. Poult. Sci. 41:881-882.

Canibe, N., O. Hojberg, S. Hojsgaard, and B. B. Jensen. 2005. Feed physical form and formic acid addition to the feed affect the gastrointestinal ecology and growth performance of growing pigs. J. Anim. Sci. 83:1287-1302.

Casa, G. D., D. Bochicchio, V. Faeti, G. Marchetto, E. Poletti, A. Rossi, A. Garavaldi, A. Panciroli, and N. Brogna. 2008. Use of pure glycerol in fattening heavy pigs. Meat Sci. 81:238-244.

Cerrate, S., F. Yan, Z. Wang, C. Coto, P. Sacakli, and P. W. Waldroup. 2006. Evaluation of glycerine from biodiesel production as a feed ingredient for broilers. Int. J. Poult. Sci. 11:1001-1007.

Cryer, A., and W. Bartley. 1973. Studies on the adaptation of rats to a diet high in glycerol. Int. J. Biochem. 4:293-308.

Dorman, D. C., J. A. Dye, M. P. Nassise, J. Ekuta, B. Bolon, and M. A. Medinsky. 1993. Acute methanol toxicity in minipigs. Fund. Appl. Toxicol. 20:341-347.

Dozier III, W. A., B. J. Kerr, and S. L. Branton. 2011. Apparent metabolizable energy of crude glycerin originating from different sources in broiler chickens. Poult. Sci. (in press).

Dozier III, W. A., B. J. Kerr, A. Corzo, M. T. Kidd, T. E. Weber, and K. Bregendahl. 2008. Apparent metabolizable energy of glycerin for broiler chickens. Poult. Sci. 87:317-322.

Duttlinger, A. W., M. D. Tokach, S. S. Dritz, J. M. DeRouchey, J. L. Nelssen, and R. D. Goodband. 2008a. Influence of glycerol and added fat on finishing pig performance. J. Anim. Sci. 86(Suppl. 2): 237 (Abstr.).

Duttlinger, A. W., M. D. Tokach, S. S. Dritz, J. M. DeRouchey, J. L. Nelssen, R. D. Goodband, and K. J. Prusa. 2008b. Effects of increasing dietary glycerol and dried distillers grains with solubles on growth performance of finishing pigs. J. Anim. Sci. 86(Suppl. 2): 607. (Abstr.).

Groesbeck, C. N., L. J. McKinney, J. M. DeRouchey, M. D. Tokach, R. D. Goodband, S. S. Dritz, J. L. Nelssen, A. W. Duttlinger, A. C. Fahrenholz, and K. C. Behnke. 2008. Effect of crude glycerol on pellet mill production and nursery pig growth performance. J. Anim. Sci. 86:2228-2236.

Guyton, A. C. 1991. Textbook of Medical Physiology. W. B. Saunders Co., Philadelphia, PA.

Hansen, C. F., A. Hernandez, B. P. Mullan, K. Moore, T. Trezona-Murray, R. H. King, and J. R. Pluske. 2009. Crude glycerol from the production of biodiesel increased plasma glycerol levels but did not influence growth performance in growing-finishing pigs or indices of meat quality at slaughter. Anim. Prod. Sci. 49:154-161.

Hanzlik, R. P., S. C. Fowler, and J. T. Eells. 2005. Absorption and elimination of formate following oral administration of calcium formate in female human subjects. Drug Metab. Disp. 23:282-286.

Hetenyi, G., G. Perez, and M. Vranic. 1983. Turnover and precursor-product relationships of nonlipid metabolites. Physiol. Rev. 63:606-667.

Hinson, R., L. Ma, and G. Allee. 2008. Use of glycerol in nursery pig diets. J. Anim. Sci. 86(E-Suppl. 3): 46 (Abstr.).

Hofmann, A. F., and B. Borgstorm. 1962. Physico chemical state of lipids in intestinal content during digestion and absorption. Fed. Proc. 21:43-50.

Hogge, D. M., K. R. Kummings, and J. L. McNaughton. 1999. Evaluation of sodium bicarbonate, chloride, or sulfate with a coccidiostat in corn-soy or corn-soy-meat diets for broiler chickens. Poult. Sci. 78:1300-1306.

Johnston, J. M. 1963. Recent developments in the mechanism of fat absorption. Adv. Lipid Res. 1:105-122. R. Paolelti and D. Kritchevsky, ed. Academic Press, NY.

Kerr, B. J., T. E. Weber, W. A. Dozier III, and M. T. Kidd. 2009. Digestible and metabolizable energy content of crude glycerin originating from different sources in nursery pigs. J. Anim. Sci. 87:4042-4049.

Khan, A., S. M. Hussain, and M. Z. Khan. 2006. Effects of formalin feeding or administering into the crops of white leghorn cockerels on hematological and biochemical parameters. Poult. Sci. 85:1513-1519.

Kijora, C., and S. D. Kupsch. 2006. Evaluation of technical glycerols from "biodiesel" production as a feed component in fattening of pigs. Lipid-Fett 98:7:240-245.

Kijora, C., R. D. Kupsch, H. Bergner, C. Wenk, and A. L. Prabucki. 1997. Comparative investigation on the utilization of glycerol, free fatty acids, free fatty acids in combination with glycerol and vegetable oil in fattening of pigs. J. Anim. Physiol. Anim. Nutr. 77:127-138.

Kijora, C., H. Bergner, R. D. Kupsch, and L. Hageman. 1995. Glycerol as feed component in diets of fattening pigs. Arch. Anim. Nutr. 47:345-360.

Lammers, P. J., B. J. Kerr, M. S. Honeyman, K. Stalder, W. A. Dozier III, T. E. Weber, M. T. Kidd, and K. Bregendahl. 2008a. Nitrogen-corrected apparent metabolizable energy value of crude glycerol for laying hens. Poult. Sci. 87:104-107.

Lammers, P. J., B. J. Kerr, T. E. Weber, W. A. Dozier III, M. T. Kidd, K. Bregendahl, and M. S. Honeyman. 2008b. Digestible and metabolizable energy of crude glycerol in pigs. J. Anim. Sci. 86:602-608.

Lammers, P. J., B. J. Kerr, T. E. Weber, K. Bregendahl, S. M. Lonergan, K. J. Prusa, D. U. Ahn, W. C. Stoffregen, W. A. Dozier, III, and M. S. Honeyman. 2008c. Growth performance, carcass characteristics, meat quality, and tissue histology of growing pigs fed crude glycerin-supplemented diets. J. Anim. Sci. 86:2962-2970.

Lessard, P., M. R. Lefrancois, and J. F. Bernier. 1993. Dietary addition of cellular metabolic intermediates and carcass fat deposition in broilers. Poult. Sci. 72:535-545.

Liesivuori, J., and H. Savolainen. 1991. Methanol and formic acid toxicity: biochemical mechanisms. Pharmacol. Toxicol. 69: 157-163

Lin, M. H., D. R. Romsos, and G. A. Leveille. 1976. Effect of glycerol on enzyme activities and on fatty acid synthesis in the rat and chicken. J. Nutr. 106:1668-1677.

Lin, E. C. C. 1977. Glycerol utilization and its regulation in mammals. Annu. Rev. Biochem. 46:765-795.

Ma, F., and M. A. Hanna. 1999. Biodiesel production: A review. Biores. Tech. 70:1-15.

Makar, A. B., T. R. Tephly, G. Sahin, and G. Osweiler. 1990. Formate metabolism in young swine. Toxicol. Appl. Pharm. 105:315-320.

Medinsky, M. A., and D. C. Dorman. 1995. Recent developments in methanol toxicity. Toxicol. Letters 82/83:707-711.

Mendoza, O. F., M. E. Ellis, F. K. McKeith, and A. M. Gaines. 2010. Metabolizable energy content of refined glycerin and its effects on growth performance and carcass and pork quality characteristics of finishing pigs. J. Anim. Sci. 88:3887-3895.

Min, Y. N., F. Yan, F. Z. Liu, C. Coto, and P. W. Waldroup. 2010. Glycerin-a new energy source for poultry. Int. J. Poult. Sci. 9:1-4.

Mourot, J., A. Aumaitre, A. Mounier, P. Peiniau, and A. C. Francois. 1994. Nutritional and physiological effects of dietary glycerol in the growing pig. Consequences on fatty tissues and post mortem muscular parameters. Livest. Prod. Sci. 38:237-244.

Normenkommission für Einzelfuttermittel im Zentralausschuss der Deutschen Landwirtschaft. 2006. Positivliste für Einzelfuttermittel, 5. Auflage, #12.07.03, p. 35.

NRC. 1998. Nutrient requirements of swine. 10th rev. Ed. Natl. Acad. Press, Washington, DC.

NRC. 1980. Mineral tolerance of domestic animals. Natl. Acad. Press, Washington, DC.

Overland, M., T. Granli, N. P. Kjos, O. Fjetland, S. H. Steien, and M. Stokstad. 2000. Effect of dietary formates on growth performance, carcass traits, sensory quality, intestinal microflora, and stomach alterations in growing-finishing pigs. J. Anim. Sci. 78:1875-1884.

Robergs, R. A., and S. E. Griffin. 1998. Glycerol: biochemistry, pharmacokinetics and clinical and practical applications. Sports Med. 26:145-167.

Roe, O. 1982. Species differences in methanol poisoning. CRC Critical Reviews in Toxicology 10:275-286.

Rosebrough, R. W., E. Geis, P. James, H. Ota, and J. Whitehead. 1980. Effects of dietary energy substitutions on reproductive performance, feed efficiency, and lipogenic enzyme activity on large white turkey hens. Poult. Sci. 59:1485-1492.

Schieck, S. J., B. J. Kerr, S. K. Baidoo, G. C. Shurson, and L. J. Johnston. 2010a. Use of crude glycerol, a biodiesel coproduct, in diets for lactating sows. J. Anim. Sci. 88:2648-2656.

Schieck, S. J., G. C. Shurson, B. J. Kerr, and L. J. Johnston. 2010b. Evaluation of glycerol, a biodiesel coproduct, in grow-finish pig diets to support growth and pork quality. J. Anim. Sci. 88:3927-3935.

Senior, J. 1964. Intestinal absorption of fats. J. Lipid Res 5:495-521.

Simon, A., H. Bergner, and M. Schwabe. 1996. Glycerol-feed ingredient for broiler chickens. Arch. Anim. Nutr. 49:103-112.

Sklan, D. 1979. Digestion and absorption of lipids in chicks fed triglycerides or free fatty acids: Synthesis of monoglycerides in the intestine. Poult. Sci. 58:885-889.

Skrzydlewska, E. 2003. Toxicological and metabolic consequences of methanol poisoning. Toxicol. Mechanisms Methods 13:277-293.

Stevens, J., A. Schinckel, M. Latour, D. Kelly, D. Sholly, B. Legan, and B. Richert. 2008. Effects of feeding increasing levels of glycerol with or without distillers dried grains with solubles in the diet on grow-finish pig growth performance and carcass quality. J. Anim. Sci. 86(Suppl. 2): 606. (Abstr.).

Sutton, A. L., V. B. Mayrose, J. C. Nye, and D. W. Nelson. 1976. Effect of dietary salt level and liquid handling systems on swine waste composition. J. Anim. Sci. 43:1129-1134.

Swiatkiewicz, S., and J. Koreleski. 2009. Effect of crude glycerin level in the diet of laying hens on egg performance and nutrient utilization. Poult. Sci. 88:615-619.

Tao, R. C., R. E. Kelley, N. N. Yoshimura, and F. Benjamin. 1983. Glycerol: Its metabolism and use as an intravenous energy source. J. Parenteral Enteral. Nutr. 7:479-488.

Thompson, J. C., and B. B. He. 2006. Characterization of crude glycerol from biodiesel production from multiple feedstocks. Appl. Eng. Agric. 22:261-265.

Van Gerpen, J. 2005. Biodiesel processing and production. J. Fuel Proc. 86:1097-1107.

Wiseman, J. and F. Salvador. 1991. The influence of free fatty acid content and degree of saturation on the apparent metabolizable energy value of fats fed to broilers. Poult. Sci. 70:573-582.

Zijlstra, R. T., K. Menjivar, E. Lawrence, and E. Beltranena. 2009. The effect of feeding crude glycerol on growth performance and nutrient digestibility in weaned pigs. Can. J. Anim. Sci. 89:85-89.